Fisheries: Techniques and Management

Fisheries: Techniques and Management

Editor: Roger Creed

www.callistoreference.com

Callisto Reference,
118-35 Queens Blvd., Suite 400,
Forest Hills, NY 11375, USA

Visit us on the World Wide Web at:
www.callistoreference.com

ISBN: 978-1-63239-912-0 (Hardback)

Cataloging-in-Publication Data

Fisheries : techniques and management / edited by Roger Creed.
p. cm.
Includes bibliographical references and index.
ISBN 978-1-63239-912-0
1. Fisheries. 2. Fishery management. I. Creed, Roger.
SH331 .F57 2018
639.2--dc23

Table of Contents

Preface

This book has been a concerted effort by a group of academicians, researchers and scientists, who have contributed their research works for the realization of the book. This book has materialized in the wake of emerging advancements and innovations in this field. Therefore, the need of the hour was to compile all the required researches and disseminate the knowledge to a broad spectrum of people comprising of students, researchers and specialists of the field.

A fishery is the business of raising fish for profit. This book on fisheries concentrates on techniques that promote sustainable amounts of fish harvesting. Some of the globally harvested species include salmon, tuna, shrimp, cod, crab, etc. Fisheries management requires concepts and techniques from fields like oceanography, marine biology, wildlife conservation, etc. Fish stocks are monitored and controlled through population dynamics. Through this book, we attempt to further enlighten the readers about the new concepts in this field. With state-of-the-art inputs by acclaimed experts of this field, this book targets students and professionals.

At the end of the preface, I would like to thank the authors for their brilliant chapters and the publisher for guiding us all-through the making of the book till its final stage. Also, I would like to thank my family for providing the support and encouragement throughout my academic career and research projects.

Editor

1

The Interaction between Water Currents and Salmon Swimming Behaviour in Sea Cages

David Johansson[1]*, Frida Laursen[1,2], Anders Fernö[2], Jan Erik Fosseidengen[1], Pascal Klebert[3], Lars Helge Stien[1], Tone Vågseth[1], Frode Oppedal[1]

1 Institute of Marine Research, Matredal, Norway, **2** Department of Biology, University of Bergen, Bergen, Norway, **3** Sintef Fisheries and Aquaculture, Trondheim, Norway

Abstract

Positioning of sea cages at sites with high water current velocities expose the fish to a largely unknown environmental challenge. In this study we observed the swimming behaviour of Atlantic salmon (*Salmo salar* L.) at a commercial farm with tidal currents altering between low, moderate and high velocities. At high current velocities the salmon switched from the traditional circular polarized group structure, seen at low and moderate current velocities, to a group structure where all fish kept stations at fixed positions swimming against the current. This type of group behaviour has not been described in sea cages previously. The structural changes could be explained by a preferred swimming speed of salmon spatially restricted in a cage in combination with a behavioural plasticity of the fish.

Editor: Z. Daniel Deng, Pacific Northwest National Laboratory, United States of America

Funding: The study was supported by Research Council of Norway (http://www.forskningsradet.no/en/Home_page/1177315753906) grant no. 207116/E40 and by the Norwegian Ministry of Fisheries. The funders had no role in study design, data collection and analysis, decision to publish, or preparation of the manuscript.

Competing Interests: The authors have declared that no competing interests exist.

* E-mail: davidjo@imr.no

Introduction

Moving sea cages to exposed sites with strong water currents is an industry-wide trend in Atlantic salmon (*Salmo salar* L.) farming [1], [2]. This could improve production efficiency through access to high water quality due to rapid transport and dilution of waste products, more stable temperatures, high levels of oxygen and less influence from terrestrial runoff [1], [3]. Other positive effects such as reduction of possible conflicts with other users in the coastal area and avoidance of the ecological carrying capacity limitations of inshore waters have been suggested [3]. One prerequisite for this progress has been the development of strong, resistant farm structures that can withstand the forces produced by strong water currents [4], [5]. However, it is not known how the fish inside the sea cages cope with high water current velocities. The fish has to cope with being forced into an environment that radically differs from the sheltered fjord sites. The question about the amount and type of stress produced by a high-energy environment and the fish capacity to cope is at least as important as the development of new resistant farming platforms. Salmon farms in sheltered localities generally experience current velocities below 20 cm s^{-1} outside the cages [6]. At such velocities the fish will often form a circular, one way directed uniform swimming pattern, possibly as a result of individuals actively avoiding collisions with each other and the cage wall [7]. At these sites salmon typically swim at speeds of 0.3–0.9 BL s^{-1}, with maximum average values of 1.9 BL s^{-1} [8], [9], [10]. The constant swimming of salmon under natural conditions has been associated with an inherent migratory tendency related to optimum cruising speed [10] [11] and in open ocean studies the speed approximates to 1 BL s^{-1}, independent of age [12]. Studies using swim tunnels indicate a critical swimming speed, U_{crit}, for small salmon (400–800 g) of 1.6–2.2 BL s^{-1} [13], [14], although one study reports values as high as 3.0 BL s^{-1} [15].

Although the exact swimming capacity of salmon is uncertain, and will vary with such factors as size, exercise level, degree of satiation [16] and individual fitness, it is evident that salmon inside sea cages must adapt their behaviour to the water current. Hence the objective of this study was to observe the general effects of high water current velocities on fish swimming behaviour at the group level, in an exposed commercial salmon cage.

Materials and Methods

The observations of schooling behaviour were performed from 11th to 13th of February 2012 at a commercial marine salmon farm near Torshavn in the Faroe Islands, Denmark (61.59° N). The farm had 8 circular cages of 41 m diameter, and 2 cages of 50 m diameter, with a depth of 12 m to the bottom ring. The depth below the cages varied from 30 to 40 m, and the total biomass at site was 1320 tonnes. The fish were fed continuously from 08:30 to 16:15 h and were exposed to continuous artificial light at 4 m depth. The observed cage (41 m diameter) was selected based on having the highest probability to be exposed to high water current velocities, due to its position at the south end of the farm. According to farm data, the stocking density in this cage was 6.2 kg m^{-3} and the average fish weight 1.54 kg, corresponding to an approximate fish length of 50 cm. During the observation period, vertical profiles of water characteristics (oxygen, temperature and salinity) showed little spatial and temporal variation: dissolved oxygen saturation levels were at 94.6±2.3% (mean±SD), temperature 6.6±0.1°C and salinity 35.0±0.1 ppt, all of which were within accepted optimal limits [6], [17], [18]. Vertical profiles of water current down to 20 m of depth were

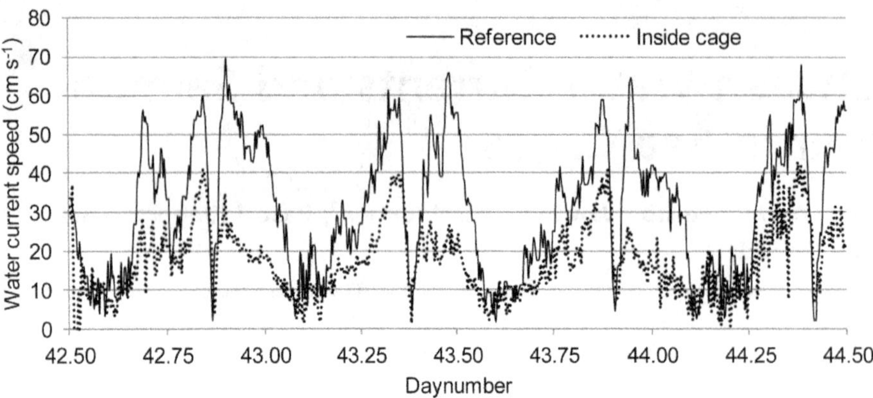

Figure 1. Water current velocity outside the cage (Reference) and inside the cage from 11th to 13th of February, 2012.

recorded 210 m south of the farm with open sea between the observed cage and the reference point using an Acoustic Wave And Current profiler (AWAC, Nortek, Oslo, Norway). In order to minimize disturbance from the fish, single point measurements were taken at 6.2 m depth in the centre of the cage using a Vector Aquadopp 3D (Nortek, Oslo, Norway). The observed water current velocities varied in a tidal pattern between 0 to 69 cm s^{-1} at the reference point, and between 0 to 42 cm s^{-1} at the single point measured inside the cage (Figure 1). The reduced current velocity inside the cage (Figure 1) is related to dampening by the net and the fish inside the cage and the cages north of the observed cage [6], [19], [20]. The vertical profiles showed little differences in current speeds and directions between 0 and 12 m depth. The tidal nature of the current produced a variable main direction between 120° and 300°. Unless otherwise specified, we refer to the current data collected at the reference point. The schooling behaviour of the salmon was observed with two remotely controlled underwater pan/tilt cameras (Orbit GMT AS, Førresfjord, Norway) connected to a recording DVD player. One camera was positioned next to the net and the other was positioned approximately 15 m from the net at the opposite side of the cage at approximately 6 m of depth to give a good representation of behaviours both up- and downwards. The 48 h period of recordings were divided into four minutes subsamples, which were post-analysed and manually classified for swimming structure (see Results). Recordings of poor quality (e.g. too low light intensity or no fish in picture) were discarded from further analysis. An average of the observed water current velocities between surface and 12 m of depth was used in the analysis. Inherently, this type of time series data produces temporal pseudo replication. The relationships between current velocity and observed swimming structure were therefore investigated using mixed effects models to resolve the non-indecencies in our data [21], with swimming structure as fixed effect and time as continuous random effect (function lme, the R software system Version 2.15.0, The R Foundation for Statistical Computing, Vienna, Austria). Model checking plots were used to check that the residuals were well behaved (function plot,~fitted(.)) and to check the normality assumption (function qqnorm).

Results and Discussion

A first screening of the videos revealed that the swimming structure could be divided into three main categories: Circle = polarized swimming in a circular movement, On Current = swimming towards the current with no forward movement and Mixed = both Circle and On Current structures present at the same time (Figure 2). Based on data from the more centralised camera (n = 155), the mixed effect model associated the Circle swimming structure with low current velocities (intercept = 22.4 cm s^{-1}, SE = 3.1, $p<0.001$), the Mixed structure with increased current velocities (+13.7 cm s^{-1}, SE = 2.2, $p<0.001$), and the On Current structure with an even higher current velocity (+24.3 cm s^{-1}, SE = 1.7, $p<0.001$). Similarly, for the camera close to the net (n = 347), the Circle structure was associated with low current velocities (intercept = 20.1 cm s^{-1}, SE = 2.6, $p<0.001$) and the Mixed and On Current swimming structures with increasing current velocities (+13.2 cm s^{-1}, SE = 2.3, $p<0.001$ and +26.5 cm s^{-1}, SE = 1.4, $p<0.001$, respectively).

Hence, at low current velocities (≈20 cm s^{-1}) the fish swam in circles (Figure 2A, Table 1) and occupied most of the cage volume. With increasing current velocities (≈35 cm s^{-1}), a shift occurred with some fish seeking a new position facing the net towards the current while other fish continued to swim in elliptic-shaped circles behind the stationary fish at the net (Figure 2B, Table 1). When the circling fish came to a position where they were exposed to

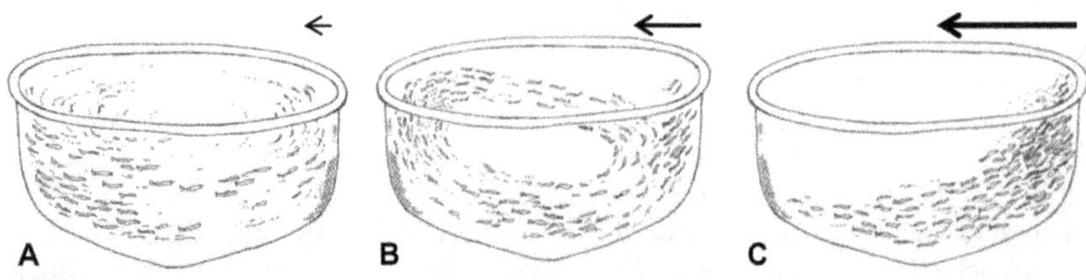

Figure 2. The three observed swimming structures Circle (A, circular movement), Mixed (B, Circle and On Current) or On Current (C, standing on current). The arrows indicate strength and direction of the water current during the different group structures. Drawings by Stein Mortensen, Institute of Marine Research.

Table 1. Modeled water current speed in cm s^{-1} at the reference point for the observed swimming categories Circle (circular movement), Mixed (Circle and On Current) or On Current (standing on current).

Camera pos.	Swimming Structure		
	Circle	Mixed	On Current
Centralized	22.4 (≈0.45)	36.1 (≈0.72)	46.7 (≈0.93)
Side	20.1 (≈0.40)	33.4 (≈0.67)	46.6 (≈0.93)

Water current velocity is given as BL s^{-1} in brackets.

current on their sides they turned inwards toward the centre of the cage and drifted with the current to the leeward side of the cage. Following this, they turned towards the current and continued to swim the remainder of the circle's distance. With further increase of current velocity, a larger proportion of the group switched from schooling to swimming towards the current next to the net wall, until all fish stood in a dense group along the side of the cage with no circling fish left (≈47 cm s^{-1}, Figure 2C, Table 1). With sudden changes of current velocities, there was a period of chaos before the fish established a stable structure again.

It is thus clear that the fish experience new challenges when exposed to strong water currents. We have for the first time observed shifts in swimming structure of salmon in sea cages connected to changes in current velocities. The shift from the traditional circular schooling to stationary swimming against the current in a group could reflect energetic optimization as a response to the increased current velocities. Fish swimming behind others have been reported to save energy [22]. However, since previously reported U_{crit} [13], [14] is higher than all the observed current velocities this is probably not the only underlying mechanism. The driving force could instead be a combination of U_{crit} and the large variation in current velocities within the cage, thereby restricting the traditional structure when swimming down- compared to upstream.

Theoretically, if a salmon cage is exposed to an increasing current speed, the typical torus shape of a salmon school within the cage will force the upstream fish to double their swimming speed in order to maintain the group structure. If this pattern is broken up by fish changing to stand on the current, the group structure is probable to collapse and move towards a new stable structure with all fish to hold a constant position against the current.

The On Current structure was observed at water current velocities of approximately 47 cm s^{-1} (Table 1). At such velocities fish in a Circle structure swimming against the current would have to swim at least 94 cm s^{-1} to maintain the group structure, which is close to the U_{crit}, for Atlantic salmon [13], [14]. However, the dampening effect of the net [20] suggests that a lower current velocity triggers the shift in group structure. Logically, the swimming speed observed in normal schooling structures, during low current velocity, can be identified as the fishes' preferred speed. This can be termed as such, since fish are able to choose their speed without influence from water current conditions. This chosen speed is assumedly a manifestation of their optimal cruising speed for minimal energy expenditure, as in migrating salmon [10]. Current velocities for Mixed structures were about 35 cm s^{-1} outside the cage; the fish swam in both Circle and On Current structures at this velocity, and this level could represent the approximate breakpoint when the swimming speed started to exceed the preferred swimming speed for some individuals. This current velocity equates to a swimming speed of $2 * 0.7$ BL $s^{-1} = 1.4$ BL s^{-1} (when fish are swimming towards the current in a circular structure), which is higher than previously reported swimming speeds of 0.3–0.9 BL s^{-1} at more sheltered sites [8], [9], [10]. Taking into account the observed dampening effect, the reduced current speed could result in swimming speeds similar to the previously reported preferred swimming speeds.

From a welfare perspective it could be argued that sites with current velocities that do not exceed the school's preferred swimming speed should provide good welfare since the animal are free to express behaviours within its natural range (item 2 of the Five Freedoms), [23]. Yet, the salmon showed a high degree of plasticity in their behaviour and adapted to the frequent challenges forced upon them by the intermittent and strong water currents. This documented adaptive capacity indicates that conclusions only based on studies performed in laboratories and at unexposed localities could be of limited value due to the different behavioural response to the variable environment. Understanding the effect of water currents on individual fish of different size, as well as on the group as whole, is therefore of utmost importance for the progress of fish farming. High-resolution studies of behaviour in relation to the environment at such sites are needed to ensure environmental conditions acceptable for animal welfare and good production performance.

Acknowledgments

The authors are greatly indebted to the personnel at the salmon farm, especially Rúni Ellingsgaard, and to Øystein Patursson and Heini Rasmussen at Fiskaaling for their kind assistance. Sincere thanks are also due to Samantha Bui for her kind help with language control in the preparation of this manuscript.

Author Contributions

Conceived and designed the experiments: FO PK DJ TV JEF FL AF. Performed the experiments: DJ JEF TV. Analyzed the data: DJ FL AF LHS. Wrote the paper: DJ FL AF JEF PK LHS TV FO.

References

1. Holmer M (2010) Environmental issues of fish farming in offshore waters: perspectives, concerns and research needs, Aquac Environ Interact 1: 57–70.
2. Perez OM, Telfer TC, Ross LG (2003) On the calculation of wave climate for offshore cage culture site selection: a case study in Tenerife (Canary Islands). Aquac Eng 29: 1–21.
3. Benetti DD, Benetti GI, Rivera JA, Sardenberg B, O'Hanlon B (2010) Site Selection Criteria for Open Ocean Aquaculture. Mar Technol Soc J 44: 22–35.
4. Fredheim A, Langan R (2009) Advances in technology for off-shore and open ocean finfish aquaculture. New technologies in aquaculture: improving production efficiency, quality and environmental management. G. Burnell, Allan, G., Woodhead Publishing in Food Science, Technology and Nutrition. 914–944.

5. Loverich GF, Gace L (1997) The effects of currents and waves on several classes of offshore sea cages. In: Helsley CE, Open Ocean Aquaculture: Charting the Future of Ocean Farming. University of Hawaii, Maui, Hawaii, USA 131–144.
6. Johansson D, Juell J-E, Oppedal F, Stiansen J-E, Ruohonen K (2007) The influence of the pycnocline and cage resistance on current flow, oxygen flux and swimming behaviour of Atlantic salmon (Salmo salar L.) in production cages. Aquaculture 265: 271–287.
7. Føre M, Dempster T, Alfredsen JA, Johansen V, Johansson D (2009) Modelling of Atlantic salmon (*Salmo salar* L.) behaviour in sea-cages: A Lagrangian approach. Aquaculture 288: 196–204.
8. Dempster T, Korsøen O, Folkedal O, Juell JE, Oppedal F (2009) Submergence of Atlantic salmon (*Salmo salar* L.) in commercial scale sea-cages: A potential short-term solution to poor surface conditions. Aquaculture 288: 254–263.
9. Juell JE (1995) The behaviour of Atlantic salmon in relation to efficient cage rearing. Rev Fish Biol Fish 5: 320–335.
10. Sutterlin AM, Jokola KJ, Holte B (1979) Swimming behaviour of salmonid fish in ocean pens. J Fish Res Board Can. 36: 948–954.
11. Brett JR (1964) The respiratory metabolism and swimming performance of young Sockeye salmon. J Fish Res Board Can 21: 1183–1226.
12. Drenner SM, Clark TD, Whitney CK, Martins EG, Cooke SJ, et al. (2012) A synthesis of tagging studies examining the behaviour and survival of anadromous salmonids in marine environments. PloS One 7: 1–13.
13. Deitch EJ, Fletcher GL, Petersen LH, Costa I, Shears MA, et al. (2006) Cardiorespiratory modifications, and limitations, in post-smolt growth hormone transgenic Atlantic salmon *Salmo salar*. J Exp Biol 209: 1310–1325.
14. McKenzie DJ, Higgs DA, Dosanjh BS, Deacon G, Randall DJ (1998) Dietary fatty acid composition influences swimming performance in Atlantic salmon (*Salmo salar*) in seawater. Fish Physiol Biochem 19: 111–122.
15. Lijalad M, Powell MD (2009) Effects of lower jaw deformity on swimming performance and recovery from exhaustive exercise in triploid and diploid Atlantic salmon *Salmo salar* L. Aquaculture, 290: 145–154.
16. Stien LH, Bracke MBM, Folkedal O, Nilsson J, Oppedal F, et al. (2013) Salmon Welfare Index Model (SWIM 1.0): a semantic model for overall welfare assessment of caged Atlantic salmon: review of the selected welfare indicators and model presentation. Rev. Aquaculture: 33–57.
17. Oppedal F, Dempster T, Stien L (2011) Environmental drivers of Atlantic salmon behaviour in sea-cages: a review. Aquaculture 311: 1–18.
18. Remen M, Oppedal F, Torgersen T, Imsland AK, Olsen RE (2012) Effects of cyclic environmental hypoxia on physiology and feed intake of post-smolt Atlantic salmon: Initial responses and acclimation. Aquaculture 326–329: 148–155.
19. Gansel LC, McClimans TA, Myrhaug D (2012) Flow around the free bottom of fish cages in a uniform flow with and without fouling. J Offshore Mech Arct Eng 134, DOI: 10.1115/1.4003695.
20. Klebert P, Gansel L, Lader P, Oppedal F (2013) Flow hydrodynamics through nets and floating cages: a review. Ocean Eng 58: 260–274.
21. Crawley MJ (2007) Mixed-effects models, in: The R Book. West Sussex, Wiley 627–660.
22. Herskin J, Steffensen JF (1998) Energy savings in sea bass swimming in a school: measurements of tail beat frequency and oxygen consumption at different swimming speeds. J Fish Biol 53: 366–376.
23. Electronic reference: UK Farm Animal Welfare Council (2005) The Five Freedoms. Available: http://www.fawc.org.uk/freedoms.htm. Accessed 2013 Apr 11.

2

Adapting to a Changing World: Unraveling the Role of Man-Made Habitats as Alternative Feeding Areas for Slender-Billed Gull (*Chroicocephalus genei*)

Francisco Ramírez[1]*, Joan Navarro[2], Isabel Afán[3], Keith A. Hobson[4], Antonio Delgado[5], Manuela G. Forero[1]

1 Estación Biológica de Doñana, Department de Biología de la Conservación, Sevilla, Spain, **2** Institut de Ciències del Mar, Barcelona, Spain, **3** Estación Biológica de Doñana, Laboratorio de SIG y Teledetección, Sevilla, Spain, **4** Environment Canada, Saskatoon, Saskatchewan, Canada, **5** Instituto Andaluz de Ciencias de la Tierra, Granada, Spain

Abstract

Current rates of wildlife habitat loss have placed increasing demands on managers to develop, validate and implement tools aimed at improving our ability to evaluate such impacts on wildlife. Here, we present a case study conducted at the Natural Area of Doñana (SW Spain) where remote sensing and stable isotope ($\delta^{13}C$, $\delta^{15}N$) analyses of individuals were combined to unravel (1) the effect of variations in availability of natural food resources (i.e. from natural marshes) on reproductive performance of a Slender-billed Gull (*Chroicocephalus genei*) population, and (2) the role of two adjacent, artificial systems (a fish farm and saltmines) as alternate anthropogenic feeding areas. Based on long-term (1983–2004) remote-sensing, we inferred the average extent of flooded area at the marshland (a proxy to natural resource availability) annually. Estimated flooded areas (ranging from extreme drought [ca. 151 ha, 1995] to high moisture [15,049 ha, 2004]) were positively related to reproductive success of gulls (estimated for the 1993–2004 period, and ranging from ca. 0 to 1.7 fledglings per breeding pairs), suggesting that habitat availability played a role in determining their reproductive performance. Based on blood $\delta^{13}C$ and $\delta^{15}N$ values of fledglings, 2001–2004, and a Bayesian isotopic mixing model, we conclude that saltmines acted as the main alternative foraging habitat for gulls, with relative contributions increasing as the extent of marshland decreased. Although adjacent, anthropogenic systems have been established as the preferred breeding sites for this gull population, dietary switches towards exploitation of alternative (anthropogenic) food resources negatively affected the reproductive output of this species, thus challenging the perception that these man-made systems are necessarily a reliable buffer against loss of natural feeding habitats. The methodology and results derived from this study could be extended to a large suite of threatened natural communities worldwide, thus providing a useful framework for management and conservation.

Editor: Zoltan Barta, University of Debrecen, Hungary

Funding: During writing, FR and JN were supported by postdoctoral contracts from FP7-REGPOT 2010-1 (Grant No. 264125 of EcoGenes project) and the Juan de la Cierva program (Ministerio de Economia y Competitividad), respectively. Research funds were provided by a grant from the Spanish Ministry REN2002-0045 (CGL2009-09801 and CGL2006-02247/BOS for historical data on inundation levels) and by funds provided by Obra Social/Fundación Cajasol. KAH benefitted from funds provided by Environment Canada. The funders had no role in study design, data collection and analysis, decision to publish, or preparation of the manuscript.

Competing Interests: The authors have declared that no competing interests exist.

* E-mail: ramirez@ebd.csic.es

Introduction

Human activities are impacting ecosystems globally to an unprecedented degree, thus leading to increasing rates of natural habitat loss [1]. Among natural ecosystems, coastal zones are of particular concern, since centuries of human influence have resulted in loss and degradation of more than 65% of natural habitats [2]. Consequently, more than 90% of species inhabiting these ecosystems are undergoing severe population declines [2], raising a grave conservation issue and placing increasing demands on managers to develop, validate and implement tools aimed at improving our ability to recognize and trace such environmental changes and evaluate their effects on free-living animals.

Simultaneously, the discipline of landscape ecology has expanded rapidly in recent years, supported by developments in spatial analysis techniques. In particular, remote sensing data is available from the last thirty years, thus providing valuable information on changes in habitat extent and heterogeneity at long spatial and temporal scales [3,4], and resulting in a powerful tool for investigating the relationship between habitat loss or fragmentation on population dynamics of species [5–7]. However, species differ tremendously in their responses to environmental changes. In particular, the extent to which populations decrease following habitat loss strongly depends on the efficiency with which they can alter their foraging behavior in order to exploit alternative food resources [8,9]. For example, several man-made structures, such as solar saltmines, fish farms or rice fields can provide complementary or alternate foraging areas for wild species naturally inhabiting and feeding on intertidal and estuarine systems when natural habitat loss occurs [9,10]. Our ability to monitor spatiotemporal variations in the use of these alternate habitats by wildlife is therefore crucial for accurate assessments of the impact of habitat loss on wild populations.

Current approaches to resource use by wildlife based on naturally occurring stable isotope measurements of carbon ($\delta^{13}C$) and nitrogen ($\delta^{15}N$) provide an exceptional opportunity to quantify how wild species use anthropogenic habitats compared to the adjacent, natural environments, given that these differ in the isotopic composition of foodwebs (e.g. [11,12]). Stable isotopes typically provide time-integrated information on animals' diet by means of isotopic measurements of various consumer tissues [13,14]. In addition, transforming isotopic information into dietary proportions of isotopically distinct dietary sources within a Bayesian or frequentist framework is now feasible through the use of multi-source isotope mixing models, thus allowing quantitative inference of nutrient pathways within natural and artificial systems [11,15]. Isotopic approaches have therefore the potential for providing key insights into the actual role of man-made structures as alternative feeding areas for wild populations when natural habitat loss occurs.

Here, we provide an instructive case study conducted at the Natural Area of Doñana (SW Spain, Figure 1) aimed at investigating the actual impact of temporal fluctuations in the extent of natural habitats on a wild avian population. Doñana is the largest protected coastal wetland in Western Europe, and natural marshes there constitute the main foraging habitat for most waterbirds and other avian inhabitants [16–18]. However, resource availability in this natural environment is strongly variable and depends on the extent of marshland, which changes both seasonally (commonly peaking in February-March and progressively decreasing thereafter up to near disappearance by July) and annually (spanning the whole continuum between extreme flood-years and extreme drought-years in which the marsh never floods). In contrast, two permanently flooded man-made systems nearby (the fish farm of Veta la Palma and the saltmines of Sanlúcar, Figure 1) provide waterbirds with two alternative and stable foraging habitats. This system (i.e. the combination of natural marshes, the fish-farm and saltmines) represents, then, a unique opportunity to investigate the impact of environmental changes affecting the extent and function of natural systems over wild populations and the role of man-made systems as alternative feeding habitats.

We used a long-term (1983–2004) remote sensing data set to infer inter-annual fluctuations in the extent of natural marshes, assumed to be a reliable surrogate of natural and local resource availability [18], and evaluate their effect over the reproductive performance of the Slender-billed Gull (*Chroicocephalus genei*) population breeding at the Natural Area of Doñana for the 1993–2004 period. Given that resource availability may constrain or influence relevant breeding parameters throughout the entire reproductive cycle: such as the likelihood of laying eggs or skipping reproduction, number and size of eggs laid, rate of hatching success, and number of nestlings raised to fledgling age (see [18]); we hypothesized that (1) inter-annual fluctuations in the extent of marshland, and therefore on natural resource availability, during different key periods (i.e. clutch production, incubation and chick-rearing periods) will result in variations in the reproductive performance of gulls; (2) individuals may cope with such variations in resource availability through dietary exploitation of the two adjacent, anthropogenic systems (i.e. the fish farm and saltmines). This second hypothesis was approached by using $\delta^{13}C$ and $\delta^{15}N$ values in tissues of fledglings and isotopic mixing models to reconstruct their use of various habitats for the 2001–2004 breeding period. Further insights into the actual role of such artificial systems as alternate feeding habitats for gulls were obtained by investigating the relationship between dietary estimates and reproductive performance (i.e. breeding success and fledgling body condition). To the best of our knowledge, this is the first study where remote sensing analyses of habitat availability and stable isotope approaches have been applied simultaneously for investigating the actual effect of habitat loss and fragmentation on population parameters. We argue that this approach provides an important advance in the way we tackle species conservation in a rapidly changing world.

Methods

Ethics Statement

The authors declare that all animals were handled in strict accordance with good animal practice as defined by the current European legislation, and all animal work was approved by the respective regional committees for scientific capture ("Consejería de Medio Ambiente de la Junta de Andalucía", Sevilla, Spain). All necessary permits were obtained for the described field studies (provided by "Dirección General de Gestión del Medio Natural-Consejería de Medio Ambiente de la Junta de Andalucía" and "Dirección Natural de Espacios Naturales y Participación Ciudadana- Consejería de Medio Ambiente de la Junta de Andalucía").

Study Area

The study was conducted at Doñana National and Natural Park (i.e the Natural Area of Doñana), a protected wetland (50,000 ha) at the Guadalquivir River estuary (SW Spain, see Figure 1). The area was declared a National Park in 1969, and, afterwards, a World Heritage Site, Biosphere Reserve (UNESCO), Important Wetland Site under the Ramsar Convention (1982), Important Bird Area (IBA) and Nature 2000 site. Strong human-induced changes, mainly due to agricultural practices and water channel drainage, occurred during the 20th Century leading to the current 27,000 ha of freshwater marshes within the National Park [19]. Natural marshes are annually flooded (between October and March) by winter rains, and then progressively dry up afterwards. Thus, the extent of the marshland radically varies both annually and seasonally according to the rainfall and climate regime [20]. Within the natural marshes, several flat islands, such as Veta de las Vaquiruelas, provide appropriate nesting sites for waterbirds. In the Natural Park, ca. 3,500 ha of the former marshland were transformed in 1992 for commercial fish farming. This private area, known as Veta la Palma (Figure 1), is formed by ca. 40 interconnected ponds with a permanent flooding regime that holds brackish water pumped from the Guadalquivir River estuary and contains re-vegetating islands for nesting waterbirds (see [16,17,21]). Within the limits of the Natural Park, ca. 3,000 ha are dominated by a commercial salt pan complex, the so-called saltmines of Sanlúcar (Figure 1). Salt production at this area goes back to the XVth century, although latest modifications of this salt pan complex were carried out during the sixties. This man-made system includes several permanently flooded ponds of increasing salinities that ultimately leads to low-diversity communities mainly consisting of brine-shrimp [16]. The three areas included in this study (i.e. the natural marshes of the Doñana National Park, the fish farm of Veta la Palma and the saltmines of Sanlúcar) show, therefore, considerable differences in type of habitat and hydrological regimens and conditions that likely affect the availability of trophic resources and their accessibility to free-living animals.

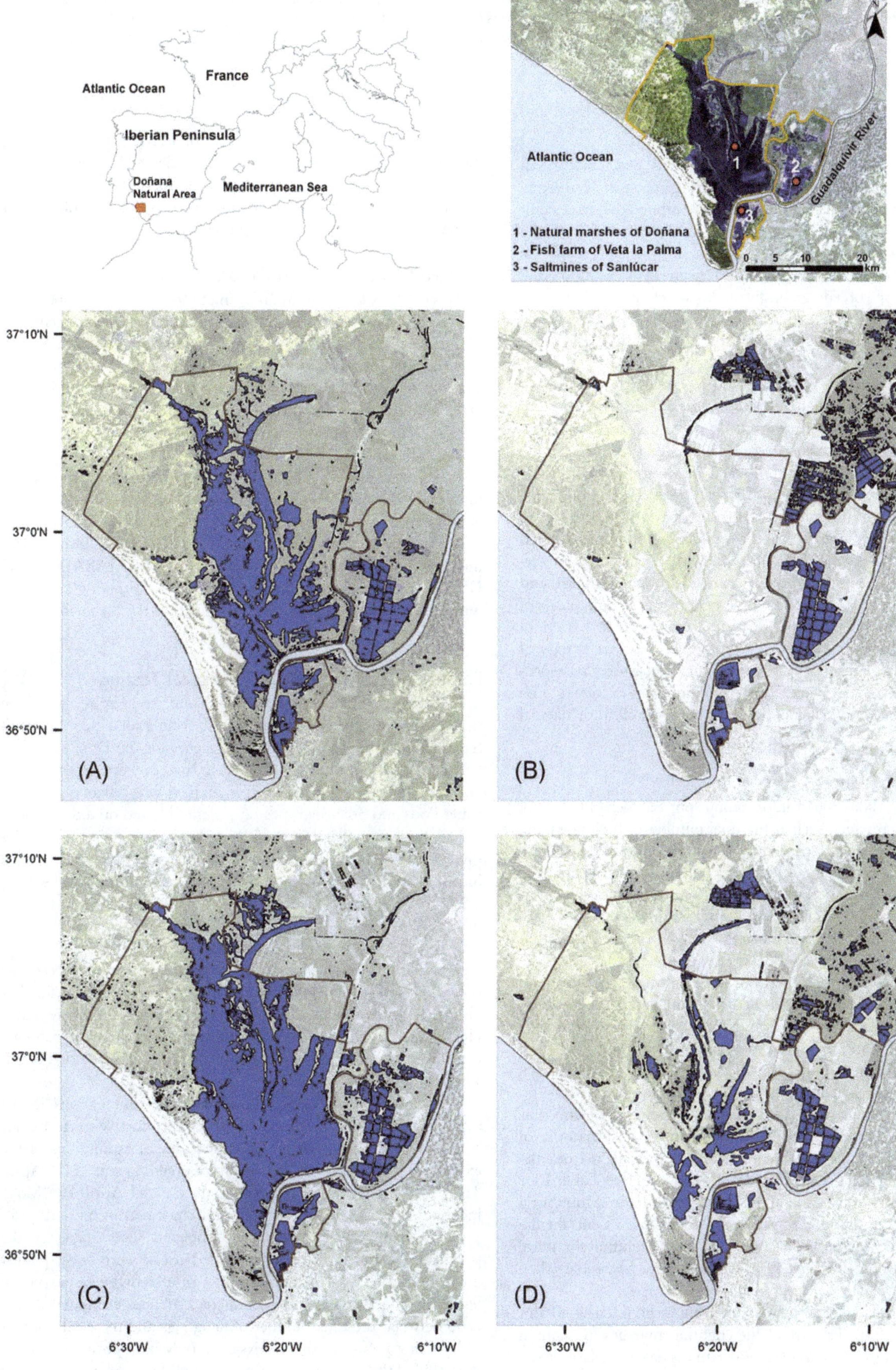
France
Atlantic Ocean
Iberian Peninsula
Doñana
Natural Area
Mediterranean Sea
Atlantic Ocean
Guadalquivir River
1 - Natural marshes of Doñana
2 - Fish farm of Veta la Palma
3 - Saltmines of Sanlúcar
0 5 10 20 km
37°10'N
37°0'N
36°50'N
(A)
(B)
(C)
(D)
6°30'W
6°20'W
6°10'W

Figure 1. Temporal trends in the flooded area extent in the marshes of Doñana. Two examples of progressive inundation rates of the marshland of Doñana National Park in two different years. Inundation masks based on Landsat 7 (ETM sensor) images acquired on 29 April 2002 (A), 18 July 2002 (B), and Landsat 5 (TM sensor) images from 26 April 2004 (C) and 15 July 2004 (D). The Landsat false color composite images include bands in 7-5-3 (SWIR, Near Infrared, Red) combination. Upper figures show geographical location of the study zone and nesting colonies of Slender-billed Gull.

Studied Species

The Slender-billed Gull is a medium-sized gull that breeds locally from Senegal and Mauritania, throughout the Mediterranean to Western India [22]. Doñana harbors an important breeding population of this species. From 1992 onward, the population has increased exponentially in parallel with the establishment of the commercial fish farm of Veta la Palma (Figure 2). For breeding, individuals select alternatively or simultaneously the above-mentioned three different areas (see Figures 1 and 2). Three different criteria made this gull species an ideal candidate for evaluating the effect of inter-annual fluctuations in the availability of natural food resources and the role of artificial systems as alternate feeding habitats: (1) individuals were expected to feed on resources both from natural and anthropogenic systems (previous dietary reports suggested this species to be mainly piscivorous, although individuals are commonly observed on saltmines consuming brine-shrimps, *Artemia spp.*, [11,23,24]), with relative contributions presumably varying according to fluctuations in availability; (2) they tend to breed on isolated colonies in high densities [24], thus allowing accurate characterization of several population parameters (e.g. population size and reproductive success); and (3) the high-energy demanding phase of chick-rearing in this species (June–July) matches the period of maximum uncertainty in the extent of the flooded area at natural marshes, so that reproductive performance (measured as reproductive success and fledglings' body condition) was expected to be particularly sensitive to fluctuations in the availability of natural resources.

Fieldwork Procedures

Long-term (1983–2004) information on Slender-billed Gull population parameters, such as the total number of breeding pairs and fledglings has been recorded by six expert ornithologists of the Monitoring Team of Natural Processes of the Biological Station of Doñana. This information was initially (1983–1992) based on non-systematic direct observations. However, banding campaigns were implemented since 1993 and all breeding colonies located within this area were visited at least once per year matching up with the end of the chick-rearing period, thus allowing the accurate recording of the total number of nests and fledglings for each breeding site. Between 2001 and 2004, fledglings were also weighed (±10 g) and measured (maximum head length; ±1 mm) using a dynamometer and a digital caliper (head lengths for the 2001 breeding season are not available), respectively. Morphometric measurements were always taken by the same person thus avoiding any collector bias. In addition, from a subsample of fledglings randomly selected, 1 ml of blood was taken from the brachial vein for isotope ($\delta^{13}C$ and $\delta^{15}N$) analyses (see Table 1 for sample sizes by year). Additionally, staff from the Monitoring Team carried out monthly censuses of Slender-billed Gulls for the 2000–2004 breeding period at the fish farm and saltmines (natural marshes were not visited as they are not easily accessible when flooded).

Based on previous information on feeding preferences of this species [11,23,24], fish from the natural marshes (Common Mummichogs *Fundulus heteroclitus* and Koi *Cyprinus carpio*) and from the fish farm (Eastern Mosquitofish *Gambusia holbrooki*), as well as brine-shrimps *Artemia* spp from the saltmines were collected during the 2004 breeding season for isotopic determination of main dietary resources. Once at the laboratory, all samples (i.e. fledglings' blood and prey items) were frozen until further preparation for stable isotope analyses.

Sample Preparation and Laboratory Analyses

All samples were dried in an oven at 60°C to constant mass and ground to a powder. Two aliquots were removed from each prey sample; one aliquot was immediately prepared for $\delta^{15}N$ analysis, whereas the other underwent lipid extraction prior to $\delta^{13}C$ analysis following the methods described at [25]. About 0.9–1 mg of each sample were combusted at 1020°C using a continuous-flow isotope-ratio mass spectrometry (CFIRMS) system consisting of a Carlo Erba 1500NC elemental analyzer interfaced with a Delta Plus XL mass spectrometer. Stable isotope ratios are expressed in the standard δ-notation (‰) relative to Vienna Pee Dee Belemnite ($\delta^{13}C$) and atmospheric N_2 ($\delta^{15}N$). Replicate assays of laboratory standards (urea and shark cartilage) previously calibrated with international reference materials (IAEA-CH-6, IAEA-N-2 and USGS-40), and routinely inserted within the sampling sequence, indicated analytical measurement errors of ±0.1 ‰ for $\delta^{13}C$ and ±0.2 ‰ for $\delta^{15}N$.

Parameter Estimations and Statistical Testing

Estimations of the extent of marshland (see Figure 1) for the periods of interest were obtained by accessing all available cloud-free Landsat MSS, TM and ETM+ scenes for the Doñana region between 1983 and 2004 (190 scenes). Images were geometrically and radiometrically corrected, transformed into reflectance values using Pons and Solé-Sugrañes [26] method based on a dark object model, and normalized to a reference image using a set of pseudo-invariant areas [27]. Inundation rate was determined using mid-infrared band 5 (1.55–1.75 μm, TM and ETM+) and band 4 (0.8–1.1 μm, MSS) to produce final inundation masks based on pixels of 30×30 m (details in [27,28]; examples in Figure 1). Since Landsat images were available approximately every 15 days (depending on cloud coverage), and because we were interested in obtaining integrative information on inundation levels for key periods comparable among years (see below), piecewise cubic Hermite interpolation of existing information was used to estimate the daily extent of the marshland for the time span included in this study.

To evaluate the influence of inter-annual fluctuations in the availability of natural resources on the reproductive performance of gulls at our study site, we obtained annual estimations of the extent of natural marshes for the entire breeding season (1st April-31th July), but also for the clutch production (1st April–15th May), incubation (15th May–15th June), and chick-rearing periods (15th June–31th July), by averaging interpolated daily data for the periods of interest. Estimated flooded areas were subsequently incorporated as explanatory variables in a stepwise regression to test their effects over gull reproductive success (estimated as the ratio between the total number of fledglings and the total number of breeding pairs). In this analysis we only incorporated data for the 1993–2004 period because of the lack of periodic and

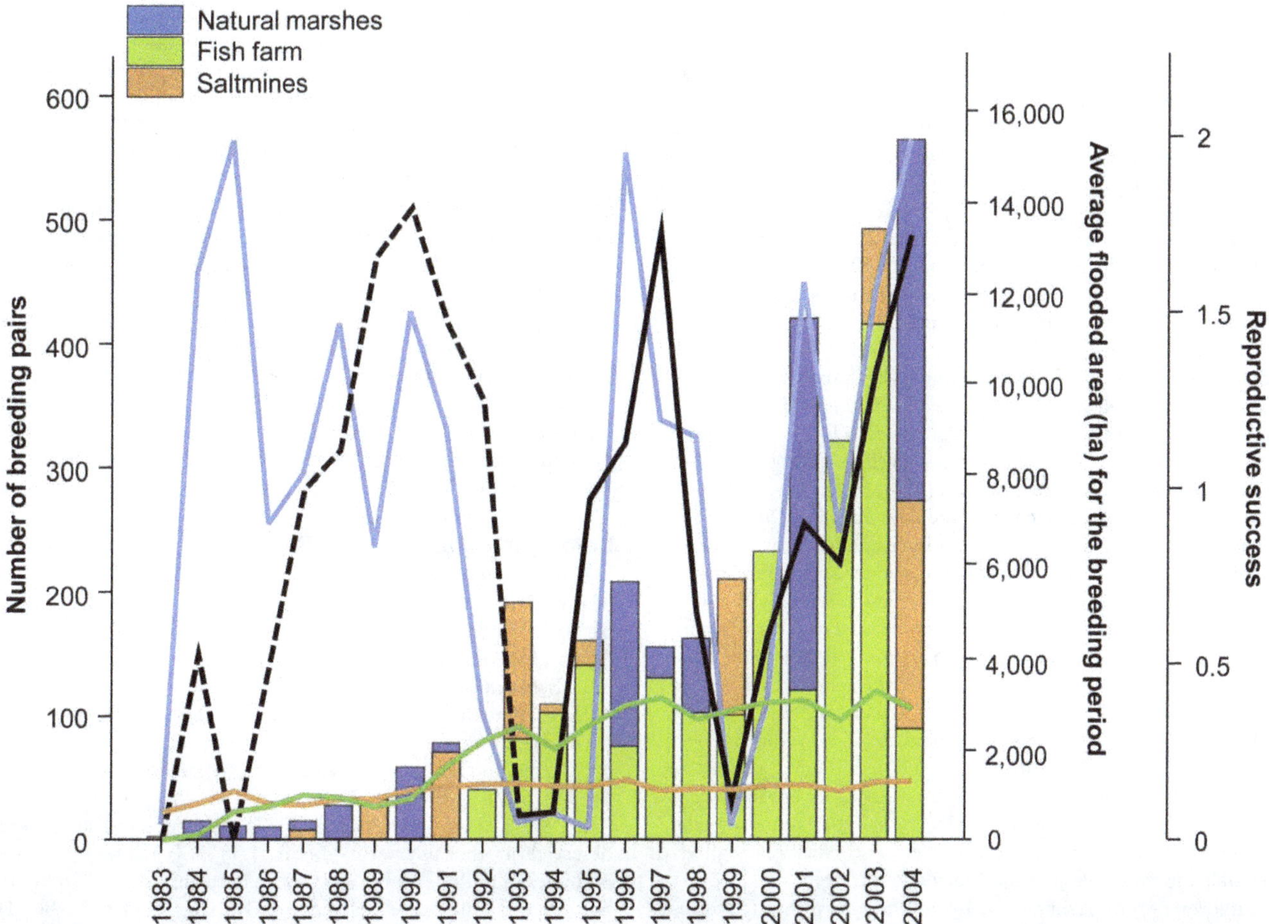

Figure 2. Linking natural resource availability with reproductive performance. Long-term (1983–2004) evolution of: bars = number of breeding pairs for the Slender-billed Gull population breeding at the Natural Area of Doñana (split by breeding site); colored lines = average extent of the flooded area for the natural marshes (blue), the fish farm (green; data for the 1983–1991 period refer to the flooded area of the former marshland that was transformed for commercial fish farming in 1992) and the saltmines (orange) during the Slender-billed gull breeding period (April-July); black line = overall reproductive success of the Slender-billed gull population (dashed black line represents the estimated reproductive success before 1993, which have been estimated based on few and non systematic direct observations and have not been considered in the analyses, see Material and Methods).

Table 1. Dietary reconstructions of Slender-billed Gull fledglings based on blood $\delta^{13}C$ and $\delta^{15}N$ values.

				Relative importance (%, 95th BCI)		
		$\delta^{13}C$ (Mean ± SD ‰)	**$\delta^{15}N$ (Mean ± SD ‰)**	**SM**	**FF**	**NM**
2001	NM (n = 14)	$-16.38\pm1.25^{2,3}$	14.52 ± 0.75^{2}	70.4–100	0–24.1	0–8.2
	FF (n = 18)	$-15.32\pm0.50^{1,2}$	14.64 ± 0.26^{1}	78.8–100	0–18.1	0–4.2
2002	FF (n = 56)	$-15.62\pm0.89^{1,2}$	13.62 ± 0.50^{2}	77.4–86.0	0–3.4	13.7–20.2
2003	FF (n = 54)	-17.43 ± 1.11^{3}	12.58 ± 0.74^{3}	56.6–68.1	0–3.4	31.3–41.6
	SM (n = 24)	-14.93 ± 1.12^{1}	13.33 ± 0.92^{1}	58.2–92.2	0–12.9	8.4–32.2
2004	NM (n = 54)	-19.02 ± 0.85^{4}	11.29 ± 0.40^{5}	27.2–35.5	0–3.6	64.0–70.3
	SM (n = 17)	-17.67 ± 0.89^{3}	11.91 ± 0.54^{4}	29.9–53.6	0–9.4	45.7–61.8

Stable isotope values for whole blood of Slender-billed Gull fledglings as a function of year (2001–2004) and breeding site (saltmines of Sanlúcar -SM-, the fish farm of Veta la Palma -FF- and natural marshes of Doñana -NM-). Pair-wise comparisons were done through Tamhane's test procedure (overall alpha level of 0.05); identical superscript numbers indicate homogeneous sub-groups. Relative importance (95% Bayesian Credibility Intervals, BCI) of food from the three sites (SM, FF and NM) in the diet of fledglings hatched at different sites and in different years was inferred through a Bayesian double-isotope ($\delta^{13}C$ and $\delta^{15}N$) isotopic mixing model (SIAR, [29]).

standardized data on population parameters before 1993 (see above).

Insights into the role of the two adjacent, anthropogenic systems as alternative feeding areas for gulls feeding chicks were obtained by investigating how availability of natural resources during the chick-rearing period affected the dietary composition of fledglings. Chick diets (split by year and breeding site) were inferred through a double-isotope ($\delta^{13}C$ and $\delta^{15}N$), three endpoint (fish from the natural marshes and from the fish farm, and brine-shrimp from the saltmines) Bayesian mixing model (SIAR, [29]). Dietary endpoints included in these models were isotopically clustered (mean ± SD) by grouping potential prey from each foraging area. One-way ANOVA (with Welch's correction to account for heteroscedasticity) and the Tamhane's tests procedure (for post-hoc pair-wise analyses; [30]) were used for testing differences in the isotopic composition of dietary endpoints, which were subsequently adjusted to account for isotopic discrimination factors (ΔX) linking diet with consumers' tissues ($\Delta^{13}C$: ~1.1 ‰; $\Delta^{15}N$: ~2.9 ‰; see [31]). The estimated extent of the marshland for the chick-rearing period was finally incorporated in a linear model to evaluate the influence of natural resource availability on the relative contribution of main dietary resources to the diet of fledglings. Complementary to our isotope approach, insights into the effect of natural resource availability on feeding habitat choice by gulls were obtained by investigating the relationship between estimated flooded areas at the natural marshes and the proportion of adult gulls (respect the total number of reproducing birds) observed at the saltmines (the main alternate foraging habitat, see Results). In particular, estimated proportions of gulls observed were fitted in a linear model in which flooded area and breeding stage (i.e. clutch production, incubation and chick-rearing) were included as covariate and fixed factor, respectively.

Information on resource use by gulls obtained throughout our isotope study was finally used to evaluated the effect of potential dietary shifts towards exploitation of alternate, anthropogenic resources on gulls' breeding performance. Mean relative contributions of food from natural marshes derived from the mixing models were incorporated in a linear model to test the effect of this explanatory variable on reproductive success (split by year and breeding site) of our gull population. We also investigated the relationship between fledglings' diet and body condition (estimated as the residuals of the linear regression of body mass, i.e. body weight, and body size, i.e. maximum head length, [32]) at the individual level through linear models in which individual body conditions was included as the response variable and raw $\delta^{13}C$ and $\delta^{15}N$ values for fledglings' blood were incorporated as explanatory variables (provided that natural resources showed depleted isotopic values with respect those ascribed to man-made systems, see Results). To account for the dependence among fledglings raised at the same colony and during the same year, the combination of year and breeding site was included as a random factor. Analyses were done using MatLab 7.11 software (MathWorks Inc., Massachusetts, USA) for interpolations and SPSS 18.0 (SPSS Inc., Chicago, Illinois, USA) for statistical testing.

Results

Marshland Extent and Reproductive Success

Estimated average extent of the marshland strongly varied among years included in this study. Overall, the average marshland extent for the Slender-billed Gull breeding period ranged from extreme flood-years like 2004, when the average extent of the marshland was 15,049 ha, to extreme drought-years like 1995, with an average extent of flooded area of 151 ha (see Figure 2 for the temporal trend in the inundation levels). In particular, and accordingly to the flooding regime at the natural marshes (see above), chick-rearing matches up with the driest period, with average flooded areas ranging from 62 ha (1983) to 9,752 ha (1985), followed by the incubation (from 129 ha in 1995 and 17,478 ha in 1996) and the clutch-production period (from 226 ha in 1995 to 21,937 ha in 1996). Similarly, great inter-annual differences in gulls' breeding performance were observed for this time-period. In particular, estimated reproductive successes ranged from ca. 0 (e.g. 1993 or 1994; values before 1993 were not considered) to ca. 1.7 chicks of fledgling age per nest (e.g. 1997 or 2004, see Figure 2), likely depending on the inundation levels at the marshland for the entire breeding season as suggested by the observed positive relationship between estimated flooded areas and reproductive success (selected model from the stepwise regression included the average extent of the natural marshes for the entire breeding period as the only explanatory variable, $F_{1, 10} = 10.6$, $p = 0.01$).

Isotopic Interpretation and Dietary Changes

Significant differences were detected in the isotopic composition of prey from the three main foraging areas ($F_{Welch\ 2,\ 6.1} = 22.9$, $p = 0.001$ for $\delta^{13}C$ and $F_{Welch\ 2,\ 6.5} = 47.3$, $p<0.001$ for $\delta^{15}N$, see Table 1 for pair-wise comparisons), thus allowing the use of mixing models to transform isotopic information into relative contributions of potential food resources to the diet of fledglings. In particular, brine-shrimp from the saltmines showed the highest $\delta^{13}C$ values (mean ± SD: −17.02±0.48 ‰), followed by fish from the fish farm (−20.13±1.77 ‰) and from the natural marshes (−21.94±2.13 ‰). Regarding $\delta^{15}N$, the highest values were observed at the fish farm (14.29±1.38 ‰), followed by the saltmines (11.48±0.54 ‰) and the natural marshes (8.28±0.81 ‰). Dietary inferences derived from mixing models (Table 1) revealed that saltmines and natural marshes constituted the main foraging areas regardless of the year and breeding site considered (with average mean contributions of ca. 67% and 28%, respectively), whereas the dietary exploitation of the fish farm was relatively low (ca. 5%), although it is noteworthy its role as breeding site (Table 1 and Figure 2). Remarkably, estimated dietary compositions were similar among breeding sites within single years, whereas great inter-annual variations in the relative contribution of the two main dietary endpoints were detected throughout the 2001–2004 time-period (Table 1).

Inter-annual variations in the diet of fledglings strongly depended on the extent of the marshland during the chick-rearing period (Figure 3). In particular, increasing flooded areas at the natural marshes resulted in a higher dietary exploitation of this natural foraging habitat ($F_{1,\ 6} = 19.8$, $p = 0.007$), in contrast with the dietary contribution of saltmine resources ($F_{1,\ 6} = 32.4$, $p = 0.002$). Relative use of resources from the fish farm stood relatively constant regardless of observed variations in the extent of the natural marshland ($F_{1,\ 6} = 0.53$, $p = 0.5$). In agreement with obtained results from our isotope approach, direct evidence based on monthly censuses also suggested that natural resource availability had a role in determining feeding habitat choice by gulls (Figure 4). In particular, the proportion of gulls observed at the saltmines diminished as extent of the natural marshland increased ($F_{1,\ 15} = 4.99$, $p = 0.047$), a trend also found for the clutch-production, incubation and chick-rearing periods ($F_{2,\ 15} = 0.36$, $p = 0.7$, for the interaction between marshland extent and breeding stage).

Use of natural foods strongly affected the reproductive performance of the species (Figure 5). In particular, a greater use of natural marshes for food resulted in higher reproductive

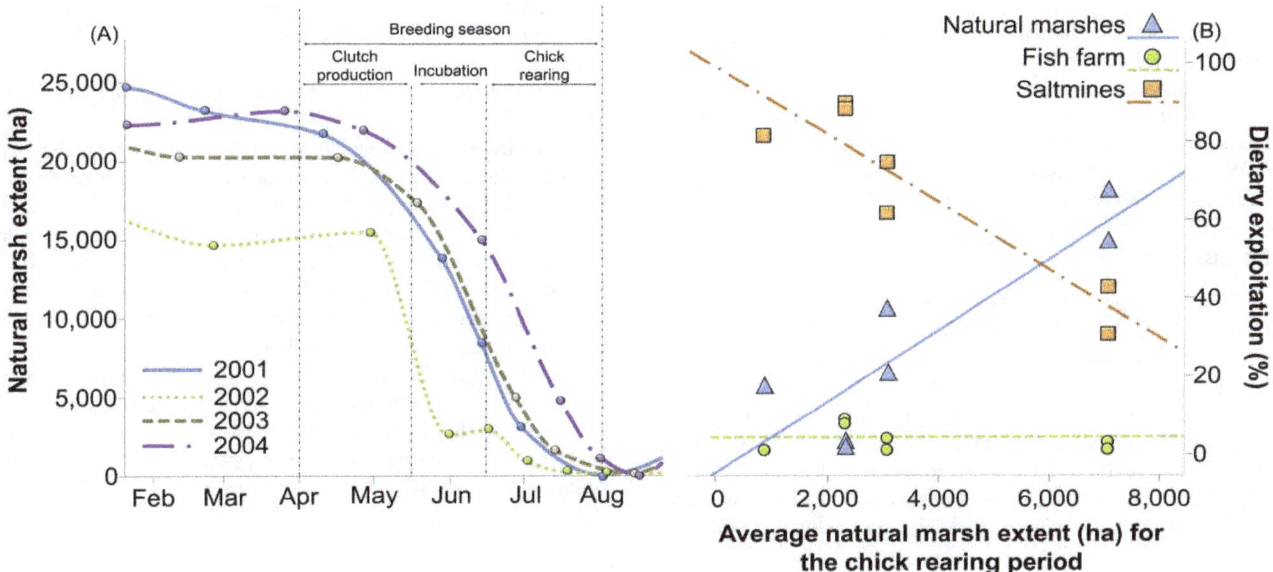

Figure 3. Linking natural resource availability with fledglings' diet. (A) Temporal trends for the flooded areas (ha) at the natural marshes for the 2001–2004 period. Dots represent real flooded areas obtained from satellite images. (B) Relationship between the estimated flooded area at the natural marshes (ha) during the chick-rearing period and the estimated relative contribution of natural resources (blue), and resources from the fish farm (green) and the saltmines (orange) to the diet of Slender-billed Gull fledglings. Dots represent dietary estimates for fledglings hatched at different sites and in different years.

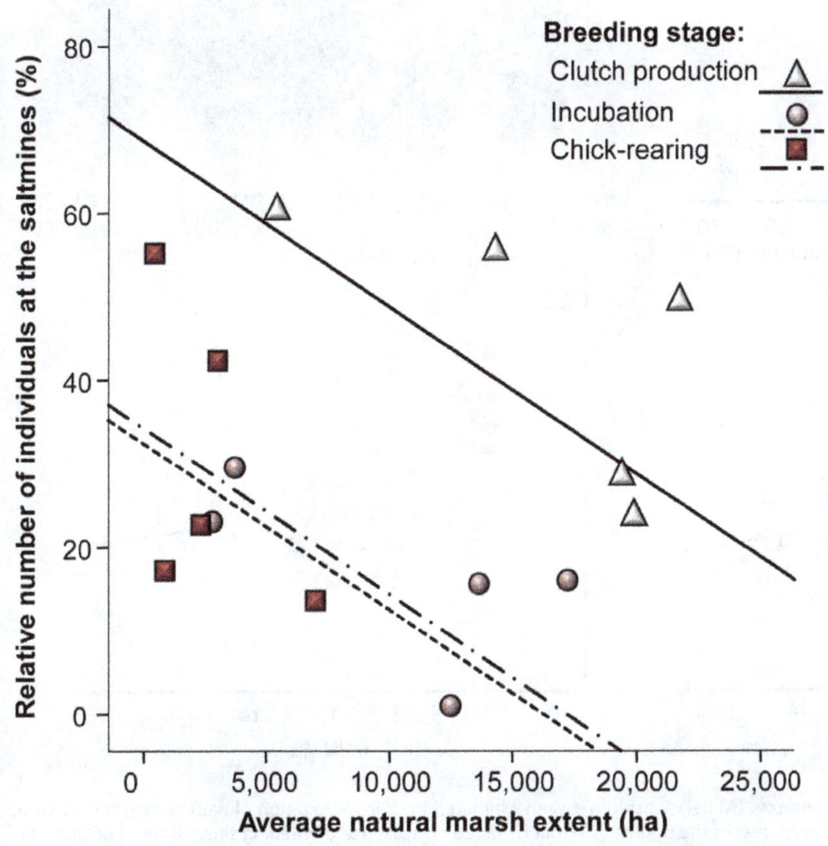

Figure 4. Linking natural resource availability with feeding habitat choice. Relationship between the estimated flooded area at the natural marshes (ha) during the clutch-production, the incubation and the chick-rearing period, and the proportion of adult gulls (i.e. respect the total number of reproducing birds) observed at the saltmines, i.e. the main alternate foraging habitat.

success ($F_{1,\ 5}=24.2$, $p=0.008$; note that Sanlúcar 2004 was excluded from the analysis since unusual rainfall levels partially flooded the colony and increased egg loss and chick mortality), and also in raising fledglings with better body condition ($F_{1,\ 106}=7.21$, $p=0.008$ and $F_{1,\ 106}=4.29$, $p=0.041$, for the observed negative relationship between fledglings' body condition and raw $\delta^{13}C$ and $\delta^{15}N$ values, respectively; data were not available from 2001, when fledgling morphometrics were not recorded).

Discussion

The connection between resource availability and population dynamics is likely mediated through the acquisition of resources and their allocation to one of the main components of animal fitness such as reproduction [33,34]. Thus, studies aimed at comparing how animals acquire resources and their consequences for population breeding performance during periods of contrasting resource availability will be informed by an understanding of the impact of global environmental change on wild populations. Here, the Natural Area of Doñana provided an exceptional opportunity for such a study, because the observed inter-annual fluctuations in the extent of natural marshland forced wild populations breeding at this area to fine-tune each reproductive event to varying availability of natural and anthropogenically derived resources [18].

Slender-billed gulls breeding at the Natural Area of Doñana showed the greatest range in reproductive success (ranging from ca. 0 to 1.7 fledglings per breeding pairs) ever reported for the Northwestern Mediterranean (breeding success for the population breeding at the Ebro Delta during the period 1992–2001 ranged from ca. 0 to 1.25; [24]), with higher values for flood-years compared to drought years. Although these results cannot prove the causality of the relationship between the extent of natural marshland and the breeding performance of this gull population, similar results have been reported for other wild species inhabiting this area (e.g. Black-crowned Night Heron *Nycticorax nycticorax*, Cattle Egret *Bubulcus ibis*, Little Egret *Egretta garzetta*, or black kite *Milvus migrans*; [18,35]). Thus, the availability of food from the

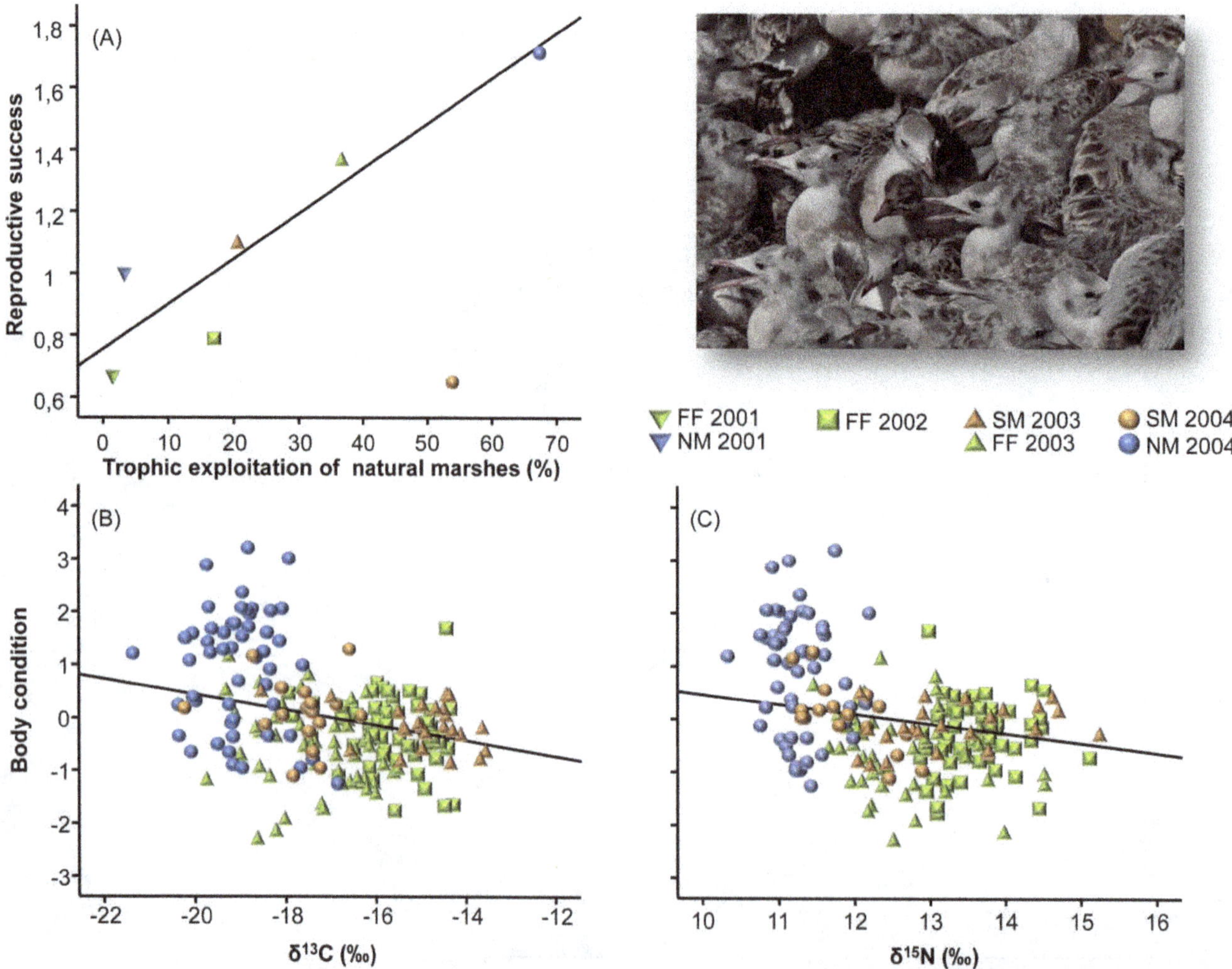

Figure 5. Linking diet with reproductive performance. (A) Relationship between estimated trophic exploitation of natural marshes (in %) by Slender-billed Gulls feeding chicks and the reproductive success of the population (split by breeding season and breeding site: saltmines of Sanlúcar -SM-, fish farm of Veta la Palma -FF- and natural marshes of Doñana -NM-). An unusual flooding episode caused a massive mortality of fledglings at SM 2004, and data on breeding success has been excluded from the regression. (B and C) relationships between raw isotopic values ($\delta^{13}C$ and $\delta^{15}N$, in ‰) for fledglings' blood and estimated body condition (residuals of the relationship body size-body weight). Morphometric data for the 2001 breeding season were not available, thus making impossible the estimation of fledglings' body condition. Picture courtesy of M. Paracuellos.

natural marshes seems to play an important role in determining the reproductive performance of species breeding here.

Food availability has been described previously as a major factor influencing the reproductive output of wild populations [36–39]. However, few studies have unequivocally demonstrated the constraining effect of food limitation on population reproductive performance [40,41]. The main reason for this appears to be the difficulty of monitoring variation in feeding preferences [41]. Indeed, investigating the relationship between food-abundances and reproductive performance is challenging, since individuals can switch to alternative resources when their staple prey becomes scarce, thus hampering rigorous analysis of the processes constraining food availability (see [18,41]). In this regard, stable isotope analysis has developed into a powerful tool for monitoring spatio-temporal variations in the diet of target populations likely associated with changes in resource availabilities (e.g. [42]), thus allowing quantitative inferences on the relationship between natural resource availability, diet-choice and the consequential effect on the reproductive performance of wild populations.

A common assumption of all retrospective isotopic studies is that isotopic baselines have either not changed through time or that such changes can be accounted for (see [43], and references therein). In our study, isotopic data for prey were only available for 2004, and so we cannot completely rule out the possibility that observed inter-annual variations in dietary estimates were caused by temporal variations in the isotopic composition of main prey types. However, several arguments suggest that this hypothesis is an unlikely explanation for our observed temporal trends. Firstly, temporal shifts in the isotopic composition of particular food resources are expected to be small compared to the considerable isotopic variations caused by changes in bird diets and foraging locations (see [42–44]). Secondly, observed relationships between marshland extent and dietary estimates clearly agree with information obtained from direct observations on feeding habitat choice by gulls, but also with observed relationship between marshland extent and reproductive success for the 1993–2004 period (provided that dietary exploitation of natural foods likely resulted in higher reproductive success). Our information on resource use by gulls therefore suggested that, when available, individuals apparently tended to prefer fish from natural marshes, whereas saltmines were the main alternative foraging area. In contrast, the minor role of the fish farm as a feeding area was noteworthy, a fact that could well be the consequence of the several mechanisms that have been implemented at this system to avoid fish predation by avian species (e.g. ponds with the highest fish densities [ca. 0.15% of the fish farm pond area] are covered with a mesh).

Brine-shrimp is a low-quality food for birds because of its low energy content [45], difficulties inherent in their consumption (i.e. the low rate of intake per unit of time [46]), and the associated energy costs related to salt excretion [46–48]. Thus, Slender-billed Gulls could tend to maximize fledgling survival, and, in turn, their breeding performance, by proactively biasing their diets towards consumption of a resource with a higher energy value, such as fish [49], whenever this resource was available for the population. Indeed, the relative contribution of fish to the diet of chicks has previously proved to positively affect the breeding success of other gull species [50–53]. This was in agreement with our inferred positive relationship between relative consumption by gulls of fish from the marshland and estimated reproductive output (reproductive success and fledglings' body condition).

In addition to inter-annual variations in the availability of natural food resources as mediated by flooding, varying predation pressure and/or inter-specific competition for food may also have a role in explaining observed temporal pattern in gulls' reproductive output. Based on historical records provided by the Monitoring Team of Natural Processes, predation episodes were sporadic (only observed during the 1985, 1993, 1999 and 2000 breeding periods) and mostly affected individuals breeding at the natural marshes and the saltmines, which are particularly susceptible to be affected by terrestrial predators. Because of the sporadic nature of these episodes, predation would be expected to have a minor role in explaining the long-term pattern in the estimated reproductive output of our gull population. However, these episodes could partially explain variations between natural resource availability and gulls' reproductive performance. For example, predation by rats (*Rattus norvegicus*) during the 1985 breeding period, when natural resources availability was relatively high (see Figure 2), caused the total failure of the gull population breeding at the natural marshes. On the other hand, and similar to Slender-billed Gulls, most of the other waterbird species inhabiting this area alternate their use of natural and man-made habitats (e.g. [17]). Variations in the availability of natural food resources may result, therefore, in varying inter-specific competition for resources ascribed to artificial systems, thus having an additive effect on the observed relationship between natural resource availability and gulls' feeding preferences and reproductive output.

In conclusion, higher energy flux to chicks during flood years, due to an increase in the nutritional quality of their diet, is the more likely explanation for the observed positive relationship between the relative contribution of natural resources to the diet of fledglings and reproductive output (i.e. breeding success and fledglings' body condition). However, the actual role of natural resource availability in constraining gull reproduction entails assessing the effect of feeding habitat choice by gulls on other fitness-related parameters, such as the likelihood of laying eggs or skipping reproduction, the number and size of eggs laid or the rate of hatching success (provided that adults segregate food for themselves from that delivered to their chicks [15]). In this regard, information derived from direct observations suggested that the availability of natural resources was also a relevant factor affecting adult gull diet, since the relationship between marshland extent and the proportion of birds feeding at the saltmines was also observed for the clutch-production and the incubation period. That the most important factor affecting reproductive success was the average extent of the natural marshes for the entire breeding period suggested that natural resources availability likely affected other reproductive parameters different from chick survival and fledgling body condition such as egg production and hatching success.

The extent of natural habitats available for wild species is currently being lost at ever increasing rates as a result of human activities [54–56]. So, there is an ongoing need to understand how wild populations adapt to the new circumstances and, particularly, the potential role of several alternative systems, such as fish farms or saltmines, as buffer areas against loss or degradation of natural feeding habitats. The benefit for wildlife of these anthropogenic structures as alternative breeding/feeding sites is undeniable as they provide individuals with a relatively safe and stable environments (see [57]). Indeed, since 1992, the fish farm of Veta la Palma has been the preferred breeding site for our Slender-billed Gull population (see Figure 2; see also [58]), and this artificial system may have played an important role in the exponential growth that this population has undergone during the last two decades. Similarly, these man-made systems have been previously considered as suitable alternative feeding areas for wild populations [9]. Here, we do not question the importance of these structures as alternative foraging habitats for free-living animals, but stress the fact that, at least for some species, they cannot completely compensate for loss of natural resources. We suggest our methodology could be extended to a large suite of natural communities that are potentially threatened as a result of habitat loss, thus providing a useful framework for management and conservation purposes.

Acknowledgments

We thank L. García, M. Mañez, F. Ibáñez, J. L. Arroyo, C. Gutiérrez, T. Roberts and several volunteers of the monitoring staff of Natural Resources of Doñana and Dirección General de Espacios Naturales (Consejería de Medio Ambiente, Junta de Andalucía) for their invaluable field assistance. We also thank our colleague R. Díaz-Delgado (Laboratorio de SIG y Teledetección, EBD-CSIC) for his helpful suggestions for dealing with remote sensing analyses. Owners of both saltmines and fish farm made always easier our work in these areas. R. Astasio and G. Jans helped us with the permit application processes. A. Granados helped us with sample preparation and stable isotope analyses.

Author Contributions

Conceived and designed the experiments: FR JN IA MGF. Performed the experiments: IA MGF. Analyzed the data: FR IA. Contributed reagents/ materials/analysis tools: KAH AD. Wrote the paper: FR JN IA KAH AD MGF.

References

1. Hoekstra JM, Boucher TM, Ricketts TH, Roberts C (2005) Confronting a biome crisis: Global disparities of habitat loss and protection. Ecology Letters 8: 23–29.
2. Lotze HK, Lenihan HS, Bourque BJ, Bradbury RH, Cooke RG, et al. (2006) Depletion, degradation, and recovery potential of estuaries and coastal seas. Science 312(5781): 1806–1809.
3. Kerr LA, Secor DH, Kraus RT (2007) Stable isotope (delta C-13 and delta O-18) and Sr/Ca composition of otoliths as proxies for environmental salinity experienced by an estuarine fish. Mar Ecol Prog Ser 349: 245–253.
4. Turner W, Spector S, Gardiner N, Fladeland M, Sterling E, et al. (2003) Remote sensing for biodiversity science and conservation. Trends Ecol Evol 18(6): 306–314.
5. Carey C, Heyer WR, Wilkinson J, Alford RA, Arntzen JW, et al. (2001) Amphibian declines and environmental change: Use of remote-sensing data to identify environmental correlates. Conserv Biol 14(4): 903–913.
6. Osborne PE, Alonso JC, Bryant RG (2001) Modelling landscape-scale habitat use using GIS and remote sensing: A case study with great bustards. J Appl Ecol 38(2): 458–471.
7. DeFries R, Hansen A, Newton AC, Hansen MC (2005) Increasing isolation of protected areas in tropical forests over the past twenty years. Ecol Appl 15(1): 19–26.
8. Dolman PM, Sutherland WJ (1995) The response of bird populations to habitat loss. Ibis 137: S38–S46.
9. Masero JA (2003) Assessing alternative anthropogenic habitats for conserving waterbirds: Salinas as buffer areas against the impact of natural habitat loss for shorebirds. Biodivers Conserv 12(6): 1157–1173.
10. Longoni V (2010) Rice fields and waterbirds in the Mediterranean region and the middle east. Waterbirds 33: 83–96.
11. Ramírez F, Abdennadher A, Sanpera C, Jover L, Hobson KA, et al. (2011) Tracing waterbird exposure to total mercury and selenium: A case study at the solar saltworks of Thyna (Sfax, Tunisia). Environ Sci Technol 45(12): 5118–5124.
12. Ramírez F, Abdennadher A, Sanpera C, Jover L, Wassenaar LI, et al. (2011) Assessing waterbird habitat use in coastal evaporative systems using stable isotopes (δ^{13}C, δ^{15}N and δD) as environmental tracers. Estuar Coast Shelf Sci 92: 217–222.
13. Hobson KA (1993) Trophic relationship among high Arctic seabirds, insights from tissues-dependent stable-isotope models. Mar Ecol Prog Ser 95: 7–18.
14. Hobson KA, Piatt JF, Pitocchelli J (1994) Using stable isotopes to determine seabird trophic relationships. J Anim Ecol 63(4): 786–798.
15. Navarro J, Oro D, Bertolero A, Genovart M, Delgado A, et al. (2010) Age and sexual differences in the exploitation of two anthropogenic food resources for an opportunistic seabird. Mar Biol 157: 2453–2459.
16. Rendón MA, Green AJ, Aguilera E, Almaraz P (2008) Status, distribution and long term changes in the waterbird community wintering in Doñana, South-West Spain. Biol Conserv 141(5): 1371–1388.
17. Kloskowski J, Green AJ, Polak M, Bustamante J, Krogulec J (2009) Complementary use of natural and artificial wetlands by waterbirds wintering in Doñana, South-West Spain. Aquatic Conserv: Mar Freshw Ecosyst 19: 815–826.
18. Sergio F, Blas J, López L, Tanferna A, Díaz-Delgado R, et al. (2011) Coping with uncertainty: Breeding adjustments to an unpredictable environment in an opportunistic raptor. Oecologia 166: 79–90.
19. Díaz-Delgado R, Aragonés D, Ameztoy I, Bustamante J (2010) Monitoring marsh dynamics through remote sensing. In: Hurford C, Schneider M, Cowx I, editors. Conservation monitoring in freshwater habitats: A practical guide and case studies. Dordrecht: Springer. 325–337.
20. Serrano L, Reina M, Martín G (2006) The aquatic systems of Doñana (SW spain): Watersheds and frontiers. Limnetica 25: 11–32.
21. Figuerola J, Green AJ, Santamaria L (2003) Passive internal transport of aquatic organisms by waterfowl in Doñana, South-West Spain. Global Ecology & Biogeography : 427–436.
22. Cramp S, Simmons KL (1983) The birds of the Western Paleartic. Oxford: Oxford University Press.
23. Fasola M, Bogliani G, Saino N, Canova L (1989) Foraging, feeding and time-activity niches of eight species of breeding seabirds in the coastal wetlands of the Adriatic Sea. Bolletino Di Zoologia 56(1): 61–72.
24. Oro D (2002) Breeding biology and population dynamics of Slender-billed Gulls at the Ebro Delta (Northwestern Mediterranean). Waterbirds 25(1): 67–77.
25. Smedes F (1999) Determination of total lipid using non-chlorinated solvents. Analyst 124: 1711–1718.
26. Pons X, Solé-Sugrañes L (1990) A simple radiometric correction model to improve automatic mapping of vegetation from multispectral satellite data. Remote Sensing of Environment 48: 191–204.
27. Bustamante J, Pacios F, Díaz-Delgado R, Aragonés D (2009) Predictive models of turbidity and water depth in the doñana marshes using landsat TM and ETM+ images. J Environ Manag 90: 2219–2225.
28. Bustamante J, Pacios F, Díaz-Delgado R, Aragonés D (2006) Predictive models of turbidity and water depth in the Doñana marshes using landsat TM and ETM+ images. Proceedings of the 1st GlobWetland Symposium, Organised by ESA and Ramsar Convention.
29. Parnell A, Inger R, Bearhop S, Jackson AL (2008) Stable isotope analysis in R (SIAR). Available: http://cran.r-project.org/web/packages/siar/index.html. Accessed 2012 Sep 19.
30. Tamhane AC (1979) Comparison of procedures for multiple comparisons of means with unequal variances. Journal of the American Statistical Association 74(366): 471–480.
31. Caut S, Angulo E, Courchamp F (2009) Variation in discrimination factors (delta N-15 and delta C-13): The effect of diet isotopic values and applications for diet reconstruction. J Appl Ecol 46(2): 443–453.
32. Alonso-Alvarez C, Velando A, Ferrer M, Veira JAR (2002) Changes in plasma biochemistry and body mass during incubation in the yellow-legged gull. Waterbirds 25(2): 253–258.
33. Stearns SC (1992) The evolution of life histories. Oxford: Oxford University Press.
34. Pinaud D, Cherel Y, Weimerskirch H (2005) Effect of environmental variability on habitat selection, diet, provisioning behaviour and chick growth in Yellow-nosed Albatrosses. Mar Ecol Prog Ser 298: 295–304.
35. Santoro S, Máñez M, Green AJ, Figuerola J (2010) Formation and growth of a heronry in a managed wetland in Doñana, Southwest Spain. Bird Study 57(4): 515–524.
36. Jouventin P, Weimerskirch H (1991) Changes in the population size and demography of southern seabirds: Management implications. In: Perrins CM, Lebreton JD, Hirons GJM, editors. Bird population studies: relevance to conservation and management. Oxford: Oxford University Press. 297–314.
37. Wanless S, Harris MP, Calladine J, Rothery P (1996) Modelling responses of Herring Gull and Lesser Black-backed Gull populations to reduction of reproductive output: Implications for control measures. J Appl Ecol 33(6): 1420–1432.
38. Phillips RA, Bearhop S, Hamer KC, Thompson DR (1999) Rapid population growth of Great Skuas *Catharacta skua* at St. Kilda: Implications for management and conservation. Bird Study 46: 174–183.
39. Oro D, Martinez-Abrain A, Paracuellos M, Nevado JC, Genovart M (2006) Influence of density dependence on predator-prey seabird interactions at large spatio-temporal scales. Proc R Soc B 273(1584): 379.
40. Pierotti R, Annett CA (1990) Diet and reproductive output in seabirds. BioScience 40(8): 568–574.
41. Rutz C, Bijlsma RG (2006) Food-limitation in a generalist predator. Proc R Soc Lond B 273(2069): 2076.
42. Ramos R, Ramírez F, Carrasco JL, Jover L (2011) Insights into the spatiotemporal component of feeding ecology: An isotopic approach for conservation management sciences. Diversity Distrib 17(2): 338–349.
43. Moody AT, Hobson KA, Gaston AJ (2012) High-arctic seabird trophic variation revealed through long-term isotopic monitoring. J Ornithol DOI: 10.1007/s10336-012-0836-0.
44. Norris DR, Arcese P, Preikshot D, Bertram DF, Kyser TK (2007) Diet reconstruction and historic population dynamics in a threatened seabird. J Appl Ecol 44(4): 875–884.

45. Verkuil Y, Van der Have TM, Van der Winden J, Chernichko II (2003) Habitat use and diet selection of northward migrating waders in the Sivash (Ukraine): The use of brine shrimp *Artemia salina* in a variable saline lagoon complex. Ardea 91(1): 71–83.
46. Rubega M, Inouye C (1994) Prey switching in Red-necked Phalaropes *Phalaropus lobatus*: Feeding limitations, the functional response and water management at Mono lake, California, USA. Biol Conserv 70(3): 205–210.
47. Klaassen M, Ens BJ (1990) Is salt stress a problem for waders wintering on the Banc d'Arguin, Mauritania? Ardea 78: 67–74.
48. Nehls G (1996) Low costs of salt turnover in Common Eiders *Somateria mollissima*. Ardea 84: 23–30.
49. Massias A, Becker PH (1990) Nutritive value of food and growth in Common Tern *Sterna hirundo* chicks. Ornis Scand 21: 187–194.
50. Pierotti R (1982) Habitat selection and its effects on the reproductive output in the Herring Gull in Newfoundland. Ecology 63: 854–868.
51. Murphy EC, Day RH, Oakley KL, Hoover AA (1984) Dietary changes and poor reproductive performance in Glaucous-winged Gulls. Auk 101: 532–541.
52. Sydeman WJ, Penniman JF, Penniman TM, Pyle P, Ainley DG (1991) Breeding performance in the Western Gull: Effect of parental age, timing of breeding and year in relation to food availability. J Anim Ecol 60: 135–149.
53. Pedrocchi V, Oro D, González-Solís J (1996) Differences between diet of adult and chick Audouin's gulls *Larus audouinii* at the Chafarinas Islands, SW Mediterranean. Ornis Fenn 73: 124–130.
54. Vitousek PM (1994) Beyong global warming: Ecology and global change. Ecology 75: 1861–1876.
55. Sutherland WJ (1996) Predicing the consequences of habitat loss for migratory populations. Proc R Soc Lond B 263: 1325–1327.
56. Walther G, Post E, Convey P, Menzel A, Parmesan C, et al. (2002) Ecological responses to recent climate change. Nature 416: 389–395.
57. Walmsley JG (2000) The ecological importance of Mediterranean salinas. In: Korovessis N, Lekkas TD, editors. Saltworks: Preserving saline coastal ecosystems. Athens: Global NEST-Hellenic Saltworks S.A. 81–95.
58. Dies JI, Dies B (2000) Breeding parameters of the Slender-billed gull *Larus genei* in a new colony located at L'Albufera de Valencia (Spain). Ardeola 47(2): 255–258.

3

Phage-Driven Loss of Virulence in a Fish Pathogenic Bacterium

Elina Laanto[1,2], Jaana K. H. Bamford[1,2], Jouni Laakso[1,3], Lotta-Riina Sundberg[1,4]*

1 Centre of Excellence in Biological Interactions, Universities of Jyväskylä and Helsinki, Finland, 2 Department of Biological and Environmental Science and Nanoscience Center, University of Jyväskylä, Jyväskylä, Finland, 3 Department of Biosciences, University of Helsinki, Helsinki, Finland, 4 Department of Biological and Environmental Science, University of Jyväskylä, Jyväskylä, Finland

Abstract

Parasites provide a selective pressure during the evolution of their hosts, and mediate a range of effects on ecological communities. Due to their short generation time, host-parasite interactions may also drive the virulence of opportunistic bacteria. This is especially relevant in systems where high densities of hosts and parasites on different trophic levels (e.g. vertebrate hosts, their bacterial pathogens, and virus parasitizing bacteria) co-exist. In farmed salmonid fingerlings, *Flavobacterium columnare* is an emerging pathogen, and phage that infect *F. columnare* have been isolated. However, the impact of these phage on their host bacterium is not well understood. To study this, four strains of *F. columnare* were exposed to three isolates of lytic phage and the development of phage resistance and changes in colony morphology were monitored. Using zebrafish (*Danio rerio*) as a model system, the ancestral rhizoid morphotypes were associated with a 25–100% mortality rate, whereas phage-resistant rough morphotypes that lost their virulence and gliding motility (which are key characteristics of the ancestral types), did not affect zebrafish survival. Both morphotypes maintained their colony morphologies over ten serial passages in liquid culture, except for the low-virulence strain, Os06, which changed morphology with each passage. To our knowledge, this is the first report of the effects of phage-host interactions in a commercially important fish pathogen where phage resistance directly correlates with a decline in bacterial virulence. These results suggest that phage can cause phenotypic changes in *F. columnare* outside the fish host, and antagonistic interactions between bacterial pathogens and their parasitic phage can favor low bacterial virulence under natural conditions. Furthermore, these results suggest that phage-based therapies can provide a disease management strategy for columnaris disease in aquaculture.

Editor: Michael A. Brockhurst, University of York, United Kingdom

Funding: This work was supported by Finnish Centre of Excellence Program of the Academy of Finland 2006–2011 CoE in Virus Research (#1129648), and CoE in Biological Interactions 2012–2017(#252411, Johanna Mappes), Academy of Finland grants 127500 (L.R.S.) and 125572 (J.L.), and a grant from the Maj and Tor Nessling Foundation (J.K.H.B.). The funders had no role in study design, data collection and analysis, decision to publish, or preparation of the manuscript.

Competing Interests: The authors have declared that no competing interests exist.

* E-mail: lotta-riina.sundberg@jyu.fi

Introduction

Over the last few years, there has been a growing concern regarding the emergence of disease outbreaks in livestock. It has become clear that environmental changes (e.g., climate warming, human intervention, enhanced transmission by transportation, use of antibiotics) have resulted in the development of new pathogens and diseases, and in addition, diseases that were previously under control have re-emerged [1–3]. In particular, intensive farming environments have been found to be evolutionary hot spots for pathogens. For example, the ecological and epidemiological features of intensive farming, including high host densities, effective transmission, and potential for serial passage, can select for high virulence of pathogens [4], [5]. Indeed, over the past 20 years, several new viral, bacterial, and eukaryotic/parasitic diseases have emerged in salmonid (*Salmo salar, S. trutta,* and *Oncorhynchus mykiss*) farming [2]. In particular, the occurrence of columnaris disease (caused by the opportunistic pathogen, *Flavobacterium columnare,* Bacteroidetes) has increased dramatically [3].

Traditionally, the evolution of pathogens is viewed as a reciprocal arms race competition between a pathogen and its host. Theories of virulence evolution predict that within-host growth of a pathogen, which is associated with virulence, is restricted by the availability of hosts [4]. This is referred to as the transmission-virulence trade-off. While obligate pathogens are dependent on their hosts and suffer from the transmission-virulence trade-off, opportunistic pathogens are able to survive and reproduce in environment, outside their hosts. In the latter case, the outside-host environment provides fluctuating selection pressures for opportunistic pathogens (e.g., predation, parasitism, and ecological changes), which may have correlated effects on pathogen virulence. Antagonistic species interactions between bacteria and their parasitic viruses (i.e. phage) are profoundly important regulators of bacterial abundance and traits [6]. Phage are the most abundant entities in the biosphere. They exceed the number of their host by ten-fold and thus have the capacity to control bacterial populations [7]. They also have a major impact on ecosystems and carbon cycling, especially in aqueous environments [8–10]. In the continuous arms race between phage and their hosts, bacteria need to evolve rapidly to avoid extinction [11].

Indeed, lytic phage have been shown to drive evolution in bacterial communities and to lead to strain diversification [12–15]. However, overlapping host-parasite interactions (e.g., bacterium-host and phage-bacterium) remain poorly understood. It is evident that theories regarding host-parasite interactions need to be expanded to involve three or more players in order to more accurately represent the evolution of virulence that is observed in complex environmental settings.

The ecological and evolutionary pressures that limit or increase the virulence of opportunistic bacterial pathogens are poorly known. In some cases, bacteria that gain resistance against lytic phage have a lower virulence, as has been demonstrated for *Serratia marcescens* [16], *Salmonella enterica* [17], and *Staphylococcus aureus* [18]. On the other hand, phage can invade the host as a prophage, and then modify and induce virulence of opportunistic bacteria by encoding virulence factors like the cholera toxin in *Vibrio cholera* [19]. This increases the pathogenicity of the host bacterium, and makes it a better competitor in the bacterial population [20], [21]. However, there is still very little information available regarding phage-host interactions that affect bacterial virulence under natural or intensive farming conditions. Moreover, the key question is, how do phage drive the evolution of opportunistic bacteria in the environment outside of a host. This aspect is especially important for understanding the emergence of opportunists, including opportunist saprotrophic pathogens which are able to replicate in, and transmit from, dead hosts. In these cases, the bacterial virulence is not necessarily limited by the virulence-transmission tradeoff.

F. columnare inhabits environmental microbial communities, but is also an opportunistic fish pathogen [22]. Disease outbreaks of columnaris disease rarely occur in nature, and the strains isolated from nature are less virulent than those isolated from fish farms [22]. Since the 1990s, columnaris infections have become increasingly more frequent and problematic in freshwater aquaculture, resulting in severe infections, increased mortality, and economic losses for fish farms producing salmonid fingerlings and fry [3]. This is mainly caused by the ability of *F. columnare* to survive in water, and outside of fish hosts for up to several months, thereby eluding antibiotic treatments [23]. *F. columnare* is also able to exploit dead fish material present in fish ponds for saprotrophic growth and as a means of transmission [23].

In nature and at fish farms, *F. columnare* encounters commensals and enemies from different trophic levels (e.g., protozoa, bacteria, viruses). However, studies of the interactions between *F. columnare* and other aquatic organisms present in the water body are scarce [24], [25]. Moreover, the impact of these interactions on bacterial virulence, and the mechanisms mediating bacterial virulence, are largely unknown. Under laboratory conditions, *F. columnare* exhibits three different colony morphologies, rhizoid (Rz), rough (R), and soft (S). The rhizoid morphotype is isolated in primary cultures from diseased fish, tank water, and biofilms. However, when the bacterium is cultured in the laboratory, or maintained under starvation, it loses its rhizoid morphology and manifests a parallel change in colony morphotype and decline in virulence and in susceptibility to phage infection [22], [26], [27], (Laanto et al., unpubl).

We have recently isolated and characterized phage capable of infecting the fish pathogen, *F. columnare* [28]. The results of previous studies, as described above, have demonstrated that lytic phage can act as a selective pressure against bacterial virulence, or can invade bacterium as a prophage and promote virulence. Based on the phage sensitivity of *F. columnare* [28], and the correlation between colony morphotype and virulence [26], [27], it is hypothesized that exposure of *F. columnare* to phage will cause a decline in bacterial virulence as a trade-off for acquiring phage resistance. Characterization of this phage-host relationship will provide novel insight into the virulence mechanisms of *F. columnare* in a fish host, as well as understanding on the infection dynamics of opportunistic pathogens outside their hosts. Moreover, studies of the complex host-parasite interactions that occur in intensive farming has the potential to facilitate our understanding of emerging new diseases, and to improve disease management in aquaculture.

Materials and Methods

Bacteria and Phage

Four previously isolated *F. columnare* strains from three different fish farms located in Central and Northern Finland were used in this study (Table 1) [28]. The bacteria were originally isolated from diseased fish and tank water obtained from fish farms as part of a disease surveillance project. To obtain isolates from fish, B67 and Os06, fish were euthanized by cutting the spinal cord prior to sampling. Isolates B185 and B245 were isolated from tank water. Standard culture methods were used to isolate bacteria, and included Shieh medium [29] for B185 and B245, Shieh medium supplemented with tobramycin [30] for B67, and AO-agar [31] for Os06. Pure cultures were stored at −80°C with 10% glycerol and 10% fetal calf serum. For the analyses performed, bacterial strains were grown in Shieh medium at room temperature (RT, approximately 24°C) with constant shaking (110 rpm). The genetic groups and phage susceptibility of B67, B185, and B245 were previously determined [28], while isolate Os06 was characterized in the present study (Table 1). Genetic grouping was based on the ribosomal intergenic spacer copy and length profile, and was performed as previously described [32]. In this study, the ancestral rhizoid morphotype is abbreviated, Rz, and the phage-induced morphotype rough is abbreviated, R.

Three lytic phage used in the present study were previously isolated from tank water from two different fish farms in Central Finland (Table 1) [28]. These *F. columnare* phage are genotype-specific [28], and therefore, phage FCL-2 were used for two bacterial strains in the genetic group G (Table 1). Phage used were enriched with bacteria using the standard double layer method. Briefly, 300 µl fresh bacterial culture was mixed with 3 ml soft agar (0.7%) and 100 µl phage suspension and poured on Shieh agar plates. After 48 h incubation at RT, phage were isolated by adding 5 ml Shieh-medium on top of the agar plates that exhibited confluent lysis. The plates were shaken at 95 rpm for 24 h at 8°C, after which the phage lysates were filtered through a 0.45 µm Supor® Membrane (PALL Corporation) and stored at 4°C.

Selection for Bacterial Colony Morphotypes with Phage

Phage that were used to select for phage-resistant morphotypes are listed in the Table 1 along with the corresponding bacterial strain. Phage lysates (50 µl of approximately 10^9 to 10^{11} plaque forming units (PFU) ml^{-1}) were spread on one half of a Shieh agar plate. Fresh bacterial culture (diluted to 10^{-1} and 10^{-2}) was then spotted on both halves of the plate (in the presence and absence of phage). After 48 h, rough (R) *F. columnare* colonies that grew in the presence of phage were selected, cultured to assure the loss of the parental Rz growth, and then stored at −80°C for further analysis.

The number of phage needed to achieve growth of only rough colonies of *F. columnare*, was analysed by taking 100 colony forming units (CFU) of bacteria (400 CFU in the case of B185) and plating the bacteria on Shieh agar with and without phage (i.e. 100 µl phage suspension with known PFU ml^{-1} at three different dilutions). The phage suspension was first spread on a plate,

Table 1. The *Flavobacterium columnare* strains and phage used in this study.

Bacterial strain*	Genetic group	Location	Source	Year Isolated	Phage used
B67	A	Farm L	Fish (*Salmo trutta*)	2007	FCL-1
B245	C	Farm V	Tank water	2009	FCV-1
B185	G	Farm L	Tank water	2008	FCL-2
Os06	G	Farm O	Fish (*Salmo salar*)	2006	FCL-2

*Bacteria and phage were previously characterized (28), except for Os06 which was genetically characterized in this study.

dried, and then inoculated with a fresh culture of *F. columnare*. Bacterial CFU was estimated from the optical density of the liquid culture.

Phage Resistance and the Presence of Phage in Rough (R) Morphotypes

Phage resistance of the rough (R) colonies that grew in the presence of phage was tested using a standard double layer method with slight modifications. Briefly, 300 µl fresh R morphotype bacterial culture was mixed with 3 ml soft agar and poured on Shieh agar plates. Five microliter aliquots from each of three dilutions of phage lysate were then spotted on the surface of the top agar. After 48 h at RT, the presence of plaques was recorded.

To determine if a phenotype change from ancestral rhizoid (Rz) to R was caused by lysogenic conversion (e.g., incorporation of phage into bacterium), the R morphotypes were tested for the release of phage into the culture medium. The lysogenic release of phage from R morphotypes into the supernatant was expected produce visible plaques on the Rz lawn. Three hundred µl of overnight grown Rz bacteria were plated with 3 ml soft agar (0.7%) on Shieh agar plates. R bacteria were centrifuged for 3 min at 13 000×g, diluted 10- and 100-fold, then 10 µl aliquots were spotted on the surface of the top agar. After 48 h at RT, the presence of plaques was recorded.

Stability of Colony Morphotypes in Serial Culture

To evaluate the stability of the colony morphotype of Rz and R bacteria, ten serial culture transfers were performed for all strains and their morphotypes. For these experiments, cultures and samplings were not replicated, since the initial bacterial strains were considered to be replicates of each other. Briefly, bacteria were cultured in Shieh broth for 22–24 h, after which 1 ml of the grown bacteria was transferred into 5 ml of fresh Shieh medium. Ten serial transfers were performed for each bacterium. From each transfer, a sample was plated onto agar plates, and the proportion of colony morphotypes present were estimated. The frequencies of the colony types observed were further analyzed using factorial analysis of variance (ANOVA), with the day of measurement, strain identity, colony class, and day*strain and strain*colony class interactions evaluated as factors. In addition, the phage resistance of a single colony of Os06 that changed colony morphotype from R to Rz was tested as described above.

Gliding Motility in Different Nutrient Conditions

Rz and R morphotypes of all bacterial strains were cultured on 2× concentrated Shieh, 1× (normal) Shieh, and 0.5× diluted Shieh agar plates. After 48 h, the colony diameter of 10 individual colonies were measured. Because the data did not fulfill the assumptions of normality, data were analysed using ANOVA on ranked data values. The sum of squares (SS) and mean squares (MS) were used to calculate test value H which was tested against chi square distribution with dfs corresponding to dfs of original treatment effects [33]. The analyses were performed using SPSS 20 (IBM).

Virulence Experiments

Fish experiments were conducted according to the Finnish Act on the Use of Animals for Experimental Purposes, under permission ESAVI-2010-05569/Ym-23 granted for L-RS by the National Animal Experiment Board at the Regional State Administrative Agency for Southern Finland.

Unsexed, adult, disease-free zebra fish (*Danio rerio*) were obtained from Core Facilities (COFA) and research services of Tampere (Tampere University, Finland). These zebra fish were infected with both colony morphotypes of the four *F. columnare* strains by bathing. The infection method and bacterial dose used were developed and optimized in preliminary experiments (Kinnula et al. unpublished). The fish were individually challenged in 100 ml water with 1×10^7 CFU ml^{-1} freshly grown bacteria for 30 min at 25–26.5°C. Each infection included eight replicates and four replicates of a negative control (i.e. fish exposed to sterile Shieh medium). After each challenge, fish were transferred into separate 3 liter aquaria, with one fish per aquarium with 1 liter bore hole water. Fish were monitored for 3 d for disease symptoms and mortality. Mortality data were analyzed using Cox regression conducted by SPSS 20 (IBM).

Results

Effect of Phage on Bacterial Colony Morphotype

When cultured in the presence of phage, all bacterial strains studied here lost their rhizoid (Rz) colony morphotype. The phage resistant bacteria growing with the phage lysate had a rough (R) colony morphotype, with solid edges and small size (Figure 1). In contrast, the spontaneously formed rough morphotypes reported in previous papers [26], [27] have irregular edges with sporadic rhizoid protrusions. Moreover, these spontaneously formed rough morphotypes also exhibit gliding motility when cultured in low nutrient media (our unpublished data). For all bacterial strains evaluated, the R colonies were confirmed to be phage-resistant, and showed no signs of lysogenic release of phage into culture supernatant.

For each bacterium-phage combination assayed, approximately 100 (B67, B245, Os06) or 400 (B185) Rz colonies were observed to convert to phage-resistant R colonies over two days of culturing on agar plates at 22°C when the number of phages on each plate was 10^6 or more. To change all the 100 ancestral Rz colonies to R colonies, 7.4×10^6 PFU ml^{-1} phage FCL-1 and 6.5×10^6 PFU ml^{-1} phage FCV-1 were needed for B67 and B245, respectively. Phage FCL-2 was not as efficient in causing a

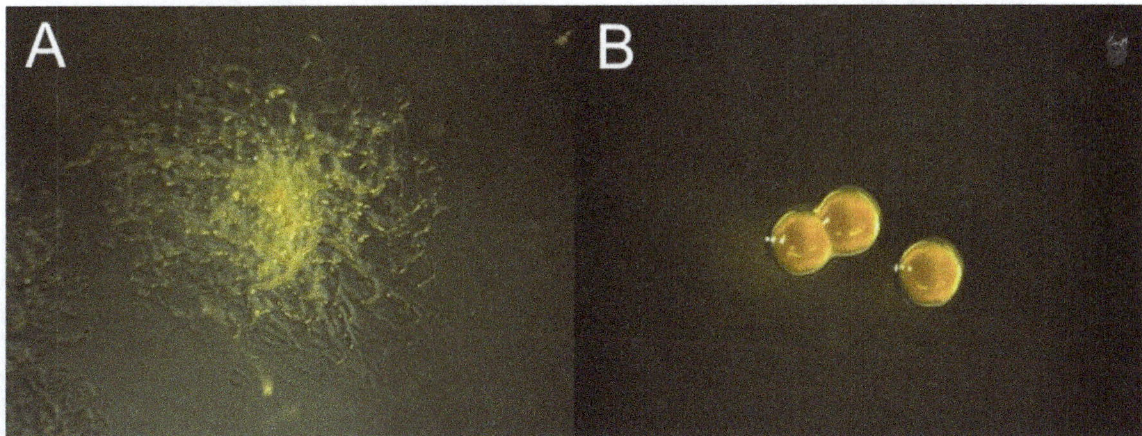

Figure 1. *Flavobacterium columnare* colony morphotypes. A) Ancestral rhizoid (Rz) colony and B) phage-resistant rough (R) colony on Shieh agar.

morphology change, with 2×10^9 PFU ml^{-1} needed for Os06, and 3×10^9 PFU ml^{-1} was needed for B185, to obtain phage-resistant colonies. For bacterial strains, B67, B245, and Os06, colony counts on plates with and without phage were similar. However, for B185, approximately 50% fewer total colonies were observed to grow in the presence of phage compared with a parental phage-free control. This may be due to the high cost of developing phage resistance in this strain, possibly by mutation.

Gliding Motility in Different Nutrient Conditions

The colony diameter of Rz and R colonies grown on agar plates with different nutrient concentrations were measured to characterize the ability of the morphotypes to move by gliding. Overall, colony diameter was found to be significantly affected by colony morphology ($H = 129.6$, $p<0.001$) and nutrient concentration ($H = 10.6$, $p<0.001$), and in addition to main effects, also significant interaction between colony morphology and nutrient concentration was found ($H = 20.1$, $p<0.001$) (Figure 2).

Stability of Colony Morphotypes in Serial Culture

The stability of the two colony types during serial culture was found to depend on the strain identity and the strain colony morphotype (strain identity: Wald $\chi2 = 364.5$, $df = 3$, $p<0.001$; strain×colony morphotype: Wald $\chi2 = 17.5$, $df = 4$, $p = 0.002$) (Figure 3a). For example, the phage-resistant strains, B67, B185, and B245, almost completely maintained their R colony morphology during the ten serial transfers. However, among the ancestral Rz morphotype spontaneous R colonies did appear. These spontaneously formed R colonies had irregular edges with some rhizoid protrusions, thereby differentiating them from the phage-resistant R colonies. The ancestral Rz and phage-resistant R morphotypes of strain Os06 were the most unstable, as they changed morphotypes in each culture. The soft (S) morphotype in particular was observed frequently (Figure 3b). Furthermore, the Os06 R morphotype did change back to the Rz morphotype and maintained its phage resistance.

Virulence Experiments

Virulence of the ancestral Rz morphotype was compared to the phage-resistant R type using a zebrafish infection model. The Rz morphotype was associated with a significantly higher mortality rate than the phage-resistant R morphotype (Cox regression, $Wald = 27.4$, $df = 1$, $p<0.001$, Table 2). Moreover, in most cases, the R morphotype bacteria were completely non-virulent, but experiments involving B185 R did result in the death of one fish. The virulence of the bacterial strains also significantly differed from each other (Cox regression, $Wald = 13.1$, $df = 1$, $p<0.001$). For example, the Rz morphotype of B185 and B67 were the most virulent, and resulted in the death of all eight fish tested within 24 h. In contrast, experiments with B245 resulted in the death of 4/8 fish, and for Os06 2/8 fish died (Table 2). The fish that died showed no obvious external signs of columnaris disease, indicating that the rapid progression of the disease can cause a symptomless death, possibly by blocking the gill surfaces. The remaining fish did not develop any disease symptoms during the three day post-infection monitoring.

Discussion

Antagonistic co-evolution between a bacterial host and its parasitic phage can represent a selective pressure for bacterial virulence. In this study, the ability of specialized parasitic phage to modulate the virulence of opportunistic pathogen outside its vertebrate host was investigated by exposing four strains of the opportunistic fish pathogen, *Flavobacterium columnare*, to corresponding lytic phage. When bacteria manifesting an ancestral rhizoid (Rz) morphotype were cultured in the presence of lytic phage, the colony morphotype changed to rough (R), and was accompanied by a loss of virulence and gliding motility. These results demonstrate that phage can cause phenotypic changes in *F. columnare* outside of the fish host, and thus, modulate its virulence.

While there are various mechanisms that can select for high virulence of pathogens, including within-host strain competition [34] and formation of persistent spores [35], opportunistic pathogens diverge fundamentally from the assumptions of epidemiological models. Opportunistic pathogenic bacteria are exposed to different environments during their life cycle, and they can have different strategies for living inside and outside a host [36]. Outside of a host, bacteria can be exposed to predators and parasitic phage, and they also need to compete for resources. On the other hand, within host growth requires traits for overcoming the immune defense of a multicellular host while growing otherwise in an enemy-free, high-resource host tissue. These conditions can either decrease or increase infectivity. Furthermore, opportunistic saprotrophic bacteria can reproduce and transmit from dead hosts, and can actively grow outside of a host as part of

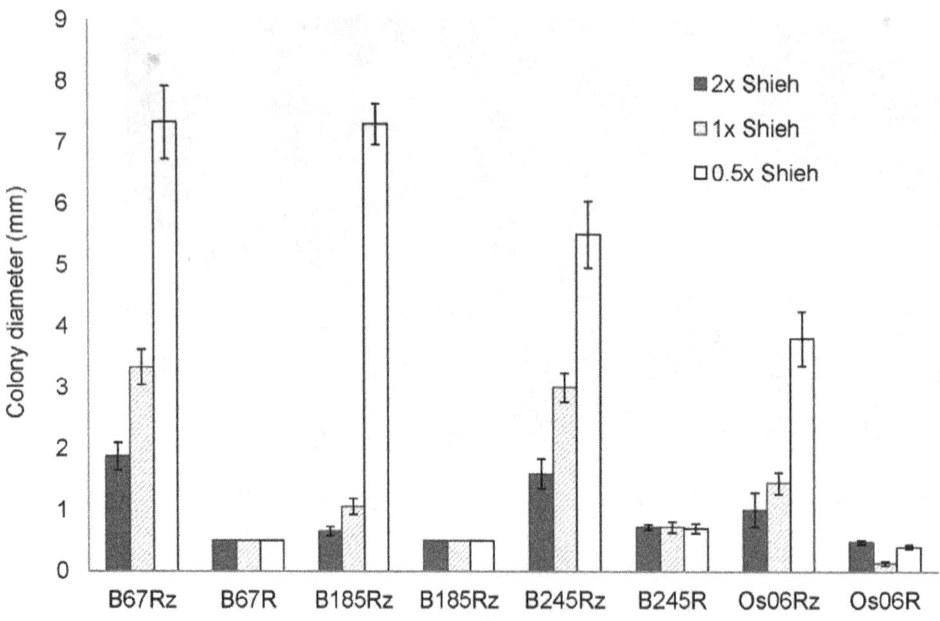

Figure 2. Gliding motility of colony morphotypes with different nutrient concentrations. Colony diameter (mm) was used to indicate the capacity for gliding motility of the *Flavobacterium columnare* rhizoid (Rz) and rough (R) colony morphotypes under different nutrient conditions. Motility is presented as the mean colony diameter (mm±SE) obtained from ten colonies grown for 48 h on 2× concentrated, normal (1×), and diluted (0.5×) Shieh agar. Colony diameter was found to be significantly affected by colony morphology ($H = 129.6$, $p < 0.001$), and by nutrient concentration ($H = 10.6$, $p < 0.001$, colony*nutrient interaction $H = 20.1$, $p < 0.001$).

the normal aquatic food web. Indeed, *F. columnare* transmits most efficiently from dead hosts (making high virulence traits beneficial), and can survive for long periods (at least five months) in water outside of a host [22], [23]. These traits may explain why columnaris disease has become problematic in fish farms [3]. However, the impact of the selective pressures generated by the environment outside of a host on the virulence of *F. columnare* has not yet been elucidated.

Based on the dramatic loss of virulence associated with phage-resistant *F. columnare* exhibiting a rough colony morphotype, we hypothesize that virulence is related to gliding motility in this bacterium. Accordingly, motility and virulence have been found to be connected in many bacteria [37], [38]. However, unlike many other bacteria, *F. columnare* does not have flagella or pili. Rather, it moves on surfaces by gliding, a motility characteristic associated with flavobacteria. Moreover, the flavobacterial gliding motility system has been shown to be orthologous to the Por secretion system (PorSS) of virulence factors in *Porphyromonas gingivalis* [39]. The corresponding genes needed for gliding/PorSS also exist in *F. columnare* [40], thereby suggesting that the motility apparatus may play a role in the secretion of virulence factors also in this species. This hypothesis would also be consistent with the motility-dependent virulence of this bacterium [26], [27]. Phage may reduce bacterial motility by using the bacterial motility apparatus as a receptor [41]–[43]. Therefore, to resist phage infection, bacteria may mutate, or down-regulate, the expression of the motility apparatus to impair motility (and virulence as well if it is coupled with motility). In other bacterial species, the negative effect of phage resistance on bacterial virulence outside a host has been observed [16]–[18]. In contrast, traditionally in the phage-host literature, more emphasis has been given to the capacity for phage to provide new genetic material and virulence factors for a host bacterium, thus increasing the fitness of the host bacterium [44], [21]. Phage resistance may, however, have trade-offs with respect to bacterial growth [18], and this may be true for *F. columnare* as well (Zhang et al., unpublished). Therefore, it is essential to recognize that phage can have both negative and positive effects on the virulence traits of their hosts, with the former highlighted in this study.

In addition to changes in colony morphotype selected by the presence of phage, spontaneous morphology changes from Rz to R in serial cultures of *F. columnare* were also detected, a phenomenon that has been previously reported [26], [27], [45], [46]. However, the appearance and biological significance of spontaneous R morphotypes are different than those of phage-resistant R types. For example, spontaneously formed R colonies appear during serial culture in the laboratory, they often exhibit weakly rhizoid or irregular edges, and they retain their gliding motility in low nutrient conditions. In contrast, phage-resistant R colonies are small with solid edges. The number of Rz morphotype colonies has a tendency to decrease in the bacterial population during culture (see Figure 3), indicating that colony morphotype is probably regulated by gene expression and phenotypic changes in response to environmental conditions. However, in the three virulent bacterial strains studied (e.g., B67, B185, and B245), the Rz morphotype was maintained at higher numbers during serial culture, suggesting that selection for the virulent Rz phenotype in these strains is strong, even under laboratory conditions. Also, the phage resistant R morphotypes of the same strains were stable, indicating that phage resistance may involve genetic effects (e.g., mutation of the phage receptor), which are not rapidly reversed under laboratory conditions, especially if phage cause a strong selective pressure as in case of B185 and FCV-2. Altogether, the general instability of the bacterial colony morphotype seems to be limited to certain strains (e.g., Os06), possibly to allow these bacteria to rapidly switch between phenotypes as necessary [47], [48]. This would increase the chances of survival in an unpredictable environment, rather than having changes in

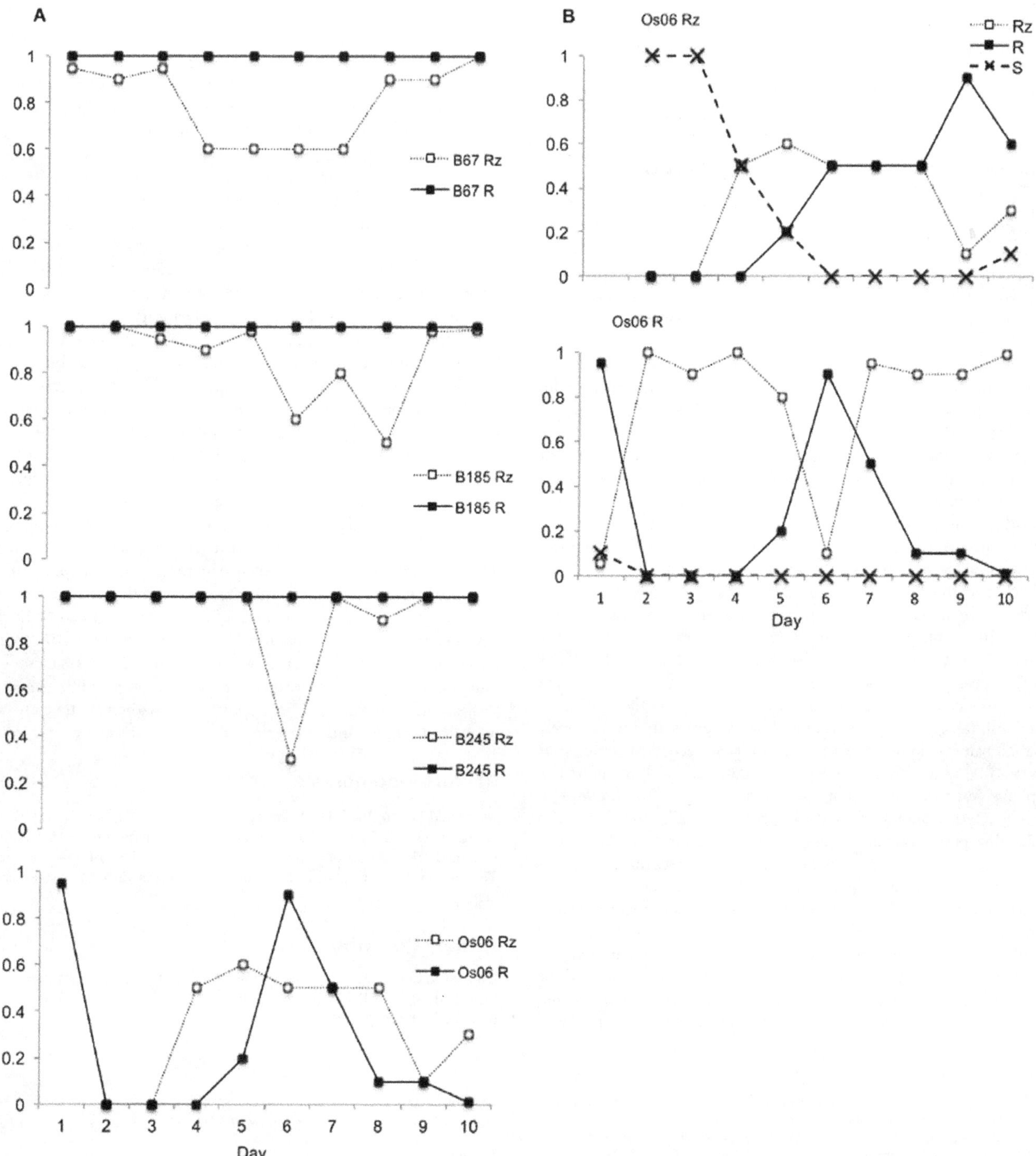

Figure 3. Proportion of *Flavobacterium columnare* colony morphotypes in serial culture. The proportion of colony morphologies (Rz = rhizoid, R = rough, and S = soft) in serial cultures of ancestral rhizoid and phage-resistant rough morphotypes of four *F. columnare* strains were estimated from pure cultures taken from each passage in Shieh broth. ND = not determined, but the proportion of Rz colonies was >50%. A) The proportion of ancestral Rz and phage-resistant R morphotypes that maintained their morphotype in serial culture, B) The fluctuation in the proportion of different morphotypes detected in Rz and R cultures of strain Os06.

phenotype/expression induced by environmental cues. Moreover, rapid shuffling between morphotypes in a low-virulence strain may be less costly than shuffling in more virulent strains, as less virulent bacteria (e.g., Os06) may not suffer from a loss of motility since they may not always be dependent on fish hosts for their survival. Furthermore, it can be assumed that less virulent strains living outside of a host are more frequently exposed to trophic interactions, and therefore the ability to change morphotype rapidly is beneficial for avoiding enemies and adapting to changing

Table 2. Zebrafish mortality caused by *Flavobacterium columnare* strains with ancestral rhizoid (Rz) versus phage-induced rough (R) morphotypes.

	Fish mortality[a] caused by morphotypes	
Strain	**Ancestral Rz**	**Phage-resistant R**
B67	100%	0%
B185	100%	12.5%
B245	50%	0%
Os06	25%	0%

[a]Zebrafish (n = 8) were individually tested for each morphotype.

environmental conditions [22]. The history of *F. columnare* as an environmental bacterium is consistent with these hypotheses.

The observation that phage resistance correlates with loss of virulence is especially important when designing novel treatment methods for intensive farming. Columnaris disease is an economically significant fish disease in freshwater aquaculture around the world. In addition, columnaris disease is an epidermal disease and transmits via water. Therefore, based on previous studies that have shown phage therapy to be successful for other fish diseases [49]–[51], and the potential for phage to be applied directly to tank water, we hypothesize that phage represent a valuable disease targeting strategy. Moreover, if phage use the gliding motility apparatus of *F. columnare* as a receptor, as suggested by the results of the present study, phage-host interactions would be mediated at the surface of the skin and gills of infected fish (or in the biofilms of the fish tank). Therefore, as fingerling fish are reared in relatively small water volumes, the amount of phage needed for treatments would be at a manageable level. In addition, the development of phage resistance would not hamper the success of the therapy, if the bacteria maintain their R morphotype as efficiently in the fish farming environments as they did in the laboratory. Additional studies will be needed to confirm these hypotheses, and to further elucidate the details of phage-host interactions.

In this study, zebrafish was used as the model organism rather than rainbow trout or other salmonids, which are common hosts of columnaris disease at fish farms. Therefore, it can be argued that the use of a model organism that is not the natural host of the studied disease could produce different outcomes [52]. However, *F. columnare* has a wide host range, and correspondingly, has been isolated from European graylings (*Thymallus thymallus*), whitefish (*Coregonus lavaretus*), bream (*Ambramis brama*), pikeperch (*Zander lucioperca*), and perch (*Perca fluviatilis*) in Fennoscandia [22], [32]. Furthermore, the use of zebrafish for *F. columnare* infections has previously been described [53], and our unpublished studies suggest that zebrafish can be used as a reliable model for studying columnaris disease instead of rainbow trout (Kinnula et al., unpublished). There are other advantages to using zebrafish as well. Zebrafish is established as model organisms for a wide range of experimental studies, disease-free fish are available year round, and they tolerate laboratory conditions better than rainbow trout. However, it will be important to confirm the phage-host interactions identified in the present study in salmonids, especially for the development of therapeutic applications.

To conclude, parasitic phage of *F. columnare* can select against bacterial virulence and motility, as indicated by colony morphology, in an environment outside a fish host. By studying phage-driven selection pressures and their impact on bacterial virulence, we have the opportunity to gain insight into the virulence factors of *F. columnare* and understand how its virulence evolves on both ecological and molecular scales. To our knowledge, this study represents the first report of the effects of phage-host interactions on a commercially important fish pathogen where phage resistance directly correlates with a decline in bacterial virulence. The phage-driven loss of virulence observed also supports the hypothesis that antagonistic co-evolution can reduce the virulence of opportunistic pathogens outside of a host due to the associated costs of defending against parasitic or predatory enemies [16]. Our results also indicate that further characterization of phage-host interactions is needed, particularly in regard to impact of phage on the evolution of virulence in opportunistic pathogens that affect intensive farming, and to develop applications for disease control.

Acknowledgments

We would like to thank Dr. Heidi Kunttu, Mr. Petri Papponen, and MSc Katja Neuvonen for skillful assistance in the laboratory, Dr. Päivi Rintamäki (University of Oulu) for kindly donating the Os06 isolate, Dr. Tarmo Ketola for help with the statistics and Kevin Slice for conceptual help.

Author Contributions

Conceived and designed the experiments: EL JKHB LRS. Performed the experiments: EL LRS. Analyzed the data: EL JKHB JL LRS. Wrote the paper: EL JKHB JL LRS.

References

1. Schrag S, Wiener P (1995) Emerging infectious disease: what are the relative roles of ecology and evolution? Trends Ecol Evol 10: 319–324.
2. Mennerat A, Nilsen F, Ebert D, Skorping A (2010) Intensive farming: evolutionary implications for parasites and pathogens. Evol Biol 37: 59–67.
3. Pulkkinen K, Suomalainen L- R, Read AF, Ebert D, Rintamaki P, et al. (2010) Intensive fish farming and the evolution of pathogen virulence: the case of columnaris disease in Finland. Proc R Soc B 277: 593–600.
4. Frank SA (1996) Models of parasite virulence. Q Rev Biol 71: 37–78.
5. Ebert D, Herre E (1996) The evolution of parasitic diseases. Parasitol Today 12: 96–101.
6. Abedon, ST (editor) (2008) Bacteriophage Ecology: Population Growth, Evolution, and Impact of Bacterial Viruses. Cambridge University Press. 526p.
7. Hendrix RW (2002) Bacteriophages: evolution of the majority. Theor. Popul Biol 61: 471–480.
8. Fuhrman JA (1999) Marine viruses and their biochemical and ecological effects. Nature 399: 541–548.
9. Suttle CA (2005) Viruses in the sea. Nature 437: 356–361.
10. Suttle CA (2007) Marine viruses-major players in the global ecosystem. Nature Rev Microbiol 5: 801–812.
11. Stern A, Sorek R (2011) The phage-host arms race: shaping the evolution of microbes. Bio Essays 33: 43–51.
12. Bohannan JMB, Lenski RE (2000) Linking genetic change to community evolution: insights from studies of bacteria and bacteriophage. Ecol Lett 3: 362–377.
13. Buckling A, Rainey PB (2002) Antagonistic coevolution between a bacterium and a bacteriophage. Proc R Soc B 269: 931–936.
14. Holmfeldt K, Middelboe M, Nybroe O, Riemann L (2007) Large variabilities in host strain susceptibility and phage host range govern interactions between lytic marine phages and their Flavobacterium hosts. Appl Env Microbiol 73: 6730–6739.
15. Lennon JT, Martiny JB (2008) Rapid evolution buffers ecosystem impacts of viruses in a microbial food web. Ecol Lett 11: 1178–1188.
16. Friman VP, Hiltunen T, Jalasvuori M, Lindstedt C, Laanto E, et al (2011) High temperature and bacteriophages can indirectly select for bacterial pathogenicity in environmental reservoirs. PloS One 6: e17651.
17. Santander J, Robeson J (2007) Phage-resistance of *Salmonella enterica* serovar Enteritidis and pathogenesis in *Caenorhabditis elegans* is mediated by the lipopolysaccharide. Electron J Biotechnol 10: 627-632.
18. Capparelli R, Nocerino N, Lanzetta R, Silipo A, Amoresano A, et al. (2010) Bacteriophage-Resistant *Staphylococcus aureus* Mutant Confers Broad Immunity against Staphylococcal Infection in Mice. PLoS One 5: e11720.

19. Waldor MK, Mekalanos JJ (1996) Lysogenic conversion by a filamentous phage encoding cholera toxin. Science 272: 1910-1914.
20. Brussow H, Canchaya C, Hardt WD (2004) Phages and the evolution of bacterial pathogens: from genomic rearrangements to lysogenic conversion. Microbiol Mol Biol Rev 68: 560-602.
21. Abedon ST, LeJeune JT (2005) Why bacteriophages encode endotoxins and other virulence factors. Evol Bioinform Online 1: 97-110.
22. Kunttu HMT, Sundberg L-R, Pulkkinen K, Valtonen ET (2012) Environment may be the source of *Flavobacterium columnare* outbreaks at fish farms. Env Microbiol Rep 4: 398-402.
23. Kunttu HMT, Valtonen ET, Jokinen IE, Suomalainen L-R (2009) Saprophytism of a fish pathogen as a transmission strategy. Epidemics 1: 96-100.
24. Revetta R, Rodgers M, Kinkle B (2005) Isolation and identification of freshwater bacteria antagonistic to *Giardia intestinalis* cysts. J Water Health 3: 83-88.
25. Rickard A, McBain A, Ledder R, Handley P, Gilbert P (2003) Coaggregation between freshwater bacteria within biofilm and planktonic communities. FEMS Microbiol Lett 220: 133-140.
26. Kunttu HMT, Suomalainen L-R, Jokinen EI, Valtonen ET (2009) F*lavobacterium columnare* colony types: connection to adhesion and virulence? Microb Pathogen 46: 21-27.
27. Kunttu HMT, Jokinen EI, Valtonen ET, Sundberg LR (2011) Virulent and nonvirulent *Flavobacterium columnare* colony morphologies: characterization of chondroitin AC lyase activity and adhesion to polystyrene. J Appl Microbiol 111: 1319-1326.
28. Laanto E, Sundberg L-R, Bamford JKH (2011) Phage specificity of the freshwater fish pathogen *Flavobacterium columnare*. Appl Env Microbiol 77: 7868-7872.
29. Shieh H (1980) Studies on the nutrition of a fish pathogen, *Flexibacter columnaris*. Microbios Lett 13: 129-133.
30. Decostere A, Haesebrouck F, Devriese LA (1997) Shieh medium supplemented with tobramycin for selective isolation of *Flavobacterium columnare (Flexibacter columnaris)* from diseased fish. J Clin Microbiol 35: 322-324.
31. Anacker R, Ordal E (1959) Studies on the myxobacterium *Chondrococcus columnaris*. I. Serological typing. J Bacteriol 78: 25-32.
32. Suomalainen L-R, Kunttu H, Valtonen ET, Hirvelä-Koski V, Tiirola M (2006) Molecular diversity and growth features of *Flavobacterium columnare* -strains isolated in Finland. Dis Aquat Org 70: 55-61.
33. Zar JH (2010) Biostatistical analysis, 5th editition. Pearson Prentice-Hall, Upper Saddle River, New Jersey. 960 p.
34. Bashey F, Reynolds C, Sarin T, Young SK (2011) Virulence and competitive ability in an obligately killing parasite. Oikos 120: 1539–1545.
35. Day T (2002) Virulence evolution via host exploitation and toxin production in spore-producing pathogens. Ecol Lett 5:471-476.
36. Brown SP, Cornforth DM, Mideo N (2012) Evolution of virulence in opportunistic pathogens: generalism, plasticity, and control. Trends Microbiol 7: 336–342.
37. Ottemann KM, Miller JF (1997) Roles for motility in bacterial-host interactions. Mol Microbiol 24: 1109-1117.
38. Josenhans C, Suerbaum S (2002) The role of motility as a virulence factor in bacteria. Int J Med Microbiol 291: 605-614.
39. Sato K, Naito M, Yukitake H, Hirakawa H, Shoji M, et al. (2010) A protein secretion system linked to bacteroidete gliding motility and pathogenesis. Proc Nat Acad Sci USA 107: 276-281.
40. Dumpala PR, Gulsoy N, Lawrence ML, Karsi A (2010) Proteomic analysis of the fish pathogen *Flavobacterium columnare*. Proteome Sci 8: 26. Available: http://www.proteomesci.com/content/8/1/26. Accessed 30th November 2012.
41. Schade SZ, Adler J, Ris H (1967) How bacteriophage chi attacks motile bacteria. J Virol 1: 599-609.
42. Heierson A, Siden I, Kivaisi A, Boman HG (1986) Bacteriophage-resistant mutants of *Bacillus thuringiensis* with decreased virulence in pupae of *Hyalophora cecropia*. J Bacteriol 167: 18-24.
43. Darzins A (1993) The pilG gene product, required for *Pseudomonas aeruginosa* pilus production and twitching motility, is homologous to the enteric, single-domain response regulator CheY. J Bacteriol 175: 5934-5944.
44. Boyd EF, Brüssow H (2002) Common themes among bacteriophage-encoded virulence factors and diversity among the bacteriophages involved. Trends Microbiol 10: 521-529.
45. Bader J, Shoemaker C, Klesius P (2005) Production, characterization and evaluation of virulence of an adhesion defective mutant of *Flavobacterium columnare* produced by beta-lactam selection. Lett Appl Microbiol 40: 123-127.
46. Olivares-Fuster O, Arias CR (2011) Development and characterization of rifampicin-resistant mutants from high virulent strains of *Flavobacterium columnare*. J Fish Dis 34: 385-394.
47. Seger J, Brockman HJ (1987) What is bet-hedging? In: Harvey PH, Partridge L, editors. Oxford Surveys in Evolutionary Biology. Oxford University Press, Oxford. pp. 182–211.
48. King OD, Base J (2007): The evolution of bet-hedging adaptations to rare scenarios. Theor Pop Biol 72: 560–575
49. Nakai T, Park SC (2002) Bacteriophage therapy of infectious diseases in aquaculture. Res Microbiol 153: 13-18.
50. Park SC, Nakai T (2002) Bacteriophage control of *Pseudomonas plecoglossicida*. infection in ayu *Plecoglossus altivelis*. Dis Aquat Org 53: 33-39.
51. Castillo D, Higuera G, Villa M, Middelboe M, Dalsgaard I. et al. (2012) Diversity of *Flavobacterium psychrophilum* and the potential use of its phages for protection against bacterial cold water disease in salmonids. J Fish Dis 35: 193-201.
52. Antonovics J, Boots M, Ebert D, Koskella B, Poss M et al. (online early) The origin of specificity by means of natural selection: evolved and non-host resistance in host-pathogen interactions. Evolution, online early. DOI: 10.1111/j.1558-5646.2012.01793.x.
53. Moyer TR, Hunnicutt DW (2007) Susceptibility of zebra fish *Danio rerio* to infection by *Flavobacterium columnare* and *F. johnsoniae*. Dis Aquat Org 76: 39–44.

4

Low Temperature-Dependent Salmonid Alphavirus Glycoprotein Processing and Recombinant Virus-Like Particle Formation

Stefan W. Metz[1,9], Femke Feenstra[1,9], Stephane Villoing[2], Marielle C. van Hulten[3], Jan W. van Lent[1], Joseph Koumans[3], Just M. Vlak[1], Gorben P. Pijlman[1]*

1 Laboratory of Virology, Wageningen University, Wageningen, The Netherlands, **2** Intervet Norbio, Bergen, Norway, **3** Intervet International BV, Boxmeer, The Netherlands

Abstract

Pancreas disease (PD) and sleeping disease (SD) are important viral scourges in aquaculture of Atlantic salmon and rainbow trout. The etiological agent of PD and SD is salmonid alphavirus (SAV), an unusual member of the *Togaviridae* (genus *Alphavirus*). SAV replicates at lower temperatures in fish. Outbreaks of SAV are associated with large economic losses of ~17 to 50 million $/year. Current control strategies rely on vaccination with inactivated virus formulations that are cumbersome to obtain and have intrinsic safety risks. In this research we were able to obtain non-infectious virus-like particles (VLPs) of SAV via expression of recombinant baculoviruses encoding SAV capsid protein and two major immunodominant viral glycoproteins, E1 and E2 in *Spodoptera frugiperda Sf9* insect cells. However, this was only achieved when a temperature shift from 27°C to lower temperatures was applied. At 27°C, precursor E2 (PE2) was misfolded and not processed by host furin into mature E2. Hence, E2 was detected neither on the surface of infected cells nor as VLPs in the culture fluid. However, when temperatures during protein expression were lowered, PE2 was processed into mature E2 in a temperature-dependent manner and VLPs were abundantly produced. So, temperature shift-down during synthesis is a prerequisite for correct SAV glycoprotein processing and recombinant VLP production.

Editor: Patricia V. Aguilar, University of Texas Medical Branch, United States of America

Funding: Intervet International provided limited funding to GP to compensate consumable costs. The funders had no role in study design, data collection and analysis, decision to publish, or preparation of the manuscript.

Competing Interests: SV, MH, and JK are employees of Intervet. JV is an employee of Wageningen University and a consultant for Intervet. All experimental work was carried out at Wageningen University, the Netherlands.

* E-mail: gorben.pijlman@wur.nl

9 These authors contributed equally to this work.

Introduction

Global fish-aquaculture has grown extensively over the past 50 years, producing yearly up to 52.5 million tons worth of an estimated US$98.5 billion. Approximately 50% of the world's fish supply is derived from fish farming, of which salmon and trout represent high value cultivates with an output of over 1.5 million tons worth US$7.2 billion per year [1]. Unlike other reared and farmed animals, fish are often cultured in open systems, exposing them to a wide variety of naturally occurring pathogens, in particular fish infecting viruses [2].

Salmonid alphavirus (SAV) is a serious pathogen for European farmed Atlantic salmon, Salmo salar and rainbow trout, Oncorhynchus mykiss and is the etiological agent of pancreas disease (PD) and sleeping disease (SD), respectively. PD is associated with lack of appetite, lethargy, pancreatic and heart lesions and increased mortality up to 48% of infected fish [3]. SD is characteristically presented by fish lying on their side on the bottom of the culture tank due to severe skeletal muscle necrosis [4]. SAV is an unusual alphavirus and differs from other alphaviruses like Sindbis virus (SINV), Semliki forest virus (SFV) or Chikungunya virus (CHIKV) as it solely replicates in fish (mainly salmonids) and has no known arthropod vector. Although SAV has successfully been detected in parasitic salmon louse species, the supposition of lice being a vector for SAV transmission remains to be proven, since active SAV replication within lice has not yet been confirmed [5,6]. In addition, it has been shown that SAV can be transmitted from fish to fish in cohabitation experiments in the absence of an arthropod vector [7] [8].

SAV belongs to the genus Alphavirus within the Togaviridae family and represents at least six closely related subtypes, of which three are of special interest: salmon pancreas disease virus (SPDV or SAV1) from Ireland, sleeping disease virus (SDV or SAV2) from France and the Norwegian salmonid alphavirus (NSAV or SAV3) [3]. SAV virions are enveloped spherical (~65 nm) particles and contain a positive sense, single-stranded RNA genome of approximately 12kb [9]. The viral RNA encodes two open reading frames (ORF); the non-structural ORF, which is directly translated from the genomic RNA, and the structural ORF, which is encoded by a 26S sub-genomic mRNA and is processed into 5 structural proteins – capsid, E3, E2, 6K, and E1. SAV transmembrane glycoproteins E1 and E2 are exposed on the virion surface as trimeric spikes, facilitating cell receptor recognition (presumably E2), cell entry via pH-dependent endocytosis (presumably E1) and support budding. E2 also serves an important role in regulating the fusion activity of E1 so that this does not occur before endocytosis [10]. The acidic endosomal environment dissociates the trimeric spike and causes E1 to initiate fusion with the endosomal membrane [11,12], thereby releasing the nucleocapsid and subsequently the viral RNA into the cytoplasm of the cell.

During translation of the structural polyprotein, the capsid protein is autocatalytically cleaved off from the structural polyprotein to encapsidate newly synthesized genomic RNA. The remaining envelope cassette (E3-E2-6K-E1) is subsequently translocated to the endoplasmic reticulum (ER) and is processed by host signalases at the N-terminal and C-terminal end of 6K, yielding E3E2 (precursor E2; PE2), 6K and E1[13]. The membrane anchored PE2 and E1 form heterodimers and three PE2-E1 dimers will eventually assemble into heterotrimers in the rough ER of infected cells [14,15,16,17]. The presence of E3 within the heterotrimers offers resistance against the acidic environment of the Golgi apparatus, to avoid premature trimer activation by homotrimerization of E1 [13]. In the trans-Golgi system, PE2 undergoes furin-dependent maturation, thereby releasing E3 from E2. However, E3 may remain associated with the E1-E2 trimers in the acidic compartments of the cell and dissociates once the trimers reach the cell surface to prime the spikes for acidic activation [18].

SAV infection in farmed salmonids can be minimized by reducing stress and good hygienic culture methods, such as sea-lice control and proper boat and transporter cleaning and disinfection. Next to these control measures, vaccination in fish has shown to be effective in protecting farmed salmon and trout from SAV infections. Few studies have focused on SAV-specific immune responses in fish, but it has been shown that infected salmonids generate a short-lasting (~9 months) protective immune response against subsequent SAV infections [19] and that, similar to other alphaviruses, immune cells are involved in dissemination of infection in the host [20,21]. Although the immune response against SAV is not fully understood, it has been demonstrated in multiple studies that salmonids can be protected against SAV challenge by immunization with inactivated virus formulations and that immunized fish elicit neutralizing antibodies, resulting in viral clearance [8,22]. Several vaccination strategies have been developed and tested, such as formalin-inactivated viral vaccines, recombinant and attenuated live vaccines [3,23,24,25,26]. Although these vaccines provide cross-protection to all SAV-subtypes, safety issues such as incomplete inactivation of active virus remain problematic. The use of a subunit vaccine may serve as an elegant alternative for inactivated or live-attenuated vaccines, in particular the use of virus-like particles (VLPs). VLPs are non-pathogenic virus look-alikes, since they are morphologically similar to virions but are non-infectious as they do not contain viral RNA. Recent studies have shown that the expression of the CHIKV-structural polyprotein in culture cells results in the formation of VLPs, which induced a protective immune response in non-human primates [27]. However, the low temperature replication and assembly of SAV are potentially major hurdles to the production of SAV VLPs.

This study focuses on the genesis of SAV VLPs using the recombinant baculovirus-insect cell expression system. This expression system has proven to be an efficient and safe way to express heterologous proteins on large scale in insect cells, which resulted in the production of several commercially available human and veterinary viral vaccines [28,29,30,31,32]. The recombinant insect cell expression system is based upon the exchange of the baculovirus polyhedrin gene for the heterologous gene of interest. Polyhedrin expression is controlled by the strong polyhedrin promoter, thereby enabling high heterologous protein expression. Next to high level protein expression, insect cells enable post-translational protein modification and accurate folding [29,33] and several studies have shown efficient alphavirus protein expression, using recombinant baculoviruses [34,35,36]. In addition, the baculovirus-insect cell expression has been shown to be an elegant method for VLP production of enveloped and non-enveloped arboviruses [37]. In this study, we show that the formation of SAV VLPs in insect cells does not take place under standard conditions, but that it is dependent on the level of processing of the envelope glycoprotein E2. In addition, we show that E2 processing is a function of temperature and is only complete at low temperatures. Finally, we present an optimized production process involving a temperature-shift regime to allow the efficient secretion of SAV VLPs in the culture fluid of baculovirus-infected insect cells.

Materials and Methods

Cells and viruses

Adherent *Spodoptera frugiperda* (*Sf9*) cells (Invitrogen) were maintained as monolayer cultures in Sf-900 II medium (Gibco), supplemented with 5% fetal calf serum. Recombinant baculoviruses were generated using a modified *Autographa californica* multicapsid nucleopolyhedrovirus (AcMNPV) backbone. Promoter sequences and large parts of the coding sequences of cathepsin and chitinase were deleted from the bacmid backbone [38]. The resulting AcΔcc was further modified by deletion of the p10 promoter and ORF. These elements were replaced by a zeocin resistance marker by Lambda Red recombination [39,40]. Modified LoxP sequences flanking the zeocin resistance marker were used for subsequent removal of the resistance marker by Cre recombinase [41]. The pFastBac1/SAV3 was constructed by cloning the SAV3 structural polyprotein (Genbank accession # AY604235) as a 3957 nt *Eco*RI-*Xba*I fragment into pFastBAC1 (Invitrogen). The recombinant baculovirus AcΔccΔp10 expressing the structural cassette of SAV3 was generated using the Bac-to-Bac baculovirus expression system (Invitrogen), resulting in Ac-SAV3. Recombinant baculovirus titers expressed as tissue culture infectious dose 50 ($TCID_{50}$) per ml, were determined by end point dilution [42]. All infections were performed in serum-free Sf-900 II medium.

Recombinant baculovirus infections and temperature-shift assay

To analyze protein expression and processing, 1.5×10^6 *Sf9*-cells were seeded in a 6-wells culture plate and infected with Ac-SAV3 at a multiplicity of infection (MOI) of 10 $TCID_{50}$ units per cell. Infections were performed at 27°C on a shaking platform. Subsequent incubations were performed at 12°C, 15°C, 18°C or 27°C for 14, 10, 7 and 3 days, respectively. The medium fraction was removed and cells were harvested, washed once in phosphate buffered saline (PBS), resuspended in 200 µl PBS and stored at −20°C. To analyze the effect of a temperature-shift on protein processing, cells were infected and harvested as previously described, but were initially incubated at 27°C. Next, cells were transferred to 12°C, 15°C or 18°C at 24hpi, 48hpi or 72hpi.

Protein analysis

Secreted protein fractions were precipitated by 7% (w/v) NaCl, 2.3% (w/v) polyethylene glycol (PEG)-6000 precipitation. Pellets were resuspended in 100 µl PBS and stored at -20°C. Whole cell lysates and medium fractions were analyzed by sodium dodecyl sulphate polyacrylamide gel electrophoresis (SDS-PAGE) and Coomassie brilliant blue (CBB) staining. Denatured proteins were transferred to an Immobilon membrane (Millipore) and examined by western analysis. Membranes were blocked with 3% skim milk in PBS-0.1% Tween-60 (PBST) for 1 h at room temperature (RT). Membranes were washed in PBST and incubated for 1h at RT with 1:2000 diluted primary monoclonal antibody (mab) α-E2 17H23 (SAV3) [43] and α-E1 (raised against the N-terminus of SAV3 E1, aa 1-26) in PBST, supplemented with 0.2% skim milk. Membranes were washed 3 times with PBST and incubated for 45 minutes at RT with 1:3000 diluted Alkaline Phosphatase (AP) conjugated secondary antibody, in PBST supplemented with 0.2% skim milk. Membranes were washed three times in PBST and incubated with AP-buffer for 10 min. Proteins were detected by NBT/BCIP staining (Roche). Antigenic mass was determined by a sandwich immune-capture ELISA using the monoclonal α-E2 17H23 (SAV3) antibody [43].

PNGase F treatment

Protein fractions were treated with PNGase F (New England Biolabs) to analyze the glycosylation status of SAV3-E2. Cell fractions and the precipitated medium fraction were treated with 1 µl glycoprotein denaturing buffer in 9 µl MilliQ for 10 min at 100°C. The denatured protein-mix was incubated for 1 h at 37°C with 2 µl 10x G7 reaction buffer, 2 µl 10% NP40, 1.5 µl PNGase F and 4.5 µl MilliQ. Treated proteins fractions were analyzed with SDS-PAGE, western blot (WB) and CBB.

Immunofluorescence assay

*Sf*9-cells were seeded on glass coverslips in 24-wells plate, infected and incubated at 12°C, 15°C, 18°C and 27°C. Cells were fixed with 4% paraformaldehyde in PBS for 5 min at RT, washed with PBS and subsequently incubated with PBS containing 1:2000 diluted primary monoclonal mouse α-E2 antibody [43] for 1 h at RT. Cells were washed carefully with PBS and treated with 1:1000 diluted goat α-mouse polyclonal Alexa 546 (Invitrogen) for 1 h at RT in dark conditions. Next, cells were washed and coverslips were fixed on glass slides with a drop of antifade fluoromount-G (Southern Biotech). Cells were analyzed by laser confocal microscopy on a Zeiss LSM 510 Meta, Axiovert 100 m.

Electron microscopy analysis

Copper 400 square mesh grids (Veco) were hydrophilized by Argon gas discharge. Next, 10 µl sample was applied for 2 min and the excess liquid carefully removed. The grid was washed 5x with MilliQ and stained with 2% uranyl acetate for 20 s. Excess dye was removed, grids were air dried and analyzed with a JEOL JEM 2100 transmission electron microscope. Analysis of fixed cell samples was performed as described [44].

Results

Expression of SAV3 structural cassette in *Sf*9 insect cells by recombinant baculoviruses

A recombinant baculovirus was generated (Ac-SAV3) to express the complete SAV3 structural cassette C-E3-E2-6K-E1 (Fig. 1A). The complete coding sequence of the structural cassette was cloned downstream of the polyhedrin promoter of an adapted AcMNPV backbone. Insect cells were infected with a MOI of 10 $TCID_{50}$/cell and were incubated for 72 h. Protein expression in the cell fraction was analyzed by CBB and WB using α-E1 and α-E2 mabs. Total protein-staining by CBB showed high levels of expression based upon the abundant band of approximately 35 kDa, which closely matches the predicted size of the SAV capsid protein (Fig. 1B, left). Western analysis using α-E1 and α-E2 mabs yielded bands of ~50 kDa and ~55 kDa, respectively (Fig. 1B, center). These sizes correspond to the predicted molecular mass of E1 (49.2 kDa) and E3E2 (54.8 kDa), suggesting that expression of the SAV3 structural cassette results in an E3E2 intermediate that is not further processed by host furin.

To analyze if SAV glycoproteins were glycosylated (Fig. 1C), expression products were treated with PNGase F, which enzymatically removes carbohydrate residues from proteins. SAV3-E2 is predicted to be N-glycosylated at N318 (Fig. 1A) [17,45]. As expected, PNGase F treatment reduced the size of E3E2 with a few kDa (Fig. 1C). The shift in molecular mass of the SAV3-E3E2 fraction after PNGase F treatment suggests that SAV3-E2 is indeed N-glycosylated, when expressed in *Sf*9 insect cell by recombinant baculoviruses. The absence of non-glycosylated protein fraction in the untreated sample indicates that SAV3-E2 is very efficiently glycosylated in insect cells, despite the fact that E3 is not released from E3E2 by furin cleavage. Infected cells displayed baculovirus-specific cytopathic effect (CPE) including decreased cell growth, enlarged nuclei and cell monolayer detachment. However, *Sf*9 cells infected with Ac-SAV3 showed additional CPE. Cell membranes fused between closely neighboring infected cells resulting in polykaryons or syncytia (Fig. 1D). This syncytia formation was most likely induced by the fusogenic activity of SAV-E1, since alphavirus E1 regulates fusion during endocytosis in wildtype infections [17]. In addition to the enlarged nuclei and syncytia, infected *Sf*9 cells contained dense nuclear bodies that were only detectable in cells expressing the SAV structural cassette (Fig. 1D and Fig. 2A, indicated with arrows) Nuclear staining using Hoechst suggested that the nuclear bodies contained nucleic acids (Fig. 1D).

To further investigate the nature of these nuclear bodies, infected cells were fixed and embedded in gelatin after which the cells were analyzed by transmission electron microscopy (TEM) (Fig. 2B). All infected cells developed the dense structures which appeared to consist of aggregated, smaller spherical structures, surrounded by a halo of viral stroma. Based on morphology and size (~30 nm) and the fact that SAV3-capsid contains a nuclear localization signal [46], the spherical structures corresponds most likely to assembled SAV nucleocapsids.

Temperature-dependent processing of SAV3 structural proteins

The analysis of the SAV3 structural cassette, expressed by recombinant baculoviruses at 27°C, showed that SAV3-E2 was efficiently glycosylated, but was not released from its E3E2 precursor by host furin-like proteases. Since SAV is an alphavirus of cold-water fish, the lack of furin-processing in E3E2 might be caused by the significant difference in environmental temperature between wildtype SAV replication (12-15°C) [3] and baculovirus expression (27°C). To investigate this putative temperature-effect on furin-dependent processing, *Sf*9-cells were infected with Ac-SAV3 and incubated at 12°C, 15°C, 18°C and 27°C for 14, 10, 7 and 3 days, respectively. Whole cell lysates were treated with PNGase F and analyzed using WB. Expression at 27°C (Fig. 3A, lane 2) resulted in similar protein patterns as seen before (Fig. 1B). The 55 kDa protein band, corresponding to glycosylated but unprocessed SAV-E3E2, was also detected at all lower expression temperatures (Fig. 3A, lane 3, 4, 5). However, in addition to E3E2, a second protein band of lower molecular mass was found. The smaller polypeptide (48 kDa) matches the predicted molecular mass of processed SAV3-E2 (46.9 kDa). This suggests that furin cleavage of E3E2 is rescued at 18°C, 15°C and 12°C. At all temperatures, PNGase I treatment resulted in an equal downward shift of both bands, suggesting that recombinant E3E2 and E2 were N-glycosylated (data not shown). From the relative intensities

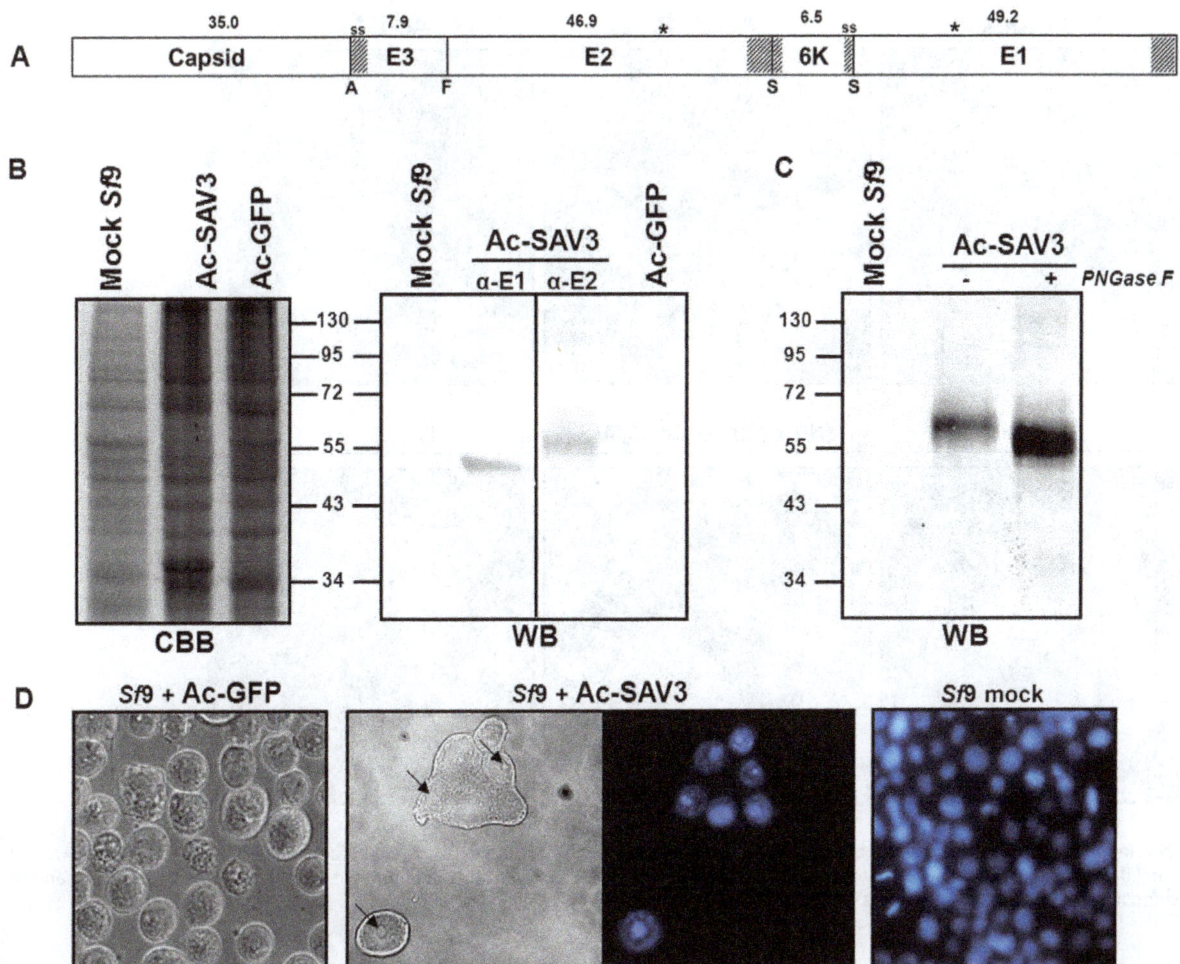

Figure 1. SAV3 structural cassette expression, using recombinant baculoviruses. A) Schematic representation of the SAV3 structural cassette as it is expressed by recombinant baculoviruses. The molecular mass of the proteins are indicated and shaded areas represent transmembrane domains or signal sequences (ss). Autocatalytic (A), furin (F) and signalase (S) cleavage sites are indicated, asterisks represent N-linked glycosylation sites. B) Protein expression in *Sf9* cells was analyzed by CBB and WB using SAV α-E1 and SAV α-E2 mabs. C) Whole cell lysates were treated with/without PNGase F and analyzed with SAV α-E2 mabs. D) Cells infected with Ac-GFP and Ac-SAV3 and stained with Hoechst. CPE was evaluated by brightfield and fluorescence microscopy. Arrows indicate dense nuclear bodies in Ac-SAV3-infected insect cells.

of the bands on the WB, it could be concluded that the intensity of the E2 fraction increased at lower temperatures. In conclusion, by decreasing the expression temperature from 27°C to 18°C, 15°C or 12°C, the ratio between E3E2 and fully processed E2 shifts towards the processed fraction (Fig. 3B). These results suggest that furin processing of the SAV3 E3E2 precursor is temperature dependent. Medium fractions of Ac-SAV3 infected Sf9 cells were PEG-precipitated to concentrate protein content and western analysis using α-E2 mabs showed that E2 can be detected in the medium fraction (Fig. 3C), but only when proteins were expressed at 12°C, 15°C or 18°C, but not at 27°C.

Surface expression of baculovirus expressed SAV3-E2 in *Sf9* cells

To generate progeny viral particles through budding, alphavirus glycoproteins assemble into heterodimers, three of which congregate into trimers, the so-called trimeric spikes. The spikes are formed in the ER and pass through the Golgi apparatus. At the end of the processing pathway, the trimeric spikes are anchored in the plasma membrane by the C-terminal transmembrane domains of E1 and E2 and are exposed on the surface of infected cells [17]. So far, it has been shown that expression of the SAV3 structural cassette by recombinant baculoviruses in insect cells results in the formation of glycosylated E3E2, which, at lower temperatures, is processed to glycosylated E2. This suggests that processing in insect cells resembles the processing as it takes place during wildtype infections in fish cells. In this case, glycoprotein spikes are exposed at the insect cell surface [13]. To analyze whether or not this is true, *Sf9* cells were infected with Ac-SAV3 at different temperatures. Next, non-permeable cells were subjected to immunofluorescence using α-E2 mabs (Fig. 4). Positive staining indicates that SAV3-E2 is exposed at the surface of the cells. Confocal microscopy revealed that the ring-like structures indicating surface expression could only be observed when Ac-SAV3 was expressed at 12°C, 15°C and 18°C. (Fig. 4). In sharp contrast, surface expression of E2 was not detected at 27°C (Fig. 4), while western analysis showed that E3E2 was expressed at high levels (Fig. 3A). No staining was observed in the mock cells (Fig. 4, left). These data comply with the previous results and demonstrate that the processing of SAV3-E2 is temperature dependent and that

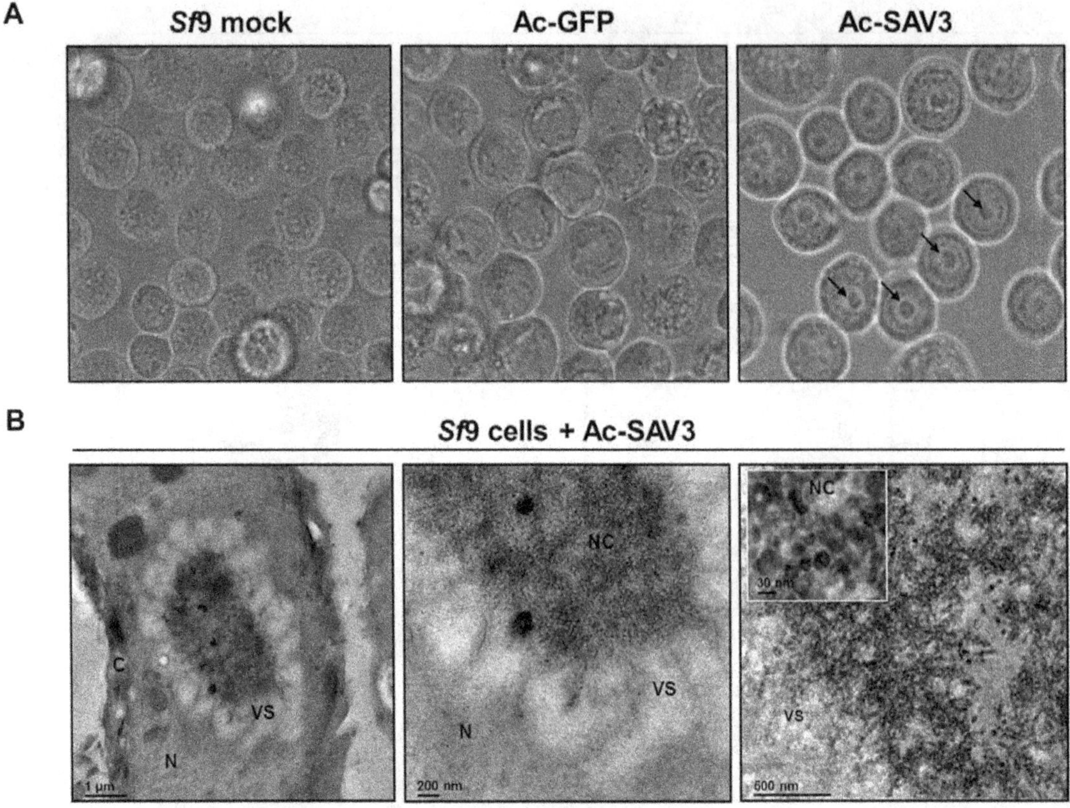

Figure 2. Nuclear localization and assembly of SAV3-nucleocapsids. *Sf9* infection with Ac-SAV3 and Ac-GFP was analyzed by A) light-microscopy at 2 d.p.i. B) Ac-SAV3-infected cells were fixed and embedded in gelatin and ultrathin coupes were analyzed by TEM, with C, N, VS and NC indicating cytoplasm, nucleus, virogenic stroma and SAV nucleocapsids, respectively.

E2 can only be detected at the surface of infected cells at lower temperatures.

SAV3 virus-like particle formation in *Sf9* cells expressing the SAV3 structural cassette

Processing of the SAV3 structural cassette, expressed by recombinant baculoviruses in insect cells is temperature dependent. Mature E2 is produced at temperatures ranging between 12°C and 18°C, where it is then translocated to the surface of Ac-SAV3 infected cells and subsequently can be detected in the medium fraction. In addition, it was shown that the capsid protein is successfully auto-cleaved from the structural polyprotein (Fig. 1B, left) and E1 is successfully released from the envelope cassette (Fig. 1B, middle). When the structural cassette was expressed at 27°C, E2 retained in the cell fraction in its E3E2 precursor form (Fig. 3A, lane 2). The effect of temperature on SAV maturation is clear, but to asses if temperature has an influence on the antigenicity of recombinant SAV-E2, which carries the major neutralizing epitopes, an immune-capture antigenic mass ELISA using the SAV neutralizing mab (17H23) was performed on total cell culture lysates and medium fraction of Sf9 cells infected with Ac-SAV3 at 27°C, 18°C, 15°C and 12°C (Fig. 5A). The highest antigenic mass was detected in the combined sonicated culture fraction and medium fraction of cells incubated at 15°C. Slightly lower antigenic mass levels were detected at cells incubated at 18°C and 12°C, whereas no antigenic mass was detected at 27°C. Detection of high antigenic mass in the medium fraction of infected cells incubated at 15°C or lower, suggested the formation of SAV VLPs. To investigate whether or not SAV VLPs could be detected, the 15°C medium fraction was evaluated by transmission electron microscopy (TEM) (Fig. 5B). VLP structures – spherical, sometimes donut-shaped, particles of 65–70 nm – that were morphologically similar to SAV3 virus particles (Fig. 5B, left) were found (Fig. 5B, middle), but these were absent in the control medium of Ac-GFP infection (Fig. 5B, right). This shows that SAV VLPs can be produced by expressing the SAV3 structural cassette, using the recombinant baculovirus-insect cell expression system. Moreover, these VLPs can be detected by a SAV-neutralizing mab, suggesting that these VLPs morphologically and antigenically resemble authentic SAV virions.

SAV3 structural cassette expression and VLP formation by temperature-shift in *Sf9* insect cells

The results so far have shown that SAV glycoprotein processing and secretion and VLP formation is dependent on the expression temperature. Although SAV-E2 processing appears to be most efficient at 12°C (Fig. 3A), expression at this temperature remains inefficient, due to the low metabolic rate of Sf9 cells and the extensive incubation time of 14 days. To optimize insect cell infection and SAV3 structural polyprotein processing, a temperature-shift experiment was performed. In this experiment, Sf9 cells were first infected with Ac-SAV3 at 27°C for 2 days to allow efficient baculovirus replication, after which the cells were transferred to 12°C for 3 days, to allow expression of properly processed SAV structural proteins. Protein expression, E2 processing, surface localization and VLP formation was analyzed 5 dpi by WB and immunostaining using α-E2 mabs and by TEM (Fig. 6). The temperature-shift resulted in increased processing of

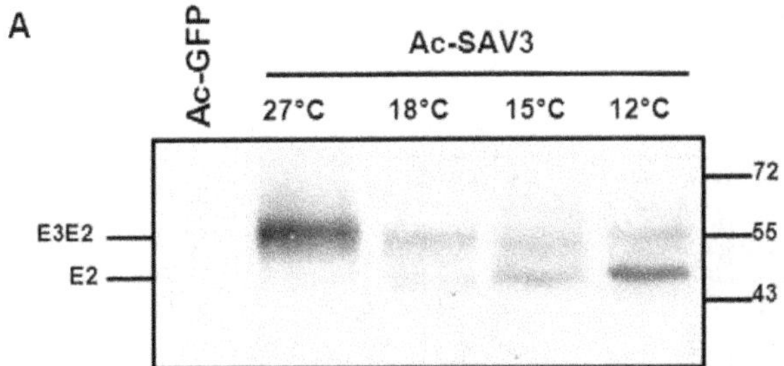

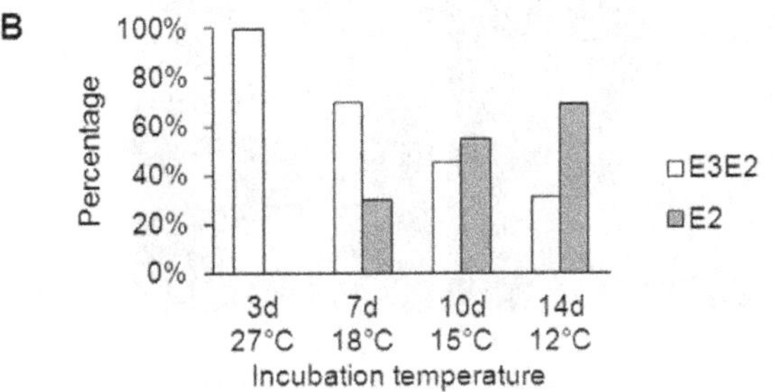

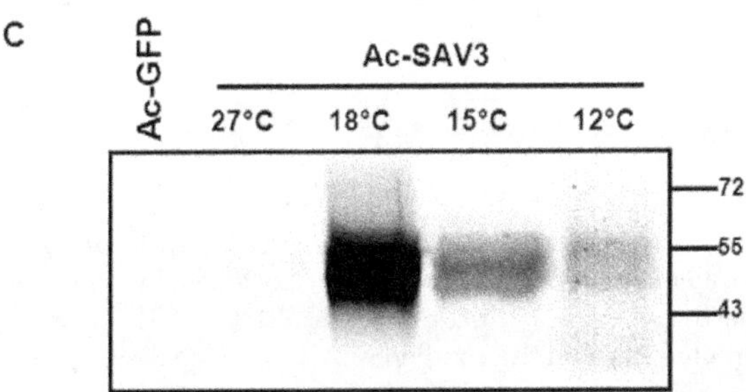

Figure 3. Temperature-dependent processing and secretion of SAV-E2. A) *Sf9* cells were infected with Ac-SAV3 with an MOI of 10 at 27°C, 18°C, 15°C and 12°C. SAV-E2 expression in the cell-fraction was analyzed by WB using SAV α-E2 mab (17H23). B) Relative percentages of SAV-E3E2 and E2, indicating more efficient processing of E3E2, with decreasing temperatures. C) Secretion of E2 as a function of the temperature. The medium fraction of infected *Sf9*-cell cultures was PEG-precipitated and analyzed by WB using SAV α-E2 mab.

SAV3-E2 (Fig. 6A, lane 3), as compared to 27°C expression (Fig. 6A, lane 2), since both E3E2 (~55 kDa) and mature E2 (~48 kDa) were detected. PNGase F treatment led to a decrease in molecular mass, showing that both E2 configurations were glycosylated (Fig. 6A, lane 4). Immunostaining on non-permeable Sf9 cells that were infected following the temperature-shift regime showed that E2 was detected on the surface of infected cells (Fig.6B), similar to results found after 12°C incubation (Fig. 4). In contrast, surface staining of the cells infected at 27°C were negative (Fig. 6B). In addition, medium of infected Sf9 cells with the temperature-shift was examined by electron microscopy (Fig. 6C). Characteristic VLP structures were again detected, which were similar to those found after expression at 15°C (Fig. 5B).

To further optimize the VLP formation in a temperature-shift regime, *Sf9* cells were infected with Ac-SAV3 at 27°C and transferred to 12°C, 15°C and 18°C at 1 dpi, 2 dpi and 3 dpi. Cells were subsequently incubated for 3 days at the indicated temperatures. As controls, cells were infected and incubated for a total of 6 days at 12°C, 15°C and 18°C. Both cell and PEG-precipitated medium fractions were analyzed by western analysis using α-E2 mabs (Fig. 7). SAV3-E2 was not detected in both the cell and medium fraction of cells that were infected at 27°C for 1 day (Fig. 7, lane 1). However, when cells were subsequently incubated for three days at lower temperatures, E2 was abundantly detected in both cell and medium fractions (Fig. 7, lane 2-4), indicating a strong increase in processing efficiency due to the temperature shift. Cells that were infected at 27°C for 2 days produced a low amount of E3E2. Although less pronounced than the shift after 1 dpi, the temperature-shift to lower temperatures 2 dpi increased processing and detection of E2 in the cell and medium fractions (Fig. 7, lane 5–8). Here, a shift to 15°C appeared optimal (Fig. 7, lane 7). Infection for 3 days at 27°C, as expected, appeared to be highly disadvantageous for the processing and secretion of E2, since only unprocessed E3E2, but no mature E2 was detected in both cell and medium fractions (Fig. 7, lane 9). It was concluded from this large temperature-shift experiment that infection for 1 day at 27°C followed by expression for 3 days at 15°C was optimal for E2 expression, processing and VLP secretion into the medium. In addition, this temperature-shift regime of in total 4 days significantly shortens infection time as compared to a 6 day infection at 15°C (Fig. 7, lane 14).

Discussion

SAV infections cause severe economic losses in European aquaculture of Atlantic salmon and rainbow trout. Next to common infection control measures, vaccination is a most effective tool to reduce SAV infections in salmonids. The presently available vaccines are based upon inactivated or attenuated-live vaccine strategies. However, a VLP-subunit approach may serve as an elegant and effective alternative to the safety issues that accompany the present way of vaccine production. The recombinant baculovirus-insect cells expression system has proven to have a high potential in the generation of VLP-subunit vaccines [29,36]. In this study, SAV VLPs were generated using recombinant baculoviruses and it was shown that the processing of SAV glycoproteins is a temperature-dependent process. VLP formation in insect cells at the normal production temperature of 27°C is not possible.

The SAV3 structural cassette (C-E3-E2-6K-E1) was cloned downstream of the polyhedrin promoter of AcMNPV (Ac-SAV3) and expressed in *Sf9* insect cells. SDS-PAGE and CBB analysis indicate that the SAV structural (glycol)proteins were expressed in high amounts and that the capsid protein was autocatalytically released from the structural polyprotein, similar as in wildtype alphavirus infection [17,47]. Alphavirus capsids are multifunctional proteins that are involved in encapsidation of viral genomic RNA, viral budding and in New-world alphaviruses induce host-cell transcription/translational shut-off [17,48]. Depending on the alphavirus species and the cell type used for expression, capsid localizes in several cellular compartments e.g. cytoplasm, nucleus/nucleoli and mitochondria [46,48,49,50]. Infection of *Sf9* cells with Ac-SAV3 resulted in the formation of dense nuclear bodies, which appeared to be specific for cells expressing the SAV structural polyprotein. In addition, Hoechst staining showed that nucleic acids co-localized with the dense nuclear bodies and TEM analysis confirmed the *in silico* predicted nuclear localization of capsid [46]. This is the first report that shows the presence of assembled alphavirus nucleocapsids of ~30 nm in the nucleus of insect cells.

The expression of the SAV-structural cassette in *Sf9* cells at 27°C clearly prevented complete processing of E2. In this regard, the efficient glycosylation of SAV glycoproteins is remarkable, since our other work showed that CHIKV glycoprotein expression

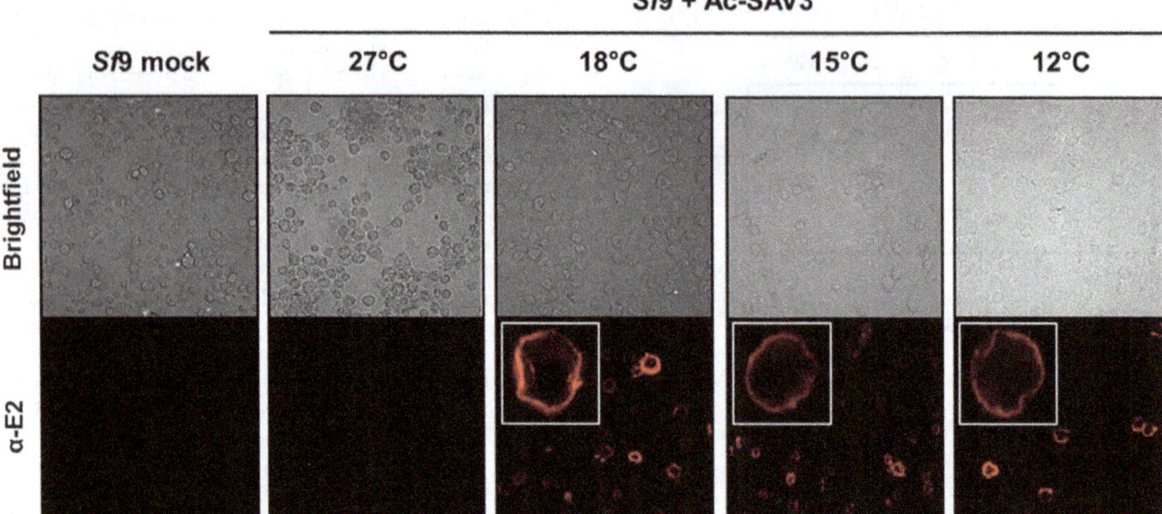

Figure 4. SAV3-E2 detection on the surface of *Sf9* cells after recombinant baculovirus expression. Cells were infected with Ac-SAV3 at 12°C, 15°C, 18°C and 27°C. Cells were fixed with 4% paraformaldehyde and subjected to immunostaining with α-E2 mabs. Cells were analyzed by confocal microscopy and positive staining indicates the presence of E2 at the surface of infected cells.

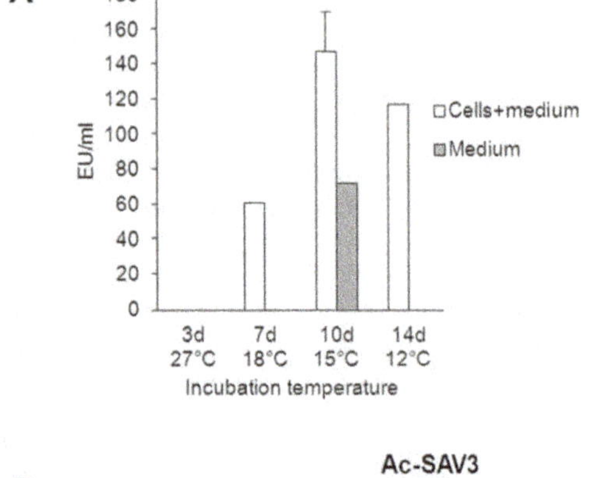

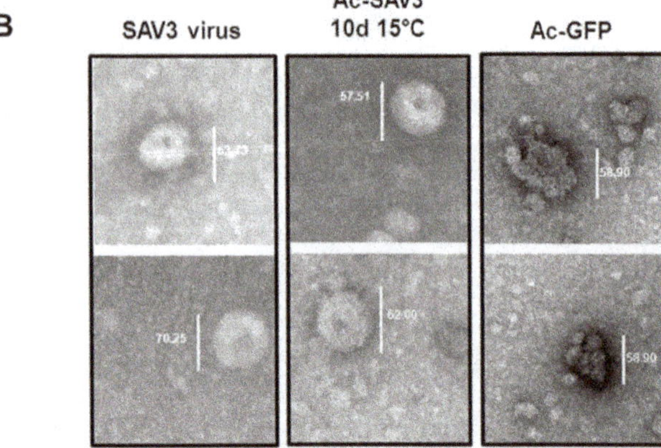

Figure 5. SAV-E2 antigenic mass determination and VLP production. A) SAV-E2 antigenic mass was determined using the SAV-neutralizing mab 17H23 on cell and medium fractions of infected *Sf9*-cells, incubated at 27°C, 18°C, 15°C and 12°C. B) The medium fraction of cells infected with Ac-SAV3 at 15°C was analyzed by TEM to analyze SAV VLP production. Medium fractions of Ac-GFP infected *Sf9* cells and SAV3 infected Chinook salmon embryo cells were used as control samples.

by recombinant baculoviruses resulted only in partial glycosylation [36].

Processing of SAV-E2 by host-furin cleavage was rescued after decreasing the infection temperatures to 18°C, 15°C or 12°C. Immunostaining indicated that only correctly folded proteins were recognized by the α-E2 mab, since at 27°C E3E2 could be detected under denaturing conditions by western analysis, but was not detected by immunostaining in permeabilized cells infected with Ac-SAV3. It has previously been shown that the 17H23 mab recognizes a discontinuous epitope (aa139-306) [43] on SAV3-E2. Incorrect folding of native PE2 might prevent the binding of mab 17H23 to the epitope, while under denaturing conditions, the conformational epitope is restored, thereby allowing antibody recognition by western analysis. Similar epitope-reformation under standard denaturing conditions has previously been described for other viral denaturation-resistant epitopes [51,52].

The incorrect folding of PE2, at 27°C that prevents binding of mab 17H23 might also render the SAV-PE2 furin-cleavage signal 68RKKR inaccessible to host furin-like proteases. An alternative explanation is that cellular furin is inactive at 27°C, however, this is highly unlikely given that baculovirus F protein is also activated by furin at similar temperatures (26°C–28°C) [53]. The apparent prevention of furin cleavage itself does not fully explain the absence of E3E2 in the medium at 27°C, because furin cleavage is not a prerequisite for alphavirus budding or E3E2 secretion [36,54,55]. The deficiency in E3E2 processing and lack of secreted E3E2 at 27°C, is therefore most likely caused by the retention of misfolded SAV-E3E2 in the ER or secretory pathway, in any case upstream of the *trans*-Golgi system where furin-dependent maturation takes place [13]. Misfolded and unfolded proteins usually accumulate in the ER, causing ER stress and thereby disrupting ER functions [56], a phenomenon often observed during overexpression of glycoproteins by recombinant baculoviruses [57]. It will be important to investigate in future experiments which structural change of E3E2 determines its intracellular retention at 27°C.

We show by electron microscopy that recombinant baculovirus expression of the SAV structural cassette in *Sf9* cells at lower

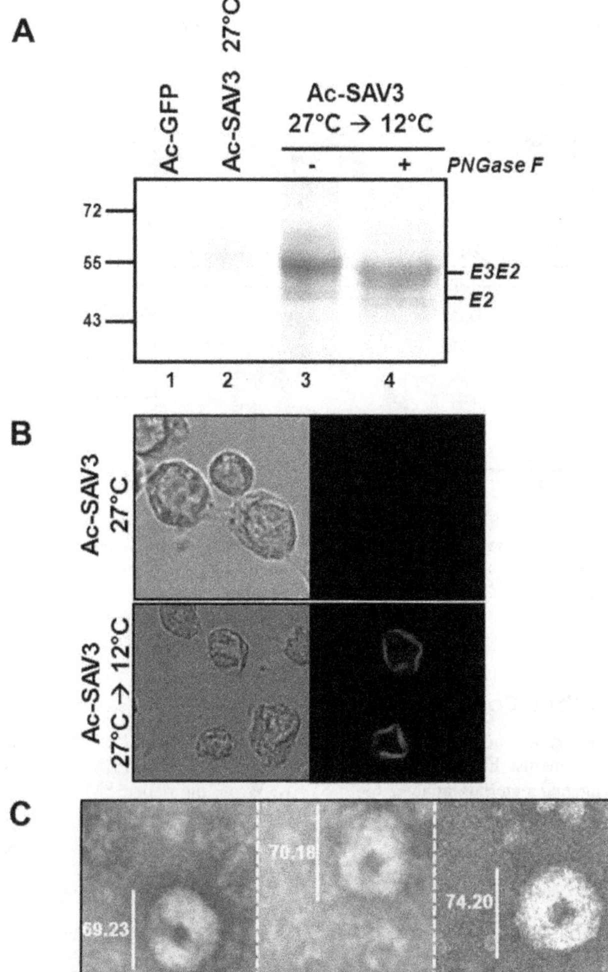

Figure 6. SAV3 structural cassette expression and VLP formation by temperature-shift in *Sf9* insect cells. *Sf9* cells were infected with Ac-SAV3, incubated for 2 days at 27°C and subsequently transferred to 12°C for 3 days. A) Cell cultures were treated with/without PNGase F and analyzed by WB using α-E2 mabs. B) Infected cells were treated with 4% paraformaldehyde and subjected to surface immunostaining with α-E2 mabs. C) The medium fraction the infected cell culture was analyzed by TEM for the presence of VLPs.

temperatures results in the formation of SAV VLPs. This result was confirmed by western analysis on the medium fraction of infected cell cultures and by sandwich immune-capture ELISA on infected cells and/or medium, in which SAV proteins were only detected at expression temperatures below 27°C. The spherical particles of ~65 nm in size, morphologically indistinguishable from SAV3 virus particles and other alphavirus VLPs [27], were found exclusively in the medium of Ac-SAV3 infected *Sf9* cells at lower temperatures, but not in the medium of control baculovirus lacking SAV sequences or mock-infected *Sf9* cells.

Since total production levels of recombinant SAV protein were the highest at 27°C, while lower temperatures were essential for PE2 processing, a hybrid protein production process, with an infection phase at 27°C followed by a production/processing phase at 12°C was developed.

Immunostaining and western analysis showed that recombinant SAV proteins produced following the temperature regime, appeared to be folded and processed correctly, and as expected, SAV VLPs were detected in the medium fraction of Ac-SAV3 infected cells. The temperature-shift production and processing regime was further optimized by varying the production time at 27°C. Both cell and medium fraction analysis showed that a 1 day infection phase at 27°C, followed by a 3 day processing phase at 15°C was optimal for combined protein expression and processing efficiency. In addition, it appeared that the vast majority of VLPs produced at 15°C have mature SAV-E2 incorporated in their envelope. However, PE2 medium detection indicated that, in a small fraction of VLPs, E3 was still associated to the trimeric spikes. This common alphavirus feature does not affect cell receptor recognition by E2 [58]. Thus, we expect VLPs carrying a minor E3E2 fraction still to be sufficiently immunogenic, especially considering that E3 from other alphaviruses harbors protective epitopes [59].

In addition to SAV VLPs, also the insect cells expressing correctly folded SAV structurals may be used in veterinary vaccine formulations. Similar to the widely used baculovirus surface display technique, based upon the expression of foreign peptides/epitopes using a chimeric baculovirus GP64 surface glycoprotein [60], SAV glycoproteins are anchored in the insect cell membrane and are displayed at its surface. However, the baculovirus surface display technique may find limited use in SAV glycoprotein production, since the preceding alphavirus E3 peptide with signal sequence is required for correct E2 folding and the use of heterologous signal peptides has recently been shown not to enhance alphavirus glycoprotein production [36].

The temperature-shift clearly rescues SAV glycoprotein processing, most likely via restoring upstream protein misfolding. A similar temperature-dependent folding phenomenon has previously been described for a temperature sensitive mutant of the vesicular stomatitis virus glycoprotein (VSV-Gmut). VSV-Gmut accumulated in the ER due to misfolding at 40°C, but was refolded when the temperature was decreased to 32°C, enabling VSV-Gmut to enter the secretory pathway into the Golgi-complex [61]. In future studies we would like to investigate the molecular mechanisms of presumed SAV glycoprotein misfolding at high temperatures, but it is unlikely that this will lead to a more efficient VLP production process at 27°C in the short term.

We clearly show that the initial incubation period for 1 day at 27°C following inoculation with baculovirus is of high importance for efficient VLP production. Nonetheless, extension of the 27°C baculovirus infection phase, negatively influenced SAV-PE2 processing and inhibited the formation of SAV VLPs. The strong CPE associated with baculovirus infection that usually occurs 2–3 dpi is most likely disabling cells to rescue SAV glycoprotein folding and processing after the shift to lower temperatures. However, a one day 27°C infection phase is highly beneficial and combined with a three day production processing phase at 15°C significantly shortens the time it takes to produce similar SAV E2 proteins levels as a six day production period at 15°C. This embodies a major advantage for large scale industrial antigen production in insect cell-bioreactors.

This study provides clear evidence that recombinant SAV VLPs can be produced in insect cells using baculovirus expression but also that SAV glycoprotein processing and folding is strictly temperature dependent and a critical determinant of VLP production. The proposed temperature-shift regime not only optimizes SAV VLP production in insect cells, but also provides a general principle for other vaccine candidates of cold-blooded infectious agents in insect cell systems. We aim to address the immunogenicity of our SAV VLPs in a follow up study involving a vaccination trial.

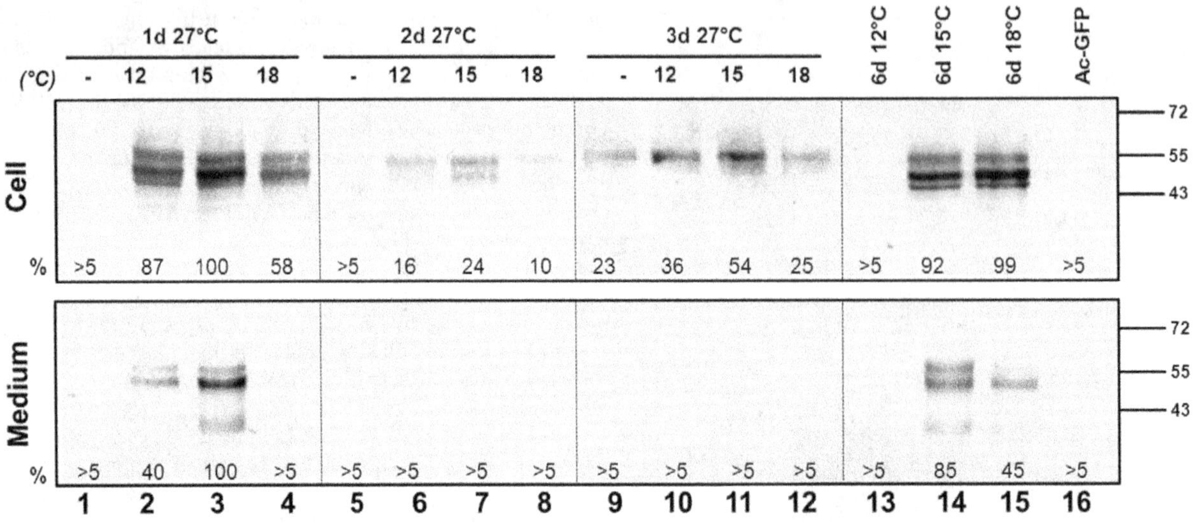

Figure 7. Comparative temperature-shift assay on *Sf*9-cells expressing the SAV3 structural polyprotein by recombinant baculoviruses. Cells were infected with Ac-SAV3 at 27°C for 1, 2 or 3 days. Next, cells were transferred for 3 days to 12°C, 15°C or 18°C. Whole cell lysates and PEG-precipitated medium fractions were analyzed by WB using α-E2 mabs. Protein fraction quantities, relative to 1 day 27°C followed by 3 days 15°C, are indicated.

Acknowledgments

We would like to acknowledge Luc Grisez and Petter Frost for continued support of this project and Corinne Geertsema for technical assistance.

Author Contributions

Conceived and designed the experiments: GP MH. Performed the experiments: FF SM JL. Analyzed the data: SM FF GP. Contributed reagents/materials/analysis tools: SV JK. Wrote the paper: SM GP JV.

References

1. Bostock J, McAndrew B, Richards R, Jauncey K, Telfer T, et al. (2010) Aquaculture: global status and trends. Philosophical Transactions of the Royal Society of London Series B, Biological Sciences 365: 2897–2912.
2. Snow M (2011) The contribution of molecular epidemiology to the understanding and control of viral diseases of salmonid aquaculture. Veterinary Research 42: 56–68.
3. McLoughlin MF, Graham DA (2007) Alphavirus infections in salmonids--a review. J Fish Dis 30: 511–531.
4. Boucher P, Laurencin FB (1996) Sleeping disease and pancreas disease: comparative histopathology and acquired cross protection. J Fish Dis 19: 303–310.
5. Petterson E, Sandberg M, Santi N (2009) Salmonid alphavirus associated with *Lepeophtheirus salmonis* (Copepoda: Caligidae) from Atlantic salmon, *Salmo salar* L. J Fish Dis 32: 477–479.
6. La Linn M, Gardner J, Warrilow D, Darnell GA, McMahon CR, et al. (2001) Arbovirus of marine mammals: a new alphavirus isolated from the elephant seal louse, *Lepidophthirus macrorhini*. J Virol 75: 4103–4109.
7. Boucher P, Raynard R, Houghton G, Baudin Laurencin F (1995) Comparative experimental transmission of pancreas disease in Atlantic salmon, rainbow trout and brown trout. Diseases of Aquatic Organisms 22: 19–19.
8. McLoughlin M, Nelson R, Rowley H, Cox D, Grant A (1996) Experimental pancreas disease in Atlantic salmon *Salmo salar* post-smolts induced by salmon pancreas disease virus (SPDV). Diseases of Aquatic Organisms 26: 117–124.
9. Nelson R, McLoughlin M, Rowley H, Platten M, McCormick J (1995) Isolation of a toga-like virus from farmed Atlantic salmon *Salmo salar* with pancreas disease. Diseases of Aquatic Organisms 22: 25–25.
10. White J, Helenius A (1980) pH-dependent fusion between the Semliki Forest virus membrane and liposomes. Proc Natl Acad Sci U S A 77: 3273–3277.
11. Gibbons DL, Ahn A, Chatterjee PK, Kielian M (2000) Formation and characterization of the trimeric form of the fusion protein of Semliki Forest Virus. J Virol 74: 7772–7780.
12. Wahlberg JM, Boere WA, Garoff H (1989) The heterodimeric association between the membrane proteins of Semliki Forest virus changes its sensitivity to low pH during virus maturation. J Virol 63: 4991–4997.
13. Kuhn RJ (2007) Togaviridae: the viruses and their replication. PhiladelphiaPA: Lippincott, Williams and Wilkins. pp 1001–1022.
14. Ziemiecki A, Garoff H, Simons K (1980) Formation of the Semliki Forest virus membrane glycoprotein complexes in the infected cell. J Gen Virol 50: 111–123.
15. Mulvey M, Brown DT (1996) Assembly of the Sindbis virus spike protein complex. Virology 219: 125–132.
16. Li L, Jose J, Xiang Y, Kuhn RJ, Rossmann MG (2010) Structural changes of envelope proteins during alphavirus fusion. Nature 468: 705–708.
17. Strauss JH, Strauss EG (1994) The alphaviruses: gene expression, replication, and evolution. Microbiology and Molecular Biology Reviews 58: 491–562.
18. Sjoberg M, Lindqvist B, Garoff H (2011) The activation of the alphavirus spike is suppressed by bound E3. J Virol 85: 5644–5650.
19. Houghton G (1994) Acquired protection in Atlantic salmon *Salmo salar* parr and post-smolts against pancreas disease. Diseases of Aquatic Organisms 18: 109–109.
20. Houghton G (1995) Kinetics of infection of plasma, blood leucocytes and lymphoid tissue from Atlantic salmon *Salmo salar* experimentally infected with pancreas disease. Diseases of Aquatic Organisms 22: 193–193.
21. Lidbury BA, Rulli NE, Suhrbier A, Smith PN, McColl SR, et al. (2008) Macrophage-derived proinflammatory factors contribute to the development of arthritis and myositis after infection with an arthrogenic alphavirus. J Infect Dis 197: 1585–1593.
22. Desvignes L, Quentel C, Lamour F, Le Ven A (2002) Pathogenesis and immune response in Atlantic salmon (*Salmo salar* L.) parr experimentally infected with salmon pancreas disease virus (SPDV). Fish & Shellfish Immunology 12: 77–95.
23. Moriette C, Leberre M, Lamoureux A, Lai TL, Bremont M (2006) Recovery of a recombinant salmonid alphavirus fully attenuated and protective for rainbow trout. J Virol 80: 4088–4098.
24. Benmansour A, de Kinkelin P (1997) Live fish vaccines: history and perspectives. Dev Biol Stand 90: 279–289.
25. Lopez-Doriga MV, Smail DA, Smith RJ, Domenech A, Castric J, et al. (2001) Isolation of salmon pancreas disease virus (SPDV) in cell culture and its ability to protect against infection by the 'wild-type' agent. Fish Shellfish Immunol 11: 505–522.
26. Sommerset I, Krossoy B, Biering E, Frost P (2005) Vaccines for fish in aquaculture. Expert Rev Vaccines 4: 89–101.
27. Akahata W, Yang ZY, Andersen H, Sun S, Holdaway HA, et al. (2010) A virus-like particle vaccine for epidemic Chikungunya virus protects nonhuman primates against infection. Nat Med 16: 334–338.
28. Kost TA, Condreay JP, Jarvis DL (2005) Baculovirus as versatile vectors for protein expression in insect and mammalian cells. Nat Biotechnol 23: 567–575.
29. van Oers MM (2006) Vaccines for viral and parasitic diseases produced with baculovirus vectors. Adv Virus Res 68: 193–253.
30. Bouma A, de Smit AJ, de Kluijver EP, Terpstra C, Moormann RJ (1999) Efficacy and stability of a subunit vaccine based on glycoprotein E2 of classical swine fever virus. Vet Microbiol 66: 101–114.
31. Paavonen J, Lehtinen M (2008) Introducing human papillomavirus vaccines - questions remain. Ann Med 40: 162–166.
32. Cox MM, Hollister JR (2009) FluBlok, a next generation influenza vaccine manufactured in insect cells. Biologicals 37: 182–189.

33. Vialard JE, Arif BM, Richardson CD (1995) Introduction to the molecular biology of baculoviruses. Methods Mol Biol 39: 1–24.
34. Hodgson LA, Ludwig GV, Smith JF (1999) Expression, processing, and immunogenicity of the structural proteins of Venezuelan equine encephalitis virus from recombinant baculovirus vectors. Vaccine 17: 1151–1160.
35. Oker-Blom C, Summers M (1989) Expression of Sindbis virus 26S cDNA in *Spodoptera frugiperda* (Sf9) cells, using a baculovirus expression vector. J Virol 63: 1256–1264.
36. Metz SW, Geertsema C, Martina BE, Andrade P, Heldens JG, et al. (2011) Functional processing and secretion of Chikungunya virus E1 and E2 glycoproteins in insect cells. Virology Journal 8: 353–365.
37. Metz SW, Pijlman GP (2011) Arbovirus vaccines; opportunities for the baculovirus-insect cell expression system. Journal of Invertebrate Pathology 107(Suppl): S16–30.
38. Kaba SA, Salcedo AM, Wafula PO, Vlak JM, van Oers MM (2004) Development of a chitinase and v-cathepsin negative bacmid for improved integrity of secreted recombinant proteins. J Virol Methods 122: 113–118.
39. Datsenko KA, Wanner BL (2000) One-step inactivation of chromosomal genes in *Escherichia coli* K-12 using PCR products. Proc Natl Acad Sci U S A 97: 6640–6645.
40. Pijlman GP, Dortmans JC, Vermeesch AM, Yang K, Martens DE, et al. (2002) Pivotal role of the non-hr origin of DNA replication in the genesis of defective interfering baculoviruses. J Virol 76: 5605–5611.
41. Suzuki N, Nonaka H, Tsuge Y, Inui M, Yukawa H (2005) New multiple-deletion method for the Corynebacterium glutamicum genome, using a mutant lox sequence. Appl Environ Microbiol 71: 8472–8480.
42. Vlak JM (1979) The proteins of nonoccluded *Autographa californica* nuclear polyhedrosis virus produced in an established cell line of *Spodoptera frugiperda*. Journal of Invertebrate Pathology 34: 110–118.
43. Moriette C, LeBerre M, Boscher SK, Castric J, Bremont M (2005) Characterization and mapping of monoclonal antibodies against the Sleeping disease virus, an aquatic alphavirus. J Gen Virol 86: 3119–3127.
44. Van Lent J, Groenen J, Klinge-Roode E, Rohrmann G, Zuidema D, et al. (1990) Localization of the 34 kDa polyhedron envelope protein in *Spodoptera frugiperda* cells infected with *Autographa california* nuclear polyhedrosis virus. Archives of Virology 111: 103–114.
45. Blom N, Sicheritz-Pontén T, Gupta R, Gammeltoft S, Brunak S (2004) Prediction of post-translational glycosylation and phosphorylation of proteins from the amino acid sequence. Proteomics 4: 1633–1649.
46. Karlsen M, Yousaf MN, Villoing S, Nylund A, Rimstad E (2010) The amino terminus of the salmonid alphavirus capsid protein determines subcellular localization and inhibits cellular proliferation. Archives of Virology. pp 1–13.
47. Saijo T, Jagdish R, Lijo J, Stephan G, Christian D, et al. (2010) Functional dissection of the alphavirus capsid protease: sequence requirements for activity. Virology Journal 7: 327–334.
48. Aguilar PV, Weaver SC, Basler CF (2007) Capsid protein of eastern equine encephalitis virus inhibits host cell gene expression. Journal of Virology 81: 3866.
49. Mitchell C, Freitas de Andrade-Rozental A, Souto-Padrón T, da Glória da Costa Carvalho M (1997) Identification of mayaro virus nucleocapsid protein in nucleus of *Aedes albopictus* cells. Virus Research 47: 67–77.
50. Michel MR, Elgizoli M, Dai Y, Jakob R, Koblet H, et al. (1990) Karyophilic properties of Semliki Forest virus nucleocapsid protein. J Virol 64: 5123–5131.
51. Frost P, Havarstein LS, Lygren B, Stahl S, Endresen C, et al. (1995) Mapping of neutralization epitopes on infectious pancreatic necrosis viruses. J Gen Virol 76: 1165–1172.
52. Wright KE, Salvato MS, Buchmeier MJ (1989) Neutralizing epitopes of lymphocytic choriomeningitis virus are conformational and require both glycosylation and disulfide bonds for expression. Virology 171: 417–426.
53. Westenberg M, Wang H, WF IJ, Goldbach RW, Vlak JM, et al. (2002) Furin is involved in baculovirus envelope fusion protein activation. J Virol 76: 178–184.
54. Ozden S, Lucas-Hourani M, Ceccaldi P, Basak A, Valentine M, et al. (2008) Inhibition of Chikungunya Virus infection in cultured human muscle cells by furin inhibitors. Journal of Biological Chemistry 283: 21899–21908.
55. Sjoberg M, Lindqvist B, Garoff H (2011) The activation of the alphavirus spike is suppressed by bound E3. J Virol 85: 5644–5650.
56. Kaufman R (1999) Stress signaling from the lumen of the endoplasmic reticulum: coordination of gene transcriptional and translational controls. Genes & Development 13: 1211–1233.
57. van Oers MM, Thomas AA, Moormann RJ, Vlak JM (2001) Secretory pathway limits the enhanced expression of classical swine fever virus E2 glycoprotein in insect cells. J Biotechnol 86: 31–38.
58. Lobigs M, Zhao HX, Garoff H (1990) Function of Semliki Forest virus E3 peptide in virus assembly: replacement of E3 with an artificial signal peptide abolishes spike heterodimerization and surface expression of E1. J Virol 64: 4346–4355.
59. Parker MD, Buckley MJ, Melanson VR, Glass PJ, Norwood D, et al. (2010) Antibody to the E3 glycoprotein protects mice against lethal Venezuelan equine encephalitis. J Virol 84: 12683–12690.
60. Makela AR, Oker-Blom C (2006) Baculovirus display: a multifunctional technology for gene delivery and eukaryotic library development. Advances in Virus Research 68: 91–112.
61. Presley JF, Cole NB, Schroer TA, Hirschberg K, Zaal KJM, et al. (1997) ER-to-Golgi transport visualized in living cells. Nature 389: 81–84.

5

Phylogenetic Evidence of Long Distance Dispersal and Transmission of Piscine Reovirus (PRV) between Farmed and Wild Atlantic Salmon

Åse Helen Garseth[1,2]*, Torbjørn Ekrem[2], Eirik Biering[1]

1 Department of Health Surveillance, Norwegian Veterinary Institute, Trondheim, Norway, 2 Department of Natural History, Norwegian University of Science and Technology University Museum, Trondheim, Norway

Abstract

The extent and effect of disease interaction and pathogen exchange between wild and farmed fish populations is an ongoing debate and an area of research that is difficult to explore. The objective of this study was to investigate pathogen transmission between farmed and wild Atlantic salmon (*Salmo salar* L.) populations in Norway by means of molecular epidemiology. Piscine reovirus (PRV) was selected as the model organism as it is widely distributed in both farmed and wild Atlantic salmon in Norway, and because infection not necessarily will lead to mortality through development of disease. A matrix comprised of PRV protein coding sequences S1, S2 and S4 from wild, hatchery-reared and farmed Atlantic salmon in addition to one sea-trout (*Salmo trutta* L.) was examined. Phylogenetic analyses based on maximum likelihood and Bayesian inference indicate long distance transport of PRV and exchange of virus between populations. The results are discussed in the context of Atlantic salmon ecology and the structure of the Norwegian salmon industry. We conclude that the lack of a geographical pattern in the phylogenetic trees is caused by extensive exchange of PRV. In addition, the detailed topography of the trees indicates long distance transportation of PRV. Through its size, structure and infection status, the Atlantic salmon farming industry has the capacity to play a central role in both long distance transportation and transmission of pathogens. Despite extensive migration, wild salmon probably play a minor role as they are fewer in numbers, appear at lower densities and are less likely to be infected. An open question is the relationship between the PRV sequences found in marine fish and those originating from salmon.

Editor: Bernhard Kaltenboeck, Auburn University, United States of America

Funding: The project was funded by The Norwegian Directorate for Nature Management (www.dirnat.no). The funders had no role in study design, data collection and analysis, decision to publish, or preparation of the manuscript.

Competing Interests: The authors have declared that no competing interests exist.

* E-mail: ase-helen.garseth@vetinst.no

Introduction

Farming of Atlantic salmon (*Salmo salar* L., 1758) is a young, fast-growing and economically important industry in Norway [1] but has not evolved without controversy. Concerns have been expressed by environmental Non-Governmental Organizations, consumers and governmental bodies with regards to animal welfare and health, area-use, pollution, exploitation of marine resources as feed ingredients and the impact of escapees and disease transmission on wild salmonid populations ([2], and references cited therein). During the last four decades when salmon farming has evolved from small scale supplementary enterprises to a multinational industry, the number of returning wild Atlantic salmon has declined [3]. These coincidental events have fed an ongoing debate concerning the potential negative effects of the growing industry on wild salmon populations. Escaped farmed salmon, sea-lice (*Lepeophtheirus salmonis* Krøyer, 1837) infestation and infectious diseases are all regarded as threats to the sustainability of wild salmon [3]. While sea-lice [4–6] and escapees [7–9] are subject of extensive research, the threat of infectious disease spreading from farmed to wild salmon has received less attention. The introduction and spread of the bacterial disease furunculosis [10,11] and the monogenean parasite *Gyrodactylus salaris* Malmberg, 1957 in Norway are a few exceptions to this rule [12–14].

Evidence of the extent and effect of disease interaction between wild and farmed Atlantic salmon populations has been difficult to obtain. Farmed Atlantic salmon have their origin in wild populations and, needless to say, so do most of the pathogens that cause diseases in farmed fish [15]. However, in contrast to the farm environment, conditions that promote epidemics and disease outbreaks, such as high host density, are rarely found in wild populations. As a consequence, farmed salmon are likely to account for higher levels of pathogen production, transmission and virulence evolution than wild salmon [15–18]. It is difficult to study the effect of pathogen transmission from farmed to wild salmon, partly due to methodological challenges as infected wild fish often die and disappear before they are detected [18,19]. The versatile life cycle of wild salmon also implies that they are affected by multiple factors, other than infectious diseases, that can cause populations to decline. These factors may act locally such as acidification [6], or at a larger scale such as climatic change [20,21] and availability of food in the ocean [20]. The outcome of these factors may camouflage potential adverse effects caused by pathogen spill from farmed salmon as they are all registered as reduced marine survival.

Molecular epidemiology has been used to investigate the dissemination and evolution of human viruses [22,23], conduct epidemiological research within the aquaculture industry [24–27], and is proposed as a tool useful for investigations of wild-farmed disease interaction [28–31]. The objective of this study was thus to investigate pathogen transmission between farmed and wild Atlantic salmon populations in Norway by means of molecular epidemiology. Piscine reovirus (PRV) was selected as the model organism as it is widely distributed in both farmed and wild Atlantic salmon in Norway. PRV is also a suitable model as infection not necessarily will lead to loss of study subjects through development of disease and mortality [32].

PRV is a reovirus associated with the development of heart and skeletal muscle inflammation (HSMI), a common and commercially important disease in farmed Atlantic salmon in Norway [33–35]. HSMI has also been found in farmed salmon in Scotland [36]. PRV is detected in both healthy and diseased salmon and appears to be ubiquitous among farmed Atlantic salmon [37]. However, the tissue distribution and increasing viral loads during an HSMI outbreak strongly support a causal relationship between PRV infection and development of HSMI [35,37–39]. Outbreaks of HSMI have so far not been associated with particular strains of PRV [37], and HSMI has not been recorded in wild Atlantic salmon, although PRV seems to be widely distributed in Atlantic salmon and to a lesser extent in sea-trout in Norwegian rivers [32,40].

Most viral agents that cause disease in salmonids in Norway have genomes consisting of RNA. Due to a higher mutation rate than DNA, the virus genome can change considerably over a relatively short period of time. This results in a high RNA-virus variability that can be used as a tool to trace spread of viral infection by the use of molecular epidemiology [41].

Reoviruses are icosahedral and non-enveloped with double-stranded RNA genomes of 10–12 segments. The *Reoviridae* consists of two subfamilies, *Spinareovirinae* and *Sedovirinae*, with altogether fifteen genera [42]. The host range of *Reoviridae* extends from insects, plants and fungi to fish, molluscs, reptiles, mammals and birds. Piscine reovirus was originally described as equally distant to genera *Orthoreovirus* and *Aquareovirus* in the subfamily *Spinareovirinae* [35]. PRV has 10 gene segments similar to *Orthoreovirus* [43], and two recent studies suggest that PRV is more closely related to *Orthoreovirus* than to *Aquareovirus* [44,45]. Hence the name *Piscine orthoreovirus* has been suggested [45]. However, a recent whole genome analysis concluded that PRV should be considered as member of a new genus within the family *Reoviridae* [46]. The same study also reports that PRV segment S1 sequences group into one genotype with two separate sub-genotypes, both found in Norway [46]. Recent research indicate that S1 is bicistronic encoding σ3 (a 330 amino acid (aa) outer capsid protein), and p13 (a 124 aa cytotoxic, nonfusogenic integral membrane protein) [44,45]. S2 is also possibly bicistronic encoding the 420 aa inner capsid protein σ2 and p8 (a 71 aa hypothetical protein) [35,45]. S4 is monocistronic encoding σ1 (a 315 aa cell attachment protein) [35,44,45].

In this study molecular epidemiology is used to investigate transmission of PRV between farmed and wild Atlantic salmon populations in Norway. Pathogen exchange can occur between Atlantic salmon stocks during marine migration, due to wild fish straying from neighboring rivers or by escapes from aquaculture. Finally, the presence of PRV in sea-trout [32] and marine species [47] raise questions regarding their role in pathogen exchange with Atlantic salmon.

Materials and Methods

Ethics statement

Samples utilized in this study and the preceding cross sectional survey of piscine reovirus infection [32] are residuals of samples originally intended for infectious pancreatic necrosis virus (IPNV) testing of brood fish as part of statutory health control in stock enhancement hatcheries and the Norwegian gene bank for wild Atlantic salmon. Additional residual samples were obtained from infectious salmon anaemia virus (ISAV) and viral hemorrhagic septicaemia virus (VHSV) surveillance conducted in wild salmonid populations. Hence, these samples represent secondary use of available material from existing health monitoring activities.

Samples from four escaped farmed salmon from river Etne (2010) were obtained during organised recapture after an escape from a nearby aquaculture site. The County Governor of Hordaland gave permission to the recapture (Fiskeløyve 23-2010).

All fish were killed in accordance with the Norwegian animal welfare act. Brood fish were anesthetized with trikainmesilat (metacaine) or benzocaine and killed by exsanguination. All other animals were stunned by a blow to their head and killed by exsanguination. No animals were killed specifically for this study. The authors have permission to use all samples.

Study sample and selection criteria

The majority of samples were from a cross-sectional survey of piscine reovirus infection described by Garseth and co-workers [32]. The survey was based on quantitative RT-PCR screening of head kidney samples from 1207 returning spawners of Atlantic salmon and 133 sea-trout captured in 36 rivers from 2007 to 2009. A total of 200 Atlantic salmon and four sea-trout (*Salmo trutta* L.) were PRV-positive. In addition, four escaped farmed salmon from river Etne (2010) were included in the study. These were caught during organised recapture after an escape from a nearby aquaculture site and are thus believed to originate from this site.

Scale-circuli patterns [48–50] and knowledge of local cultivation and release practices were used to determine the origin (*life-history*) of the Atlantic salmon. Hence, the term *wild* describes individuals that are the result of natural spawning and recruitement in the river, the term *escaped farmed* describes individuals displaying scale-circuli patterns of salmon escaped from commercial aquaculture, while the term *hatchery-reared* describes individuals that are offspring of wild parents but reared in hatcheries and released for stock enhancement or restoration purposes [32].

The selection criteria were chosen to agree with the objective of the study; to investigate pathogen transmission between wild and farmed salmon populations. Hence, salmon from all counties and life-histories were included. In addition, only samples from PRV-positive salmon with cycle threshold (C_t) values below 30 were included to ensure good sequence quality. However, all four PRV-positive sea-trout (C_t -values 25.9–39.5) were included in the initial amplification step. Sequences generated in this study are deposited in the European Nucleotide Archive with accession numbers HG329842 to HG330021 (http://www.ebi.ac.uk/ena/data/view/HG329842-HG330021).

All tissue-samples and RNA-extracts used in this study are deposited in the collections of the Norwegian University of Science and Technology (NTNU) University Museum and at the Norwegian Veterinary Institute, section for environmental and biosecurity measures.

RNA-extraction, RT-PCR amplification and sequencing

RNA was extracted from head kidney tissue as described by Garseth and co-workers [32]. RNA was isolated from approximately 20 mg of tissue with MagMAX TM-96 Total RNA Isolation Kit (cat #1830, Ambion). The subsequent RNA extraction was performed according to the manufacturers' recommendations with the same kit. A KingFisher (Labsystems Oy) was used in the magnetic-based separation. After elution, RNA concentration and purity was measured by use of NanoDrop ND-1000 spectrophotometer (NanoDrop Technologies). All samples had OD260/280 ratios between 1.97 and 2.12 (mean 2.06). Four aliqots à 15 µl eluated RNA were produced from each sample, one of these were used in the initial qRT-PCR PRV screening and three were frozen at −70°C. Altogether 91 samples were selected for amplification and transferred to NTNU University Museum on dry ice. Piscine reovirus genome segments S1, S2 and S4 were selected for amplification and sequencing (based on recommendations from Espen Rimstad and Torstein Tengs, coauthors of [35]).

An overview of analysed gene segments, primer combinations and primer sequences is shown in Table 1.

Reverse transcription and PCR amplification of S1, S2 and S4 were carried out in one step with QIAGEN OneStep RT-PCR kit (QIAGEN AB) using the primer combinations in Table 1. S1 was initially amplified using primer set 3 enabeling a near full length amplification. This approach was abandoned as sequence quality was improved by the amplification of S1 in two overlapping fragments (using primer sets 1 and 2). 2 µl template (2–10 ng total RNA), 1.5 µl forward primer and 1.5 µl reverse primer (final concentration 4 pmol/µl) was denaturated for 5 min at 95°C before 19.85 µl primer free Mastermix (QIAGEN OneStep RT-PCR kit) and 0.15 µl RNAse Out (Invitrogen) were added. The following PCR conditions were used: 30 min at 50°C (reverse transcription): 15 min at 95°C (inactivation of reverse transcriptase and activation of hot-start PCR DNA polymerase): 30 sec at 94°C (template denaturation): 30 sec at 55°C (primer annealing): 1 min at 72°C (fragment elongation). Steps 3–5 were repeted 40 times followed by a final elongation step of 3 min at 72°C.

Gel electrophoresis in 1% agarose gel with SYBR Safe stain (Invitrogen) was used to test the success of the amplification and served as an additional criterion for selecting samples for sequencing.

PCR products selected for sequencing were purified with ExoSAP-IT (USB Products) to remove excess nucleotides and unincorporated primers. Selected samples were sequenced bi-directionally by cycle sequencing technology using dideoxy chain termination/cycle sequencing on ABI 3730XL sequencing machines at Eurofins.

Amplification and sequencing was conducted twice for a proportion of the samples as a test of lab routine quality. For S1, 8 sequences were run twice, 15 were run twice for S2 and 19 were run twice for S4. Altogether 42 sequences were run twice, and of these 40 were identical while 2 had too low quality in the second run to be compared with the sequences from the first run. In total, 27 of the 180 sequences (15%) selected for the final dataset were included in this quality control.

Sequence editing and alignment

DNA sequences were assembled and edited with DNABaser Sequence Assembler v3.5.0 2011 (Heracle BioSoft SRL, http://www.DnaBaser.com). Sequences were assembled automatically and inspected and edited manually. In cases of ambiguity of base calls, the appropriate International Union of Pure and Applied Chemistry (IUPAC) code was inserted. Edited nucleotide contigs were imported to MEGA5 [51] and aligned as codons by MUSCLE [52] under default settings. Alignment was trivial since no internal indels were observed. Both ends of the alignments were trimmed to remove primers and parts with low sequence quality and indistinct base calls. Translation of nucleotides to amino acid sequences gave complete coding sequences.

For all three segments a standard nucleotide NCBI BLAST search (blastn) was conducted to identify and add available sequences of aquaculture origin to the alignments. Altogether 10 sequences were obtained from GenBank, whereof three were consensus sequences deposited in GenBank by Palacios and co-workers [35]. These were not included in the alignments as geographic origin was a key selection criterion. The remaining seven sequences (accessions JN991006-JN991012) were PRV S1 sequences derived from an industry based study conducted in Norway [37]. Information with regards to geographic origin of these samples was obtained from the authors and anonymized by limiting information to county of origin.

Phylogenetic analysis

Description of marker composition and initial maximum likelihood phylogenetic analyses were made on all three genomic segments in MEGA5 using 1000 bootstrap replicates and the Kimura 2-Parameter model (K2) on S1 and S2 [53] and K2 with

Table 1. Primers and their combination used in amplification and sequencing of segments S1, S2 and S4 of the piscine reovirus genome.

Genome segment	Primer set	Forward primer	Primer sequence	Reverse primer	Primer sequence
S1	S1 No 1	S1_39F	AAACCCAAATGGCGAACCA	S1_621R	TGCTCCACTGGGTTCAGCTC
	S1 No 2	S1_460F	TTGAAGCTAAGCGACGCCTT	S1_1036R	ACAGTAGGCTCCCCATCACG
	S1 No 3	S1_39F	AAACCCAAATGGCGAACCA	S1_1036R	ACAGTAGGCTCCCCATCACG
S2	S2 No 1	S2_43F	TGGCTAGAGCAATTTTCTCGG	S2_720R	GCCATTCCATGTCATCGTTG
	S2 No 2	S2_603F	TCGGTGCACGATATGAAAGC	S2_1304R	GTGGTCAGTCCCGGCTAGAG
	S2 No 3	S2_43F	TGGCTAGAGCAATTTTCTCGG	S2_1304R	GTGGTCAGTCCCGGCTAGAG
S4	S4 No1	S4_30F	TTAACCGCAGCGACATCTCA	S4_591R	TTGGTGCCGTCCCAACA
	S4 No 2	S4_456F	ACTGACCTGCTTGGACACACTG	S4_1005R	GACACGTGGCTCTTCCACG
	S4 No 3	S4_30F	TTAACCGCAGCGACATCTCA	S4_1005R	GACACGTGGCTCTTCCACG

Table 2. Basic statistics on the reverse transcribed genome segments used in the phylogenetic analyses of Norwegian piscine reovirus strains.

Genome segment		S1	S2	S4	Total
Nucleotides	Length of segment (bp)	1081	1329	1040	3450
	Length of segment used in analyses (bp)	837	1182	879	2898
	Conserved sites (bp)	785	1140	843	2768
	Variable sites (bp)	52	42	36	130
	Parsimony informative sites (bp)	43	28	28	99
	A (%)	27.4	23.8	25.4	25.4
	C (%)	23.9	24.0	23.2	23.7
	G (%)	25.7.	24.5	26.1	25.3
	T (%)	23.0	27.7	25.3	25.5

Gamma correction on segment S4. However, rigorous phylogenetic analyses were performed on S1 and the concatenated dataset only, as these were the alignments containing sequences derived from farmed Atlantic salmon. Moreover, S1 was the most phylogenetically informative segment in our dataset (Table 2).

The best fit substitution model and partition scheme was found using PartitionFinder 1.0.1 [54] testing for all substitution models and all possible combinations of markers and nucleotide positions. The best partition scheme according to the Bayesian Information Criterion (BIC) on the concatenated dataset contained three partitions, consisting of nucleotides from 1st position, 2nd position and 3rd position for all markers. The best substitution models on these partitions were the Hasegawa, Kishino and Yano model [55] with a proportion of invariable sites (HKY+I) on partition 1 and 2, and HKY+G+I (including gamma corrections for rate variation among sites) on partition 3. The best partition scheme for the marker S1 alone contained two partitions: 3rd position and 1st + 2nd position, both with the Kimura 2-Parameter model [53] as the best fit substitution model. Maximum likelihood (ML) analyses with 1000 bootstrap replicates were run on the partitioned dataset in RAxML 7.4.2 [56] utilizing the software raxmlGUI [57]. Since the best fit substitution models are not implemented in RAxML, we used the GTR+G model in our analyses.

Phylogenetic analyses by Bayesian inference were performed on the partitioned datasets in MrBayes 3.2.1 [58,59]. The Metropolis-Coupled Monte Carlo Markov Chain method with default four independent chains (nchains = 4) was run for 3,000,000 generations (ngen = 3,000,000). The frequency with which the chains were swapped was set to 0.2 (temp = 0.2). Every 200 generations a tree and corresponding parameter values were sampled and recorded to file (samplefreq = 200). The first 25% of sampled trees were discarded as the burnin fraction (relburnin = yes burninfrac = 0.25). Effective sample size (ESS) estimated with Tracer v1.5.0 [60] and standard deviation of split frequencies (≤0.01) were used as convergence diagnostic. For S1 and the concatenated dataset 50% majority rule consensus trees (contype = halfcompat) were constructed from the tree output files. Phylogenetic analyses on reduced datasets that only contained information from synonymous sites were also conducted. These were run with the same setup as the full datasets described above but without partitioning. The estimated phylogenetic trees were visualised in Figtree v1.3.1 [61] and MEGA5.

Results

Sequence composition and description of alignment

For all three genomic segments the final alignment matrix comprised sequences from 27 rivers with wild (N = 45) and hatchery-reared (N = 6) Atlantic salmon, one anadromous trout (sea-trout) and eight escaped farmed salmon whereof four were captured in river Etne in 2010 during an escape from a nearby aquaculture site. In addition, S1 and the concatenated alignment also comprised the seven sequences from GenBank derived from six cohorts of farmed Atlantic salmon from five counties. The final matrix of aligned sequences is described in Tables 2 and 3.

Table 3. Overview of origin of samples used in phylogenetic analyses.

Sample category		Samples	PRV-positive	Phylogeny	Rivers/sites represented
Sea-trout		133	4	1	1
Atlantic salmon	Wild	1008	134	45	24
	Hatchery reared	124	30	6	2
	Escaped farmed	61	33	4	4
	Uncertain	14	3	-	-
	Escaped farmed Etne*	38	37	4	1
GenBank	Farmed	-	-	7	6 (5 counties)

The final alignment comprised PRV protein coding sequences S1, S2 and S4 from sea-trout, wild, hatchery-reared and farmed Atlantic salmon.
*From Etne; believed to come from the same aquaculture site.

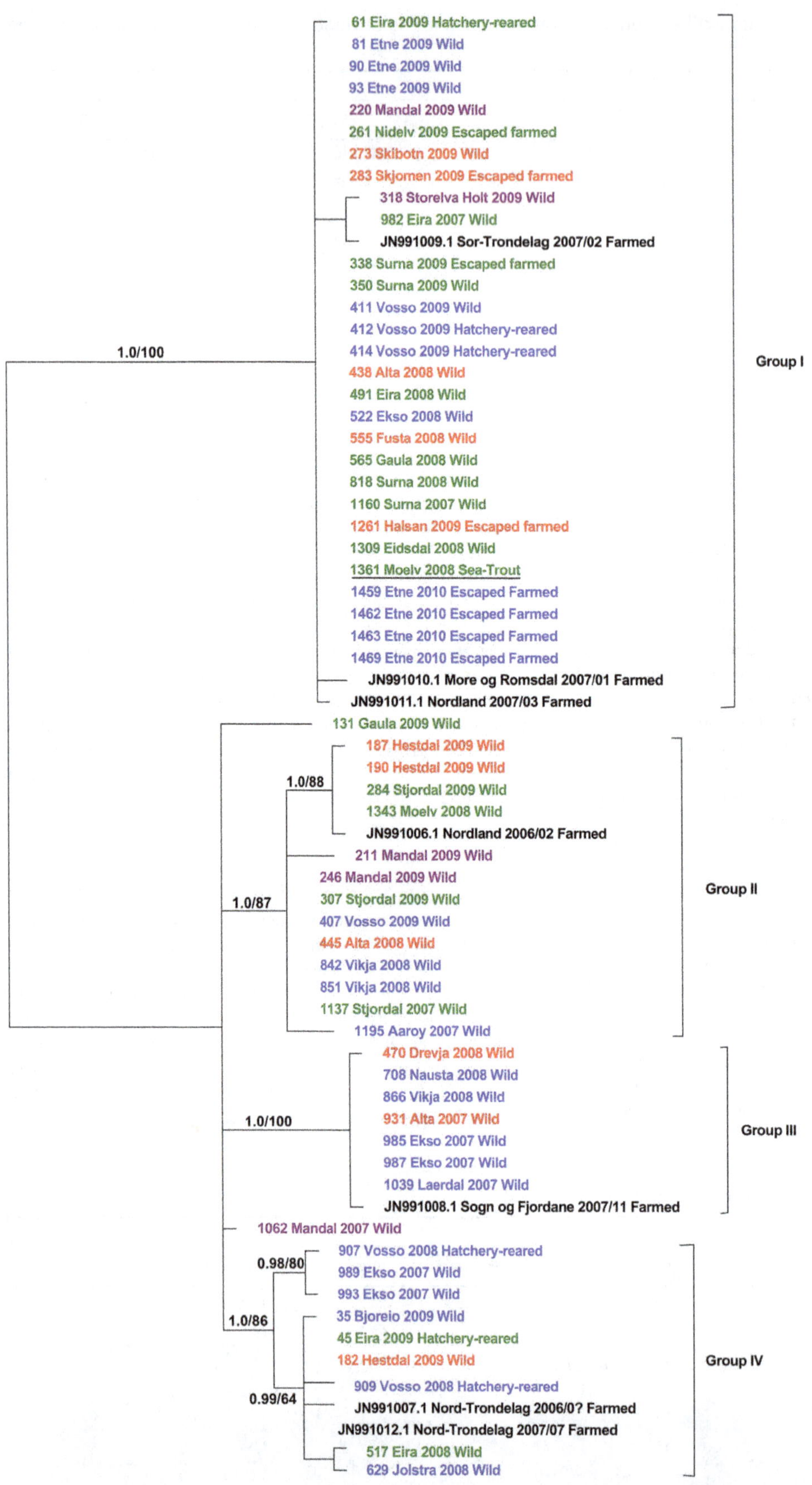
61 Eira 2009 Hatchery-reared
81 Etne 2009 Wild
90 Etne 2009 Wild
93 Etne 2009 Wild
220 Mandal 2009 Wild
261 Nidelv 2009 Escaped farmed
273 Skibotn 2009 Wild
283 Skjomen 2009 Escaped farmed
318 Storelva Holt 2009 Wild
982 Eira 2007 Wild
JN991009.1 Sor-Trondelag 2007/02 Farmed
338 Surna 2009 Escaped farmed
350 Surna 2009 Wild
411 Vosso 2009 Wild
412 Vosso 2009 Hatchery-reared
414 Vosso 2009 Hatchery-reared
1.0/100
Group I
438 Alta 2008 Wild
491 Eira 2008 Wild
522 Ekso 2008 Wild
555 Fusta 2008 Wild
565 Gaula 2008 Wild
818 Surna 2008 Wild
1160 Surna 2007 Wild
1261 Halsan 2009 Escaped farmed
1309 Eidsdal 2008 Wild
1361 Moelv 2008 Sea-Trout
1459 Etne 2010 Escaped Farmed
1462 Etne 2010 Escaped Farmed
1463 Etne 2010 Escaped Farmed
1469 Etne 2010 Escaped Farmed
JN991010.1 More og Romsdal 2007/01 Farmed
JN991011.1 Nordland 2007/03 Farmed
131 Gaula 2009 Wild
187 Hestdal 2009 Wild
190 Hestdal 2009 Wild
1.0/88
284 Stjordal 2009 Wild
1343 Moelv 2008 Wild
JN991006.1 Nordland 2006/02 Farmed
211 Mandal 2009 Wild
246 Mandal 2009 Wild
Group II
1.0/87
307 Stjordal 2009 Wild
407 Vosso 2009 Wild
445 Alta 2008 Wild
842 Vikja 2008 Wild
851 Vikja 2008 Wild
1137 Stjordal 2007 Wild
1195 Aaroy 2007 Wild
470 Drevja 2008 Wild
708 Nausta 2008 Wild
866 Vikja 2008 Wild
931 Alta 2007 Wild
1.0/100
Group III
985 Ekso 2007 Wild
987 Ekso 2007 Wild
1039 Laerdal 2007 Wild
JN991008.1 Sogn og Fjordane 2007/11 Farmed
1062 Mandal 2007 Wild
907 Vosso 2008 Hatchery-reared
0.98/80
989 Ekso 2007 Wild
993 Ekso 2007 Wild
1.0/86
35 Bjoreio 2009 Wild
45 Eira 2009 Hatchery-reared
182 Hestdal 2009 Wild
Group IV
909 Vosso 2008 Hatchery-reared
0.99/64
JN991007.1 Nord-Trondelag 2006/0? Farmed
JN991012.1 Nord-Trondelag 2007/07 Farmed
517 Eira 2008 Wild
629 Jolstra 2008 Wild
0.005

Figure 1. Resulting phylogenetic tree derived from Bayesian analysis of protein-coding PRV genome segment S1. Numbers above branches refer to Bayesian posterior probabilities and bootstrap support from corresponding maximum likelihood, respectively. Samples are identified with ID-numbers, geographical origin, year of sampling and life-history. Colours are corresponding to geographical regions in Figure 3. Sequences representing farmed Atlantic salmon are in black and marked with their respective GenBank accessions, county of origin and life history. Sequence from *Salmo trutta* is underlined.

None of the three nucleotide alignments (S1, S2 and S4) contained insertions or deletions. As described in Table 2, 837 of 1081 (77.4%) nucleotides were used in the analyses of PRV S1. This genome segment was the most variable segment with 52 (6.2%) variable sites whereof 43 (5.1%) were parsimony informative. For S2 1182 of 1329 (88.9%) nucleotides and for S4 879 of 1040 (84.5%) nucleotides were used in the analyses.

Phylogenetic analysis

The initial phylogenetic analysis of S1 returned a result nearly identical to the tree in Figure 1.

The result from rigorous Bayesian and ML analyses of S1 and the concatenated dataset are concordant and support the same groups (Figures 1 & 2). For the concatenated dataset, three major groups and several minor clades are well supported, with the exception of Group II in ML analyses (Figure 2). The same major groups are evident in the result based on analysis of the S1 dataset, but here an additional Group IV is also well supported. The initial analyses of segments S2 and S4 presented largely non-conflicting patterns to S1, but groups as defined in the phylogenetic analysis of S1 are not recovered in the same degree. Generally branches have lower support and are shorter (Figure S1 & Figure S2).

With the exception of a few smaller groupings, for instance the Vosso-Ekso clade in Group IV, there is very poor geographical structuring in our trees. All 3-4 main groups include samples from wild stocks (i.e. rivers) situated geographically far apart. For instance, wild salmon from rivers Alta (69°N, red) and Mandal (58°N, purple) appear together in Group I and II even if they are situated 1800 km apart (see Figure 3 for geographic location). The rivers Storelva Holt (purple southern region) and Skibotn (red northern region) are both present in Group I, and river Hestdal (red northern region) appear together with samples from river Vosso (blue western region) and Bjoreio (blue western region) in Groups II and IV respectively (Figures 1 & 2).

Further investigation revealed that S1 sequences were identical in multiple salmon from the same rivers and that in a subset of these S1, S2 and S4 were identical. In two of these rivers, sequences were obtained from salmon that according to the local stock enhancement hatcheries were cohabitants in the same tank before stripping and sampling. In a third river, identical sequences came from a hatchery where salmon were moved between several tanks. Hence, the identical sequences could be caused by infection during cohabitation.

The PRV S1 sequences from farmed Atlantic salmon, representing six aquaculture cohorts from five different counties, are dispersed among all four main groups along with sequences obtained from wild, hatchery-reared and escaped farmed salmon (Figures 1 & 2). In Group I, three sequences from farmed salmon representing three aquaculture cohorts from three counties appear together (Figures 1 & 2). In Group IV, two sequences from farmed salmon representing the same aquaculture cohort (fresh water and sea-water phase) group together (Figure 1). Groups II and III each have one sequence originating from farmed Atlantic salmon. PRV obtained from escaped farmed salmon captured in river Etne in 2010 group together in Group I indicating that at the point of escape there was limited within-site variation. Finally, PRV from sea-trout group together with PRV from Atlantic salmon in Group I (1361 Moelv 2008 Sea-trout).

Phylogenetic analyses of the synonymous sites from the S1 and concatenated datasets resulted in trees with the same general pattern (Figure S3 & Figure S4).

Discussion

Phylogenetic evidence of pathogen transmission between populations

Due to the assumed functions of the genome segments analysed, it is likely that their protein products are subject to natural selection, especially from the hosts' immune systems. This violates the assumption of neutral markers in phylogenetic reconstructions and could potentially strongly influence the relationship between virus strains. To investigate if variation in non-synonymous sites influenced our results, we ran the same analyses on reduced datasets only incorporating synonymous sites. Although some resolution was lost and some groups received slightly lower support values (Figure S3 & Figure S4), the results of these analyses were concordant with the results from the full datasets. We therefore conclude that the relationships seen between PRV strains in our data is not significantly influenced by converging or parallel evolution.

Geography was a key criterion in planning and conducting the study. This was based on the hypothesis that if distinct host populations are isolated geographically and there is no pathogen exchange between them; pathogen sequences will group according to the geographic origin of the host. Our results strongly indicate pathogen exchange between distant populations of Atlantic salmon, as PRV sequences from these populations are placed together in well supported genetic clusters.

Pathogen exchange can occur between Atlantic salmon stocks during marine migration, and be caused by straying from other rivers or escapes from aquaculture. Finally, the presence of PRV in sea-trout and marine species raise the question regarding their role in pathogen exchange with Atlantic salmon.

Pathogen exchange between wild Atlantic salmon stocks

Most Atlantic salmon spawners return to the river they left as smolts. This has led to genetically distinct salmon stocks or even several distinct populations in each river [62]. Contact between wild salmon stocks can occur during migration, straying and within live gene banks for wild Atlantic salmon (see below).

The marine feeding migration is the least studied phase of Atlantic salmon life, and information regarding the spatial and temporal distribution during this period is scarce. Dadswell and co-workers [63] reviewed data accumulated during the last five decades and concluded that the most probable marine migration model is the "Merry-Go-Round"- hypothesis. This hypothesis proposes that North-American and European stocks enter the North Atlantic Sub-polar Gyre from their respective sides of the Atlantic and migrate counter clockwise until they return to their native river [63]. It is difficult to estimate the extent of interaction and the potential for pathogen transmission between individual salmon and stocks during migration. Catch rates from Faroese long-line fisheries from November 1981 to May 1982 showed that

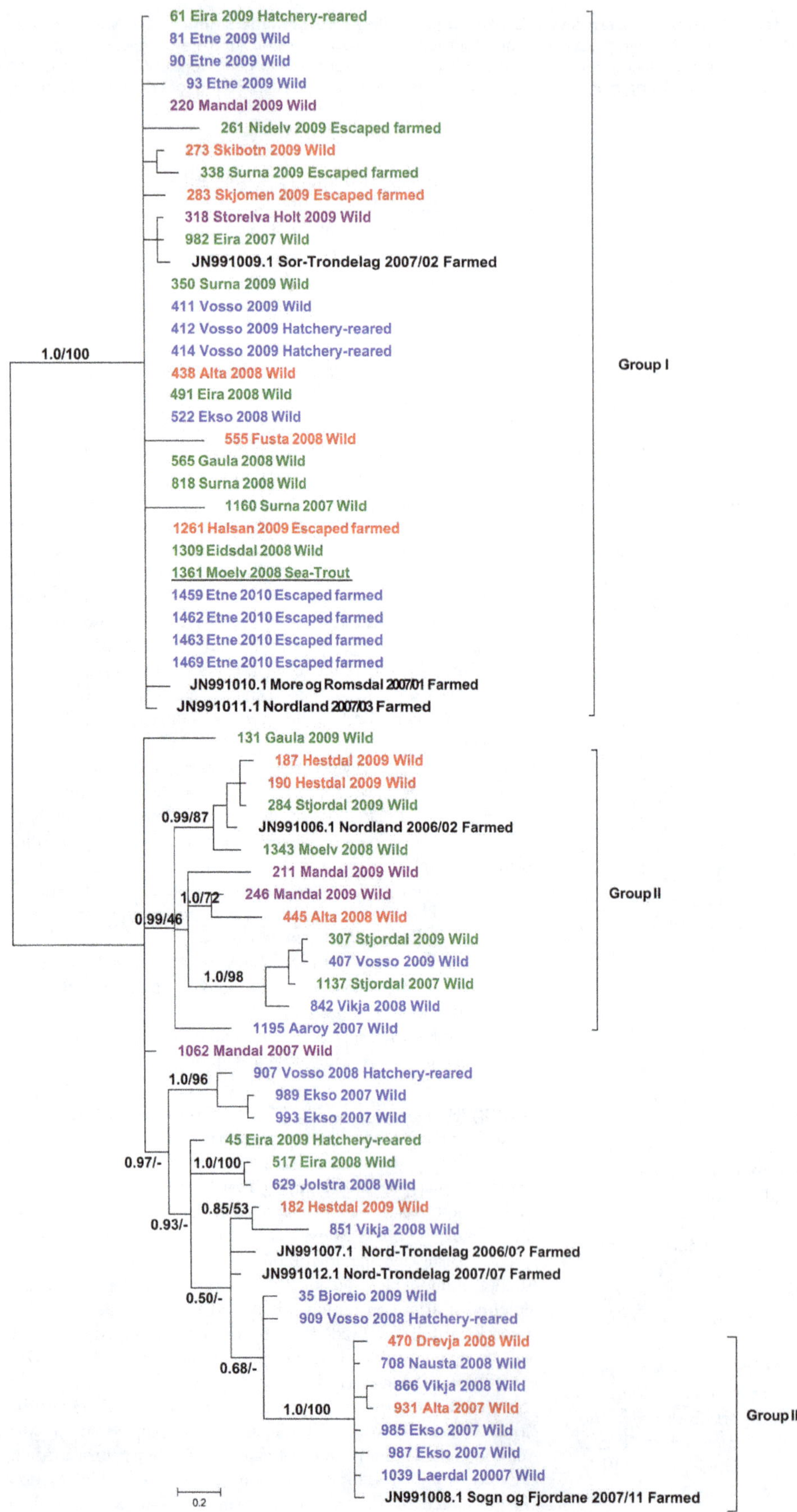
61 Eira 2009 Hatchery-reared
81 Etne 2009 Wild
90 Etne 2009 Wild
93 Etne 2009 Wild
220 Mandal 2009 Wild
261 Nidelv 2009 Escaped farmed
273 Skibotn 2009 Wild
338 Surna 2009 Escaped farmed
283 Skjomen 2009 Escaped farmed
318 Storelva Holt 2009 Wild
982 Eira 2007 Wild
JN991009.1 Sor-Trondelag 2007/02 Farmed
350 Surna 2009 Wild
411 Vosso 2009 Wild
412 Vosso 2009 Hatchery-reared
414 Vosso 2009 Hatchery-reared
1.0/100
438 Alta 2008 Wild
491 Eira 2008 Wild
522 Ekso 2008 Wild
555 Fusta 2008 Wild
565 Gaula 2008 Wild
818 Surna 2008 Wild
1160 Surna 2007 Wild
1261 Halsan 2009 Escaped farmed
1309 Eidsdal 2008 Wild
1361 Moelv 2008 Sea-Trout
1459 Etne 2010 Escaped farmed
1462 Etne 2010 Escaped farmed
1463 Etne 2010 Escaped farmed
1469 Etne 2010 Escaped farmed
JN991010.1 More og Romsdal 2007/01 Farmed
JN991011.1 Nordland 2007/03 Farmed
Group I
131 Gaula 2009 Wild
187 Hestdal 2009 Wild
190 Hestdal 2009 Wild
284 Stjordal 2009 Wild
0.99/87
JN991006.1 Nordland 2006/02 Farmed
1343 Moelv 2008 Wild
211 Mandal 2009 Wild
1.0/72
246 Mandal 2009 Wild
0.99/46
445 Alta 2008 Wild
307 Stjordal 2009 Wild
407 Vosso 2009 Wild
1.0/98
1137 Stjordal 2007 Wild
842 Vikja 2008 Wild
1195 Aaroy 2007 Wild
Group II
1062 Mandal 2007 Wild
1.0/96
907 Vosso 2008 Hatchery-reared
989 Ekso 2007 Wild
993 Ekso 2007 Wild
45 Eira 2009 Hatchery-reared
0.97/-
1.0/100
517 Eira 2008 Wild
629 Jolstra 2008 Wild
0.85/53
182 Hestdal 2009 Wild
0.93/-
851 Vikja 2008 Wild
JN991007.1 Nord-Trondelag 2006/0? Farmed
JN991012.1 Nord-Trondelag 2007/07 Farmed
0.50/-
35 Bjoreio 2009 Wild
909 Vosso 2008 Hatchery-reared
470 Drevja 2008 Wild
0.68/-
708 Nausta 2008 Wild
866 Vikja 2008 Wild
931 Alta 2007 Wild
1.0/100
985 Ekso 2007 Wild
987 Ekso 2007 Wild
1039 Laerdal 20007 Wild
JN991008.1 Sogn og Fjordane 2007/11 Farmed
Group III
0.2

Figure 2. Resulting phylogenetic tree derived from Bayesian analysis of the concatenated dataset. The dataset contained protein-coding PRV genome segments S1, S2 and S4. Numbers above branches refer to Bayesian posterior probabilities and bootstrap support from corresponding maximum likelihood analysis, respectively. Samples are identified with ID-numbers, geographical origin, year of sampling and life-history. Colours are corresponding to geographical regions in Figure 3. Sequences representing farmed Atlantic salmon are in black and marked with their respective GenBank accessions, county of origin and life history. Sequence from *Salmo trutta* is underlined.

from 0 to 286 salmon were caught per 1000 hooks, and during the west Greenland fisheries (before 1980) the highest catches were 20–70 salmon km^{-1}. This indicates that salmon occur in small shoals and provides some insight about density. However, these data were generated before the major decline of wild populations had occurred. Scale discrimination and marking studies indicate a mixed stock structure, i.e. representation from both European and North-American stocks among catches throughout the North Atlantic Sub-polar Gyre. Tagged adult salmon from the Faroes were recovered in Canadian rivers while marked North American smolt were captured off the Faroes, west and east Greenland and Norway [63]. This indicates that pathogen exchange may occur between individuals and stocks of diverse origin. Still, extensive transmission during migration seems unlikely when the low host density in wild salmon populations is taken into account.

About 3–6% of wild salmon and 15% of hatchery-reared salmon may stray to other rivers during homeward spawning migration [64,65]. Studies show that most of them enter nearby rivers. For instance, 96% of straying Imsa salmon entered streams within 420 km, and 80% entered streams within 60 km of the mouth of the River Imsa [65]. Hence, pathogen exchange between wild stocks from nearby rivers can happen due to straying.

Norway has established three gene bank stations for live Atlantic salmon to facilitate conservation and restoration of endangered wild stocks. Each gene bank harbour several stocks mainly of regional origin, and a biosecurity strategy has been implemented to minimize the risk of horizontal and vertical pathogen transmission. Founder stocks are established and maintained in the gene bank by importing disinfected, fertilized eggs from wild brood fish that have been subject to pathogen testing and health control. Stocks from different rivers are kept in separate tanks throughout the lifespan, and only disinfected, fertilized eggs from these are exported back to the river. Since stocks within each gene bank are of regional origin, the phylogenetic pattern caused by pathogen transmission within the gene banks cannot be distinguished from the pattern caused by straying. A phylogenetic pattern derived from pathogen dissemination through straying or gene banking cannot be excluded in any of the groups (Figure 2). Still, the pattern is systematically violated by sequences from farmed and wild salmon from other geographic regions.

Pathogen exchange between wild and farmed Atlantic salmon

Grouping of PRV S1 sequences from farmed Atlantic salmon together with sequences obtained from wild, hatchery reared and escaped farmed salmon indicates that wild and farmed salmon harbour the same virus strains and that virus have been exchanged between populations of different origin. Pathogen exchange between farmed and wild Atlantic salmon can occur by interaction between wild and *escaped* farmed salmon during marine migration or spawning, but also when wild salmon pass through areas with aquaculture production during sea-ward migration as post-smolt or homeward migration as spawners.

Simulated escapes of farmed Atlantic salmon show that migratory behavior of escaped farmed salmon depends on the development stage at point of release. Post-smolt released during early summer migrates out of the fjord whereas post-smolt escaping during late summer and autumn were recaptured in the fjord [66]. Escaped large salmon have the capacity for long distance migration and are recaptured in rivers [67].

Wild and escaped salmon stay in the same parts of the river and interbreeding is known to occur [7,68]. In addition, congregation of salmon under natural or man-made migration barriers can facilitate pathogen exchange as demonstrated by furunculosis induced mass mortality in Norwegian rivers [11].

Investigation of transmission of HSMI, conducted prior to the description of PRV, confirm horizontal transmission as an important route [69]. Horizontal transmission between sites can occur through virus dispersal by ocean currents, sharing of personnel and equipment, but also through wild fish movements if these are susceptible to the pathogen in question [70].

PRV is ubiquitous in salmon farms [37] and higher odds of PRV-infection in escaped farmed salmon than wild salmon (odds ratio 7.3 $p<0{,}001$) is considered both plausible and expected based on the number of HSMI-outbreaks and the PRV-prevalence in farmed fish [71]. Since reoviruses are hydrophilic, non-enveloped viruses and considered relatively robust outside the host [72,73] it is plausible to suggest that PRV can be abundant in sea-water near aquaculture sites. This makes it possible for wild salmonids to contract PRV-infection as they pass aquaculture sites. Research with salmonids carrying acoustic tags show that while sea-trout stay near aquaculture sites and move between them, wild Atlantic salmon post-smolt and Atlantic salmon x sea-trout hybrids pass the same sites without delay or inter-site movement [74].

The fact that PRV-sequences from wild Atlantic salmon in rivers Mandal and Alta appear together in Groups I and II (Figures 1 & 2), could be explained by transportation of PRV-carrier fish within the Atlantic salmon industry. The structure of the Norwegian salmon farming industry is to a high degree dependent upon transportation of live fish. Fertilized eggs are moved from broodfish stations to hatcheries. Smolt are moved from smolt production sites to on-growing sites in sea-water, and full-grown salmon are transported to the abattoirs for slaughtering. Some of these movements represent long distance transportation. A public record of live fish movement has not been established in Norway; hence detailed information is not available. However, the discrepancy between smolt production and input to sea in most counties is an indication of trans-county movement [75]. Likewise is the discrepancy between production and slaughter capacity in the different counties. The production of smolt in the northern part of Norway (counties Nordland, Troms and Finnmark) has so far not been able to meet the demand of the local industry. This imbalance has been solved by moving smolt from the southern part of Norway. Cases of disease outbreaks have been known to occur after such fish movements; the first cases of pancreas disease in the northern part of Norway occurred in smolt imported from the endemic area in south-western Norway [76,77].

Farmed salmon far outnumber wild salmon in Norway. At the end of 2011, a total of 366 million individuals (679 398 metric tons) farmed Atlantic salmon were in cages along the coast of Norway [75]. The same year 500 000 wild salmon returned to Norwegian rivers [78], whereof 45% were captured and killed in rivers and fjords. In the rivers escapees constituted approximately 4% of salmon caught during angling season in 2011 (compared to

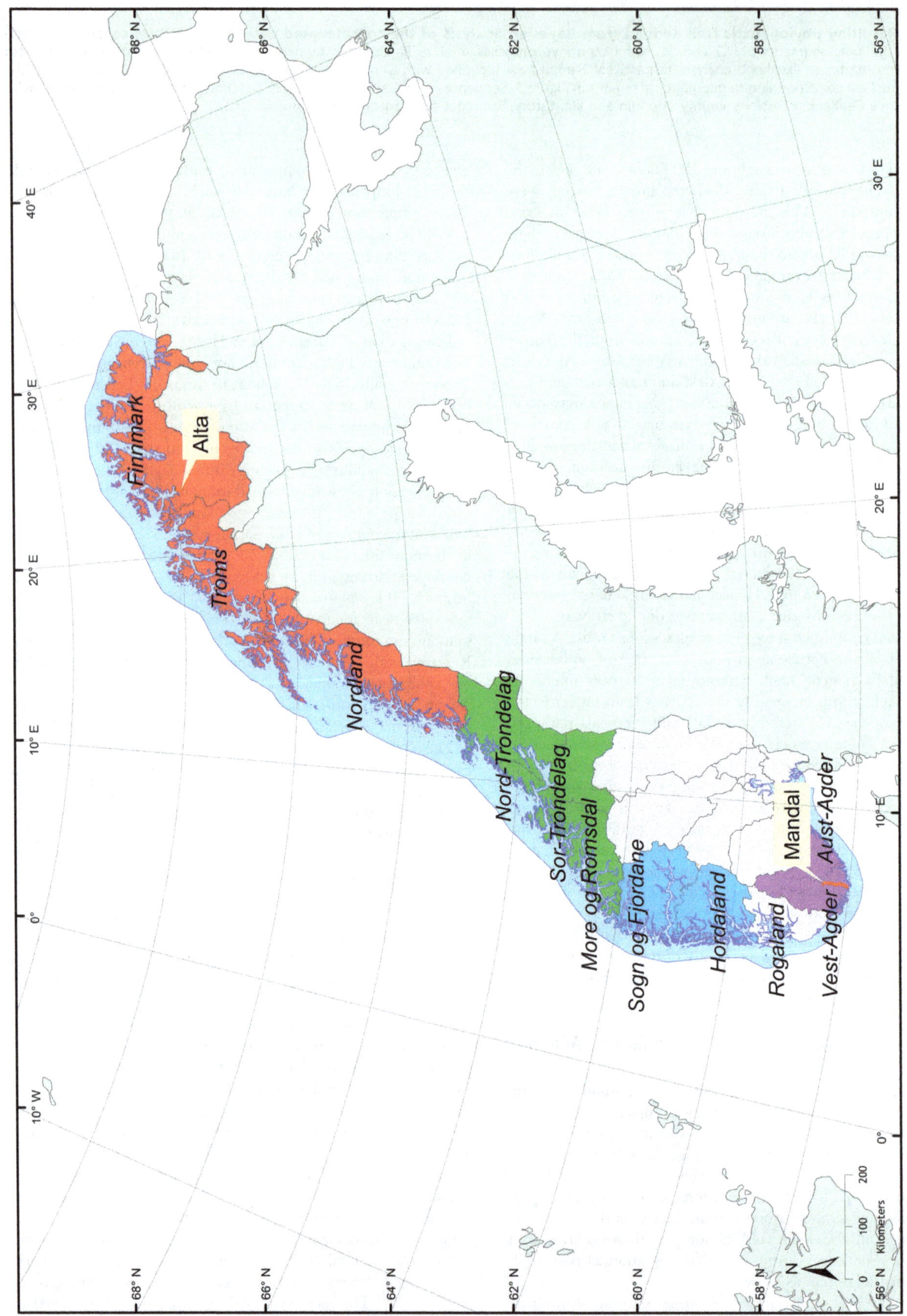

Figure 3. Map of Norway showing rivers Alta and Mandal and counties. Counties are color coded according to geographic region.

6–9% the previous 10 years). During the spawning season later that year 12% were escapees (compared to 11–18% the previous 13 years) [78].

The Norwegian salmon farming industry experiences large disease related losses [79] and a considerable proportion of sites experience disease outbreaks with potential pathogen spill over to the environment. In 2011 there were 440 recorded disease outbreaks caused by viral agents [79] in approximately 1000 licensed sites for grow-out production of Atlantic salmon [75].

Fish-farming constitutes a favorable environment for within- and between sites transmission of pathogens. Fish are held at high stocking densities in open cages during the sea-water phase, and sites are connected to nearby sites by coastal currents, movement of fish and sharing of equipment and personnel. The near endless access to susceptible hosts in high densities in the farm environment will keep infections alive over an extended period. This will not only increase the likelihood of pathogen spill over to the environment, but also increase the potential for evolution of more virulent strains [17]. Thus, farmed salmon seem to outnumber wild salmon not only in sheer numbers, but also in the potential for propagation and spread of infectious agents. While diseased farmed salmon are fed and protected against predators in cages, diseased wild fish will strive to catch their prey and to avoid predators. Hence by implication, diseased wild fish will to a greater extent succumb to infections.

The role of sea-trout

Between 1.9 and 3.0% of sea-trout are PRV-infected [32,40], and PRV obtained from one sea-trout group together with PRV obtained from Atlantic salmon (Group I). Although only one sequence was available for phylogenetic analyses, this may indicate that sea-trout can play a role in pathogen exchange with and between Atlantic salmon. Although some individuals migrate out of the fjords, most sea-trout spend their entire marine phase in the fjords. Research conducted by the Norwegian Institute for Water Research (NIVA) show that they stay temporarily around aquaculture sites, but also connect different sites by moving between them [74]. Sea-trout are also in close contact with wild Atlantic salmon in the rivers. As most sea-trout limit their marine migration to the fjord, and few of them are infected, sea-trout will *per se* not contribute to the lack of geographic pattern in the phylogenetic tree. However, they can be a link between farmed salmon, which may have been transported long distances, and the local wild salmon stock. Further sequencing, phylogenetic analyses and research are needed to conclude on the role of sea-trout.

Pathogen exchange between marine fish species and wild Atlantic salmon

Wiik-Nielsen and co-workers [47] screened a total of 1627 fish (379 pools) from 37 different wild marine species using a PRV-specific RT-qPCR assay. Pools from four species yielded positive results; *Argentina silus* (Atlantic Argentine, 1 of 38 pools), *Trachurus trachurus* (Atlantic horse mackerel, 1 of 1 pools), *Mallotus villosus* (Capelin, 1 of 16 pools) and *Clupea harengus* (Atlantic herring, 1 of 37 pools). The highest viral load was detected in herring. To this point virus from marine species have not been sequenced, and it is not known if they represent a marine genotype. Caplin and 0+ herring are important prey for Atlantic salmon post-smolt [80], while 1+ herring often compete with post-smolt for food and occur in high densities in the same habitat in both fjords and the open ocean. Atlantic salmon are hence caught as by-catch in herring surface trawls [81]. Accordingly, the possibility of pathogen exchange between these marine species and Atlantic salmon cannot be excluded and will be better understood when PRV-sequences from marine fish are available.

Conclusion

This study pinpoints the complex nature of research concerning pathogen exchange between farmed and wild Atlantic salmon. Many factors influence the life and survival of wild salmon and should be accounted for before a conclusion is drawn.

In the present study, PRV serves as a model of pathogen exchange between different wild and farmed populations of Atlantic salmon. PRV is a suitable model organism because it is widely distributed in both populations and because it doesn't necessarily lead to loss of study subjects through development of disease and mortality. We conclude that the lack of a geographical pattern in the phylogenetic trees is caused by extensive exchange of PRV. In addition, the detailed topography of the trees indicates long distance transportation of PRV.

Through its size, structure and infection status, the Atlantic salmon farming industry has the capacity to play a central role in both long distance transportation and transmission of pathogens. Despite extensive migration, wild salmon probably play a minor role as they are fewer in numbers, appear at lower densities and are less likely to be infected. An open question is the relationship between PRV-sequences found in marine fish and those from salmon.

In this study we have used PRV as a model for pathogen dissemination, and the study strongly supports the existence of pathways for pathogen transmission between farmed and wild salmon. We have so far no indications that PRV-infection leads to disease in wild salmon, this remains to be shown. But, as transmission of PRV is possible, it is not unlikely that other more virulent agents are transferred. If this occurs, and if it has an impact on wild fish, are important questions for future research.

Supporting Information

Figure S1 Phylogeny from initial maximum likelihood analysis of protein-coding PRV genome segment S2 in MEGA5. Numbers above branches refer to bootstrap support based on 1000 random replicates. Samples are identified with ID-numbers, geographical origin, year of sampling and life-history. Colours are corresponding to geographical regions in Figure 3. Sequence from *Salmo trutta* is underlined. Groups I and III as defined in the phylogenetic analysis of S1 are indicated. Sample sequences not belonging to groups as defined by S1 are marked with an asterisk. Groups II and IV are not recovered in the same degree and therefore not indicated.

Figure S2 Phylogeny from initial maximum likelihood analysis of protein-coding PRV genome segment S4 in MEGA5. Numbers above branches refer to bootstrap support based on 1000 random replicates. Samples are identified with ID-numbers, geographical origin, year of sampling and life-history. Colours are corresponding to geographical regions in Figure 3. Sequence from *Salmo trutta* is underlined. Groups I, III and IV as defined by phylogenetic analysis of S1 are indicated. Sample sequences not belonging to groups as defined by S1 are marked with an asterisk. Group II is not recovered in the same degree and therefore not indicated.

Figure S3 Maximum likelihood phylogeny of synonymous sites from the concatenated dataset. The dataset contained protein-coding PRV genome segments S1, S2 and S4. Numbers above branches refer to Bayesian posterior probabilities and maximum likelihood bootstrap support, respectively. Samples are identified with ID-numbers, geographical origin and year of sampling. Sequences representing farmed Atlantic salmon are marked with their respective GenBank accessions. Colours are corresponding to geographical regions in Figure 3. Sequence from *Salmo trutta* is underlined.

Figure S4 Maximum likelihood phylogeny of synonymous sites from PRV genome segment S1. Numbers above branches refer to Bayesian posterior probabilities and maximum likelihood bootstrap support, respectively. Samples are identified with ID-numbers, geographical origin and year of sampling. Sequences representing farmed Atlantic salmon are marked with their respective GenBank accessions. Colours are corresponding to geographical regions in Figure 3. Sequence from *Salmo trutta* is underlined.

Acknowledgments

The authors would like to thank Torbjørn Forseth (NINA), Torstein Tengs (NVI) and Turhan Markussen (NVI) for valuable information and discussions. Thanks to Attila Tarpai for constructing the map in Figure 3.

Author Contributions

Conceived and designed the experiments: AHG EB TE. Performed the experiments: AHG. Analyzed the data: AHG TE EB. Contributed reagents/materials/analysis tools: AHG. Wrote the paper: AHG TE EB.

References

1. Anonymous (2011) Effectiv and sustainable area use in aquaculture. The Norwegian Ministry of Fisheries and Coastal Affairs. 190 p. (In Norwegian). Available: http://www.regjeringen.no/nb/dep/fkd/dok/rapporter_planer/rapporter/2011/arealutvalget.html?id = 647391. Accessed 20 August 2011
2. Olesen I, Myhr AI, Rosendal GK (2011) Sustainable aquaculture: Are we getting there? Ethical perspectives on salmon farming. J Agri Environ Ethic 24(4): 381–408.
3. Anonymous. (2009) The status of Norwegian salmon stocks in 2009 and catch advice. Scientific Advisory Committee for Atlantic Salmon Management in Norway. 230 p. (In Norwegian, English summary). Available: http://www.vitenskapsradet.no/Publikasjoner/RapportfraVitenskapeligr%C3%A5dforlakse forvaltning.aspx. Accessed 20 August 2010
4. Heuch PA, Bjørn PA, Finstad B, Holst JC, Asplin L, et al. (2005) A review of the Norwegian 'National action plan against salmon lice on salmonids': The effect on wild salmonids. Aquaculture 250(1–2): 535–535.
5. Heuch PA, Bjørn PA, Finstad B, Asplin L, Holst JC (2009) Salmon lice infection of farmed and wild salmonids in Norway: an Overview. Integ Comp Biol 49: 74.
6. Finstad B, Kroglund F, Bjørn PA, Nilsen R, Pettersen K, et al. (2012) Salmon lice-induced mortality of Atlantic salmon postsmolts experiencing episodic acidification and recovery in freshwater. Aquaculture 2012. 362: p. 193–199.
7. Hindar K, Ryman N, Utter F (1991) Genetic-effects of cultured fish on natural fish populations. Can J Fish Aquat Sci 48(5): 945–957.
8. Hindar K, Fleming IA, McGinnity P, Diserud O (2006) Genetic and ecological effects of salmon farming on wild salmon: modelling from experimental results. Ices J Mar Sci 63(7): 1234–1247.
9. Jensen Ø, Dempster T, Thorstad EB, Uglem I, Fredheim A (2010) Escapes of fishes from Norwegian sea-cage aquaculture: causes, consequences and prevention. Aquacult Environ Interact 1(1): 71–83.
10. Heggberget TG, Johnsen BO, Hindar K, Jonsson B, Hansen LP, et al. (1993) Interactions between wild and cultured Atlantic salmon - A review of the Norwegian experience. Fish Res 18(1–2): 123–146.
11. Johnsen BO, Jensen AJ (1994) The spread of furunculosis in salmonids in Norwegian rivers. J Fish Biol 45(1): 47–55.
12. Peeler EJ, Thrush MA (2004) Qualitative analysis of the risk of introducing *Gyrodactylus salaris* into the United Kingdom. Dis aquat organ 62(1–2): 103–113.
13. Mo TA (1994) Status of *Gyrodactylus salaris* problems and research in Norway. In: Lewis JW, editor. Parasitic diseases of fish. Dyfed: Samara Publishing: pp. 43–48.
14. Johnsen BO, Jensen AJ (1991) The gyrodactylus story in Norway. Aquaculture 98(1–3): 289–302.
15. Raynard RS, Wahli T, Vatsos I, Mortensen S (2007) Review of disease interaction and pathogen exchange between farmed and wild finfish and shellfish in Europe Available: http://www.revistaaquatic.com/DIPNET/. Accessed 15 January 2012.
16. Frank SA (1996) Models of parasite virulence. Q Rev Biol 71(1): 37–78.
17. Mennerat A, Nilsen F, Ebert D, Skorping A (2010) Intensive farming: Evolutionary implications for parasites and pathogens. Evol Biol 37(2–3): 59–67.
18. Johansen LH, Jensen I, Mikkelsen H, Bjørn PA, Jansen PA, et al. (2011) Disease interaction and pathogens exchange between wild and farmed fish populations with special reference to Norway. Aquaculture 315(3–4): 167–186.
19. Bergh O (2007) The dual myths of the healthy wild fish and the unhealthy farmed fish. Dis aquat organ 75(2): 159–164.
20. Beaugrand G, Reid PC (2003) Long-term changes in phytoplankton, zooplankton and salmon related to climate. Glob Change Biol 9(6): 801–817.
21. Peyronnet A, Friedland KD, Maoileidigh NÓ (2008) Different ocean and climate factors control the marine survival of wild and hatchery Atlantic salmon *Salmo salar* in the north-east Atlantic Ocean. J Fish Biol 73(4): 945–962.
22. Garten RJ, Davis CT, Russell CA, Shu B, Lindstrom S, et al. (2009) Antigenic and genetic characteristics of swine-origin 2009 A(H1N1) influenza viruses circulating in humans. Science 325(5937): 197–201.
23. Hungnes O, Jonassen TØ, Jonassen CM, Grinde B (2000) Molecular epidemiology of viral infections - How sequence information helps us understand the evolution and dissemination of viruses. Apmis 108(2): 81–97.
24. Lyngstad TM, Hjortaas MJ, Kristoffersen AB, Markussen T, Karlsen ET, et al. (2011) Use of molecular epidemiology to trace transmission pathways for infectious salmon anaemia virus (ISAV) in Norwegian salmon farming. Epidemics 3(1): 1–11.
25. Wiik-Nielsen J, Alarcón M, Fineid B, Rode M, Haugland Ø (2013) Genetic variation in Norwegian piscine myocarditis virus in Atlantic salmon, *Salmo salar* L. J Fish Dis 36(2): 129–139.
26. Fringuelli E, Rowley HM, Wilson JC, Hunter R, Rodger H, et al. (2008) Phylogenetic analyses and molecular epidemiology of European salmonid alphaviruses (SAV) based on partial E2 and nsP3 gene nucleotide sequences. J Fish Dis 31(11): 811–823.
27. Nylund A, Karlsbakk E, Nylund S, Isaksen TE, Karlsen M, et al. (2008) New clade of betanodaviruses detected in wild and farmed cod (*Gadus morhua*) in Norway. Arch Virol 153(3): 541–547.
28. Mladineo I, Šegvić T, Grubišić L (2009) Molecular evidence for the lack of transmission of the monogenean *Sparicotyle chrysophrii* (Monogenea, Polyopisthocotylea) and isopod *Ceratothoa oestroides* (Crustacea, Cymothoidae) between wild bogue (*Boops boops*) and cage-reared sea bream (*Sparus aurata*) and sea bass (*Dicentrarchus labrax*). Aquaculture 295(3-4): 160–167.
29. Ruane NM, McCarthy LJ, Swords D, Henshilwood K (2009) Molecular differentiation of infectious pancreatic necrosis virus isolates from farmed and wild salmonids in Ireland. J Fish Dis 32(12): 979–987.
30. Vendramin N, Patarnello P, Toffan A, Panzarin V, Cappellozza E, et al. (2013) Viral encephalopathy and retinopathy in groupers (*Epinephelus spp.*) in southern Italy: a threat for wild endangered species? Bmc Vet Res 9: 20.
31. Anderson ED, Engelking HM, Emmenegger EJ, Kurath G (2000) Molecular epidemiology reveals emergence of a virulent infectious hematopoietic necrosis (IHN) virus strain in wild salmon and its transmission to hatchery fish. J Aquat Anim Health 12(2): 85–99.
32. Garseth ÅH, Fritsvold C, Opheim M, Skjerve E, Biering E (2013) Piscine reovirus (PRV) in wild Atlantic salmon, *Salmo salar* L., and sea-trout, *Salmo trutta* L., in Norway. J Fish Dis 36(5): 483–493.
33. Kongtorp RT, Kjerstad A; Taksdal T, Guttvik A, Falk K (2004) Heart and skeletal muscle inflammation in Atlantic salmon, *Salmo salar* L.: a new infectious disease. J Fish Dis 27(6): 351–358.
34. Kongtorp RT, Taksdal T, Lyngoy A (2004) Pathology of heart and skeletal muscle inflammation (HSMI) in farmed Atlantic salmon *Salmo salar*. Dis aquat organ 59(3): 217–224.
35. Palacios G, Lovoll M, Tengs T, Hornig M, Hutchison S, et al. (2010) Heart and skeletal muscle inflammation of farmed salmon is associated with infection with a novel reovirus. PloS One 5(7): e11487. DOI: 10.1371/journal.pone.0011487.
36. Ferguson HW, Kongtorp RT, Taksdal T, Graham D, Falk K (2005) An outbreak of disease resembling heart and skeletal muscle inflammation in Scottish farmed salmon, *Salmo salar* L., with observations on myocardial regeneration. J Fish Dis 28(2): 119–123.
37. Lovoll M, Alarcón M, Bang Jensen B, Taksdal T, Kristoffersen AB, et al. (2012) Quantification of piscine reovirus (PRV) at different stages of Atlantic salmon *Salmo salar* production. Dis Aquat Organ 99(1): 7–12.
38. Finstad ØW, Falk K, Løvoll M, Evensen Ø, Rimstad E (2012) Immunohistochemical detection of piscine reovirus (PRV) in hearts of Atlantic salmon coincide with the course of heart and skeletal muscle inflammation (HSMI). Vet Res 43:27. doi:10.1186/1297-9716-43-27.
39. Mikalsen AB, Haugland O, Rode M, Solbakk IT, Evensen O (2012) Atlantic salmon reovirus infection causes a CD8 T Cell myocarditis in Atlantic salmon (*Salmo salar* L.). PloS One 7(6): e37269. doi:10.1371/journal.pone.0037269.
40. Biering E, Madhun AS, Isachsen CH, Omdal LM, Bårdsgjære Einen AC, et al. (2013) Annual report on health monitoring of wild anadromous salmonids in

Norway. Norwegian Veterinary Institute and Institute of Marine research. Available: www.vetinst.no. Accessed 10 June 2013.
41. Grenfell BT, Pybus OG; Gog JR, Wood JLN, Daly JM, et al. (2004) Unifying the epidemiological and evolutionary dynamics of pathogens. Science 303(5656): 327–332.
42. Carstens EB, Ball LA (2009) Ratification vote on taxonomic proposals to the International Committee on Taxonomy of Viruses (2008). Arch Virol 154(7): 1181–1188.
43. Day JM (2009) The diversity of the orthoreoviruses: Molecular taxonomy and phylogentic divides. Infection Gen Evol 9(4): 390–400.
44. Key T, Read J, Nibert ML, Duncan R (2013) Piscine reovirus encodes a cytotoxic, non-fusogenic, integral membrane protein and previously unrecognized virion outer-capsid proteins. J Gen Virol 94(94): 1039–1050.
45. Markussen T, Dahle MK, Tengs T, Løvoll M, Finstad ØW, et al. (2013) Sequence analysis of the genome of piscine orthoreovirus (PRV) associated with heart and skeletal muscle inflammation (HSMI) in Atlantic salmon (Salmo salar). PLoS ONE 8(7): e70075. doi:10.1371/journal.pone.0070075
46. Kibenge M JT, Iwamoto T, Wang Y, Morton A, Godoy MG, et al. (2013) Whole-genome analysis of piscine reovirus (PRV) shows PRV represents a new genus in family *Reoviridae* and its genome segment S1 sequences group it into two separate sub-genotypes. Virol J 10: 230. doi: 10.1186/1743-422X-10-230.
47. Wiik-Nielsen CR, Løvoll M, Sandlund N, Faller R, Wiik-Nielsen J, et al. (2012) First detection of piscine reovirus (PRV) in marine fish species. Dis Aquat Organ 97(3): 255–258.
48. Antere I, Ikonen E (1983) A method of distinguishing wild salmon from those originating from fish farms on the basis of scale structure. Ices J Mar Sci 26: 1–7.
49. Lund RA, Hansen LP (1991) Identification of wild and reared Atlantic salmon, *Salmo salar* L., using scale characters. Aquacult Fish Manag 22: 499–508.
50. Fiske P, Lund RA, Hansen LP. (2004) Identifying fish farm escapees. In: Cadrin S, Friedland K, Waldman J, editors. Stock identification methods, Elsevier. pp. 659–680.
51. Tamura K, Peterson D, Peterson N, Stecher G, Nei M, et al., (2011) MEGA5: Molecular evolutionary genetics analysis using maximum likelihood, evolutionary distance, and maximum parsimony methods. Mol Biol Evol 28(10): p. 2731–2739.
52. Edgar RC (2004) MUSCLE: multiple sequence alignment with high accuracy and high throughput. Nucleic Acids Res 32(5): 1792–1797.
53. Kimura M (1980) A simple method for estimating evolutionary rates of base substitutions through comparative studies of nucleotide sequences. J Mol Evol. 16(2): 111–120.
54. Lanfear R, Calcott B, Ho SYW, Guindon S (2012) PartitionFinder: Combined selection of partitioning schemes and substitution models for phylogenetic analyses. Mol Biol Evol 29(6): 1695–1701.
55. Hasegawa M, Kishino H, Yano T (1985) Dating of human-ape splitting by a molecular clock of mitochondrial DNA. J Mol Evol 22(2): 160–174.
56. Stamatakis A (2006) RAxML-VI-HPC: Maximum likelihood-based phylogenetic analyses with thousands of taxa and mixed models. Bioinformatics 22(21): 2688–2690.
57. Silvestro D, Michalak I (2012) raxmlGUI: a graphical front-end for RAxML. Org Diver Evol 12(4): 335–337.
58. Huelsenbeck JP, Ronquist F (2001) MRBAYES: Bayesian inference of phylogenetic trees. Bioinformatics 17(8): 754–755.
59. Ronquist F, Huelsenbeck JP (2003) MrBayes 3: Bayesian phylogenetic inference under mixed models. Bioinformatics 19(12): 1572–1574.
60. Rambaut A (2009) Molecular evolution, phylogenetics and epidemiology: Tracer. http://tree.bio.ed.ac.uk/software/tracer/ Accessed 11 November 2012.
61. Rambaut A (2009) Molecular evolution, phylogenetics and epidemiology: FigTree http://tree.bio.ed.ac.uk/software/figtree/ Accessed 11 November 2012.
62. Verspoor E, Beardmore JA, Consuegra S, Garcìa De Leàniz C, Hindar K, et al. (2005) Population structure in the Atlantic salmon: insights from 40 years of research into genetic protein variation. J Fish Biol 67: 3–54.
63. Dadswell MJ, Spares AD, Reader JM, Stokesbury MJW (2010) The North Atlantic subpolar gyre and the marine migration of Atlantic salmon *Salmo salar*: the 'Merry-Go-Round' hypothesis. J Fish Biol 77(3): 435–467.
64. Stabell OB (1984) Homing and olfaction in salmonids: a critical review with special reference to the Atlantic salmon. Biol Rev 59: 333–388.
65. Jonsson B, Jonsson N, Hansen LP (2003) Atlantic salmon straying from the River Imsa. J Fish Biol 62(3): 641–657.
66. Skilbrei OT (2010) Reduced migratory performance of farmed Atlantic salmon post-smolts from a simulated escape during autumn. Aquacult Environ Interact 1(2): 117–125.
67. Hansen LP, Youngson AF (2010) Dispersal of large farmed Atlantic salmon, *Salmo salar*, from simulated escapes at fish farms in Norway and Scotland. Fisheries Manag Ecol 17(1): 28–32.
68. Naylor R, Hindar K, Fleming IA, Goldburg R, Williams S, et al. (2005) Fugitive salmon: Assessing the risks of escaped fish from net-pen aquaculture. Bioscience 55(5): 427–437.
69. Aldrin M, Storvik B, Frigessi A; Viljugrein H, Jansen PA, et al. (2010) A stochastic model for the assessment of the transmission pathways of heart and skeleton muscle inflammation, pancreas disease and infectious salmon anaemia in marine fish farms in Norway. Prev Vet Med 93(1): 51–61.
70. Uglem I, Dempster T, Bjørn PA, Sanchez-Jerez P, Økland F (2009) High connectivity of salmon farms revealed by aggregation, residence and repeated movements of wild fish among farms. Mar Ecol-Prog Ser 384: 251–260.
71. Garseth AH, Biering E, Aunsmo A (2013) Associations between piscine reovirus infection and life history traits in wild-caught Atlantic salmon *Salmo salar* L. in Norway. Pre Vet Med 112: 138–146.
72. Rivas C, Bandìn I, Cepeda C, Dopazo CP (1994) Efficacy of chemical disinfectants against turbot aquareovirus. Appl Environ Microb 60(6): 2168–2169.
73. Jones RC (2000) Avian reovirus infections. Revue Scientifique et Technique de l Office International des Epizooties 19(2): 614–625.
74. Urke HA, Kristensen T, Arnekleiv JV, Haugen TO, Kjærstad G, et al. (2013) Seawater tolerance and post-smolt migration of wild Atlantic salmon *Salmo salar* X brown trout *S. trutta* hybrid smolts. J Fish Biol 82(1): 206–227.
75. The Norwegian Directorate of Fisheries (2013) Statistics. Available: http://www.fiskeridir.no/fiskeridirektoratets-statistikkbank. Accessed 10 April 2013
76. Karlsen M, Hodneland K, Endresen C, Nylund A (2006) Genetic stability within the Norwegian subtype of salmonid alphavirus (family *Togaviridae*). Arch Virol 151(5): 861–874.
77. Sæther PA (2004) PD in Norway and in particular in the north of Norway. The Norwegian Veterinary Bulletin. No 7/2004(116): 548–550. (In Norwegian).
78. Anonymous (2012) The status of Norwegian salmon stocks in 2012. In: Thorstad EB, Forseth T, Editors. Report from the scientific advisory committee for Atlantic salmon management in Norway. p. 103 (In Norwegian) Available: www.vitenskapsradet.no Accessed 10 April 2013
79. Alarcon M, Biering E, Colquhoun D, Dale OB, Falk K, et al. (2012) The health situation in farmed salmonids 2011. In: Olsen AB, Hellberg H, Editors. The health situation in Norwegian aquaculture 2011. Norwegian Veterinary Institute. Available: http://www.vetinst.no/eng/Publications/Fish-Health-Report. Accessed 10 April 2013.
80. Haugland M, Holst JC, Holm M, Hansen LP (2006) Feeding of Atlantic salmon (*Salmo salar* L.) post-smolts in the Northeast Atlantic. Ices J Mar Sci 63(8): 1488–1500.
81. Rikardsen AH, Hansen LP, Jensen AJ, Vollen T, Finstad B (2008) Do Norwegian Atlantic salmon feed in the northern Barents Sea? Tag recoveries from 70 to 78 degrees N. J Fish Biol 72(7): 1792–1798.

6

Quantitative Trait Loci (QTL) Associated with Resistance to a Monogenean Parasite (*Benedenia seriolae*) in Yellowtail (*Seriola quinqueradiata*) through Genome Wide Analysis

Akiyuki Ozaki[1*❍], Kazunori Yoshida[2], Kanako Fuji[3], Satoshi Kubota[3], Wataru Kai[1], Jun-ya Aoki[1], Yumi Kawabata[1], Junpei Suzuki[3], Kazuki Akita[3], Takashi Koyama[3], Masahiro Nakagawa[2], Takurou Hotta[2], Tatsuo Tsuzaki[2], Nobuaki Okamoto[3], Kazuo Araki[1], Takashi Sakamoto[3❍]

1 National Research Institute of Aquaculture, Fisheries Research Agency, Nakatsuhamaura, Minamiise-cho, Watarai-gun, Mie, Japan, **2** Seikai National Fisheries Research Institute, Fisheries Research Agency, Nunoura, Tamanoura-machi, Goto-shi, Nagasaki, Japan, **3** Faculty of Marine Science, Tokyo University of Marine Science and Technology, Konan, Minato-ku, Tokyo, Japan

Abstract

Benedenia infections caused by the monogenean fluke ectoparasite *Benedenia seriolae* seriously impact marine finfish aquaculture. Genetic variation has been inferred to play a significant role in determining the susceptibility to this parasitic disease. To evaluate the genetic basis of Benedenia disease resistance in yellowtail (*Seriola quinqueradiata*), a genome-wide and chromosome-wide linkage analyses were initiated using F_1 yellowtail families (n = 90 per family) based on a high-density linkage map with 860 microsatellite and 142 single nucleotide polymorphism (SNP) markers. Two major quantitative trait loci (QTL) regions on linkage groups Squ2 (*BDR-1*) and Squ20 (*BDR-2*) were identified. These QTL regions explained 32.9–35.5% of the phenotypic variance. On the other hand, we investigated the relationship between QTL for susceptibility to *B. seriolae* and QTL for fish body size. The QTL related to growth was found on another linkage group (Squ7). As a result, this is the first genetic evidence that contributes to detailing phenotypic resistance to Benedenia disease, and the results will help resolve the mechanism of resistance to this important parasitic infection of yellowtail.

Editor: Rongling Wu, Pennsylvania State University, United States of America

Funding: The work was supported by ministry of Fisheries Research Agency (FRA) grants-in-aid for scientific research "New Technology of Fish Breeding". And funded by the Program for Promotion of Basic and Applied Researches for Innovations in Bio-oriented Industry (BRAIN). The funders had no role in study design, data collection and analysis, decision to publish or preparation of the manuscript.

Competing Interests: The authors have declared that no competing interests exist.

* E-mail: aozaki@affrc.go.jp

❍ These authors contributed equally to this work.

Introduction

The production of cultured species of yellowtail in Japan was approximately 152,800 tons in 2009, which accounts for 59% of marine finfish aquaculture in Japan [1]. Yellowtail has been cultured in southern areas of Japan using juveniles caught from natural stock. But in recent years the harvest quantity has declined probably because of decreasing wild populations [2]. Capture-based aquaculture however negatively impacts wild stocks of the targeted species as well as non-targeted species. Therefore, it is expected that artificially produced seed will eventually replace seeds caught from the natural source [3,4]. Although research on disease, nutrition and pond management has supported the development of the yellowtail aquaculture industry, genetic improvement programs leading to improve yellowtail lines are only at the beginning.

Genetic linkage maps play a prominent role in many areas of genetics, including quantitative trait locus (QTL) analysis, marker-assisted selection (MAS), positional candidate or positional cloning of genes approach, and comparative genomics. The first genetic linkage map for yellowtail was conducted by Ohara et al. [5]. Recently, a second generation map that spans the genome at a higher resolution has been constructed for *Seriola quinqueradiata* [6]. The map contains several hundred markers with microsatellites associated with candidate genes. The map will facilitate genome mapping efforts in *S. quinqueradiata*, and other related species. The mapping data can be compared to reference species and utilized for QTL analyses and further MAS breeding programs of yellowtail.

Benedenia disease caused by infection by *Benedenia seriolae* is a serious parasitic disease for yellowtail in aquaculture, leading to secondary infection due to viral or bacterial disease. This is because fish rub their bodies against the fish cage to remove the parasite, in certain conditions the mortality is quite high especially in juvenile fishes. Although the way of removing the parasite is generally to soak the fish in a freshwater bath, this method requires a high cost and is labor intensive. Thus, Benedenia disease is difficult to prevent in marine aquaculture systems. Besides from the point of view of wildlife conservation, yellowtail aquaculture is considered as a hotbed for enhancement of parasite transmission [7,8]. Risk management is an important consideration for the long-term sustainability of the aquaculture industry.

S. quinqueradiata is generally regarded to have a higher inherent resistance to *B. seriolae* than other kinds of yellowtail, such as *S. lalandi* and *S. dumerili* [9,10]. The levels of infestation among individuals in *S. quinqueradiata* have been observed to have some degree of heritable variation. These results confirmed earlier evidence of genetic variation to susceptibility to *B. seriolae* in *S. quinqueradiata* and indicate that the host genes play a significant role in determining infection levels against the parasite [9,10].

Genetic studies of parasitic infections have been reported. A study of *Myxobolus cerebralis* infection in rainbow trout phenotype primarily focused on QTL [11] and *Lepeophtheirus salmonis* in Atlantic salmon phenotype focused on candidate genes [12]. Based on QTL and candidate gene changes in response to infection, these studies have allowed us to gain insights into potential genes and pathways that may be differentially regulated between resistant and susceptible strains. This is an important step towards understanding host responses to infection, however, much remains to be learned about the genetic basis controlling the immunological response to parasitic infection.

In this study, we performed QTL analyses using wild F_1 strains of *S. quinqueradiata* to elucidate the genetic evidence of resistance to Benedenia disease. By using the high-density linkage map with microsatellite and SNP markers, we identified two major QTL regions contributing to the Benedenia disease resistance. The discovery of a large QTL effect for Benedenia disease resistance has broad implications for improving our general understanding of external parasitic diseases and host pathogen interactions.

Results

Phenotypic Trait Correlation with Fish Size and Number of Parasites (*B. seriolae*) in Family A and B

Pearson correlation coefficients for total length, body length, body weight, surface area and number of pathogens are shown in Table 1. Weight and length were positively correlated with each other and, to a lesser degree, the number of pathogens and fish size variables were negatively correlated in family A. However in family B, the number of pathogens was marginally correlated with the fish size variables ($P = 0.001$).

1st Screening by Kruskal-Wallis Analysis (K-W test) of Family A for Benedenia Disease Resistance

In analysis of family A, twelve markers were significant ($P<0.01$) on linkage group corresponding to chromosome Squ2, twenty-five markers were significant on linkage group Squ8, thirty-one markers were significant on linkage group Squ20, as 1st screening about Benedenia disease resistance QTL candidates using Kruskal-Wallis analysis (K-W test) (Table 2). A total of sixty-eight markers were informative and indicative of only one family. All of these markers achieved QTL possible using the MapQTL 5 software.

Simple Interval Mapping Results about All Linkage Groups in Family A

We show the interval mapping results for Benedenia disease resistance in family A for all linkage groups in Figure 1. Three regions of the chromosomes were identified to be significantly associated with Benedenia desease for family A (Table 3). The QTL at Squ2 identified by simple interval mapping were also found by using K-W test. The peak LOD value of sequ1295BAC (LOD = 4.71) was substantially higher than the genome-wide LOD significance threshold value of 2.9 determined by permutation testing (*Pg* <0.05; Pg: P value genome-wide LOD). Linkage group in Squ2 QTL region (tentatively termed *BDR-1*) was observed as a high single peak as genome-wide LOD significance level (*Pg* <0.001) in interval mapping (Figure 2A). The markers of chromosomal region of Squ8 linkage group, example sequ0670-BAC (LOD = 2.45), was less than the genome-wide LOD significance level (*Pg* <0.05). The markers of chromosomal region of Squ20 linkage group, example Sequ0808TUF (LOD = 2.98), had slightly exceeded the genome-wide LOD significance level (*Pg* <0.05) (Figure 2B). About one of the peaks, it can tentatively be called as the *BDR-2* significant region, based on the rules of QTL nomenclature [13,14].

Each of the LOD peaks, Squ2 (Sequ1295BAC), Squ8 (Sequ0670BAC), Squ20 (Sequ1071TUF), and Squ20 (Sequ0808-TUF), can explain the phenotypic variance ranging from 11.8 to 21.4% by simple interval mapping. When LOD peaks were combined into simple interval mapping results, two loci (Sequ1295BAC; Squ2, Sequ0808T TUF; Squ20) could explain the phenotypic variance ranging up to 35.5%. If the other LOD peak of Squ20 (Sequ1071TUF) and the LOD peak of Squ8 (Sequ0670-BAC) are added, these four loci were responsible for a significant portion 60.8% of the total phenotypic variation in family A.

Table 1. Between fish size and number of parasites in pairwise Pearson correlations.

Family A		**body length**	**body weight**	**surface area**	**number of *B. seriolae***
	total length	**0.967**	**0.932**	**0.999**	0.050
	body length		**0.947**	**0.963**	0.041
	body weight			**0.933**	0.094
	surface area				0.050
Family B		body length	body weight	surface area	number of *B. seriolae*
	total length	**0.964**	**0.933**	**0.999**	**0.389**
	body length		**0.946**	**0.964**	**0.386**
	body weight			**0.935**	**0.429**
	surface area				**0.390**

Values in bold are different from a significance level $P = 0.001$.

Table 2. Significant markers for Benedenia disease resistance using Kruskal-Wallis analysis with A and B families.

		Family A		Family B		
Linkage group	Locus	K-W test	Signif.	Linkage group	K-W test	Signif.
Squ2F	Sequ0941TUF	0.005	NS	Squ2M	2.772	*
	Sequ0020TUF	0.005	NS		2.893	*
	Sequ2216BAC	0.005	NS		3.403	*
	Sequ0603TUF	0.041	NS		4.106	**
	Sequ0832TUF	0.041	NS		2.743	*
	Sequ0648TUF	0.041	NS		3.725	*
	Sequ3121BAC	0.041	NS		3.725	*
	Sequ0174TUF	0.249	NS		3.725	*
	Sequ0171BAC	6.901	***		3.725	*
	Sequ0172TUF	8.966	****		3.725	*
	Sequ0672TUF	12.926	******		3.725	*
	Sequ0125TUF	16.83	*******		3.725	*
	Sequ1065TUF	16.83	*******		–	
	Sequ1295BAC	18.998	*******		–	
	Sequ1066TUF	17.478	*******		–	
	Sequ0979BAC	17.478	*******		3.725	*
	Sequ1067TUF	14.393	******		–	
	Sequ1068TUF	7.961	****		–	
	Sequ1069TUF	9.115	****		–	
	Sequ1070TUF	7.136	***		–	
	Sequ0788TUF	5.404	**		3.76	*
	Sequ0549TUF	2.939	*		3.725	*
	Sequ1026TUF	2.939	*		–	
	Sequ1749BAC	2.939	*		3.725	*
	Sequ0827TUF	2.866	*		3.725	*
	Sequ0828TUF	2.866	*		3.725	*
Squ8F	Sequ0074TUF	9.852	****	Squ8M	0	NS
	Sequ0409TUF	9.852	****		0.728	NS
	Sequ0431TUF	9.852	****		0.322	NS
	Sequ0851TUF	9.852	****		0.605	NS
	Sequ0906TUF	9.852	****		0.728	NS
	Sequ3175BAC	9.852	****		0.103	NS
	Sequ2078BAC	9.852	****		0.911	NS
	Sequ1769BAC	10.247	****		0.029	NS
	Sequ0670BAC	10.398	****		0.064	NS
	Sequ0503TUF	10.398	****		0.485	NS
	Sequ0610TUF	10.398	****		–	
	Sequ0101TUF	10.398	****		1.804	NS
	Sequ2536BAC	10.398	****		1.804	NS
	Sequ2198BAC	10.398	****		3.506	*
	Sequ0985TUF	10.398	****		1.973	NS
	Sequ1013TUF	10.398	****		1.973	NS
	Sequ0507BAC	10.398	****		–	
	Sequ0575BAC	10.398	****		–	
	Sequ2218BAC	10.398	****		1.973	NS
	Sequ3193BAC	10.398	****		–	

Table 2. Cont.

		Family A		Family B		
Linkage group	Locus	K-W test	Signif.	Linkage group	K-W test	Signif.
	Sequ3288BAC	10.398	****		–	
	Sequ00955SNP	10.398	****		–	
	Sequ01036SNP	10.398	****		–	
	Sequ02608SNP	10.398	****		–	
	Sequ02777SNP	10.398	****		–	
Squ20F	Sequ1071TUF	12.821	******	Squ20M	–	
	Sequ0439TUF	12.821	******		2.721	*
	Sequ2134BAC	12.821	******		–	
	Sequ1100BAC	12.821	******		0.808	NS
	Sequ01056SNP	12.038	*****		–	
	Sequ02679SNP	12.821	******		–	
	Sequ1072TUF	11.438	*****		–	
	Sequ3071BAC	11.438	*****		8.263	****
	Sequ0938TUF	11.438	*****		9.786	****
	Sequ0719TUF	9.914	****		–	
	Sequ1074TUF	9.914	****		–	
	Sequ1075TUF	9.914	****		–	
	Sequ00695SNP	10.844	*****		–	
	Sequ02734SNP	9.371	****		–	
	Sequ2569BAC	9.914	****		8.802	****
	Sequ1073TUF	9.914	****		–	
	Sequ2645BAC	9.914	****		3.928	**
	Sequ1076TUF	9.914	****		–	
	Sequ0537TUF	9.914	****		7.914	****
	Sequ0596BAC	9.914	****		9.011	****
	Sequ0829TUF	9.914	****		–	
	Sequ0836TUF	9.914	****		8.338	****
	Sequ1989BAC	9.914	****		7.161	***
	Sequ2312BAC	9.914	****		–	
	Sequ0017BAC	9.914	****		6.647	***
	Sequ0730TUF	9.914	****		6.647	***
	Sequ1077TUF	10.211	****		–	
	Sequ1078TUF	11.544	*****		–	
	Sequ0808TUF	13.494	******		6.647	***
	Sequ1079TUF	13.494	******		–	
	Sequ0288TUF	10.037	****		6.647	***
	Sequ1702BAC	0.036	NS		6.647	***

Signif.; Significance levels:
*<0.1
**<0.05
***<0.01
****<0.005
*****<0.001
******<0.0005
*******<0.0001.
NS; not significant, -; not informative in this locus.
Squ(linkage group)F; F is dam allele in female linkage group. Squ(linkage group)M; M is sire allele in male linkage group.

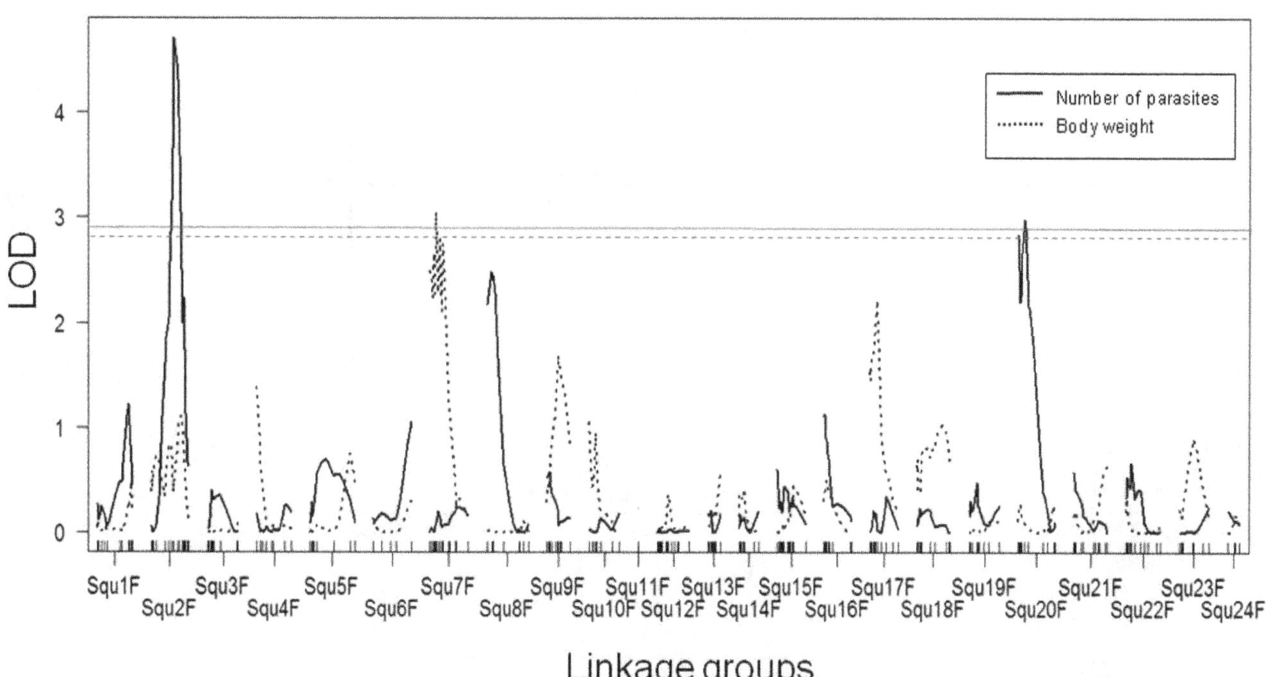

Figure 1. Simple interval mapping results for Benedenia disease resistance and body weight in all linkage groups with family A. Squ(linkage group)F; marker distance in female map. This figure is described using R/qtl. Number of parasites; *Pg* <0.05 significant threshold is indicated as solid line. Body weight; *Pg* <0.05 significant is indicated as dashed line.

Multiple QTL Model Mapping about Significant Loci in Linkage Groups Squ2 and Squ20

After simple interval mapping consideration of analysis results of family A, Multiple QTL model was performed. Multiple QTL model mapping was applied to detect significant loci with the exception of ghost QTL, and is based on backward elimination. Therefore significant regions in linkage group Squ8 (ex. Sequ0670BAC) were rejected as QTL region in this step. Map positions and LOD scores are based on multiple QTL model analysis using the software MapQTL 5. The results of the multiple QTL model mapping are shown in Table 4 and Figure 3. Peaks of LOD score were higher than the simple interval mapping results, example *BDR-1* on Squ2 was indicated as LOD = 5.21, and *BDR-2* on Squ20 was indicated as LOD = 3.47. But the marker locus Sequ1071TUF (LOD = 2.89) was less than the genome-wide LOD significance level (*Pg* <0.05) in edge of Squ20 linkage group. LOD peaks were combined into multiple QTL model mapping results, the two loci (Sequ1295BAC; Squ2, Sequ0808T TUF; Squ20) could explain phenotypic variance ranging up to 32.9%.

K-W test of Benedenia disease resistance in family B

About candidate marker loci in family A, we collected genotype data in family B. A total of six markers of linkage group Squ20 showed consistent significant results (*P*<0.005) (Table 2). But one markers of linkage group Squ2 (Sequ0603TUF) were marginally significant (*P*<0.05), while all markers of linkage group Squ8 were not significant (*P*<0.05) in K–W test.

Simple Interval Mapping Results on Chromosomal-wide Analysis in Family B

After K-W test in family B, simple interval mapping was performed to identify the location of significant QTL regions on Squ2, Squ8, and Squ20 (Table 3). About the candidate QTL region in Squ20 linkage group, loci of which were confirmed to have a significant value. Also these loci were observed to have a significant LOD score 2.24 in family B, which was confirmed in reproducible families chromosome-wide LOD significance level (*Pc* <0.05, Pc: P value chromosome-wide LOD) by interval mapping (Figure S1). Results for both A and B families as two peaks about significant region in linkage group Squ20.

Linkage Analysis Estimation of Other Phenotype Fish Size QTL Regions

We show the interval mapping for body weight QTL in family A about all linkage groups in Figure 1. Significant loci about fish size (total length, body length, body weight, surface area) in linkage groups Squ7, Squ17 are shown in Table 5. The markers of chromosomal region of Squ7 linkage group, for example body weight, Sequ0582TUF (LOD = 3.04) had values exceeding the genome-wide LOD significance level (LOD 2.8, *Pg* <0.05) in family A (Figure S2). The region of LOD maximum locus (Sequ0582TUF) could explain phenotypic variance ranging 14.4% of the trait body weight. Furthermore in family B analysis about fish size, the region of Squ7 linkage group was significant (*P*<0.05) by K-W test (data not shown). For the fish size QTL candidate region in these families, the number of pathogens was negatively correlated with fish size, the same as the correlation coefficient results in this study.

A

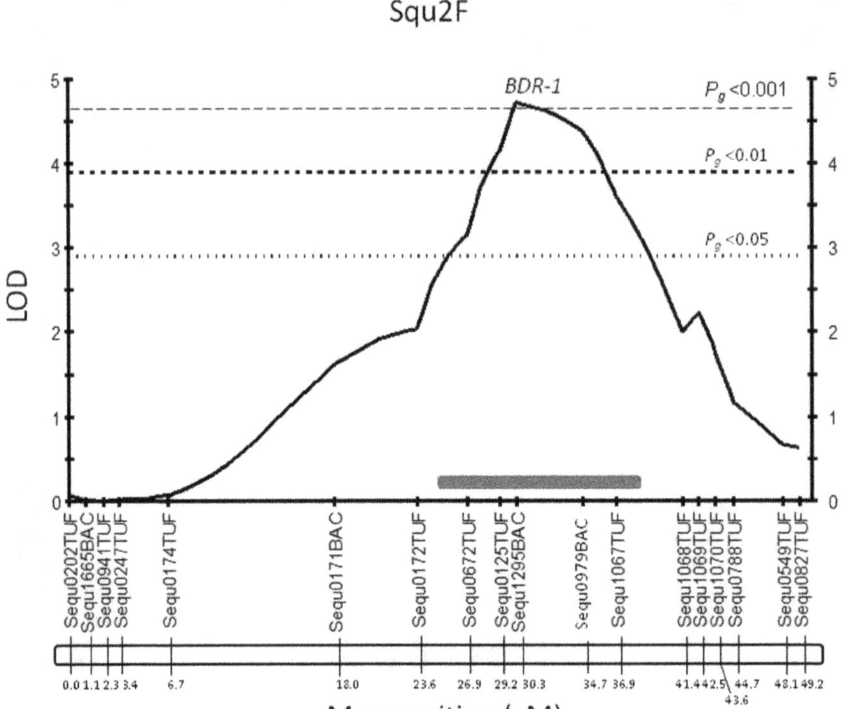

B

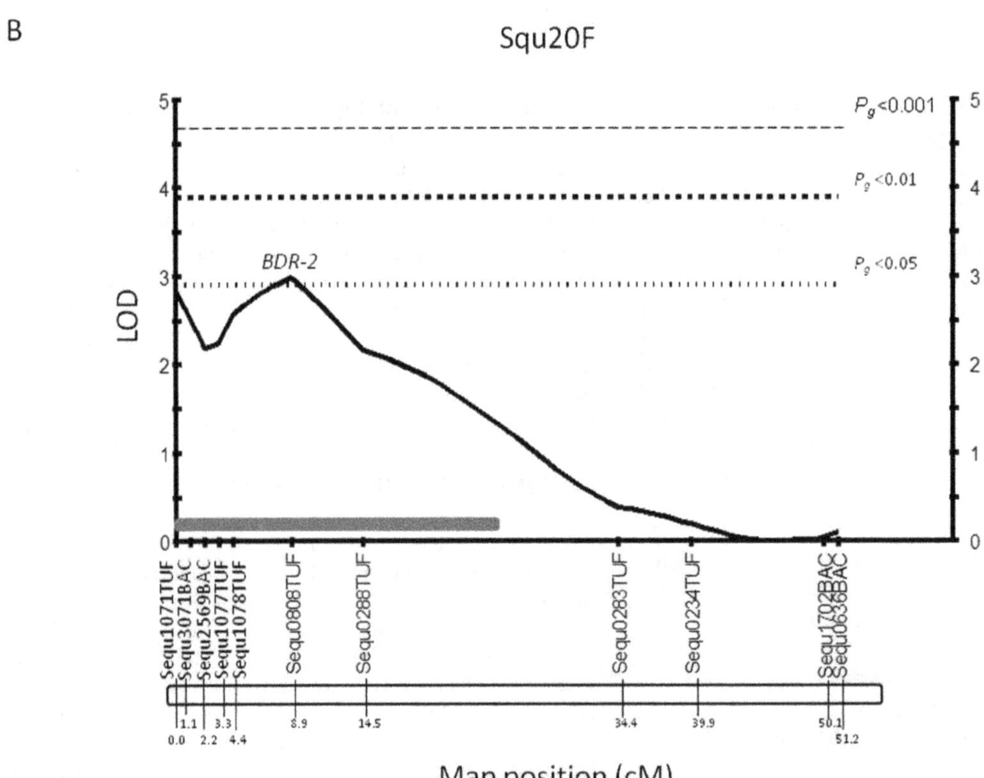

Figure 2. Localization of significant markers for Benedenia disease resistance in linkage group Squ2F and Squ20F with family A. Squ(linkage group)F; marker distance in female map. (A) Squ2F, (B) Squ20F. Map positions and LOD scores are based on a simple interval mapping QTL analysis using the software MapQTL 5. Marker absolute map distances are given in (cM). 95% confidence probability LOD support interval was indicated as Gray bold line. Horizontal lines across each plot indicate LOD siginificance threshold, P_g; genome-wide significance threshold.

Table 3. Simple interval mapping results of the significant markers for Benedenia disease resistance in linkage group 2, 8, and 20 with two families.

		Family A				Family B		
Linkage Group	Locus	LOD	% Var.	Effect	Linkage Group	LOD	% Var.	Effect
Squ2F	Sequ0171BAC	1.61	7.9	0.60	Squ2M	NS		
	Sequ0172TUF	2.04	9.9	0.66		NS		
	Sequ0672TUF	3.15	14.9	0.82		NS		
	Sequ0125TUF	4.15	19.1	0.94		NS		
	Sequ1065TUF	4.15	19.1	0.94		NS		
	Sequ1295BAC	**4.71**	**21.4**	**1.00**		NS		
	Sequ1066TUF	4.38	20.1	0.96		NS		
	Sequ0979BAC	4.38	20.1	0.96		NS		
	Sequ1067TUF	3.60	16.8	0.89		NS		
	Sequ1068TUF	2.00	9.7	0.66		NS		
	Sequ1069TUF	2.24	10.8	0.70		NS		
	Sequ1070TUF	1.74	8.5	0.62		NS		
	Sequ0788TUF	1.17	5.8	0.51		NS		
Squ8F	Sequ0074TUF	2.17	10.5	0.68	Squ8M	NS		
	Sequ0409TUF	2.17	10.5	0.68		NS		
	Sequ0431TUF	2.17	10.5	0.68		NS		
	Sequ0851TUF	2.17	10.5	0.68		NS		
	Sequ0906TUF	2.17	10.5	0.68		NS		
	Sequ3175BAC	2.17	10.5	0.68		NS		
	Sequ2078BAC	2.17	10.5	0.68		NS		
	Sequ1769BAC	2.41	11.6	0.72		NS		
	Sequ0670BAC	**2.45**	**11.8**	**0.72**		NS		
	Sequ0503TUF	**2.45**	**11.8**	**0.72**		NS		
	Sequ0610TUF	**2.45**	**11.8**	**0.72**		NS		
	Sequ0101TUF	**2.45**	**11.8**	**0.72**		NS		
	Sequ2536BAC	**2.45**	**11.8**	**0.72**		NS		
	Sequ2198BAC	**2.45**	**11.8**	**0.72**		NS		
	Sequ0985TUF	**2.45**	**11.8**	**0.72**		NS		
	Sequ1013TUF	**2.45**	**11.8**	**0.72**		NS		
	Sequ0507BAC	**2.45**	**11.8**	**0.72**		NS		
	Sequ0575BAC	**2.45**	**11.8**	**0.72**		NS		
	Sequ2218BAC	**2.45**	**11.8**	**0.72**		NS		
	Sequ3193BAC	**2.45**	**11.8**	**0.72**		NS		
	Sequ3288BAC	**2.45**	**11.8**	**0.72**		NS		
	Sequ00955SNP	**2.45**	**11.8**	**0.72**		NS		
	Sequ01036SNP	**2.45**	**11.8**	**0.72**		NS		
	Sequ02608SNP	**2.45**	**11.8**	**0.72**		NS		
	Sequ02777SNP	**2.45**	**11.8**	**0.72**		NS		
Squ20F	Sequ1071TUF	**2.83**	**13.5**	**0.77**	Squ20M	NS		
	Sequ0439TUF	**2.83**	**13.5**	**0.77**		NS		
	Sequ2134BAC	**2.83**	**13.5**	**0.77**		NS		
	Sequ1100BAC	**2.83**	**13.5**	**0.77**		NS		
	Sequ01056SNP	**2.83**	**13.5**	**0.77**		NS		
	Sequ02679SNP	**2.83**	**13.5**	**0.77**		NS		
	Sequ1072TUF	2.51	12.0	0.73		1.91	9.0	0.61
	Sequ3071BAC	2.51	12.0	0.73		**2.24**	**10.5**	**0.66**
	Sequ0938TUF	2.51	12.0	0.73		**2.24**	**10.5**	**0.66**

Table 3. Cont.

		Family A			Family B			
Linkage Group	**Locus**	**LOD**	**% Var.**	**Effect**	**Linkage Group**	**LOD**	**% Var.**	**Effect**
	Sequ0719TUF	2.19	10.6	0.68		1.96	9.3	0.62
	Sequ1074TUF	2.19	10.6	0.68		1.96	9.3	0.62
	Sequ1075TUF	2.19	10.6	0.68		1.96	9.3	0.62
	Sequ00695SNP	2.19	10.6	0.68		1.96	9.3	0.62
	Sequ02734SNP	2.19	10.6	0.68		1.96	9.3	0.62
	Sequ2569BAC	2.19	10.6	0.68		1.96	9.3	0.62
	Sequ1073TUF	2.19	10.6	0.68		1.96	9.3	0.62
	Sequ2645BAC	2.19	10.6	0.68		NS		
	Sequ1076TUF	2.19	10.6	0.68		NS		
	Sequ0537TUF	2.19	10.6	0.68		1.66	7.9	0.57
	Sequ0596BAC	2.19	10.6	0.68		**1.91**	**9.0**	**0.61**
	Sequ0829TUF	2.19	10.6	0.68		**1.91**	**9.0**	**0.61**
	Sequ0836TUF	2.19	10.6	0.68		**1.91**	**9.0**	**0.61**
	Sequ1989BAC	2.19	10.6	0.68		1.48	7.1	0.54
	Sequ2312BAC	2.19	10.6	0.68		1.48	7.1	0.54
	Sequ0017BAC	2.19	10.6	0.68		1.39	6.6	0.53
	Sequ0730TUF	2.19	10.6	0.68		1.39	6.6	0.53
	Sequ1077TUF	2.25	10.9	0.69		1.39	6.7	0.53
	Sequ1078TUF	2.56	12.3	0.74		1.39	6.7	0.53
	Sequ0808TUF	**2.98**	**14.1**	**0.79**		1.39	6.6	0.53
	Sequ1079TUF	**2.98**	**14.1**	**0.79**		1.39	6.6	0.53
	Sequ0288TUF	2.17	10.5	0.68		1.39	6.6	0.53

Locus; marker name, LOD; Lod scores, % Var; percent of variance explained, Effect; estimated effect, NS; not significant. Squ(linkage group)F; F is dam allele in female linkage group. Squ(linkage group)M; M is sire allele in male linkage group. Values in bold are LOD max in peak of each QTL marker position and each value.

Discussion

This study is the first to report the detection and positioning of major QTLs affecting resistance to external parasites in yellowtail. We identified in yellowtail two chromosomal regions containing QTL (*BDR-1, BDR-2*) that are associated with Benedenia disease resistance. Two putative QTL associations, of intermediate to large effect on Benedenia disease resistance, were localized to linkage groups Squ2 and Squ20.

On Squ2, the example marker loci Sequ1295BAC, which can explain phenotypic variance ranging from 20.1 to 21.4% by simple interval and multiple model interval mapping results. On Squ20, the example marker loci Sequ0808TUF, which can explain phenotypic variance ranging from 12.8 to 14.1% by both mapping methods. These two loci were responsible for a range from 32.9 to 35.5% of the total phenotypic variation in family A. If other peaks which the marker Sequ1071TUF on Squ20 can be considered as significant loci, it will explain the phenotypic variance range from 10.8% to 13.5%. In addition, suggested level marker Sequ0670-BAC on Squ8 potentially explains about 11.8% of the phenotypic variance. In total, these four loci can explain phenotypic variance ranging from 55.5% to 60.8% in family A. However, these two additional QTL were only significant at the chromosome-wide level and should be regarded as tentative until other family results are confirmed. Besides QTL interaction was found not to occur between Squ2 region and Squ20 region when considered from the 2D-QTL scan function in R/qtl in family A (data not shown).

Family B was not used in the main genome-wide family analysis. The number of pathogens was marginally significantly correlated with fish size. The family B was analyzed in limited linkage groups with chromosome-wide significance levels to confirm that the candidate QTL regions are reproducible. For the candidate QTL regions in the Squ20 linkage group, which were confirmed to have a significant value in family B. However the highly significant region in family A on linkage group Squ2 was rejected as in family B. Also the suggested region in family A on linkage group Squ8 was rejected in family B. We consider that the reasons for these differences are due to the different parental fish of family B that were selected, and their F_1 progeny showed more susceptibility to the parasite than that of family A (Figure S3).

The most important finding of this study was detected as a single peak of QTL (*BDR-1*) associated with Benedenia infection resistance within the proximal region of linkage group Squ2. The QTL peak (Sequ1295BAC) was located at position 30.3 cM, with a 95% confidence interval that the QTL region lies within 10 cM of the most proximal marker from Sequ0672TUF to Sequ1067-TUF by simple interval mapping. Furthermore the results of multiple QTL model mapping indicated the QTL region within 5.5 cM interval (Figure 3A). The 5.5 cM interval is narrow as QTL candidate region and should be considered as a fine approximation, given the large QTL effect and high recombination rate found in yellowtail females. Besides the QTL peak marker Sequ1295BAC isolated from BAC library end sequencing [6], and adjacent markers Sequ0979BAC were also isolated from

A

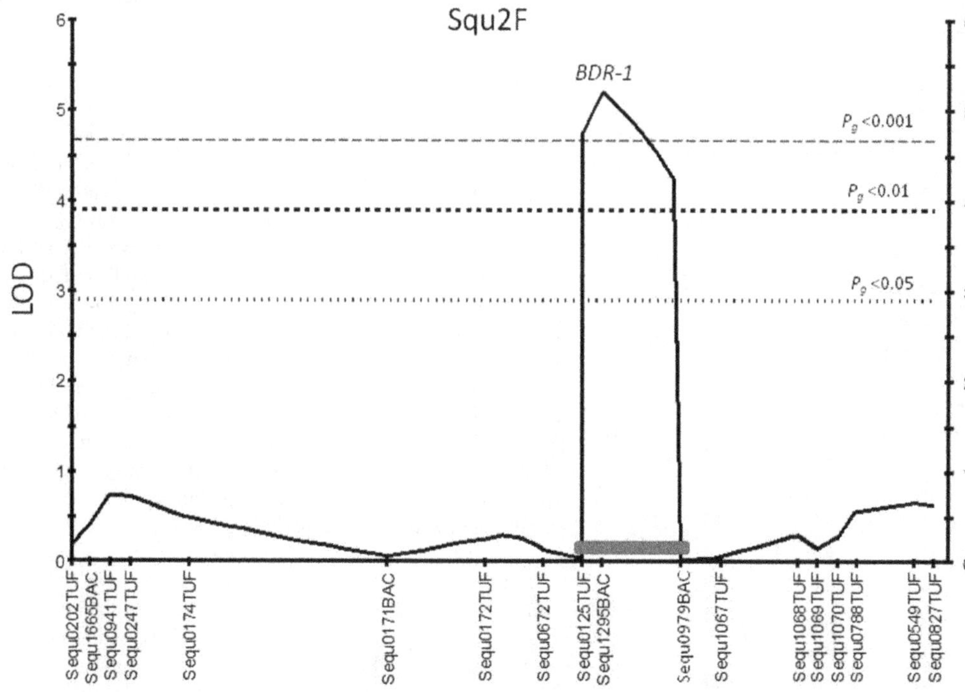

B

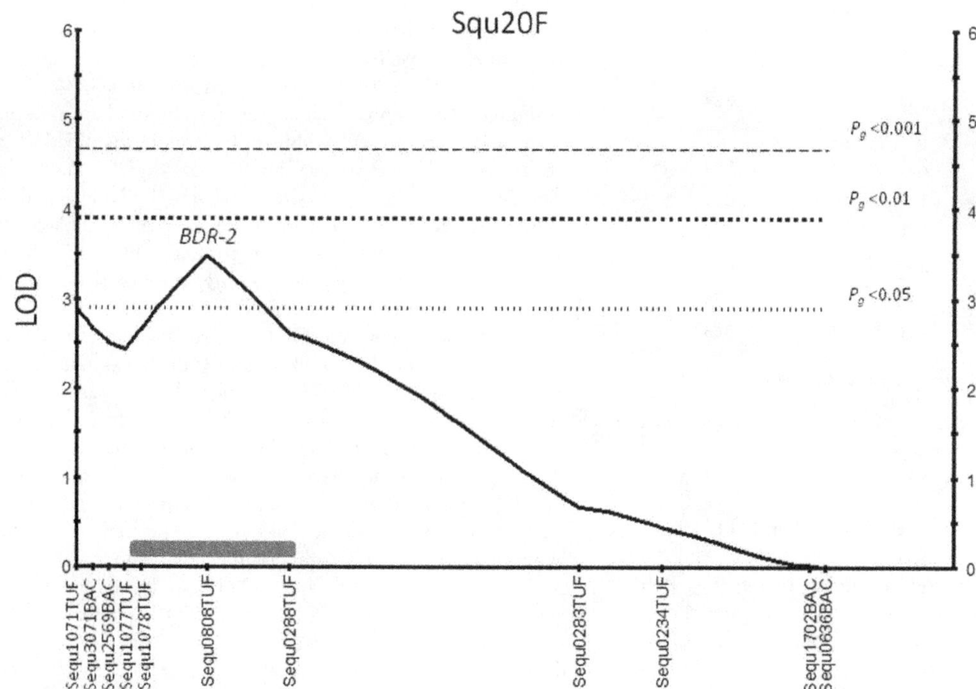

Figure 3. Significant markers for Benedenia disease resistance multiple-QTL model mapping in linkage group Squ2F and Squ20F with family A. Squ(linkage group)F; marker distance in female map. (A) Squ2F, (B) Squ20F. This figure was described using by MapQTL 5.

BAC library end sequencing. These physical sequences based on BAC library were developed for future identification of positional candidate genes or positional cloning regarding external parasitism disease resistant genes. However before initiative sequences of contignation from BAC library, it would be beneficial to further refine the QTL region by increasing the marker density around the QTL peak, likely with EST based SNP markers. Furthermore it would be necessary to map additional families using near-

Table 4. Multiple QTL model mapping results of the significant markers for Benedenia disease resistance in linkage group 2 and 20 in family A.

		Family A		
Linkage Group	**Locus**	**LOD**	**% Var.**	**Effect**
Squ2F	Sequ1065TUF	4.75	18.5	0.92
	Sequ1295BAC	**5.21**	**20.1**	**0.97**
Squ20F	Sequ1071TUF	**2.89**	**10.8**	**0.69**
	Sequ1072TUF	2.66	10.0	0.66
	Sequ3071BAC	2.66	10.0	0.66
	Sequ0938TUF	2.66	10.0	0.66
	Sequ0719TUF	2.50	9.4	0.64
	Sequ1074TUF	2.50	9.4	0.64
	Sequ1075TUF	2.50	9.4	0.64
	Sequ00695SNP	2.50	9.4	0.64
	Sequ02734SNP	2.50	9.4	0.64
	Sequ2569BAC	2.50	9.4	0.64
	Sequ1073TUF	2.50	9.4	0.64
	Sequ2645BAC	2.50	9.4	0.64
	Sequ1076TUF	2.50	9.4	0.64
	Sequ0537TUF	2.50	9.4	0.64
	Sequ0596BAC	2.50	9.4	0.64
	Sequ0829TUF	2.50	9.4	0.64
	Sequ0836TUF	2.50	9.4	0.64
	Sequ1989BAC	2.50	9.4	0.64
	Sequ2312BAC	2.50	9.4	0.64
	Sequ0017BAC	2.50	9.4	0.64
	Sequ0730TUF	2.50	9.4	0.64
	Sequ1077TUF	2.43	9.2	0.64
	Sequ1078TUF	2.66	10.0	0.67
	Sequ0808TUF	**3.47**	**12.8**	**0.75**
	Sequ1079TUF	3.47	12.8	0.75
	Sequ0288TUF	2.61	9.8	0.66

Locus; marker name, LOD; Lod scores, % Var; percent of variance explained, Effect; estimated effect. Squ(linkage group)F; F is dam allele in female linkage group. Values in bold are LOD max in peak of each QTL marker position and each value.

isogenic lines, which are separated from other QTL effects on other linkage groups, and successive generations with larger progeny sizes to increase the total recombination events. Further QTL studies should therefore focus on fine-mapping the QTL identified on Squ2 as well as searching for additional QTL on other linkage groups in yellowtail.

On the other hand, the QTL (*BDR-2*) significant region would exist as multiple QTL on linkage group Squ20 (Figure 2B, Figure 3B). This is because results of analyses of family A and family B were detected as two peaks. In addition, this QTL significant region was confirmed for Benedenia disease resistance in hybrid lines between *S. quinqueradiata* and *S. lalandi* [15]. Therefore, it is possible that this linkage group is enriched for external parasitism disease resistance genes, although without a sequenced genome, this remains highly speculative. Genetic variation will contribute to the Benedenia disease resistance phenotype in different families or cross species results. Finding this QTL region strongly supports the potential for success of marker-assisted selection (MAS) for this disease. But it is difficult to evaluate each separate QTL effect from these two peaks. Utilization of this QTL has to pay attention to the broad region on Squ20, which should be integrated into the next generation for MAS.

There is another possibility about the identification of the positional candidate gene or utilization for MAS. The synteny study among vertebrate fish species, medaka, zebra, pufferfish, and yellowtail can be considered. The physical BAC clones and EST based SNPs will help to gain new information about both QTLs (*BDR-1, BDR-2*). Still due to the insufficient data about the yellowtail genome, it is difficult to compare the orthologus region or gene based on homological sequence. This approach is now progressing. Some studies on salmonid species have shown significant regions of disease resistance are common in separated genetic research or ortholog regions in cross-species [11,16,17,18,19,20]. These research results indicate the necessity of synteny analysis in fish species to determine the orthologous candidate gene about disease resistance. Also the immune-related genes on the same linkage group are correlated with significant QTL about disease resistance. Especially some studies have reported that QTLs are located on the same linkage group with the MHC class I, II or toll-like receptor regions [12,21,22,23,24]. Linkage groups on QTL region were specifically targeted in these studies because they carry the classical MHC class I and II genes previously linked to differences in susceptibility to viral, bacterial, and parasitic diseases.

Effect of fish size in relation to the number of *B. seriolae*, our results suggest that fish size is not responsible for the resistance of *B. seriolae* in yellowtail, because the number of pathogens and fish size were negatively correlated in family A (Table 1). Although we cannot exclude the possibility that multiple fish growth loci are present within the currently identified *B. seriolae* resistance QTL region in Squ2, and Squ20, the results of significant loci about fish size (total length, body length, body weight, surface area) were identified in different linkage groups in Squ7 by the results of the three month old stage. The results from our current study suggest a negative correlation between growth traits and the number of parasites. However, given the effect of the discovered QTL region has on the Benedenia disease phenotype, it is unlikely that growth has a major role in conferring parasite disease resistance. In the case of whirling disease, caused by the pathogen *Myxobolus cerebralis*, fish age and size were found to be key factors influencing the severity of whirling disease in experimentally infected rainbow trout [25,26]. In this disease, rainbow trout growing at faster rates may more rapidly become resistant to both clinical symptoms of whirling disease and high numbers of myxospores in their skeletal elements. But our QTL study targeted external parasitism, and as such the case of whirling disease is difficult to compare with Benedenia disease. Fish size QTLs were identified for different linkage groups in Squ7. However, this QTL possibly affects significant locus at the genome-wide level ($P<0.05$) by simple interval mapping results in the three month old stage, about 185 mm size and 65 g weight. For determining growth QTL, it might be necessary to measure the phenotype using different development stages by time series analysis.

Sex was not directly correlated with the number of *B. seriolae* in the fish used in this study. Actually we could not morphologically distinguish males and females at the three month old stage. But sex-linked markers have already been identified on the linkage group Squ12 within several families [27]. Our results indicated

Table 5. Simple interval mapping results of the fish sizes in linkage group 7 and 17 in family A.

		total length			body length			total weight			surface area		
Linkage Group	**Locus**	**LOD**	**% Var.**	**Effect**	**LOD**	**% Var.**	**Effect**	**LOD**	**% Var.**	**Effect**	**LOD**	**% Var.**	**Effect**
Squ7F	Sequ1041TUF	2.46	11.8	0.70	2.62	12.6	0.72	2.50	12.0	0.70	2.46	11.8	0.70
	Sequ1041BAC	2.65	12.7	0.73	2.67	12.8	0.73	2.43	11.7	0.70	2.65	12.7	0.73
	Sequ3208BAC	2.06	10.0	0.65	2.50	12.0	0.72	2.24	10.8	0.68	2.06	10.0	0.65
	Sequ0447TUF	2.48	11.9	0.71	2.90	13.8	0.77	2.66	12.7	0.73	2.48	11.9	0.71
	Sequ0582TUF	**2.57**	**12.3**	**0.72**	**2.92**	**13.9**	**0.76**	**3.04**	**14.4**	**0.77**	**2.57**	**12.3**	**0.72**
	Sequ0623TUF	2.12	10.3	0.65	2.22	10.7	0.67	2.27	11.0	0.67	2.12	10.3	0.65
	Sequ2990BAC	2.50	12.0	0.70	2.62	12.5	0.72	2.78	13.3	0.74	2.50	12.0	0.70
	Sequ00662SNP	1.83	8.9	0.60	2.02	9.8	0.63	2.10	10.2	0.64	1.83	8.9	0.60
	Sequ0416TUF	2.49	12.0	0.70	2.72	13.0	0.73	2.74	13.1	0.73	2.49	12.0	0.70
	Sequ0784TUF	1.39	6.8	0.53	1.62	7.9	0.57	1.22	6.1	0.50	1.39	6.8	0.53
	Sequ1421BAC	1.23	6.1	0.50	1.47	7.2	0.54	1.15	5.7	0.48	1.23	6.1	0.50
	Sequ0382BAC	0.65	3.3	0.36	0.81	4.1	0.41	0.48	2.5	0.31	0.65	3.3	0.36
	Sequ1628BAC	0.44	2.2	0.30	0.53	2.7	0.33	0.27	1.4	0.24	0.44	2.2	0.30
	Sequ00355_1SNP	0.52	2.6	0.33	0.66	3.3	0.37	0.34	1.7	0.26	0.52	2.6	0.33
	Sequ1700BAC	0.44	2.2	0.30	0.45	2.3	0.30	0.15	0.8	0.18	0.44	2.2	0.30
Squ17F	Sequ1016TUF	1.30	6.4	0.51	1.84	9.0	0.61	1.53	7.6	0.56	1.30	6.4	0.51
	Sequ0025TUF	1.10	5.5	0.47	1.68	8.2	0.58	1.46	7.2	0.54	1.10	5.5	0.47
	Sequ0895TUF	1.30	6.4	0.51	1.82	8.9	0.60	1.68	8.3	0.58	1.30	6.4	0.51
	Sequ0228TUF	1.35	6.7	0.52	1.78	8.7	0.60	1.70	8.3	0.58	1.35	6.7	0.52
	Sequ3088BAC	1.59	7.8	0.56	1.94	9.4	0.62	1.78	8.7	0.59	1.59	7.8	0.56
	Sequ1690BAC	1.73	8.5	0.59	2.15	10.4	0.65	2.01	9.8	0.63	1.73	8.5	0.59
	Sequ01964SNP	**1.79**	**8.8**	**0.60**	**2.30**	**11.1**	**0.67**	**2.22**	**10.8**	**0.66**	**1.79**	**8.8**	**0.60**
	Sequ0716TUF	0.86	4.3	0.42	1.05	5.2	0.46	0.92	4.6	0.43	0.86	4.3	0.42
	Sequ2051BAC	0.45	2.3	0.30	0.60	3.0	0.35	0.56	2.8	0.34	0.45	2.3	0.30
	Sequ2269BAC	0.12	0.6	0.16	0.18	0.9	0.20	0.34	1.7	0.27	0.12	0.6	0.16
	Sequ2942BAC	0.02	0.1	0.06	0.02	0.1	0.06	0.16	0.8	0.18	0.02	0.1	0.06

Locus; marker name, LOD; Lod scores, % Var; percent of variance explained. Effect; estimated effect. Squ(linkage group)F; F is dam allele in female linkage group. Values in bold are LOD max in peak of each QTL, marker position and each value.

that there are no differences between males and females. And we can support the QTL study results about *Lepeophtheirus salmonis* in Atlantic salmon [12]. Hence, no differences were observed between males and females, although fish size, which is known to vary between sexes has been found to be an important factor in determining lice abundance [28].

There are no previous studies that investigated whether host genes are important determinants of susceptibility to *B. seriolae* in yellowtail. However, further research is required to elucidate the functional basis of the differential susceptibility to infection among individuals. Possible mechanisms include a variety of host factors, such as specific and non-specific immune-related genes [29], and skin epithelial extract biochemistry [30]. Perhaps the most significant instances are the inflammatory differences associated with variation in susceptibility to initial infection among yellowtail species [9,10]. For example, Japanese flounder, exhibit a peripheral blood leucocyte response to the presence of *Neoheterobothrium hirame*, which has been linked to the earlier onset and significant greater up- and down-regulation of pro-inflammatory genes, such as matrix metalloproteinase MMP-9, MMP-13, leukotriene B4 receptor, CD20 receptor, major histocompatibility complex MHC Class I, MHC Class II beta-chain, immunoglobulin light chain and immunoglobulin heavy chain and unknown genes using microarray analysis [29]. In yellowtail, there is a possibility that *B. seriolae* also induces changes in the expression of pro-inflammatory and other immune-related genes. The EST based gene-expression research and usage of EST-SNP polymorphism will help to determine the functional mechanisms as the key to a solution about parasite disease in yellowtail.

The ancestral chromosome and syntenic gene, or expression candidate gene information has been used to assess fine QTL mapping and to determine candidate gene loci for economically important traits in domestic animals and crops research. But the discovery of major QTL influencing susceptibility to *B. seriolae* represents a significant step forward for a better understanding of the host response to parasite infection, which in turn will assist fine QTL mapping and MAS studies for future yellowtail breeding programs.

MAS will greatly increase the efficiency and effectiveness of breeding compared to traditional breeding programs. The fundamental advantage of MAS compared to conventional phenotypic selection is that it is faster since the early selection could be made without collecting the phenotypic data, and fish will be selected with high reliability, and successive generations will enable avoidance of inbreeding depression.

In addition, the potential of MAS and marker-assisted introgression (MAI) use for disease risk management of marine aquaculture finfish can be considered. Wild aquatic species are not selected and still maintain high genetic diversities. Individuals have high potential for genetic breeding regarding phenotypic variation. Natural populations will be more appropriate to contribute to those genetic resources to find large QTL effects than strain populations, as breeding materials. Therefore, there is a real possibility that marker-assisted selection can be a success for aquatic species.

Materials and Methods

Ethics Statement

Field permits are not required for this species in Japan. The fish handling, husbandry and sampling methods were approved by Institutional Animal Care and Use Committee of National Research Institute of Aquaculture (IACUC-NRIA No. 03).

Fish Families and Samples for QTL Analysis

Juvenile fish of *S. quinqueradiata* (100–120 mm in total length) were captured in the coastal waters of Goto Fukue island (Tsushima Strait, Nagasaki Prefecture). The yellowtail juveniles were purchased from commercial fishermen who catch wild juvenile yellowtail to supply fish farms for culture (artificial method of rearing from eggs for this species still does not allow a stable supply for practical use, and the aquaculture industry relies on wild caught juveniles for on-growing). One thousand juveniles were kept in a growing fish pen for two years. Two hundred of the three year old fish were pit tagged, and the number of *B. seriolae* counted six times to select the parents (data not shown). The selected fish were reared until maturity. We prepared two F_1 families (family A and family B) for QTL analysis. Both families are pair-mating from different parents. To evaluate Benedenia disease resistance, artificial infections of *B. seriolae* were performed in about 100 progeny of each F_1 family. For the genome-wide linkage analyses, 90 fishes of family A were used for finding candidate QTL, and 93 fishes of family B were used to confirm the QTL which are reproducible in the other family.

Parasite Collection and Artificial Infection Experiment

B. seriolae used for artificial infection were collected from adult fish stock in growing pens. Mesh nets were hung in fish pens in which adult fish were parasitized by *B. seriolae*. Eggs of *B. seriolae* stuck to the net and a method to collect only *B. seriolae* was established and confirmed under experimental conditions several times for reproducibility [10]. Collected *B. seriolae* eggs could be induced to hatch by a 15 minute exposure to fluorescent light. Hatched larvae were kept in a shaded tank before infection. In preliminary experiments, it was confirmed that the parasite did not exist on their the test fishes bodies after they were soaked in a freshwater bath for five minutes. And the fish were allowed to recover from freshwater stress for two days before artificial infection. Hatched *B. seriolae* larvae were introduced to experimental fish tanks, and individual fish were exposed to two hundred larvae of *B. seriolae*. Water temperature was kept at 25.3–25.6°C during the infection experiment for 10 days until the larvae grew to the countable adult stage.

Phenotypic Measurement of External Parasitism Resistance Traits Recorded

Ten days after exposure to *B. seriolae*, each fish was individually dipped in a tank of freshwater to remove the parasites. Thus, in the context of this study we recorded the total number of *B. seriolae* per fish. Benedenia disease frequency conformed to a normal phenotypic distribution as shown in Figure S3. Phenotypic information was used as quantitative trait values after linkage analysis (File S1). Total number of *B. seriolae* per fish was used as phenotypic information for QTL analysis. Number of pathogens for both families was normally distributed (Shapiro–Wilk test). Number of parasites was converted as Z score in each fish, in this case Z score was calculated as: number of parasites on each fish – average number of parasites/*standard deviation*.

Phenotypic Measurement of Total Length, Body Length, Body Weight, and Surface Area Recorded

We measured the fish total length, body length, body weight, and surface area, which conformed to a normal phenotypic distribution as shown in Figure S4. The average length and weight across all F_1 progeny were 185±5 mm and 65±10 g, and the surface area was calculated as (surface area) = $0.109\times2\times(\text{total length})^{2.113}$ [31]. Furthermore we have to know the correlation of these factors with the number of *B. seriolae* parasites. The phenotypic information of fish size was used as quantitative trait values for genome-wide analysis (File S1).

Data Collection and Genotyping

Microsatellite genotyping was performed in a 10 µl reaction volume containing 0.5 pmol/µl of unlabeled primer, 0.05 pmol/µl of fluorescence-end-labeled primer with [5′-TET], plus 1×buffer, 2.0 mM $MgCl_2$, 0.2 mM dNTP, 1% BSA, 0.025 U of Taq DNA polymerase (Takara: Ex-Taq) and 25 ng template DNA. Suitable annealing temperatures for each microsatellite marker were used. PCR was performed on a MJ PTC-100 (Bio-Rad), and the program conditions were 95°C for 2 min for initial denaturation, followed by 35 cycles of 30 s at 95°C, 1 min at the annealing temperature (56–58°C), 1 min at 72°C and 3 min at 72°C for final extension. Amplification products were mixed with an equal volume of loading buffer [98% formamide, 10 mM EDTA (pH 8.0), 0.05% bromophenol blue], heated for 5 min at 95°C and then immediately cooled on ice. The mixture was loaded onto 6% PAGE-PLUS gel (Amresco, OH, USA) containing 7 M urea and 0.5×TBE buffer. Electrophoresis was performed in 0.5×TBE buffer at 1800 V constant voltage for 1.5 h. After electrophoresis, the gel was scanned and imaged using an FMBIO III Multi-View fluorescence image analyzer (Hitachi-soft, Tokyo, Japan).

Single Nucleotide Polymorphism (SNP) loci from new expressed sequence tag (EST) sequences of *S. quinqueradiata* (Kai *et al.*, unpublished data) were used. To identify the polymorphism on SNP sites, we directly sequenced the PCR products of SNP regions in the parent of mapping family A using Sanger sequencer. The regions of polymorphic SNPs in the parents were also sequenced in 90 progeny of family A. Details of the microsatellite and SNP markers included in the linkage map are shown in File S1 and Figure S5.

A total of 1002 polymorphic markers in 24 linkage groups were used for family A using 860 microsatellite and 142 SNP markers. These marker locations and groups likely encompass all chromosomes for this cross. Nomenclature for the linkage groups was based on chromosome names using the linkage map [6]. Marker sex-specific map positions and genotypes for family A can be found in Figure S5 and File S1.

Linkage and QTL Analyses

Genotype scoring was performed by using LINKMFEX ver. 2.3 application package [32]. The application can separate the allele genotypes which originate from males or females, and check the accuracy of genotypes in their progeny from parental male and female alleles to avoid genotype scoring errors. Linkage analysis was performed using genotype data converted to a backcross format. As grandparent genotypes were unknown, pairwise analyses were performed, and markers were sorted in linkage groups at LOD threshold of 5.0. Linkage phases were determined retrospectively by examining the assortment of alleles among linked markers. Then the allele was tested for goodness-of-fit for Mendelian segregation distortion using χ^2-analysis. Also the order of the marker loci was confirmed to be correctly positioned, and was checked by double recombination events with the software application program in Map Manager QTX [33]. Graphic representations of linkage groups were generated with MAP-CHART version 2.1 [34] using raw recombination fractions as estimates of map distances (Figure S5).

The estimated total genome-length of the female map is 1054.5 (cM) by Kosambi function, with an average of 34 markers per linkage group. In the male map, the estimated total genome length is 1054.0 (cM) by Kosambi function, with an average of 32 markers per linkage group in this analysis. The female checked marker average distance was 1.29 cM in total length, and the male average distance was 1.35 cM in total length. The present map would be useful in some molecular studies for genome-wide linkage analysis, because the average inter-marker distances of each map were calculated on the basis of one marker for one cluster. Thus, the female and male maps had 249 and 281 unique positions, with estimated average intervals of 4.2 cM and 3.7 cM in the female and male maps, respectively, and offer sufficient marker density for QTL studies.

Phase-corrected genotypes using linkage group marker orders were established using LINKMFEX and GENOVECT [32] prior to QTL analysis (File S1). QTL effects were tested for the segregating alleles from the sires and dams separately for each interval or markers [example: Squ(linkage group)F; F is dam allele in female linkage group. Squ(linkage group)M; M is sire allele in male linkage group].And the results of sires and dams are reported in tabular and figure form. Further QTL modeling was carried out by analysis of variance (ANOVA) using the fitqtl function in R/qtl [35,36]. Putative QTL genotypes were added to the model using conditional genotype probabilities calculated at 1 cM intervals, and a genotyping error rate of 1%. Estimates of QTL effects and percentage of variance explained were obtained by comparing the full model to the sub-model without QTL. A similar approach was used for estimating covariate effects.

Estimation of QTL Region of Benedenia Disease Resistance

QTL analysis was conducted using MapQTL 5 software [37]. In the first genome-wide screening to find candidate loci, we used the Kruskal–Wallis (K–W) test initially to determine the significance level of all marker loci associated with the disease resistance phenotype for the family A. Limited significant loci associated with the disease resistance in the family A, which were analyzed in the linkage groups from chromosome-wide assessment in family B, because the candidate QTL region was reproducible in other families.

Then, all QTL analyses were carried out using simple interval mapping which was also used to identify the location of significant LOD max position. Simple interval mapping models were fitted in each parent using the Haley & Knott regression method [38] with conditional genotype probabilities calculated at 1 cM intervals and constant genotype error rate 1%. Multiple QTL model mapping was then used to reduce the background genetic noise and the influence of other QTLs from other chromosomes [39,40]. Although the results of both analyses are reported in tabular form for completeness, the results from the multiple QTL model analysis are used in the figures and discussion.

Significance Thresholds and Confidence Intervals

Experiment-wide "genome-wide and chromosome-wide" significance thresholds were derived from permutation estimates by dividing the nominal p-value by the total number of chromosomes examined in the study [41]. Permutation tests were performed (1,000 replicates) to determine the threshold for LOD using a type I error rate of $P = 0.05$, $P = 0.01$ and $P = 0.001$. Significant QTLs and regions were graphically visualized using the software MapChart 2.1 and MapQTL 5. QTL confidence intervals were estimated by 1.8-LOD support interval with 95% confidence interval probability coverage [42].

Estimation of Other QTL Regions Related to Total Length, Body Length, Body Weight, and Surface Area

QTL analyses for total length, body length, body weight, and surface area were also performed using K–W test and simple interval mapping, i.e. the same method of linkage analysis was performed to detect QTL region for Benedenia disease resistance.

Conclusion

We have discovered the first genetic evidence that contributes to detailing the phenotypic resistance to Benedenia disease in yellowtail. Furthermore we identified two chromosomal regions containing QTL (*BDR-1, BDR-2*) that were associated with Benedenia disease resistance. Two putative QTL associations, of medium to large effect of with Benedenia disease resistance, were localized to linkage groups Squ2 and Squ20. These two loci were responsible for ranging from 32.9 to 35.5% of the total phenotypic variation. The important finding of this study was detected as a single peak of QTL (*BDR-1*) associated with Benedenia disease resistance within the proximal region of linkage group Squ2. The QTL peak (Sequ1295BAC) was located at position 30.3 cM, with a 95% confidence interval that the QTL region lies within 10 cM of the most proximal marker from Sequ0672TUF to Sequ1067-TUF by simple interval mapping. The results of multiple QTL model mapping indicated QTL region within 5.5 cM interval. Furthermore the QTL (*BDR-2*) significant region would exist as multiple QTL on Squ20, because results of A and B families are detected as two peaks about the significant region in linkage group Squ20. Finding the QTL region strongly supports the potential for success of MAS for this disease. Moreover, we also identified the QTL on Squ7 associated with fish size (total length, body length, body weight, surface area) in yellowtail. The results from our current study suggested a negative correlation between growth traits and the number of parasites. The results will help resolve the mechanism of resistance to this important disease of yellowtail.

Supporting Information

Figure S1 Localization of significant markers for Benedenia disease resistance in linkage group Squ20M with family B. Squ(linkage group)M; marker distance in male map. Map positions and LOD scores are based on a simple interval mapping QTL analysis using the software MapQTL 5. Horizontal lines across each plot indicate LOD significance threshold, P_c; chomosome-wide significance threshold.

Figure S2 Significant markers for body weight simple interval mapping in linkage group Squ7F with family A. Squ(linkage group)F; marker distance in female map. Map positions and LOD scores are based on a simple interval mapping QTL analysis using the software MapQTL 5. Horizontal lines across each plot indicate LOD significance threshold, P_g; genome-wide significance threshold.

Figure S3 Benedenia disease frequency conformed to a normal phenotypic distribution (Shapiro-Wilk test).

Figure S4 Fish total length, body length, body weight and surface area conformed normal to a phenotypic distribution (Shapiro-Wilk test).

Figure S5 Details of microsatellite and SNP markers included in the linkage map. The marker positions of linkage map are identified in male and female sex-specific location.

File S1 Genotype of microsatellite and SNP markers included in family A and B. And the phenotypic information of fish was used as quantitative trait values for genome-wide analysis.

Acknowledgments

Authors would like to thank Kishiko Kubo for assistance with genotyping.

Author Contributions

Conceived and designed the experiments: AO KY K. Araki NO TS. Performed the experiments: AO KY KF SK JS WK YK. Analyzed the data: AO KF SK WK K. Akita TK. Contributed reagents/materials/analysis tools: JA MN TH TT. Wrote the paper: AO.

References

1. MAFF Japan (2009) Ministry of Agriculture. Forestry and Fisheries. Annual Statistics of Fishery and Aquaculture Production. Statistics Department. (in Japanese).
2. Nakada M (2008) Capture-based aquaculture of yellowtail. In Lovatelli A and Holthus PF (eds). Capture-based aquaculture. Global overview. FAO Fisheries Technical Paper 508: 199–215.
3. Mushiake K, Kawano K, Sakamoto W, Hasegawa I (1994) Effects of extended daylength on ovarian maturation and HCG-induced spawning in yellowtail fed moist pellets. Fish Sci 60: 647–651.
4. Mushiake K, Kawano K, Kobayashi T, Yamazaki T (1998) Advanced spawning in yellowtail, *Seriola quinqueradiata*, by manipulations of the photoperiod and water temperature. Fish Sci 64: 727–731.
5. Ohara E, Nishimura T, Nagakura T, Sakamoto T, Mushiake K, et al. (2005) Genetic linkage maps of two yellowtails (*Seriola quinqueradiata* and *Seriola lalandi*). Aquaculture 244: 41–48.
6. Fuji K, Kai W, Kubota S, Yoshida K, Ozaki A, et al. (2012) A genetic linkage map of yellowtail (*Seriola quinqueradiata*) constructed with microsatellite isolated from BAC-end sequences. BMC Genomics: submitted.
7. Ogawa K (1996) Marine parasitology with special reference to Japanese fisheries and mariculture. Veterinary Parasitology 64: 95–105.
8. Hutson K, Ernst I, Whittington ID (2007) Risk assessment for metazoan parasites of yellowtail kingfish (*Seriola lalandi*) (Perciformes: Carangidae) in South Australian sea-cage aquaculture. Aquaculture 271: 85–99.
9. Nagakura Y, Nakano S, Mushiake K, Ohara E, Okamoto N, et al. (2006) Differences in susceptibility to the monogenean parasite *Benedenia seriolae* among *Seriola quinqueradiata, S. lalandi* and their hybrid. Aquaculture Science 54: 335–340. (in Japanese).
10. Nagakura Y, Yoshinaga T, Sakamoto T, Hattori K, Okamoto N (2010) Susceptibility of four families derived from two *Seriola* species to the monogenean parasite (*Benedenia seriolae*) using a new challenge method. Journal of Fisheries Technology 3: 21–26. (in Japanese).
11. Baerwald MR, Petersen JL, Hedrick RP, Schisler GJ, May B (2011) A major effect quantitative trait locus for whirling disease resistance identified in rainbow trout (*Oncorhynchus mykiss*). Heredity 106: 920–926.
12. Gharbi K, Glover KA, Stone LC, MacDonald ES, Matthews L, et al. (2009) Genetic dissection of MHC-associated susceptibility to *Lepeophtheirus salmonis* in Atlantic salmon. BMC Genet 10: 20.
13. Ishikawa A (2010) Nagoya Repository, version 5. The Japanese Association for Laboratory Animal Science 51: 51–61 (in Japanese). Available: http://hdl.handle.net/2237/6779.
14. Members of the Complex Trait Consortium (2003) The nature and identification of quantitative trait loci: a community's view. Nat Rev Genet 4: 911–916.
15. Kubota S, Ohara E, Nishimura T, Nagakura Y, Ozaki A, et al. (2009) QTL analysis for Benedeniasis resistance in hybrids between two yellowtails (*Seriola quinqueradiata* and *Seriola lalandi*). The 10th International Symposium on Genetics in Aquaculture, Bangkok, Thailand, PP093: 194.
16. Ozaki A, Sakamoto T, Khoo SK, Nakamura K, Coimbra MR, et al. (2001) Quantitative trait loci (QTLs) associated with resistance/susceptibility to infectious pancreatic necrosis virus (IPNV) in rainbow trout (*Oncorhynchus mykiss*). Mol Genet Genomics 265: 23–31.
17. Nichols KM, Bartholomew J, Thorgaard GH (2003) Mapping multiple genetic loci associated with *Ceratomyxa shasta* resistance in *Oncorhynchus mykiss*. Dis Aquat Org 56: 145–154.
18. Moen T, Sonesson AK, Hayes B, Lien S, Munck H, et al. (2007) Mapping of a quantitative trait locus for resistance against infectious salmon anemia in Atlantic salmon (*Salmo salar*): comparing survival analysis with analysis on affected/resistant data. BMC Genet 8: 13.
19. Houston DR, Haley SC, Hamilton A, Guy RD, Tinch EA, et al. (2008) Major quantitative trait loci affect resistance to infectious pancreatic necrosis in Atlantic salmon (*Salmo salar*). Genetics 178: 1109–1115.
20. Moen T, Baranski M, Sonesson AK, Kjøglum S (2009) Confirmation and fine-mapping of a major QTL for resistance to infectious pancreatic necrosis in Atlantic salmon (*Salmo salar*): population-level associations between markers and trait. BMC Genomics 10: 368.
21. Ozaki A, Khoo SK, Yoshiura Y, Ototake M, Sakamoto T, et al. (2007) Identification of additional quantitative trait loci (QTL) responsible for susceptibility to infectious pancreatic necrosis virus (IPNV) in rainbow trout (*Oncorhynchus mykiss*). Fish Pathology 42: 131–140.
22. Glover KA, Grimholt U, Bakke HG, Nilsen F, Storset A, et al. (2007) Major histocompatibility complex (MHC) variation and susceptibility to the sea louse *Lepeophtheirus salmonis* in Altantic salmon (*Salmo salar*). Dis Aquat Org 76: 57–65.
23. Fuji K, Kobayashi K, Hasegawa O, Coimbra MRM, Sakamoto T, et al. (2006) Identification of a single major genetic locus controlling the resistance to lymphocystis disease in Japanese flounder (*Paralichthys olivaceus*). Aquaculture 254: 203–210.
24. Hwang SD, Fuji K, Takano T, Sakamoto T, Kondo H, et al. (2011) Linkage mapping of toll-like receptors (TLRs) in Japanese flounder, *Paralichthys olivaceus*. Mar Biotechnol 13: 1086–1091.
25. Ryce EKN, Zale AV, MacConnell E (2004) Effects of fish age and development of whirling parasite dose on the disease in rainbow trout. Dis Aquat Org 59: 225–233.
26. Ryce EKN, Zale AV, MacConnell E, Nelson M (2005) Effects of fish age versus size on the development of whirling disease in rainbow trout. Dis Aquat Org 63: 69–76.
27. Fuji K, Yoshida K, Hattori K, Ozaki A, Araki K, et al. (2010) Identification of the sex-linked locus in yellowtail, *Seriola quinqueradiata*. Aquaculture 308: S51–S55.
28. Glover KA, Aasmundstad T, Nilsen F, Storset A, Skaala O (2005) Variation of Atlantic salmon families (*Salmo salar*) in susceptibility to the sea lice *Lepeophtheirus salmonis* and *Caligus elongatus*. Aquaculture 245: 19–30.
29. Matsuyama T, Fujiwara A, Nakayasu C, Kamaishi T, Oseko N, et al. (2007) Microarray analyses of gene expression in Japanese flounder *Paralichthys olivaceus* leucocytes during monogenean parasite *Neoheterobothrium hirame* infection. Dis Aquat Org 75: 79–83.
30. Yoshinaga T, Nagakura T, Ogawa K, Fukuda Y, Wakabayashi H (2002) Attachment-inducing capacities of fish skin epithelial extracts on oncomiracidia of *Benedenia seriolae* (Monogenea: Capsalidae). Int J Parasitol 32: 381–384.
31. Ohno Y, Kawano F, Hirazawa N (2008) Susceptibility by amberjack (*Seriola dumerili*), yellowtail (*S. quinqueradiata*) and Japanese flounder (*Paralichthys olivaceus*) to *Neobenedenia girellae* (Monogenea) infection and their acquired protection. Aquaculture 274: 30–35.
32. Danzmann RG (2006) Linkage analysis package for outcrossed families with male or female exchange of the mapping parent, version 2.3. Available: http://www.uoguelph.ca/~rdanzman/> software/LINKMFEX.
33. Manly KF, Cudmore RH Jr, Meer JM (2001) Map Manager QTX, cross-platform software for genetic mapping. Mamm Genome 12: 930–932.
34. Voorrips RE (2002) MapChart: software for the graphical presentation of linkage maps and QTLs. J Hered 93: 77–78.
35. Broman KW, Wu H, Sen S, Churchill GA (2003) R/qtl: QTL mapping in experimental crosses. Bioinformatics 19: 889–890.
36. Arends D, Prins P, Jansen RC, Broman KW (2010) R/qtl: high-throughput multiple QTL mapping. Bioinformatics 26: 2990–2992.
37. Van Ooijen JW (2004) MapQTL 5. Software for the mapping of quantitative trait loci in experimental populations of diploid species. Kyazma BV: Wageningen, Netherlands.

38. Haley CS, Knott SA (1992) A simple regression method for mapping quantitative trait loci in line crosses using flanking markers. Heredity 69: 315–324.
39. Jansen RC, Stam P (1994) High resolution of quantitative traits into multiple loci via interval mapping. Genetics 136: 1447–1455.
40. Jansen RC, Van Ooijen JW, Stam P, Lister C, Dean C (1995) Genotype-by-environment interaction in genetic mapping of multiple quantitative trait loci. Theor Appl Genet 91: 33–37.
41. Churchill GA, Doerge RW (1994) Empirical threshold values for quantitative trait mapping. Genetics 138: 963–971.
42. Manichaikul A, Dupuis J, Sen S, Broman KW (2006) Poor performance of bootstrap confidence intervals for the location of a quantitative trait locus. Genetics 174: 481–489.

7

Effect of Hydrogen Peroxide on Immersion Challenge of Rainbow Trout Fry with *Flavobacterium psychrophilum*

Maya Maria Mihályi Henriksen*, Lone Madsen, Inger Dalsgaard

Technical University of Denmark, National Veterinary Institute, Bülowsvej 27, Frederiksberg C, Denmark

Abstract

An experimental model for immersion challenge of rainbow trout fry (*Oncorhynchus mykiss*) with *Flavobacterium psychrophilum*, the causative agent of rainbow trout fry syndrome and bacterial cold water disease was established in the present study. Although injection-based infection models are reliable and produce high levels of mortality attempts to establish a reproducible immersion model have been less successful. Various concentrations of hydrogen peroxide (H_2O_2) were evaluated before being used as a pre-treatment stressor prior to immersion exposure to *F. psychrophilum*. H_2O_2 accelerated the onset of mortality and increased mortality approximately two-fold; from 9.1% to 19.2% and from 14.7% to 30.3% in two separate experiments. Clinical signs observed in the infected fish corresponded to symptoms characteristically seen during natural outbreaks. These findings indicate that pre-treatment with H_2O_2 can increase the level of mortality in rainbow trout fry after exposure to *F. psychrophilum*.

Editor: Dongsheng Zhou, Beijing Institute of Microbiology and Epidemiology, China

Funding: This work was performed as a part of the Danish Fish Immunology Research Centre and Network (DAFINET; www.dafinet.dk) supported by the Danish Council of Strategic Research (grant 2101-080017). The funders had no role in study design, data collection and analysis, decision to publish, or preparation of the manuscript.

Competing Interests: The authors have declared that no competing interests exist.

* E-mail: mmah@vet.dtu.dk

Introduction

Flavobacterium psychrophilum is a Gram-negative fish pathogen and the causative agent of bacterial cold water disease (BCWD), also called rainbow trout fry syndrome (RTFS). Since first being discovered in USA in the 1940s, the disease has spread worldwide and was identified in Western Europe in the mid 1980s. Today it has a significant impact on rainbow trout (*Oncorhynchus mykiss*) and other salmonid fish in aquaculture worldwide [1]. Clinical signs of the disease include a dark coloration of the skin, pale gills and organs (kidney, intestine and liver) due to anemia. A loss of appetite and lethargic behavior is also observed.

Injection-based experimental challenges with *F. psychrophilum* have been standardized and the resulting mortality depends on several factors, including the number of colony forming units (CFU) injected, size of the fish, batch differences and the number of fish in each tank [2]. However, infection by injection is not a suitable approach for all investigations since the first line of defense, consisting of non-specific barriers, are bypassed. It has proven difficult to produce high and consistent mortality rates using cohabitation and immersion models unless stress or scarification has been applied [2–7], while immersion exposure to *F. psychrophilum* in the logarithmic phase of growth reportedly results in significant mortality rates [8].

Bath-treatment with various non-medical compounds, such as copper sulphate, chloramine-T, sodium carbonates, sodium chloride, formalin and hydrogen peroxide (H_2O_2), are routinely used against pathogens in aquaculture [9]. One previous immersion model for *F. psychrophilum* used formalin as stressor, which raised the mortality rate significantly [2]. Since the use of formalin is to be phased out by Danish fish farmers due to human health considerations, another less harmful stressor is needed. H_2O_2 is a useful and environmentally friendly alternative to formalin and has been used against skin parasites, bacterial gill infections and mold on eggs in Denmark for over ten years [10,11]. The use of H_2O_2 has been shown to accelerate *Tenacibaculum maritimum* infections in turbot [12] and *Flavobacterium columnaris* infections in channel catfish [13]. Accordingly, H_2O_2 was an obvious candidate for a stressor in a *F. psychrophilum* immersion model.

The aim of this study was to establish a reproducible method for increasing mortality of RTFS in immersion challenge of rainbow trout fry. A reliable model is needed for studies on pathogen invasion, host immune response and efficacy of treatments. First, various concentrations of H_2O_2 were tested in experiment 1 before being combined with immersion exposure to a virulent *F. psychrophilum* strain at two different phases of bacterial growth in experiment 2. The most favorable H_2O_2 concentration and bacterial growth phase were combined in experiment 3, after which the temperature and number of animals were adjusted in experiment 4 to increase the power of the model.

Materials and Methods

Fish and Rearing Conditions

Three batches of eggs from the same stock and family of fish were used in all four experiments (see Table 1). The eggs originated from Fousing Trout Farm (Denmark) and disinfection and hatching was carried out at AquaBaltic (Bornholm, Denmark). The fish were acclimatized for at least 3 weeks in 200 L tanks containing 15°C recirculated tap water (non-chlorinated), while nitrate, nitrite and ammonia levels were monitored continuously.

The fish were fed dry commercial feed daily (INICIO Plus, BioMar A/S, Denmark). All tanks, except the 1 and 2 L tanks used during challenge and pre-treatment, were opaque in order to minimize external stress due to movement in the experimental facility. Upon arrival, at least 10 fish from each batch where sacrificed by an overdose of MS-222 (Ethyl 3-aminobenzoate methanesulfonate, Sigma), examined for *F. psychrophilum* and proven to be free of the pathogen. Samples from brain and kidney were streaked on tryptone yeast extract salts (TYES) and blood agar plates, while spleen samples were placed in TYES broth (110 rpm). Samples were incubated for 7 d at 15°C for TYES broth/plates and at 20°C for blood plates. Yellow colonies were further examined by species specific PCR. First, a pure culture was obtained by streaking and after 5–7 d of growth, a single colony was suspended in 50 μL distilled water and boiled for 5 min to lyse the bacterial cells and placed on ice. Then, species specific primers, referred to as PSY1 and PSY2, amplifying the 16S rRNA genes of *F. psychrophilum* were used [14] for PCR amplifications, which were carried out according to a previously described method [15]. In brief, Ready to Go PCR beads (cat. no. 27955901, Amersham Pharmacia Biotech, Millwaukee, USA), including all reagents, were used. Ten pmol of each primer and 1 μL of the lysed bacteria were added to each reaction. Distilled water was added to a total volume of 25 μL. Amplifications were carried out using a Biometra T- 3000 thermocycler amplifier (Biometra, Göttingen, Germany): 95°C for 5 min followed by 35 cycles consisting of 30 s at 95°C, 30 s at 57°C and 60 s at 72°C.

Before randomly selecting the experimental groups, the largest (approx. >1.5 g) and smallest (approx. <0.6 g) fish were removed in order to minimize size variation. The fish were inspected at least once every 24 h and dead fish were examined for *F. psychrophilum* as previously described. Fish surviving until the completion of each experiment were euthanized using an overdose of MS-222.

The described work was carried out in accordance with the internationally accepted guidelines for care and use of laboratory animals in research and procedures were approved by the Committee for Animal Experimentation, Ministry of Justice, Copenhagen, Denmark (J.nr. 2006/561-1204 and 2011/561-51). During experiments, the fish were inspected several times a day and precautions were taken to minimize stress by keeping the fish in opaque tanks, minimizing handling and monitoring water chemistry.

Bacterial Strain and Challenge Dose

A Danish *F. psychrophilum* strain, 950106-1/1, was used for all challenges (serotype Fd and elastin degrading). It was isolated from a clinical outbreak of RTFS in a freshwater farm in 1995, and has previously been used for several i.p. and immersion challenges [2]. The bacterial strain was stored at -80°C in TYES broth with 15–20% glycerol and pre-cultivated in 10 mL TYES broth at 15°C for 3 d (110 rpm). Then, 100 mL TYES broth was inoculated using 0.5 mL of the pre-culture and optical density (OD) was measured at 525 nm using a spectrophotometer (Shimadzu UV-1201) at regular intervals to establish a growth curve. Logarithmic and stationary phases were determined after 24 or 48 h of growth, respectively. Before being used for challenge, each culture was examined microscopically to assert purity and CFU counts were carried out by spreading 0.1 mL of 10-fold dilutions (10^{-4} to 10^{-7}) on TYES agar in duplicates.

Experiment 1: H_2O_2 as Pre-treatment Stressor

Three groups of 6 fish in duplicates (n = 36) were immersed for 60 min in aerated 1 L tanks containing final concentrations of 50, 150 or 300 mg H_2O_2 L^{-1} (prod. no. 1072090250, Merck KGaA, Germany). H_2O_2 was added directly into the tanks already containing the fish. Subsequently, the fish were removed and placed in 20 L aquaria and observed for 14 d. The temperature during the experiment was 15±0.5°C, while H_2O_2 treatment was carried out at approximately 12°C. Based on the results, a concentration of 200 mg H_2O_2 L^{-1} was chosen as stressor for experiment 2.

Experiment 2: H_2O_2 Combined with Bacterial Growth Phases (24 and 48 h)

Four groups of 10 fish in duplicates (n = 80) were immersed for 60 min in aerated 1 L tanks containing 200 mg H_2O_2 L^{-1} or kept under similar conditions in water. H_2O_2 was added directly into the aquaria already containing the fish. Afterwards, the fish were removed and placed for 30 min in a tank containing a 1:9 dilution of cultures grown for either 24 h or 48 h. This resulted in bacterial concentrations of 10^5 and 10^7 CFU mL^{-1} water, respectively. Finally, the fish were placed in 20 L aquaria containing tap water. The fish were observed for 30 d. The temperature during the experiment was 14.5±0.5°C, while H_2O_2 treatment and bacterial challenge were carried out at a temperature below 14°C. The animals dying after exposure to the pathogen were sampled as previously described for the bacteriological examination. Since no significant difference regarding either mortality or reisolation of the pathogen was observed, the 48 h culture was chosen for the following experiments, since it has been used successfully in previous investigations [2].

Experiment 3:150 mg L^{-1} H_2O_2 as Experimental Stressor

Four groups of 50 fish in duplicates (n = 400) were immersed for 60 min in aerated 2 L tanks containing 150 mg H_2O_2 L^{-1} or kept under similar conditions in tap water. H_2O_2 was diluted in 300 mL tap water, before being added to the aquaria already containing the fish. Subsequently, the fish were removed and immersed in a 1:9 diluted broth culture (48 h) containing 10^7 CFU mL^{-1} for 30 min, while controls were immersed in

Table 1. Experimental parameters.

Exp.	Batch	Weight	Length	H_2O_2 mg L^{-1}	*F.p* CFU mL^{-1}	°C	n (pr. replicate)
1	1	0.69 g (±0.21)	4.4 cm (±0.3)	50/150/300	-	15.5±1.0	36 (6)
2	1	0.77 g (±0.18)	4.4 cm (±0.3)	–/200	10^5(24 h)/10^7(48 h)	16.0±0.5	80 (10)
3	2	1.2 g (±0.50)	5.0 cm (±0.8)	–/150	–/10^7	16.8±0.5	400 (50)
4	3	1.1 g (±0.34)	4.7 cm (±0.5)	–/200	–/10^7	14.0±0.5	500 (50)

Fish batches and mean values for weight, length and temperature are stated along with standard deviation (SD). The applied H_2O_2 concentrations are given in mg L^{-1}, while the final bacterial concentrations in broth used for immersion challenge are stated in CFU mL^{-1}. The symbol '–' designates the absence of *F.p.* and H_2O_2.

sterile broth. Finally, the fish were placed in 30 L aquaria and mortality observed for 50 d. The temperature during the experiment was 16.8±0.5°C, while H_2O_2 treatment and bacterial challenge were carried out at a temperature below 14°C. To increase mortality in experiment 4, the temperature was lowered and the concentration of H_2O_2 increased.

Experiment 4:200 mg L^{-1} H_2O_2 as Experimental Stressor

Four groups of 50 fish in duplicates for uninfected or triplicates for infected (n = 500) were immersed for 60 min in aerated 2 L tanks containing 200 mg H_2O_2 L^{-1} or kept under similar conditions in tap water. H_2O_2 was diluted in 300 mL tap water, before being added to the treatment tanks already containing the fish. Subsequently, the fish were removed and immersed in a 1:9 diluted broth culture (48 h) containing 10^7 CFU mL^{-1} for 30 min, while controls were immersed in sterile broth. Finally, the fish were placed in 30 L aquaria and observed for 40 d. The temperature during the experiment was 14±0.5°C, while H_2O_2 treatment and bacterial challenge were carried out at a temperature below 14°C. Conditions during treatments of the various groups were kept as identical as possible with the same number of nettings and amount of retention time in the 2 L tanks during treatments.

Statistical Analysis

Mortality data was analyzed using the generalized linear model (GLM) on the R software platform [16] and figures have been produced using GraphPad Prism 5. The relative standard deviation (RSD = (standard deviation/mean)*100) was also calculated for mortality rates of infected groups in experiment 3 and 4. Weight and length of the fish was compared between groups (where a sufficient n was available) by using 1-way ANOVA combined with Tukey's test.

In tables and figures, significance levels are denoted by * for p = 0.05, ** for p = 0.01 and *** for p = 0.001.

Results

Hydrogen peroxide pre-treatment was found to elevate the mortality of subsequent immersion exposure to *F. psychrophilum* in the stationary phase of growth.

Experiment 1: H_2O_2 as Pre-treatment Stressor

An apparent escape response was observed in all groups immediately after exposure to H_2O_2; the majority of fish clumped together and swam towards the tank wall. This was most prominent in tanks treated with medium and high concentrations, where the escape response was followed by fish laying on bottom of the tank. Furthermore, a few fish from the high dose group showed erratic swimming behavior for shorter periods of time. All visual changes in behavior ceased after a few min at the lowest concentration, while persisting longer in groups treated with higher concentrations of H_2O_2. All fish survived the treatment, although one death occurred in the high dosage group within the first 24 h.

Experiment 2:200 mg L^{-1} H_2O_2 Combined with Bacterial Growth Phases (24 and 48 h)

The experiment included four groups in duplicates: (1) 24 h culture, (2) 24 h culture with H_2O_2 pre-treatment, (3) 48 h culture and (4) 48 h culture with H_2O_2 pre-treatment (Table 2).

Nine of the 40 fish pre-treated with H_2O_2 died during treatment and were accounted for as acute mortality. It was only possible to re-isolate *F. psychrophilum* from dead fish belonging to groups pre-treated with H_2O_2.

Table 2. H_2O_2 combined with bacterial growth phases.

Group	H_2O_2	*F.p* (CFU mL^{-1})	n	Re-isolation/ dead fish	% mortality
1	-	**24 h**	20	0/3	15%
2	+	(10^5)	12	3/3	25%
3	-	**48 h**	20	0/1	5%
4	+	(10^7)	19	3/4	21%

Experiment 2. The symbols '+' and '-' designate addition or absence of the two factors, H_2O_2 (200 mg L^{-1} for 60 min). and *F.p.* ($10^5/10^7$ CFU mL^{-1} *F. psychrophilum* for 30 min). Successful reisolations of *F. psychrophilum* from dead fish are given as a fraction of the total number of dead fish. Finally, cumulative mortality is stated in percent.

Experiment 3:150 mg L^{-1} H_2O_2 as Experimental Stressor

The experiment consisted of four groups in duplicates: (I) untreated control, (II) H_2O_2, (III) *F. psychrophilum* and (IV) H_2O_2 and *F. psychrophilum* (Table 3).

Seven of the 200 fish pre-treated with 150 mg H_2O_2 L^{-1} died during the treatment or subsequent transfer to tap water and were accounted for as acute mortality. Seven days after challenge, technical problems lead to an increased temperature in the room and flooding of tank III.B. This led to the loss of 23 fish, which were excluded from the experiment, since the event took place before onset of mortality.

The cumulative mortality (Table 3) was higher in both infected groups compared to their respective controls (p = 0.05 for III and p = 0.001 for IV). The cumulative mortality for infected groups was 9.1% for *F. psychrophilum* challenge alone and 19.2% in combination with H_2O_2 (p = 0.001). Relative standard deviation for the two groups was 23% and 39%, respectively. Furthermore, H_2O_2 accelerated the onset of mortality post challenge from 9 to 3 d (Fig. 1).

Experiment 4:200 mg L^{-1} H_2O_2 as Experimental Stressor

The experiment consisted of four groups, which were set up in either duplicates for uninfected or triplicates for infected groups: (A) untreated control and (B) H_2O_2, (C) *F. psychrophilum* and (D) H_2O_2 and *F. psychrophilum* (Table 4).

Five of the 250 fish pre-treated with 200 mg H_2O_2 L^{-1} died during the treatment and subsequent transfer to tap water and were accounted for as acute mortality.

The cumulative mortality was higher in both infected groups compared to their respective controls (p = 0.001). The cumulative mortality (Table 4) of infected groups was 14.7% (n = 22) for *F. psychrophilum* challenge alone and rose to 30.3% (n = 54) in combination with H_2O_2 (p = 0.001). RSD for the two groups was 27% and 38%, respectively. Furthermore, H_2O_2 accelerated the onset of mortality post challenge from 10 to 4 d (Fig. 2). A statistically significant difference in mortality was found between replicate tanks D1 (20.0%) and D2 (42.9%) from group D (p = 0.05), which was taken into consideration in data processing.

Discussion

Several studies have investigated the potential adverse effects of H_2O_2 on salmonids to establish recommendations for safe treatment dosages [10,11,17–20]. Possible damage caused by

Table 3. Challenge after 150 mg L^{-1} H_2O_2.

Group	Replicate	H_2O_2	*F.p.*	n	% mortality (n)	Cumulative % mortality	SD
I	A	–	–	45	2.0% (1)	**2.0%**	
	B	–	–	47	2.0% (1)	(a*)	
II	A	+	–	49	4.1% (2)	**4.0%**	
	B	+	–	44	4.4% (2)	(b***)	
III	A	–	+	49	8.0% (4)	**9.1%**	2.19
	B	–	+	27	11.1% (3)	(a*)(c**)	(23%)
IV	A	+	+	49	14.0% (7)	**19.2%**	7.42
	B	+	+	48	24.5% (12)	(b***)(c**)	(39%)

Experiment 3. The fish weighed 1.2 g and the temperature was 16.8±0.5°C. The symbols '+' and '-' designate addition or absence of the two factors, H_2O_2 (150 mg L^{-1} for 60 min) and *F.p.* (10^7 CFU mL^{-1} *F. psychrophilum* for 30 min). The % cumulative mortality is stated for both replicates and groups along with SD and RSD in brackets. Statistical differences are denoted by a letter and * for p = 0.05, ** for p = 0.01 and *** for p = 0.001.

H_2O_2 depends on several factors, including the applied concentration, exposure time, frequency of treatment, life stage of the fish and temperature during treatment. Damage generally consists of injury to the gills and larger fish are more susceptible [19]. Pathological changes may include an increase in epithelial cell granularity, edemas, lamellar fusion, epithelial hyperplasia as well as swelling and lifting of the gill epithelium. The gills are a frequent target in stress responses and of the mentioned lesions can, with the existing knowledge, be induced by many types of environmental stressor; not just H_2O_2 [23]. Treatment is generally discouraged at temperatures above 14–15°C [11,21,22], but temperatures approaching 14°C during H_2O_2 treatment did not result in increased mortality under the conditions applied in the present experiments.

Mortality due to H_2O_2 has been shown to occur within the first days after treatment; predominantly during or within hours of exposure [17,18]. It is difficult to directly compare the published studies regarding H_2O_2, and the performance of pilot studies on small groups of fish are recommended before treating an entire population [18,19].

H_2O_2 as Pre-treatment Stressor

Previous studies have demonstrated that H_2O_2 has a fast short-term stress effect in both Atlantic salmon and sea bass [24,25]. In the present study, an escape response was observed in all fish treated with H_2O_2, although most prominent in the high dose group, in which the only mortality occurred within 24 h after exposure. Based on these results, a concentration of 200 mg H_2O_2 L^{-1} was chosen as stressor for experiment 2, where 9 of the 40 treated fish died from the pre-treatment alone. The increased mortality is likely due to the direct administration of undiluted 35% H_2O_2 into the water. Since the density of fish was higher in experiment 2, the probability of contact with high local concentrations of H_2O_2 before dispersal in the water was

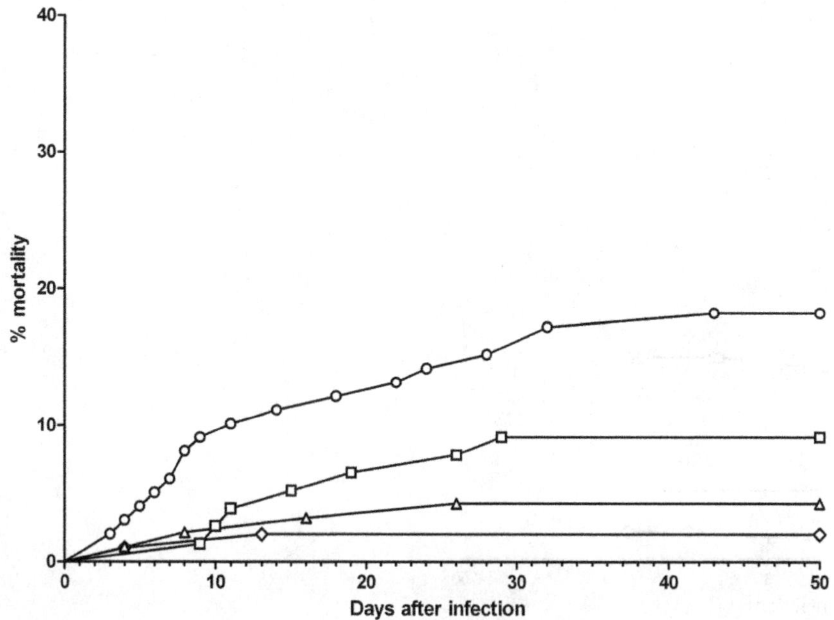

Figure 1. Challenge after 150 mg L^{-1} H_2O_2. Cumulative mortality of the merged replicates is shown in the figure. The fish weighed 1.2 g and the temperature was 16.8±0.5°C. The experiment ran for 50 days. ◇: control, △: H_2O_2, □: *F. psychrophilum*, ○: H_2O_2+ *F. psychrophilum*.

Table 4. Challenge after 200 mg L^{-1} H_2O_2.

Group	Replicate	H_2O_2	*F.p.*	n	% mortality (n)	Cumulative % mortality	SD
A	1	–	–	50	0.0% (0)	**3%**	
	2	–	–	50	6.0% (3)	(a***)	
B	1	+	–	50	8.0% (4)	**5%**	
	2	+	–	50	2.0% (1)	(b***)	
C	1	–	+	50	12.0% (5)	**14.7%**	4.19
	2	–	+	50	14.3% (7)	(a***)(c***)	(27%)
	3	–	+	50	20.0% (10)		
D	1	+	+	50	20.0% (10)	**30.3%**	11.57
	2	+	+	49	42.9% (21)	(b***)(c***)	(39%)
	3	+	+	46	28.3% (13)		

Experiment 4. The fish weighed 1.1 g and the temperature was 14±0.5°C. The symbols '+' and '–' designate addition or absence of the two factors, *F.p.* (10^7 CFU mL^{-1} *F. psychrophilum* for 30 min) and H_2O_2 (200 mg L^{-1} for 60 min). The % cumulative mortality is stated for both replicates and groups along with SD and RSD. Statistical differences are denoted by a letter and * for p = 0.05, ** for p = 0.01 and *** for p = 0.001.

increased. In experiment 3 and 4, H_2O_2 was diluted at in least 300 mL water before being added to the treatment tanks. This approach resulted in 2–3.5% acute mortality in addition to 4–5% mortality during the experiments, compared to a total of 2–3% in the untreated control groups. In a previous test, healthy juvenile rainbow trout were treated twice a week for seven weeks (200 mg L^{-1} for 60 min) reported lower feed conversion ratios in the first weeks of treatment but no noteworthy mortality [26], while the same dosage given only once resulted in high mortalities within a few hours under slightly different conditions [17]. The experimental design was not intended to investigate the potential long-term consequences of H_2O_2 treatment, but the results indicate the possibility of excess mortality due to as little as a single treatment but the subject should be investigated further.

H_2O_2 Combined with Bacterial Growth Phases (24 and 48 h)

Experiment 2 was a pilot test used to indicate, whether there was a difference in mortality for fish infected with bacteria in either the logarithmic or stationary phase of growth. Furthermore, it was relevant to conduct a small scale test of H_2O_2 in combination with *F. psychrophilum*. Based on previous results, an increased mortality was expected for the 24 h culture [8], but no differences were seen. Reisolation of *F. psychrophilum* from spleen, kidney and brain of dead fish was only successful in H_2O_2 treated groups. Although the portals of entry have not been determined, *F. psychrophilum* has been found in mucus, fins, gills and stomach of infected fish and it has been speculated, that sub-optimal environmental conditions may allow the bacterium to get across skin and gills [27]. Besides

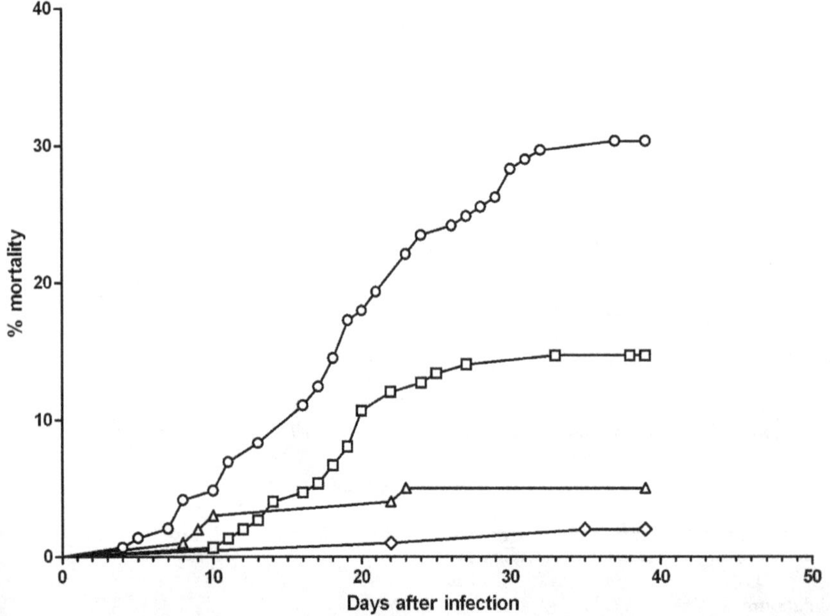

Figure 2. Challenge after 200 mg L^{-1} H_2O_2. Cumulative mortality of the merged replicates is shown in the figure. The fish weighed 1.1 g and the temperature was 14±0.5°C. The experiment ran for 40 days. ◇: control, △: H_2O_2, □: *F. psychrophilum*, ○: H_2O_2+ *F. psychrophilum*.

stressing the host, treatment with H_2O_2 may have resulted in better access to the blood stream, possibly via the gills or by damaging mucus on either skin or gastrointestinal tract, hence allowing for a rapid spread inside the host.

Based on these findings, it was decided to proceed using H_2O_2 as a stressor and the 48 h culture. The 48 h culture has been used successfully in previous investigations [2] and resulted in a higher number of CFU. Cumulative mortality was relatively low for all treatments in experiment 2. The small size of the experimental groups could have affected the outcome in several ways and the results of this experiment cannot be considered to be conclusive. Firstly, the dynamics between healthy, infected and dead fish were influenced by the low density. Secondly, the low number of fish also decreases the statistical robustness, especially since reproducibility is a known problem regarding immersion challenge with *F. psychrophilum*. Choice of strain might have played a role, although 950106-1/1, which was used in the present study, is known to cause mortality in rainbow trout fry [2].

H_2O_2 as Experimental Stressor in Immersion Challenge

The cumulative mortality was significantly increased by pre-treatment with H_2O_2 in both experiment 3 and 4. In experiment 3, pre-treatment with H_2O_2 increased mortality due to *F. psychrophilum* from 9.1% to 19.2%. In experiment 4, the temperature was lowered and the pre-treatment dosage increased slightly, resulting in mortality rates of 14.7% for *F. psychrophilum* alone and 30.3% in combination with H_2O_2. The increased dosage of H_2O_2 may have stressed the fish more or resulted in more damage to gills or other tissues in contact with the surrounding water, such as gastrointestinal tract or skin. The lowered temperature is also likely to have played a role, since even smaller changes have been shown to have significant consequences on survival of infections [27]. At lower temperatures, the immune response of the fish is delayed [28,29], while the psychrophilic bacteria's physiological functions are impaired to a lesser degree [1]. Another explanation is a suppression of the immune system due to cortisol, which is a corticosteroid associated with stress [23]. Finally, batch differences can also have contributed to the difference in mortality seen between the two experiments.

Variation was consistent around 40% for the pre-treated and infected group, and 25% for the infected group for both experiment 3 and 4. Thus, variation did not seem to be influenced by smaller differences in handling through the experiments, since strictly uniformed management was applied in experiment 4. Both variation and proportional change in mortality induced by H_2O_2 was comparable between the two experiments and to the results obtained in a similar model using formalin [2]. Despite the variation, H_2O_2 pre-treatment consistently elevated the mortality after subsequent immersion exposure to *F. psychrophilum*. Conversely, the results underline the importance of maximizing the statistical power of the experimental model by using larger experimental groups or more replicates.

Onset of mortality was accelerated, a pattern which was also observed for H_2O_2 treatment in combination with *T. maritimum* and *F. columnaris* [12,13]. The clinical signs in both experiment 3 and 4 corresponded to the symptoms seen in natural outbreaks and included anemia (resulting in pale gills and organs), splenomegaly (which could in some cases be seen through the skin as a red coloration of the abdomen) and dark pigmentation of the skin [1,30,31].

A number of previous studies have also focused on developing a reliable model for infection, which mimics natural transmission, since it is essential to gain more knowledge regarding transmission and host-pathogen interactions. Previous studies have highlighted the limited pathogenicity of *F. psychrophilum* without using various forms of stress or scarification. A number of studies have not resulted in mortality [4,5,32,33], while infections have been successfully established in other studies [2,7,8,34]. Even when mortality occurred, reproducibility has been a problem [35].

It is hard to directly compare the various studies due to differences in setup; including origin, size and health status of the fish, differences in bacterial strains and growth conditions, method of exposure and diversity in the experimental setup, including the number replicates and animals in each group. Since it is difficult to induce *F. psychrophilum* mortality without using stressors or compromising the outer barriers, it seems unlikely that immersion challenge will be standardized on the same level as seen with injections. Although mortality did not reach 60%, which is the desired goal for testing the potency of fish vaccines [36], the current model seems reproducible, since mortality was elevated in both full scale experiments.

In conclusion, the investigated model seems to be a good alternative to injections for studies requiring natural transmission of the pathogen without bypassing the outer barriers. Although the present study does not deal with the cause for the increased mortality induced by H_2O_2, it emphasizes and allows further investigation of the potential connection between routine non-medical treatments and pathogen outbreaks in aquaculture.

Acknowledgments

The technical assistance provided by Lene Gertman has been greatly appreciated as well as advice given regarding statistics by both Henrik Spliid and Claes Enø. Thanks to Kurt Buchmann and Per Walther Kania from the University of Copenhagen, Department of Veterinary Disease Biology, for allowing the use of their experimental facilities during the first three experiments. Finally, comments and suggestions given by Rune Lindgreen were also greatly appreciated.

Author Contributions

Proofreading: LM ID. Conceived and designed the experiments: MH LM ID. Performed the experiments: MH. Analyzed the data: MH. Contributed reagents/materials/analysis tools: LM ID. Wrote the paper: MH.

References

1. Nematollahi A, Decostere A, Pasmans F, Haesebrouck F (2003) *Flavobacterium psychrophilum* infections in salmonid fish. Journal of Fish Diseases 26: 563–574.
2. Madsen L, Dalsgaard I (1999) Reproducible methods for experimental infection with *Flavobacterium psychrophilum* in rainbow trout *oncorhynchus mykiss*. Diseases of Aquatic Organisms 36: 169–176.
3. Borg AF (1960) Studies on myxobacteria associated with diseases in salmonid fishes. American Association for the Advancement of Science 8: 1–85.
4. Madetoja J, Nyman P, Wiklund T (2000) Flavobacterium psychrophilum, invasion into and shedding by rainbow trout *Oncorhynchus mykiss*. Diseases of Aquatic Organisms 43: 27–38.
5. Busch S, Dalsgaard I, Buchmann K (2003) Concomitant exposure of rainbow trout fry to *Gyrodactylus derjavini* and *Flavobacterium psychrophilum*: Effects on infection and mortality of host. Veterinary Parasitology 117: 117–122.
6. Holt RA (1988) *Cytophaga psychrophila*, the causative agent of bacterial cold-water disease in salmonid fish. PhD thesis, Oregon State University, Corvallis.
7. Rangdale RE (1995) Studies on rainbow trout fry syndrome (RTFS). PhD thesis, University of Stirling, Stirling.
8. Aoki M, Kondo M, Kawai K, Oshima S (2005) Experimental bath infection with *Flavobacterium psychrophilum*, inducing typical signs of rainbow trout *Oncorhynchus mykiss* fry syndrome. Diseases of Aquatic Organisms 67: 73–79.

9. Jokumsen A, Svendsen LM (2010) Farming of freshwater rainbow trout in denmark. Charlottenlund. National Institute of Aquatic Resources, Technical University of Denmark. DTU Aqua report no.219–2010.
10. Pedersen LF (2010) Investigations of environmental benign aquaculture therapeutics replacing formalin (in Danish). DTU Aqua, Charlottenlund. DTU Aqua report, no.218–2010.
11. Sortkjær O, Bovbjerg P, Steenfeldt SJ, Bruun MS, Dalsgaard I, et al. (2000) Undersøgelse af eventuelle miljøpåvirkninger ved anvendelse af hjælpestoffer og medicin i ferskvandsdambrug samt metoder til at reducere/eliminere sådanne påvirkninger [in Danish]. DFU report 79–00: 206.
12. Avendaño-Herrera R, Magariños B, Irgang R, Toranzo AE (2006) Use of hydrogen peroxide against the fish pathogen *Tenacibaculum maritimum* and its effect on infected turbot (*Scophthalmus maximus*). Aquaculture 257: 104–110.
13. Thomas-Jinu S, Goodwin AE (2004) Acute columnaris infection in channel catfish, Ictalurus punctatus (Rafinesque): Efficacy of practical treatments for warmwater aquaculture ponds. Journal of Fish Diseases 27: 23–28.
14. Toyama T, Kita-Tsukamoto K, Wakabayashi H (1994) Identification of *Cytophaga psychrophila* by PCR targeted 16S ribosomal RNA. Fish Pathology 29: 271–275.
15. Wiklund T, Madsen L, Bruun MS, Dalsgaard I (2000) Detection of *Flavobacterium psychrophilum* from fish tissue and water samples by PCR amplification. Journal of Applied Microbiology 88: 299–307.
16. R Development Core Team (2011) R: A language and environment for statistical computing, reference index version 2.13.0. R Foundation for Statistical Computing, Vienna.
17. Tort MJ, Jennings-Bashore C, Wilson D, Wooster GA, Bowser PR (2002) Assessing the effects of increasing hydrogen peroxide dosage on rainbow trout gills utilizing a digitized scoring methodology. Journal of Aquatic Animal Health 14: 95–103.
18. Gaikowski MP, Rach JJ, Ramsay RT (1999) Acute toxicity of hydrogen peroxide treatments to selected lifestages of cold-, cool-, and warmwater fish. Aquaculture 178: 191–207.
19. Rach JJ, Schreier TM, Howe GE, Redman SD (1997) Effect of species, life stage, and water temperature on the toxicity of hydrogen peroxide to fish. Progressive Fish-Culturist 59: 41–46.
20. Arndt RE, Wagner EJ (1997) The toxicity of hydrogen peroxide to rainbow trout *Oncorhynchus mykiss* and cutthroat trout *Oncorhynchus clarki* fry and fingerlings. Journal of the World Aquaculture Society 28: 150–157.
21. Harper C, Wolf JC (2009) Morphologic effects of the stress response in fish. Ilar Journal 50: 387–396.
22. Kiemer MCB, Black KD (1997) The effects of hydrogen peroxide on the gill tissues of atlantic salmon, *Salmo salar* L. Aquaculture 153: 181–189.
23. Bruno DW, Raynard RS (1994) Studies on the use of hydrogen peroxide as a method for the control of sea lice on atlantic salmon. Aquaculture International 2: 10–18.
24. Bowers JM, Speare DJ, Burka JF (2002) The effects of hydrogen peroxide on the stress response of Atlantic salmon (*salmo salar*). J Journal of Veterinary Pharmacology and Therapeutics 25: 311–313.
25. Roque A, Yildiz HY, Carazo I, Duncan N (2010) Physiological stress responses of sea bass (*Dicentrarchus labrax*) to hydrogen peroxide (H_2O_2) exposure. Aquaculture 304: 104–107.
26. Speare DJ, Arsenault GJ (1997) Effects of intermittent hydrogen peroxide exposure on growth and columnaris disease prevention of juvenile rainbow trout (*Oncorhynchus mykiss*). Canadian Journal of Fisheries and Aquatic Sciences 54: 2653–2658.
27. Nematollahi A, Decostere A, Pasmans F, Haesebrouck F (2003) *Flavobacterium psychrophilum* infections in salmonid fish. Journal of Fish Disease 26: 563–574.
28. Covert JB, Reynolds WW (1977) Survival value of fever in fish. Nature 267: 43–45.
29. Raida MK, Buchmann K (2007) Temperature-dependent expression of immune-relevant genes in rainbow trout following *Yersinia ruckeri* vaccination. Diseases of Aquatic Organisms 77: 41–52.
30. Gräns A, Rosengren M, Niklasson L, Axelsson M (2012) Behavioural fever boosts the inflammatory response in rainbow trout *Oncorhynchus mykiss*. Journal of Fish Biology 81: 1111–1117.
31. Lorenzen E, Dalsgaard I, From J, Hansen EM, Hørlyck V, et al. (1991) Preliminary investigations of fry mortality syndrome in rainbow trout. Bulletin of the European Association of Fish Pathologists 11: 77–79.
32. Bruno DW (1992) *Cytophaga psychrophila* (= '*Flexibacter psychrophilus*') (borg), histopathology associated with mortalities among farmed rainbow trout, *Oncorhynchus mykiss* (Walbaum) in the UK. Eur Assoc Fish Pathol 12: 215–216.
33. Decostere A, Lammens M, Haesebrouck F (2000) Difficulties in experimental infection studies with *Flavobacterium psychrophilum* in rainbow trout (*Oncorhynchus mykiss*) using immersion, oral and anal challenges. Research in Veterinary Science 69: 165–169.
34. Chua FHC (1991) A study on the rainbow trout fry syndrome. MSc thesis, Institute of Aquaculture, University of Stirling, Stirling.
35. Garcia C, Pozet F, Michel C (2000) Standardization of experimental infection with *Flavobacterium psychrophilum*, the agent of rainbow trout *Oncorhynchus mykiss* fry syndrome. Diseases of Aquatic Organisms 42: 191–197.
36. Amend DF (1981) Potency testing of fish vaccines. Developments in Biological Standardization 49: 447–454.

8

Transcriptomic Analyses of Sexual Dimorphism of the Zebrafish Liver and the Effect of Sex Hormones

Weiling Zheng[1], Hongyan Xu[1], Siew Hong Lam[1], Huaien Luo[2], R. Krishna Murthy Karuturi[2], Zhiyuan Gong[1]*

1 Department of Biological Sciences, National University of Singapore, Singapore, Singapore, **2** Computational and Systems Biology, Genome Institute of Singapore, Singapore, Singapore

Abstract

The liver is one of the most sex-dimorphic organs in both oviparous and viviparous animals. In order to understand the molecular basis of the difference between male and female livers, high-throughput RNA-SAGE (serial analysis of gene expression) sequencing was performed for zebrafish livers of both sexes and their transcriptomes were compared. Both sexes had abundantly expressed genes involved in translation, coagulation and lipid metabolism, consistent with the general function of the liver. For sex-biased transcripts, from in addition to the high enrichment of vitellogenin transcripts in spawning female livers, which constituted nearly 80% of total mRNA, it is apparent that the female-biased genes were mostly involved in ribosome/translation, estrogen pathway, lipid transport, etc, while the male-biased genes were enriched for oxidation reduction, carbohydrate metabolism, coagulation, protein transport and localization, etc. Sexual dimorphism on xenobiotic metabolism and anti-oxidation was also noted and it is likely that retinol x receptor (RXR) and liver x receptor (LXR) play central roles in regulating the sexual differences of lipid and cholesterol metabolisms. Consistent with high ribosomal/translational activities in the female liver, female-biased genes were significantly regulated by two important transcription factors, Myc and Mycn. In contrast, Male livers showed activation of transcription factors Ppargc1b, Hnf4a, and Stat4, which regulate lipid and glucose metabolisms and various cellular activities. The transcriptomic responses to sex hormones, 17β-estradiol (E2) or 11-keto testosterone (KT11), were also investigated in both male and female livers and we found that female livers were relatively insensitive to sex hormone disturbance, while the male livers were readily affected. E2 feminized male liver by up-regulating female-biased transcripts and down-regulating male-biased transcripts. The information obtained in this study provides comprehensive insights into the sexual dimorphism of zebrafish liver transcriptome and will facilitate further development of the zebrafish as a human liver disease model.

Editor: Zhanjiang Liu, Auburn University, United States of America

Funding: This work was supported by the Singapore National Research Foundation under its Environmental & Water Technologies Strategic Research Programme and administered by the Environment & Water Industry Programme Office (EWI) of the PUB, grant number R-154-000-328-272. The funders had no role in study design, data collection and analysis, decision to publish, or reparation of the manuscript.

* E-mail: dbsgzy@nus.edu.sg

Introduction

The liver plays a critical role in the coordination of various physiological processes including digestion, metabolism, detoxification, biosynthesis of serum proteins, endocrine and immune response, etc. Because of the different metabolic needs for male and female reproduction, the liver is one of the most sexually dimorphic organs in terms of gene expression [1]. This is particularly prominent in oviparous species as the female liver is the main organ for production of yolk protein precursors (vitellogenins) and some zona pellucida proteins.

Recently, the zebrafish has emerged as models for liver diseases such as steatosis [2], alcoholic liver disease [3], polycystic liver disease [4], and tumorigenesis [5,6,7,8,9,10,11] as well as liver regeneration [12,13] and environmental hepatotoxicity [14,15]. Sex differences in the zebrafish transcriptome have been studied previously with the whole organism [16,17], gonads [17,18,19] or other organs [20,21]. Sexual dimorphism of gene expression in the liver has also been investigated in other fish species, including tilapia [22] and turbot [23]. Sexual dimorphism of hepatic response to dietary carbohydrate manipulation [24], brominated flame retardants [25] and perfluorononanoic acid (PFNA) [26] has also been reported in the zebrafish. One microarray-based study in zebrafish has indicated that female livers have higher levels of transcripts associated with translation, while the male up-regulated genes are associated with oxidative metabolism, carbohydrate metabolism, energy production, and amelioration of oxidative stress [24].

The available evidence indicates that sexual dimorphism in the liver is mediated via the sex hormones in both oviparous and viviparous animals [27]. In the present study, we intend to compare the transcriptomic difference between female and male livers in zebrafish using the deep sequencing technology. Our comparative transcriptomic analyses indicated functional differences in translation, carbohydrate metabolism, lipid and cholesterol metabolism, and xenobiotic metabolism between female and male zebrafish livers. Different gene expression regulatory networks for causing these differences were also identified.

Furthermore, we also used female and male sex hormones to treat both male and female zebrafish and found that male liver transcriptome was readily responsive to both female and male hormones while female livers were relatively resistant to the sex hormone perturbation. Thus, our transcriptomic data presented here should provide a molecular basis for a better understanding of the sexual dimorphism of zebrafish and facilitate proper experimental design in future studies.

Materials and Methods

Ethics Statement

All experimental protocols were approved by Institutional Animal Care and Use Committee (IACUC) of National University of Singapore (Protocol 079/07).

Zebrafish Treatment and Sample Collection

Four-month-old adult zebrafish were purchased from a local fish farm (Mainland Tropical Fish Farm, Singapore) and were acclimated for one week prior to experimental treatments. Fish were maintained based on the standard methods [28] and water quality was monitored daily in the zebrafish aquarium of Department of Biological Sciences, National University of Singapore. For experimental treatment, male and female fish were kept separately at room temperature (28±0.5°C) under 14 h of light and 10 h of dark cycle. 15 fish were placed in a 3-liter tank and were exposed for 48 hours in a static condition with 5 μg/L 17β-estradiol (E2) or 5 μg/L 11-keto testosterone (KT11) (Sigma-Aldrich); the relatively high concentrations of sex hormones used in the experiment was to ensure the response of the fish in the short, acute treatment. Water was changed daily with fresh sex hormone added. The same treatment was also conducted with 0.01% (v/v) DMSO as a vehicle control. A previous report indicated that 0.01% DMSO did not cause any developmental defect or induce stress protein expression in zebrafish embryos and it is within the range of recommended concentration for solvent controls in zebrafish experiments [29] There was no feeding and aeration during the short 48-hour of acute chemical exposure experiment and at the end of the exposure, the liver samples were collected for RNA isolation.

RNA-SAGE (Serial Analysis of Gene Expression) Sequencing and Data Processing

Total RNA was extracted from individual zebrafish livers using TRIzol® Reagent (Invitrogen) and treated with DNase I (Invitrogen) to remove genomic DNA contamination. Male and female fish were discerned by morphological appearance and confirmed by presence of testis and ovary after dissection. To ascertain the correct sex, RNA samples were further confirmed by real-time RT-PCR with *vtg1* primers (Forward: 5′-GGATTCCAGAGATCACAATGT-3′; Reward: 5′-CAGTACAGCAGTGGTCTAAT-3′) as female livers contained extraordinarily high level of vitellogenin mRNA. For RNA-SAGE sequencing, equal amount of total RNA from seven individual fish liver were pooled within the same experimental group for construction of 3′ SAGE libraries. The construction of SAGE libraries and sequencing

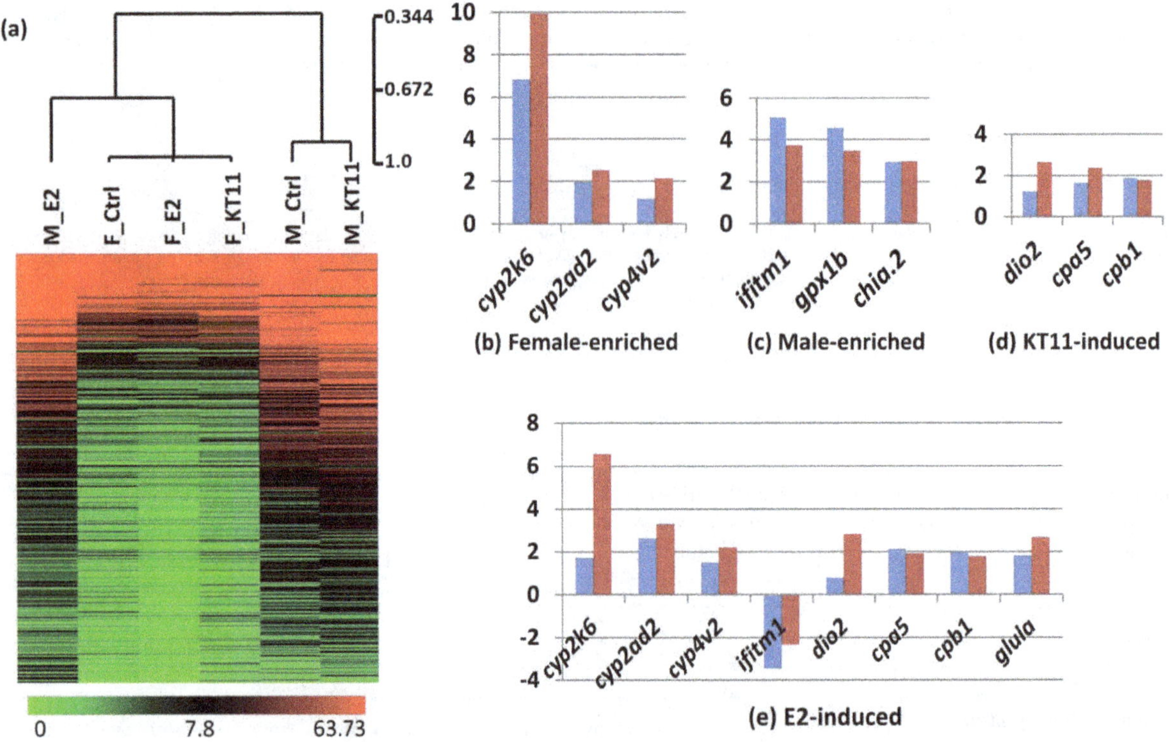

Figure 1. Hierarchical clustering of the six liver transcriptomes and real-time RT-PCR validation of RNA expression of selected genes. (a) Hierarchical clustering of the six liver transcriptomes from control males and females, E2-treated males and females, and KT11-treated males and females. The three female samples were closely clustered together, and E2-treated males was clustered with the female samples. Control males and KT11-treated males were clustered together and were distinct from the others. Scale bar represented the pearson correlation score. Heatmap was constructed with transcripts that showed significant differences in at least one comparison. (b–e) Real-time RT-PCR validation of transcripts enriched in the female liver (b), male liver (c), induced by KT11 (d) or E2 (e) in male livers. Fold changes (log2 base) measured by real-time RT-PCR (blue bars) are compared with those measured by RNA-SAGE sequencing (red bars).

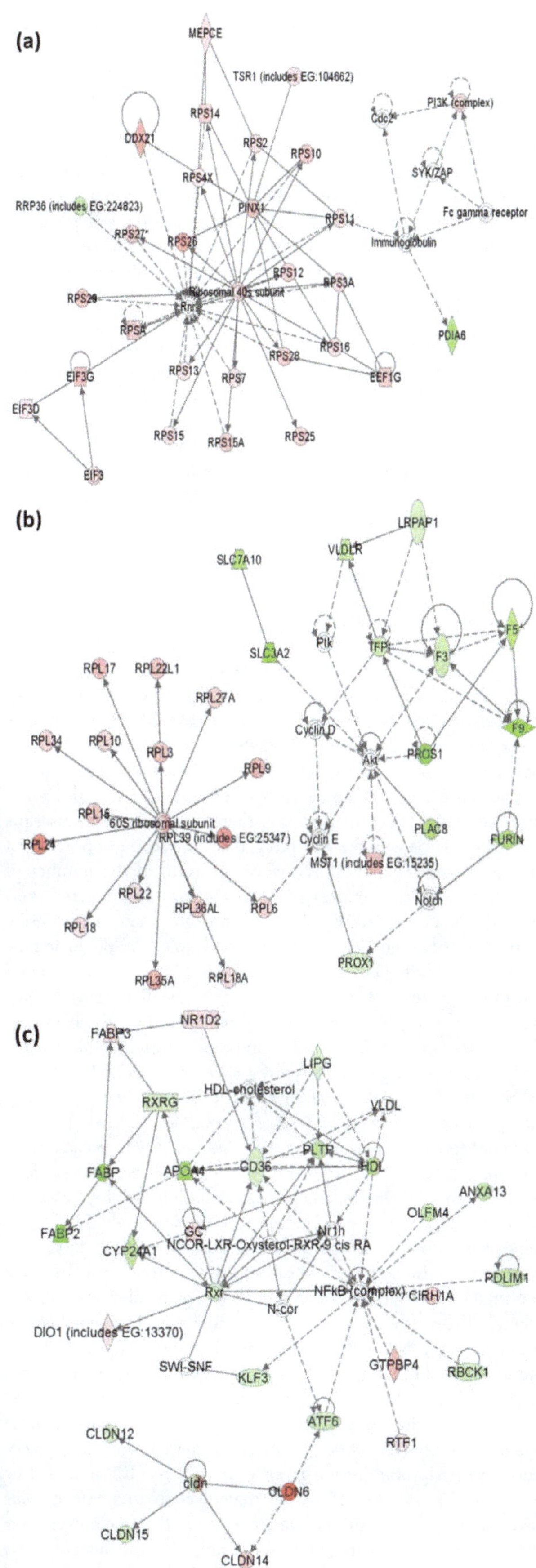

Figure 2. The top three networks as revealed by knowledge-based functional analyses of sex-biased genes. (a) The top network featured mainly by female-biased transcripts involving in gene expression, protein synthesis, RNA post-transcriptional modification (Score = 46). (b) The second network focuses on hematological system development and function, organismal functions, infectious disease (Score = 44). (c) The third network is associated with lipid metabolism, molecular transport, and small molecule biochemistry (Score = 39). Female-biased transcripts are indicated in red and male-biased transcripts in green.

were performed using SOLiD™ Analyzer 4 (Applied Biosystems) by Mission Biotech Co. Ltd, Taiwan according to manufacturer's protocol (Applied Biosystems SOLiD SAGE Guide). Briefly, mRNA was purified using Dynabeads® Oligo(dT) EcoP (Invitrogen) and subjected to cDNA synthesis. Synthesized cDNA was digested by NlaIII and EcoP15I, and sequencing adapters were added to the cDNA fragments after the digestion. A total of six SAGE libraries were sequenced: F_DMSO (control female), F_E2 (E2-treated female), F_11KT (11KT-treated female), M_DMSO (control male), M_E2 (E2-treated male), and M_11KT (11KT-treated male).

Sequence Tag Mapping and Annotation

50 nucleotides were sequenced from the SOLiD system and 27 nucleotide starting from CATG (based on the distance of EcoP15I recognition and cut sites) were used to map to the zebrafish Reference Sequence database (http://www.ncbi.nlm.nih.gov/RefSeq) with criteria of allowing maximum 2 nucleotide mismatches by taking account of sequencing errors and sequence polymorphism [30]. Uniquely mapped tag counts for each transcript were normalized to TPM (transcript per million) to facilitate comparison among different samples. In some cases, SAGE tags mapped to the same sequence ID were pooled.

Identification of Differentially Expressed Genes

Differentially expressed genes were identified using edgeR, a Bioconductor package for differential expression analysis of digital gene expression data [31,32,33]. edgeR estimates the genewise dispersions by conditional maximum likelihood, conditioning on the total count for that gene. This method models tag counts as negative binomial (NB) distributed to account for overdispersion in the digital gene expression data. Then a common dispersion is estimated for all tags. Finally, an exact test similar to Fisher's exact test is carried out to assess the differential expression for each gene. As we did not have replicates in this experiment, we followed the recommendation in the edgeR user guide based on the assumption that tag counts are not too small and a relatively small number of genes are differentially expressed. Genes with sum of tag counts in six samples less than 30 were removed and p value cut-off was set at 0.05.

Real-time RT-PCR Validation

To validate RNA-SAGE data, zebrafish from the same experiment but not used for RNA-SAGE sequencing were used for real-time RT-PCR analyses. For each condition, equal amount of total RNA from three fish livers were pooled and three biological replicates from a total of nine fish were used. For each biological replicate, five technical replicates were performed. Coefficient variance for the five technical replicates was smaller than 0.05 for each group. *Ef1a* was used as the housekeeping gene. Real-time PCR was performed using SYBR Green I Master (Roche) on the LightCycler 480 (Roche) following the manufacture's protocol.

(a)

	Transcription Regulator	Regulation z-score	P value
Female	MYCN	5.886	7.22E-19
	MYC	3.963	1.48E-05
Male	STAT4	2.693	3.86E-02
	HNF4A	2.017	2.17E-02
	PPARGC1B	2.006	1.42E-02

(b)

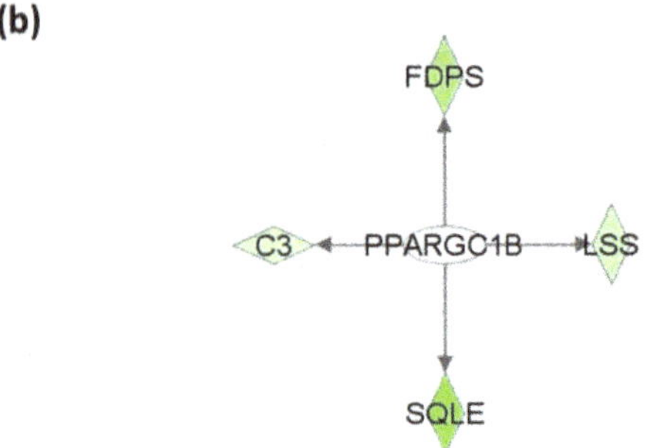

(c)

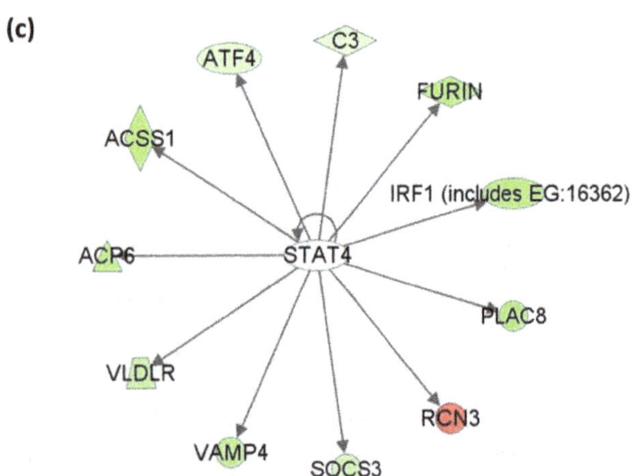

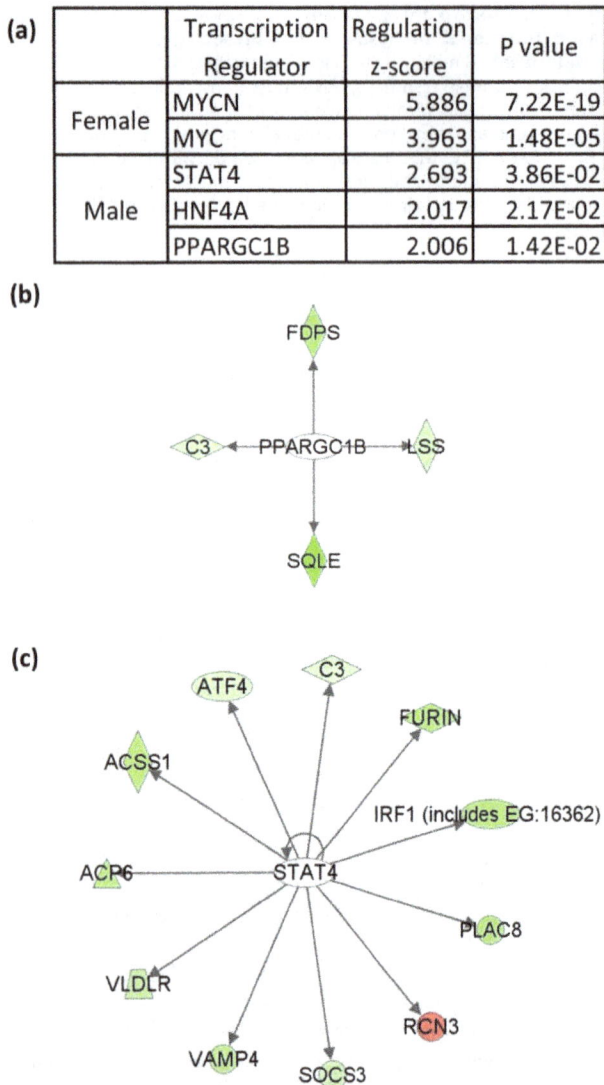

Figure 3. Significantly deregulated transcription factor networks. (a) List of sex-biased transcription factors in the zebrafish liver transcriptome. Regulation z-score indicates the degree of enrichment and p value indicates the level of significance. (b) Transcriptional targets network of *ppargc1b.* (c) Transcriptional targets network of *stat4.* Female- and male-biased genes are indicated in red and green, respectively. Non-colored genes are not in the sex-biased gene lists but are associated with the sex-biased genes and are introduced by the software to link up the network.

Gene Ontology Enrichment Analysis

Gene ontology enrichment analysis was performed using DAVID (The Database for Annotation, Visualization and Integrated Discovery) [34] with the total zebrafish genome information as the background and p-values representing a modified Fisher's exact t-test. Gene Ontology Fat categories were used for this analysis and the GO Fat attempts to filter out the broadest terms so that they will not overshadow the more specific terms. The p value cut-off was 0.05.

Ingenuity Pathway Analysis (IPA)

The sex-biased transcripts ($p<0.1$) were ranked by logarithm-transformed (base 10) p values. Positive values were given to female-biased transcripts, and negative values were given to male-biased transcripts. The ranked list was then analysed with IPA software (www.ingenuity.com) for pathway and transcription factors (TFs). The algorithm of IPA also incorporates other genes from the database to maximize the connectivity with the sex-biased transcripts to assemble a 'focus gene network'. Networks are limited to 35 molecules each to keep them to a functional size. Network scores were generated based on the hypergeometric distribution and were calculated with the right-tailed Fisher's exact test. In the networks, nodes represent biological entities (e.g. genes, proteins, and complexes) and edges represent interactions (e.g. induction, inhibition, binding, regulation, and phosphorylation) between nodes in the pathway. TF prediction by IPA was based on information compiled only from literature with experimental evidence. First, known targets of each transcription factor in the sex-biased transcript list were examined and compared with the direction of altered target gene expression, and finally a prediction for each transcription factor based on the direction of change was generated. The regulation z-score algorithm was used to make predictions. Only TFs with absolute z-score>2 and p-value<0.05 were considered as significantly enriched.

Results and Discussion

Overview of SAGE Sequencing Data

A total of six liver SAGE libraries were generated, including control female and male, E2-treated female and male, and KT11-treated female and male. For each library, between 11 and 20 million sequence reads were generated (**Table S1**). The read tags were mapped to the zebrafish RefSeq database with mapping efficiency from 29.0% to 43.4%. A total of 8,154 transcript entries were detected from the female liver, while 12,183 expressed genes from the male liver. In general, there were fewer transcripts detected in female livers probably due to the high portion of vitellogenin mRNA (78 %). However, in term of the number of genes more robustly expressed with at least 30 sequence tags detected in both SAGE libraries, the numbers were quite comparable between female and male samples (4,386 in female and 4,548 in male) (**Table S1**).

Among the top 50 expressed genes in the female and male livers, there were 34 genes overlapping, including three vitellogenin genes (*vtg1, vtg4, vtg5*), 11 ribosomal protein genes (*rplp0, rps25, rps15a, rps10, rps27a, rpl7, rps29, rpl31, rpl6, rps27.1, rpl10*), six complement factor genes (*fga, fgg, c3a, LOC100334885, cfb, plg*), four lipoprotein genes (*fabp10a, apobl, apoa2, apoc1l*), two glycoprotein genes (*tfa, ahsg*), and a few others (*il7r, a2m, tpt1, mibp2, gpx4a* and three unannotated genes). In the female liver, the top four transcripts were all vitellogenin genes, including *vtg1, vtg4, vtg5* and *vtg2*. In the top 50 expressed transcripts in the female liver, 23 of them were ribosomal protein genes. On the other hand, in the top 50 expressed transcripts in the male liver, there were only 12 ribosomal protein genes, indicating a strong translational activity in female livers. It is also interesting to note that three vitellogenin transcripts (*vtg1, vtg4, vtg5*) also appeared in the top 50 list, implying potentially yolk-unrelated functions of vitellogenins, such as lipid transportation [35] and anti-microbes [36].

Hierarchical clustering of the six samples showed that the three female samples were closely clustered together, indicating very minor transcriptome perturbations caused by E2 and KT11 treatments (**Figure 1a**). However, male liver treated with E2 was clustered with the female samples, suggesting that male liver after E2 treatment showed a higher similarity with the female liver. Control male liver and KT11-treated male liver were clustered together and were distinct from the others.

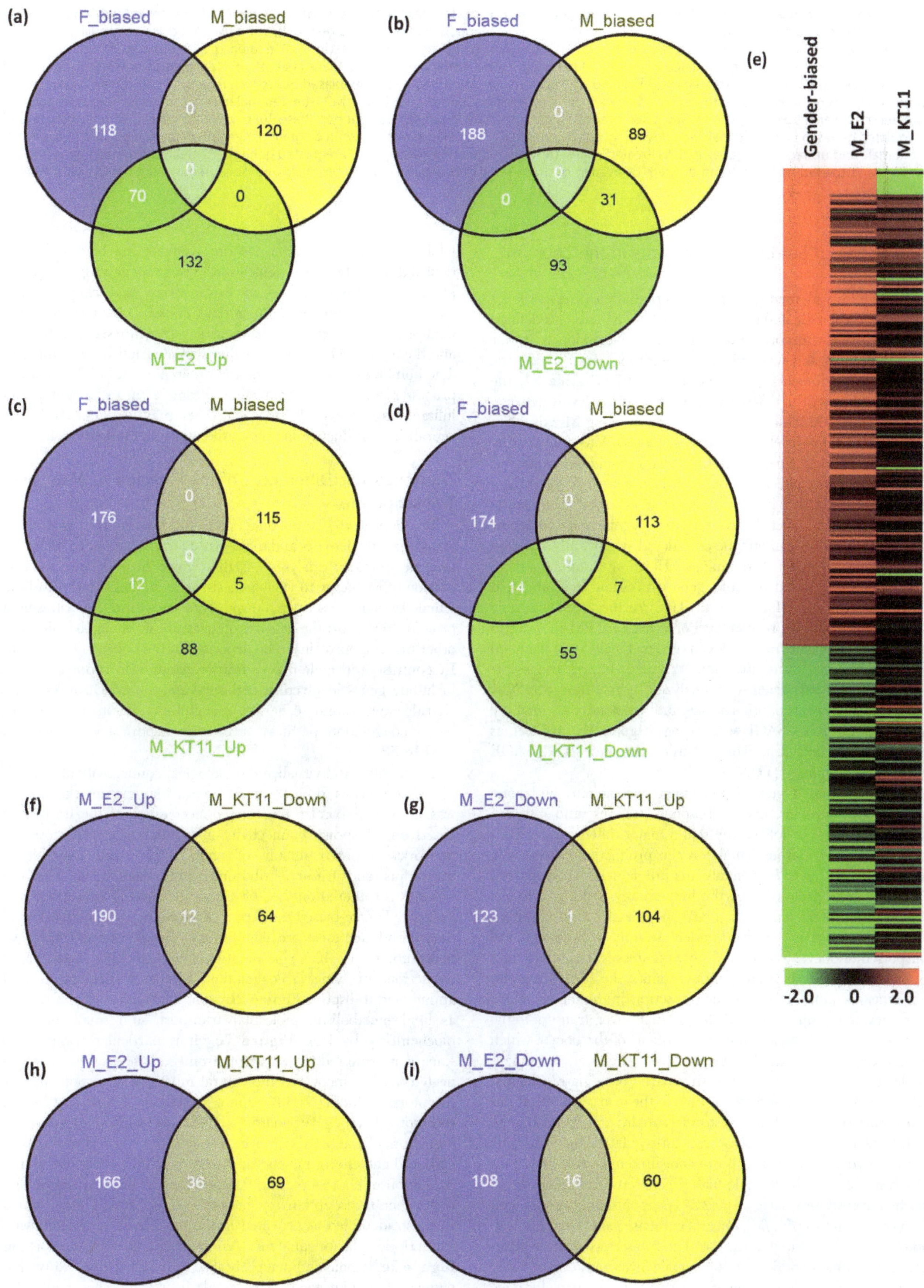
(a)
F_biased
M_biased
118
0
120
0
70
0
132
M_E2_Up
(b)
F_biased
M_biased
188
0
89
0
0
31
93
M_E2_Down
(c)
F_biased
M_biased
176
0
115
0
12
5
88
M_KT11_Up
(d)
F_biased
M_biased
174
0
113
0
14
7
55
M_KT11_Down
(e)
Gender-biased
M_E2
M_KT11
-2.0
0
2.0
(f)
M_E2_Up
M_KT11_Down
190
12
64
(g)
M_E2_Down
M_KT11_Up
123
1
104
(h)
M_E2_Up
M_KT11_Up
166
36
69
(i)
M_E2_Down
M_KT11_Down
108
16
60

Figure 4. Intersections of differentially expressed genes in the male zebrafish livers after E2 or KT11 treatment with sex-biased genes. (a) Venn diagram of up-regulated genes in E2-treated males overlapped with sex-biased genes. (b) Venn diagram of down-regulated genes in E2-treated males overlapped with sex-biased genes. (c) Venn diagram of up-regulated genes in KT11-treated males overlapped with sex-biased genes. (d) Venn diagram of down-regulated genes in KT11-treated males overlapped with sex-biased genes . (e) Heat map of expression changes of sex-biased genes in E2- and KT11-treated male livers. For the column corresponding to the sex-biased transcripts, red represents female-biased genes and green represents male-biased genes. For the two columns corresponding to the E2- and KT11-treated male livers, red represents up-regulation and green represents down-regulation. The color intensity is calculated by logarithm-transformed (base 10) p-value. (f) Venn diagram of overlap of up-regulated genes in E2-treated male and down-regulated genes in KT11-treated males. (g) Venn diagram of overlap of down-regulated genes in E2-treated male and up-regulated genes in KT11-treated males. (h) Venn diagram of overlap of up-regulated genes in E2-treated male and up-regulated genes in KT11-treated males. (i) Venn diagram of overlap of down-regulated genes in E2-treated male and down-regulated genes in KT11-treated males.

Identification of Sex-biased Transcripts in the Zebrafish Liver

An MA plot was used to display differentially expressed genes between control male and female livers, where M-axis was defined as the logarithm-transformed fold change of expression levels for each gene and A-axis was defined as logarithm-transformed gene expression level for each gene. As illustrated in **Figure S1**, the log2-ratio-range of the differential gene expression levels between control female and male livers was decreased along M-axis when the gene expression levels were increased along A-axis, indicating that the fold change of gene expression levels should not be used as a sole significant indicator for identification of differentially expressed genes. This is because a small fold change of abundantly expressed genes can alter the transcriptome significantly while low abundant transcripts can only be considered significantly changed when having a higher fold change. Here we used edgeR to calculate p value based on both gene expression level and fold change and the cut-off was set at 0.05. Furthermore, we only included genes with expression level of at least 2 TPM in order to capture physiologically more relevant genes. Finally, we identified 186 female-biased transcripts and 121 male-biased transcripts (**Table S2**). Selected genes were analyzed by real-time RT-PCR from independent sets of fish samples and our data confirmed the accuracy of the RNA-SAGE sequencing (**Figure 1b and 1c**), as we previously reported in zebrafish using the same RNA-SAGE sequencing technology [**11**].

The female-biased gene list included vitellogenins and zona pellucida glycoproteins, many ribosomal proteins, and estrogen receptor 1 (*esr1*). It is noteworthy that a major difference between oviparous and viviparous females is the production of egg yolk proteins in the liver. In sexually mature females of oviparous animals, a large proportion of the liver energy is devoted to the synthesis of vitellogenins, which are specific glycolipo-phosphoproteins produced in the liver under estradiol stimulation, and transported to oocyte as maternal stored materials to support early development. In our transcriptome data, the 8 vitellogenin transcripts account for 78% of the transcripts in the female livers, while they make up only 0.3% of the male liver transcriptome. Zona pellucidas are membrane glycoproteins of the oocyte which are essential for oocyte fertilization [37]. Another prominent female biased gene, *nots* encodes an aspartic proteinase, which is synthesized in the liver and transported to the ovary [38,39]. It has been reported as one of the most differentially up-regulated gene by E2 treatment [40] and is also up-regulated by polycyclic aromatic hydrocarbons and alkylphenols in female fish [41].

In contrast, the list of male-biased gene contains many liver-enriched transcripts, such as *fabp2*, apolipoprotein (*apoa4*), and complement factors (*f9b*, *f3b*). Several chitinase genes (*chia.2*, *chia.1*, and *chia.3*) also appeared in the list. The transcripts of these genes were either not identified in the female liver or at marginally detectable level. It is interesting to note that chitinase proteins have been previously identified in zebrafish ovarian follicle and testis, while the corresponding transcript counterparts have not been detected in fully-grown follicles or testis, suggesting that these proteins were synthesized in extraovarian and extratesticular tissues and transported to the gonads [39,42]. Our transcriptome data provided evidence that chitinases were synthesized in the liver and the male fish might have higher synthetic activity of chitinases. The human homologs of chitinases are important in inflammatory processes and tissue remodeling during arthritis, asthma, and inflammatory bowel disease [43]. They may act similarly in the zebrafish reproductive organs in the anti-inflammatory defense.

Transcriptomic Differences of the Female and Male Liver Transcriptome

To depict the functional constitutions of the sex-biased transcripts, the female- and male- biased transcripts were subjected to gene ontology analyses. As the female liver devotes a large portion of its energy to vitellogenesis, gene ontology analysis of the female-biased transcripts showed high enrichment of ribosomal proteins to support the active protein synthesis of vitellogenins and other maternal proteins to be deposited to the oocytes (**Table S3**). In contrast, the male-biased transcripts showed more diversity, including genes involved in carbohydrate, aminoglycan, polysaccharide and monosaccharide metabolism, chitin metabolism, blood coagulation pathway, protein transport and localization (**Table S3**).

To discover and visualize the biological connectivity of the sex-biased transcripts ($p<0.1$), the IPA network generating algorithm was used to maximize the interconnectedness of the transcripts based on all known connectivity in the database. The top two networks consisted mainly of only female- and male-biased transcripts and featured ribosome and coagulation. The top network is centered on the 40S ribosomal subunit with essentially exclusive female-biased transcripts (**Figure 2a**). The network with the second highest score featured the female-biased transcripts constituting the 40S ribosomal subunit and the male-biased transcripts involving in coagulation, fatty acid metabolism, and amino acid transport (**Figure 2b**). The third network was named as lipid metabolism, molecular transport, and small molecule biochemistry by IPA (**Figure 2c**). It is particularly interesting since it is consisted of transcripts from both sexes well interconnected, and it uncovered the central functions of several nuclear receptors, including NFkB, retinol x receptor (RXR) and liver x receptor (LXR). Particularly, RXR and LXR are critical regulators of the sex difference of liver transcriptome in terms of lipid and cholesterol metabolism [44,45]. RXR participates in the regulation of cholesterol and fatty acid metabolism by interacting with other transcription regulators such as PPARG [46], which was also identified as activated in the male liver in the following transcription factor analysis. As indicated in the network in **Figure 2c**, female and male zebrafish liver has distinct expression pattern of claudin genes. The female liver is enriched with *cldn6* and *cldn14*, while the male liver is enriched with *cldn12* and *cldn15*.

Table 1. Top 50 expressed genes in the female and male livers.

	Female			Male		
	GI	Gene Symbol	TPM	GI	Gene Symbol	TPM
1	166795886	*vtg1*	475563.6	303304953	*fga*	117924.7
2	160420305	*vtg4*	188213.4	164698522	*il7r*	69007.6
3	68448529	*vtg5*	91869.6	227430294	*itln3*	26027.1
4	113678457	*vtg2*	24067.4	288856245	*LOC100003647*	25947.3
5	23308626	*fabp10a*	11708.9	292610076	*c3a*	22169.3
6	18859322	*rplp0*	8113.4	23308626	*fabp10a*	21888.8
7	62632716	*tfa*	7990.9	62632716	*tfa*	20925.9
8	303304953	*fga*	6546.7	292611118	*LOC100329302*	19307.8
9	41387125	*rps25*	4438.9	47085768	*fgg*	18839.7
10	164698522	*il7r*	4201.1	55925455	*gpx4a*	18297.0
11	47086524	*rps15a*	4142.4	166795886	*vtg1*	16449.5
12	41055645	*rps10*	3875.2	18859322	*rplp0*	12649.9
13	41055515	*rps27a*	3845.6	292626490	*LOC100334885*	11125.8
14	47271397	*rpl7*	3825.0	80751158	*zgc:123103*	10854.0
15	37700236	*tpt1*	3691.2	71834285	*apobl*	10589.9
16	47086132	*rps29*	3657.0	115529392	*zgc:152945*	9281.5
17	50540043	*rpl35a*	3525.7	189526818	*si:ch211-270n8.1*	8857.7
18	91176309	*rpl31*	3081.7	292621451	*a2m*	8826.5
19	51467910	*rpl6*	2970.3	18858246	*cfb*	8774.4
20	229335594	*rpl39*	2903.1	160420305	*vtg4*	8281.4
21	292610076	*c3a*	2869.9	47086792	*ahsg*	7600.9
22	41152198	*rps26l*	2712.4	221307487	*si:ch1073-126c3.2*	7303.7
23	80751158	*zgc:123103*	2704.1	41055515	*rps27a*	6890.0
24	47086792	*ahsg*	2623.0	41393104	*plg*	6781.0
25	226958508	*rps27.1*	2442.6	37700236	*tpt1*	6168.5
26	41053336	*rpl10*	2342.7	41387125	*rps25*	6006.3
27	48597013	*rpl3*	2314.2	47086524	*rps15a*	5513.9
28	41152463	*rps14*	1946.1	194578932	*apoa2*	5457.1
29	38564414	*mibp2*	1917.0	292622269	*apoc1l*	5214.0
30	55925455	*gpx4a*	1849.3	42734413	*igfbp1a*	4639.7
31	51010974	*rpl15*	1791.9	41055645	*rps10*	4622.5
32	189526818	*si:ch211-270n8.1*	1742.3	91176309	*rpl31*	4480.5
33	71834285	*apobl*	1724.2	31340752	*krt18*	4476.4
34	229335607	*rps28*	1669.9	300934727	*ambpl*	4421.7
35	41054304	*rpl5a*	1590.7	41152438	*rpl10a*	4332.7
36	50053845	*rps12*	1544.6	51467910	*rpl6*	4324.0
37	91176291	*rpl32*	1507.5	226958508	*rps27.1*	4133.9
38	160333704	*vtg2*	1476.9	47271397	*rpl7*	4085.6
39	292626490	*LOC100334885*	1457.1	121583580	*serpinf2b*	4047.3
40	292622269	*apoc1l*	1442.1	47086132	*rps29*	3994.1
41	61657912	*serp1*	1384.5	292622268	*LOC570585*	3894.7
42	288856245	*LOC100003647*	1378.8	38564414	*mibp2*	3870.6
43	61806481	*rps17*	1266.9	54261746	*ucp1*	3790.1
44	47085768	*fgg*	1255.5	41053336	*rpl10*	3596.7
45	292621451	*a2m*	1235.5	68448529	*vtg5*	3570.2
46	47086532	*ppia*	1126.1	162287364	*hpx*	3546.2
47	18858246	*cfb*	1103.2	50539723	*uox*	3452.2
48	194578932	*apoa2*	1062.8	40538763	*cp*	3441.9

Table 1. Cont.

	Female			Male		
	GI	Gene Symbol	TPM	GI	Gene Symbol	TPM
49	50344933	*rpl11*	1020.0	113951770	*crp*	3213.3
50	41393104	*plg*	1013.9	33504508	*serpinc1*	3011.7

doi:10.1371/journal.pone.0053562.t001

Claudins are critical components of tight junctions which regulate paracellular permeability and polarity. CLDN6 could be up-regulated by ERα activation [47], which is consistent with our observation that *cldn6* is highly female-biased.

Sexual Dimorphism of Zebrafish Liver on Xenobiotic Metabolism and Anti-oxidation

It has been known that sexual dimorphism of the liver differentially affects the metabolism of drugs and toxins in both men and women [48]. CYPs are a large and diverse group of enzymes to catalyze the oxidation of organic substances and it has been found that the difference in CYP expression pattern is one of the major reasons resulting in differential drug metabolism between men and women [48,49]. In the present study, we also identified several *cyp* transcripts biased either in the female or male zebrafish liver, including female-biased *cyp2k6, cyp2ad2, cyp4v2* and *cyp2aa4* and male-biased *cyp2aa2* and *cyp3c1l2*. Interestingly, all the four female-biased cyp genes were up-regulated in E2-treated male liver (**Table S4**), suggesting that these genes are E2-responsive. *Cyp2k6,* which is the most significant female-biased *cyp*, was demonstrated to catalyze the activation of the myco-toxin aflatoxin B1 (AFB1) to the carcinogenic exo-8,9-AFB1epoxide, and it is only expressed in the liver and ovary of the zebrafish [50]. It has been shown that the zebrafish is much more resistant to AFB1-induced carcinogenesis than the rainbow trout and the female fish is more susceptible to AFB1-induced DNA damage [51]. Our data probably provide a basis for this as male fish basically has no *cyp2k6* expression and they would be naturally resistant to AFB1, while female fish are more sensitive to AFB1 due to the higher level of *cyp2k6* expression.

Besides cyps, an antioxidant gene, *gpx1,* is expressed in male liver significantly higher than in female liver. Reactive oxygen species, such as superoxide and hydrogen peroxide, are generated in the mitochondrions and during enzymatic reactions. It can cause oxidative damage to DNA, proteins, and membrane lipids. Gpx1 is an intracellular antioxidant enzyme that enzymatically reduces hydrogen peroxide to water to limit its harmful effects. The higher level of *gpx1* in the male liver may be correlated with the higher mitochondrial activities and oxidoreduction. The gene encoding another enzyme which is known to reduce oxidative stress, *txnl4a,* was also enriched in the male liver.

Differential Transcriptional Networks in the Female and Male Livers

The intrinsic difference between female and male livers is also reflected by their different regulatory networks. Transcription factor (TF) analysis was conducted using female- and male-biased transcripts ($p<0.1$) to identify the transcription factors that may be responsible for the sex-biased gene expression profiles (**Figure 3a**). Different from previous methods which use predictive binding sites, IPA TF prediction is based on experimentally derived relationships from published literature, thus enabling to focus on important interactions with greater confidence in prediction. A transcription factor was predicted to be activated if z-score≥2 and inhibited if z-score≤-2, and the significance was defined by p value≤0.05. Our analysis indicated that Myc and Mycn are the only activated TFs in the female-biased transcripts. The majority of Myc-regulated female-biased transcripts are involved in ribosomal function [52]. Estrogen has been known to induce Myc expression by activating an upstream enhancer [53,54,55]. The activated TFs in male livers included Ppargc1b, Hnf4a and Stat4. Ppargc1b is involved in fat oxidation, non-oxidative glucose metabolism, and the regulation of energy expenditure [56,57,58] (**Figure 3b**). Hnf4a is best known as a master regulator of liver-specific gene expression, especially those involved in lipid transport and glucose metabolism [59,60]. It has also been reported to contribute to the sex-specific expression of mouse liver *Cyp* genes by positively regulating male-specific genes [61]. The targets of Stat4 include Jak-Stat signaling transducer Irf1 and negative regulator Socs3 (**Figure 3c**). STATs (signal transducers and activators of transcription) are a family of TFs which could be activated by JAK (Janus kinase). The Jak-STAT signaling pathway has been implicated in a variety of cellular functions in the liver, such as liver metabolism and differentiation, antiviral defense in hepatitis virus infection, as well as liver injury and regeneration [62]. However, the function of Stat4 in liver is thus far poorly studied.

Transcriptomic Changes Induced by Sex Hormone Treatment

The liver is highly sensitive to sex hormone disturbance. Previous studies have been conducted using either estrogen or androgen to examine the effect of sex hormone on liver transcriptome [40,63,64,65,66,67,68,69,70], this study is the first to directly compare the effect of estrogen and androgen in female and male. Genes whose expression levels have been significantly altered by E2 or KT11 treatment in the female or male zebrafish livers were identified using edgeR using cutoff at p value<0.05. The alterations of RNA expression of several genes were further confirmed by real-time PCR from independent RNA samples after the same sex hormone treatment (**Figure 1d and 1e**).

Overall, the female zebrafish liver transcriptome did not show much change after E2 or KT11 treatment, as judged from a small number of genes significantly affected (**Table S4**). In E2-treated female livers, only five genes were up-regulated and 11 down-regulated. In KT11-treated female livers, 10 genes were up-regulated and 12 down-regulated. In contrast, both E2 and KT11 treatment induced dramatic transcriptome changes in male livers. In E2-treated male livers, 202 genes were up-regulated, 70 of which belonged to the female-biased transcripts and none of them was male-biased gene; 124 genes were down-regulated, 31 of which were male-biased transcripts and none of them was female-biased gene. Thus, 37.2% of female-biased transcripts and 25.8% of male-biased transcripts were responsive to E2 treatment of the

	Functions	GO Categories	M_E2	M_KT11
Female	Ribosome/ Translation	GO:0006412~translation		
		GO:0006414~translational elongation		
		GO:0003735~structural constituent of ribosome		
		GO:0003743~translation initiation factor activity		
		GO:0003746~translation elongation factor activity		
		GO:0005840~ribosome		
		GO:0030529~ribonucleoprotein complex		
		GO:0043232~intracellular non-membrane-bounded organelle		
		GO:0043228~non-membrane-bounded organelle		
		GO:0005198~structural molecule activity		
		GO:0003723~RNA binding		
		GO:0008135~translation factor activity, nucleic acid binding		
		GO:0033279~ribosomal subunit		
		GO:0051726~regulation of cell cycle		
	Estrogen pathway	GO:0043627~response to estrogen stimulus		
		GO:0032355~response to estradiol stimulus		
		GO:0048545~response to steroid hormone stimulus		
		GO:0009725~response to hormone stimulus		
		GO:0009719~response to endogenous stimulus		
	Lipid transportation	GO:0005319~lipid transporter activity		
	Pigment biosynthesis	GO:0046148~pigment biosynthetic process		
		GO:0042440~pigment metabolic process		
		GO:0009113~purine base biosynthetic process		
	Sugar binding	GO:0005529~sugar binding		
Male	Oxidation reduction	GO:0055114~oxidation reduction		
		GO:0016052~carbohydrate catabolic process		
		GO:0044275~cellular carbohydrate catabolic process		
		GO:0046164~alcohol catabolic process		
		GO:0005996~monosaccharide metabolic process		
	Polysaccharide matabolism	GO:0006030~chitin metabolic process		
		GO:0006032~chitin catabolic process		
		GO:0008061~chitin binding		
		GO:0004568~chitinase activity		
		GO:0000272~polysaccharide catabolic process		
		GO:0005976~polysaccharide metabolic process		
		GO:0030247~polysaccharide binding		
		GO:0006026~aminoglycan catabolic process		
		GO:0006022~aminoglycan metabolic process		
		GO:0009057~macromolecule catabolic process		
		GO:0001871~pattern binding		
		GO:0030246~carbohydrate binding		
	Coagulation	GO:0050817~coagulation		
		GO:0007596~blood coagulation		
		GO:0050878~regulation of body fluid levels		
		GO:0007599~hemostasis		
		GO:0042060~wound healing		
		GO:0010038~response to metal ion		
		GO:0010035~response to inorganic substance		
	Protein localization	GO:0015031~protein transport		
		GO:0008104~protein localization		
		GO:0045184~establishment of protein localization		

Figure 5. Changes of sex-biased GO categories in E2- and KT11-treated male zebrafish liver. Female-biased GO categories are in yellow shade and male-biased GO categories are in blue shade. GO categories under different classification (biological process, molecular function, and cellular component) are combined into functional categories to facilitate the interpretation. Red color represents up-regulation of the GO category and green color represents down-regulation of the GO category. It is clear that most of the female-biased GO categories were up-regulated in E2-treated males and down-regulated in KT11-treated males, while most of the male-biased GO categories were not affected.

male fish (**Figure 4a and 4b**), indicating that E2 treatment strongly feminized male liver. In the KT11-treated male liver, 105 transcripts were up-regulated and 76 transcripts were down-regulated. However, the overlapping of these differentially expressed genes with sex-biased genes was quite small (**Figure 4c and 4d**), suggesting that the KT11-induced liver transcriptomic changes are not sex-orientated. We also examined the change of all sex-biased transcripts in E2- and KT11-treated male livers (**Figure 4e**). The heat map showed that most of the female-biased transcripts were up-regulated (86.4%) and male-biased transcripts were down-regulated (58.3%) in the E2-treated male livers, while they did not show a very clear trend in the KT11-treated male livers. Furthermore, there were 12 genes up-regulated in E2-treated males but down-regulated in KT11-treated males (**Figure 4f**), including all the eight *vtgs*, *nots*, *gc*, *rpl39* and *rpl23a*. However, only one gene (*fabp6*) was down-regulated by E2 and up-regulated by 11KT (**Figure 4g**). Surprisingly, there were larger numbers of transcripts showing expression change at the same direction with either E2 or KT11 treatment; 36 genes were up-regulated by both E2 and KT11 (**Figure 4h**), and 16 genes were down-regulated by both E2 and KT11 (**Figure 4i**). These genes could be common targets of E2 and KT11 and several of these are involved in xenobiotic metabolism (*cpa1, cpa4, cpa5* and *cpb1*) and glutamate metabolism (*glub, sat2*).

E2 treatment up-regulated *vtg* expression in both female and male zebrafish livers. In the control female liver, the total number of *vtg* (*vtg1–7*) transcripts constitutes 78% of the transcriptome body. This high expression level was further up-regulated by E2 treatment to 85%. In the male liver, *vtgs* were only expressed at the level of 0.3 % of total transcriptome, while E2 treatment dramatically increased this by over 300 fold to 18.6%. Vitellogenin has been long recognized as a biomarker for environmental estrogenic compounds [71,72]. However, *vtg* level in the female liver after KT11 treatment was not affected. Similar results have been reported by Hoffman *et. al.*, where female zebrafish treated with 17α-methyldihydro-testosterone (MDHT), a model androgen, showed no significant change in the level of *vtg* RNA [66]. In contrast to the female liver, the male liver is very sensitive to sex hormonal disturbance. Besides estrogenic compounds, the male liver could be feminized by various acute injuries [73,74] and disruption of certain signaling pathways including Wnt signaling and the cytochrome P450 reductase system [75,76]. Therefore, the male liver is predisposed to sexual dimorphic gene alterations in response to a wide variety of perturbations. It is also worthy to note that the current set of experiments employed fish directly from a local fish farm and these fish may had been exposed to a low level of estrogenic compounds from water and/or feed, as indicated by a relatively high level of *vtg* mRNAs in male control fish (**Table 1**). Nevertheless, we still observed a strong response of the male fish to E2 treatment in our current experiment.

As summarized in **Figure 5**, the female-biased genes are mainly involved in ribosome/translation, estrogen pathway, lipid transportation, pigment biosynthesis and sugar binding. In contrast, the male-biased genes are involved in Oxidation reduction, Polysaccharide metabolism, Coagulation and Protein localization. Gene ontology analysis of E2 up-regulated genes in male livers showed that the most abundant category was oxidation reduction, including several *cyps*, mitochondrial genes and various enzymes involved in carbohydrate, amino acid, and hydroxysteroid metabolisms (**Table S5**). In addition, many other prominent up-regulated categories were female-biased categories, such as response to estrogen stimulus, translation, and lipid transport (**Figure 5, Table S5**). Meanwhile, another small group of oxidoreduction-related transcripts was also down-regulated, including several *cyps*, heme xygenase (*hmox1*), eosinophil peroxidase (*epx*), and several enzymes involving in retinol and hydroxysteroid metabolisms. The bi-directional change of genes involved in oxidoreduction indicates that there were major shifts in the cellular metabolisms induced by E2 treatment.

Gene ontology enrichment analysis showed that up-regulated categories in KT11-treated male livers included proteolysis, immune response, lipid binding and catecholamine metabolic process (**Table S5**). Genes involved in immune response include major histocompatibility complex (MHC) genes and chemokines. In contrast, down-regulated categories were translation, lipid transport and response to estrogen stimulus, suggesting that KT11 further down-regulated female-biased categories (**Figure 5**). Interestingly, MHC genes showed both up- and down-regulation by KT11, indicating an alteration of antigen processing and presentation in the liver.

As indicated in our transcriptomic data, the effect of androgen treatment on the male livers was not sex-oriented. It is believed that liver masculinization is regulated predominantly by indirect androgen effects rather than by direct androgen receptor-ligand binding [77]. The sexual difference of liver transcriptome is controlled by the hypothalamic-pituitary-adrenal system. Catecholamines, especially epinephrine, play a crucial role in stress response and can modulate the basic function of immune cells in terms of proliferation, differentiation and production of cytokines. Catecholamines can influence the hepatic inflammatory response by altering hepatic blood flow through vasospasm and centrilobular hypoxia. Plasma catecholamine level is elevated in case of acute liver failure and acute and chronic hepatic inflammation [78]. Our result showed that administration of extra KT11 to the male fish up-regulated the catecholamine metabolism, which may provide a positive feedback cycle.

In summary, this study presented a comprehensive description of liver transcriptomes of the female and male zebrafish. We identified a list of sex-biased transcripts, which provides comprehensive information on sexual dimorphism of the zebrafish livers in term of functional categories, interactive networks and regulatory molecules. Our results from the sex hormone treatment experiments showed that the female zebrafish liver is relatively insensitive to sex hormone perturbation, while the male liver is highly responsive to both E2 and KT11 treatments. E2 feminizes the male liver by up-regulating many female biased genes that cause increase of vitellogenins synthesis and change of cellular metabolisms. In contrast, KT11 modulates the hypothalamic-pituitary-adrenal system, which imposes secondary effect on the liver.

Supporting Information

Figure S1 0MA plotting of the transcripts from control male and female samples. M-axis is defined as the logarithm-transformed fold change of expression levels for each gene between control female and male liver while A-axis is defined as logarithm-transformed gene expression level for each gene. Red dots represented statistically significant differentially expressed genes between female and male liver (p-value<0.05).

Table S1 Summary of RNA-SAGE sequencing results.

Table S2 Lists of female- and male-biased genes in the zebrafish liver transcriptome.

Table S3 Gene ontology enrichment analysis of female- and male-biased transcripts.

Table S4 Lists of differently expressed transcripts in the female and male liver by E2 or KT11 treatment.

Table S5 Gene ontology enrichment analysis of E2- and KT11-induced transcriptome changes in the male liver.

Author Contributions

Conceived and designed the experiments: HX SHL ZG. Performed the experiments: HX WZ. Analyzed the data: WZ SHL ZG. Contributed reagents/materials/analysis tools: HL RKMK. Wrote the paper: WZ ZG.

References

1. Yang X, Schadt EE, Wang S, Wang H, Arnold AP, et al. (2006) Tissue-specific expression and regulation of sexually dimorphic genes in mice. Genome Res 16: 995–1004.
2. Cinaroglu A, Gao C, Imrie D, Sadler KC (2011) Activating transcription factor 6 plays protective and pathological roles in steatosis due to endoplasmic reticulum stress in zebrafish. Hepatology.
3. Howarth DL, Passeri M, Sadler KC (2011) Drinks like a fish: using zebrafish to understand alcoholic liver disease. Alcohol Clin Exp Res 35: 826–829.
4. Gao H, Wang Y, Wegierski T, Skouloudaki K, Putz M, et al. (2010) PRKCSH/80K-H, the protein mutated in polycystic liver disease, protects polycystin-2/TRPP2 against HERP-mediated degradation. Hum Mol Genet 19: 16–24.
5. Ung CY, Lam SH, Gong ZY (2009) Comparative Transcriptome Analyses Revealed Conserved Biological and Transcription Factor Target Modules Between the Zebrafish and Human Tumors. Zebrafish 6: 425–431.
6. Lam SH, Gong Z (2006) Modeling liver cancer using zebrafish: a comparative oncogenomics approach. Cell Cycle 5: 573–577.
7. Lam SH, Wu YL, Vega VB, Miller LD, Spitsbergen J, et al. (2006) Conservation of gene expression signatures between zebrafish and human liver tumors and tumor progression. Nat Biotechnol 24: 73–75.
8. Li Z, Huang X, Zhan H, Zeng Z, Li C, et al. (2011) Inducible and repressable oncogene-addicted hepatocellular carcinoma in Tet-on xmrk transgenic zebrafish. J Hepatol 56: 419–425.
9. Nguyen AT, Emelyanov A, Koh CH, Spitsbergen JM, Lam SH, et al. (2011) A high level of liver-specific expression of oncogenic Kras(V12) drives robust liver tumorigenesis in transgenic zebrafish. Dis Model Mech 4: 801–813.
10. Nguyen AT, Emelyanov A, Koh CH, Spitsbergen JM, Parinov S, et al. (2012) An inducible kras(V12) transgenic zebrafish model for liver tumorigenesis and chemical drug screening. Dis Model Mech 5: 63–72.
11. Li Z, Zheng W, Wang Z, Zeng Z, Zhan H, et al. (2012) An inducible Myc zebrafish liver tumor model revealed conserved Myc signatures with mammalian liver tumors. Disease models & mechanisms.
12. Curado S, Stainier DY (2010) deLiver'in regeneration: injury response and development. Semin Liver Dis 30: 288–295.
13. Curado S, Ober EA, Walsh S, Cortes-Hernandez P, Verkade H, et al. (2010) The mitochondrial import gene tomm22 is specifically required for hepatocyte survival and provides a liver regeneration model. Dis Model Mech 3: 486–495.
14. Ung CY, Lam SH, Hlaing MM, Winata CL, Korzh S, et al. (2010) Mercury-induced hepatotoxicity in zebrafish: in vivo mechanistic insights from transcriptome analysis, phenotype anchoring and targeted gene expression validation. BMC Genomics 11: 212.
15. Lam SH, Mathavan S, Tong Y, Li H, Karuturi RK, et al. (2008) Zebrafish whole-adult-organism chemogenomics for large-scale predictive and discovery chemical biology. PLoS Genet 4: e1000121.
16. Wen C, Zhang Z, Ma W, Xu M, Wen Z, et al. (2005) Genome-wide identification of female-enriched genes in zebrafish. Dev Dyn 232: 171–179.
17. Small CM, Carney GE, Mo Q, Vannucci M, Jones AG (2009) A microarray analysis of sex- and gonad-biased gene expression in the zebrafish: evidence for masculinization of the transcriptome. BMC Genomics 10: 579.
18. Zeng S, Gong Z (2002) Expressed sequence tag analysis of expression profiles of zebrafish testis and ovary. Gene 294: 45–53.
19. Li Y, Chia JM, Bartfai R, Christoffels A, Yue GH, et al. (2004) Comparative analysis of the testis and ovary transcriptomes in zebrafish by combining experimental and computational tools. Comp Funct Genomics 5: 403–418.
20. Sreenivasan R, Cai M, Bartfai R, Wang X, Christoffels A, et al. (2008) Transcriptomic analyses reveal novel genes with sexually dimorphic expression in the zebrafish gonad and brain. PLoS One 3: e1791.
21. Santos EM, Kille P, Workman VL, Paull GC, Tyler CR (2008) Sexually dimorphic gene expression in the brains of mature zebrafish. Comp Biochem Physiol A Mol Integr Physiol 149: 314–324.
22. Davis LK, Pierce AL, Hiramatsu N, Sullivan CV, Hirano T, et al. (2008) Gender-specific expression of multiple estrogen receptors, growth hormone receptors, insulin-like growth factors and vitellogenins, and effects of 17 beta-estradiol in the male tilapia (Oreochromis mossambicus). Gen Comp Endocrinol 156: 544–551.
23. Taboada X, Robledo D, Del Palacio L, Rodeiro A, Felip A, et al. (2012) Comparative expression analysis in mature gonads, liver and brain of turbot (Scophthalmus maximus) by cDNA-AFLPS. Gene 492: 250–261.
24. Robison BD, Drew RE, Murdoch GK, Powell M, Rodnick KJ, et al. (2008) Sexual dimorphism in hepatic gene expression and the response to dietary carbohydrate manipulation in the zebrafish (Danio rerio). Comp Biochem Physiol Part D Genomics Proteomics 3: 141–154.
25. Kling P, Norman A, Andersson PL, Norrgren L, Forlin L (2008) Gender-specific proteomic responses in zebrafish liver following exposure to a selected mixture of brominated flame retardants. Ecotoxicol Environ Saf 71: 319–327.
26. Zhang W, Zhang Y, Zhang H, Wang J, Cui R, et al. (2012) Sex differences in transcriptional expression of FABPs in zebrafish liver after chronic perfluorononanoic acid exposure. Environ Sci Technol 46: 5175–5182.
27. Roy AK, Chatterjee B (1983) Sexual dimorphism in the liver. Annu Rev Physiol 45: 37–50.
28. Westerfield M (2000) The zebrafish book. A guide for the laboratory use of zebrafish (Danio rerio). University of Oregon Press, Eugene.
29. Hallare A, Nagel K, Kohler HR, Triebskorn R (2006) Comparative embryotoxicity and proteotoxicity of three carrier solvents to zebrafish (Danio rerio) embryos. Ecotoxicology and environmental safety 63: 378–388.
30. Hashimoto S, Qu W, Ahsan B, Ogoshi K, Sasaki A, et al. (2009) High-resolution analysis of the 5′-end transcriptome using a next generation DNA sequencer. PLoS One 4: e4108.
31. Robinson MD, McCarthy DJ, Smyth GK (2010) edgeR: a Bioconductor package for differential expression analysis of digital gene expression data. Bioinformatics 26: 139–140.
32. Robinson MD, Smyth GK (2007) Moderated statistical tests for assessing differences in tag abundance. Bioinformatics 23: 2881–2887.
33. Robinson MD, Smyth GK (2008) Small-sample estimation of negative binomial dispersion, with applications to SAGE data. Biostatistics 9: 321–332.
34. Huang da W, Sherman BT, Lempicki RA (2009) Systematic and integrative analysis of large gene lists using DAVID bioinformatics resources. Nat Protoc 4: 44–57.
35. Baker ME (1988) Is vitellogenin an ancestor of apolipoprotein B-100 of human low-density lipoprotein and human lipoprotein lipase? Biochem J 255: 1057–1060.
36. Li ZJ, Zhang SC, Zhang J, Liu M, Liu ZH (2009) Vitellogenin is a cidal factor capable of killing bacteria via interaction with lipopolysaccharide and lipoteichoic acid. Molecular Immunology 46: 3232–3239.
37. Liu X, Wang H, Gong Z (2006) Tandem-repeated Zebrafish zp3 genes possess oocyte-specific promoters and are insensitive to estrogen induction. Biol Reprod 74: 1016–1025.
38. Cheng W, Guo L, Zhang Z, Soo HM, Wen C, et al. (2006) HNF factors form a network to regulate liver-enriched genes in zebrafish. Dev Biol 294: 482–496.
39. Groh KJ, Nesatyy VJ, Segner H, Eggen RI, Suter MJ (2011) Global proteomics analysis of testis and ovary in adult zebrafish (Danio rerio). Fish Physiol Biochem 37: 619–647.
40. Levi L, Pekarski I, Gutman E, Fortina P, Hyslop T, et al. (2009) Revealing genes associated with vitellogenesis in the liver of the zebrafish (Danio rerio) by transcriptome profiling. BMC genomics 10: 141.
41. Holth TF, Nourizadeh-Lillabadi R, Blaesbjerg M, Grung M, Holbech H, et al. (2008) Differential gene expression and biomarkers in zebrafish (Danio rerio) following exposure to produced water components. Aquat Toxicol 90: 277–291.
42. Knoll-Gellida A, Andre M, Gattegno T, Forgue J, Admon A, et al. (2006) Molecular phenotype of zebrafish ovarian follicle by serial analysis of gene expression and proteomic profiling, and comparison with the transcriptomes of other animals. BMC Genomics 7: 46.
43. Correale J, Fiol M (2011) Chitinase effects on immune cell response in neuromyelitis optica and multiple sclerosis. Mult Scler 17: 521–531.
44. Zhao C, Dahlman-Wright K (2010) Liver X receptor in cholesterol metabolism. J Endocrinol 204: 233–240.
45. Zhao S, Li R, Li Y, Chen W, Zhang Y, et al. (2012) Roles of vitamin A status and retinoids in glucose and fatty acid metabolism. Biochem Cell Biol 90: 142–152.
46. Desvergne B, Michalik L, Wahli W (2006) Transcriptional regulation of metabolism. Physiol Rev 86: 465–514.
47. Yafang L, Qiong W, Yue R, Xiaoming X, Lina Y, et al. (2011) Role of Estrogen Receptor-alpha in the Regulation of Claudin-6 Expression in Breast Cancer Cells. J Breast Cancer 14: 20–27.
48. Waxman DJ, Holloway MG (2009) Sex differences in the expression of hepatic drug metabolizing enzymes. Mol Pharmacol 76: 215–228.

49. Mode A, Gustafsson JA (2006) Sex and the liver - a journey through five decades. Drug Metab Rev 38: 197–207.
50. Wang-Buhler JL, Lee SJ, Chung WG, Stevens JF, Tseng HP, et al. (2005) CYP2K6 from zebrafish (Danio rerio): cloning, mapping, developmental/tissue expression, and aflatoxin B1 activation by baculovirus expressed enzyme. Comp Biochem Physiol C Toxicol Pharmacol 140: 207–219.
51. Troxel CM, Reddy AP, O'Neal PE, Hendricks JD, Bailey GS (1997) In vivo aflatoxin B1 metabolism and hepatic DNA adduction in zebrafish (Danio rerio). Toxicol Appl Pharmacol 143: 213–220.
52. van Riggelen J, Yetil A, Felsher DW (2010) MYC as a regulator of ribosome biogenesis and protein synthesis. Nat Rev Cancer 10: 301–309.
53. Shang Y, Hu X, DiRenzo J, Lazar MA, Brown M (2000) Cofactor dynamics and sufficiency in estrogen receptor-regulated transcription. Cell 103: 843–852.
54. Shang Y, Brown M (2002) Molecular determinants for the tissue specificity of SERMs. Science 295: 2465–2468.
55. Wang C, Mayer JA, Mazumdar A, Fertuck K, Kim H, et al. (2011) Estrogen induces c-myc gene expression via an upstream enhancer activated by the estrogen receptor and the AP-1 transcription factor. Mol Endocrinol 25: 1527–1538.
56. Wolfrum C, Stoffel M (2006) Coactivation of Foxa2 through Pgc-1beta promotes liver fatty acid oxidation and triglyceride/VLDL secretion. Cell Metab 3: 99–110.
57. Handschin C, Spiegelman BM (2006) Peroxisome proliferator-activated receptor gamma coactivator 1 coactivators, energy homeostasis, and metabolism. Endocr Rev 27: 728–735.
58. Gao M, Wang J, Lu N, Fang F, Liu J, et al. (2011) Mitogen-activated protein kinase kinases promote mitochondrial biogenesis in part through inducing peroxisome proliferator-activated receptor gamma coactivator-1beta expression. Biochim Biophys Acta 1813: 1239–1244.
59. Dell H, Hadzopoulou-Cladaras M (1999) CREB-binding protein is a transcriptional coactivator for hepatocyte nuclear factor-4 and enhances apolipoprotein gene expression. J Biol Chem 274: 9013–9021.
60. Stoffel M, Duncan SA (1997) The maturity-onset diabetes of the young (MODY1) transcription factor HNF4alpha regulates expression of genes required for glucose transport and metabolism. Proc Natl Acad Sci U S A 94: 13209–13214.
61. Wiwi CA, Gupte M, Waxman DJ (2004) Sexually dimorphic P450 gene expression in liver-specific hepatocyte nuclear factor 4alpha-deficient mice. Mol Endocrinol 18: 1975–1987.
62. Gao B (2005) Cytokines, STATs and liver disease. Cell Mol Immunol 2: 92–100.
63. Lam SH, Lee SG, Lin CY, Thomsen JS, Fu PY, et al. (2011) Molecular conservation of estrogen-response associated with cell cycle regulation, hormonal carcinogenesis and cancer in zebrafish and human cancer cell lines. BMC medical genomics 4: 41.
64. Goetz FW, Rise ML, Rise M, Goetz GW, Binkowski F, et al. (2009) Stimulation of growth and changes in the hepatic transcriptome by 17beta-estradiol in the yellow perch (Perca flavescens). Physiol Genomics 38: 261–280.
65. Pham CH, Park KS, Kim BC, Kim HN, Gu MB (2011) Construction and characterization of Japanese medaka (Oryzias latipes) hepatic cDNA library and its implementation to biomarker screening in aquatic toxicology. Aquat Toxicol 105: 569–575.
66. Hoffmann JL, Thomason RG, Lee DM, Brill JL, Price BB, et al. (2008) Hepatic gene expression profiling using GeneChips in zebrafish exposed to 17alpha-methyldihydrotestosterone. Aquat Toxicol 87: 69–80.
67. Moens LN, van der Ven K, Van Remortel P, Del-Favero J, De Coen WM (2007) Gene expression analysis of estrogenic compounds in the liver of common carp (Cyprinus carpio) using a custom cDNA microarray. J Biochem Mol Toxicol 21: 299–311.
68. Hook SE, Skillman AD, Small JA, Schultz IR (2006) Gene expression patterns in rainbow trout, Oncorhynchus mykiss, exposed to a suite of model toxicants. Aquat Toxicol 77: 372–385.
69. Ruggeri B, Ubaldi M, Lourdusamy A, Soverchia L, Ciccocioppo R, et al. (2008) Variation of the genetic expression pattern after exposure to estradiol-17beta and 4-nonylphenol in male zebrafish (Danio rerio). General and comparative endocrinology 158: 138–144.
70. Kausch U, Alberti M, Haindl S, Budczies J, Hock B (2008) Biomarkers for exposure to estrogenic compounds: gene expression analysis in zebrafish (Danio rerio). Environmental toxicology 23: 15–24.
71. Wang H, Tan JT, Emelyanov A, Korzh V, Gong Z (2005) Hepatic and extrahepatic expression of vitellogenin genes in the zebrafish, Danio rerio. Gene 356: 91–100.
72. Tong Y, Shan T, Poh YK, Yan T, Wang H, et al. (2004) Molecular cloning of zebrafish and medaka vitellogenin genes and comparison of their expression in response to 17beta-estradiol. Gene 328: 25–36.
73. Rogers AB, Boutin SR, Whary MT, Sundina N, Ge Z, et al. (2004) Progression of chronic hepatitis and preneoplasia in Helicobacter hepaticus-infected A/JCr mice. Toxicol Pathol 32: 668–677.
74. Deaciuc IV, Doherty DE, Burikhanov R, Lee EY, Stromberg AJ, et al. (2004) Large-scale gene profiling of the liver in a mouse model of chronic, intragastric ethanol infusion. J Hepatol 40: 219–227.
75. Tan X, Behari J, Cieply B, Michalopoulos GK, Monga SP (2006) Conditional deletion of beta-catenin reveals its role in liver growth and regeneration. Gastroenterology 131: 1561–1572.
76. Weng Y, DiRusso CC, Reilly AA, Black PN, Ding X (2005) Hepatic gene expression changes in mouse models with liver-specific deletion or global suppression of the NADPH-cytochrome P450 reductase gene. Mechanistic implications for the regulation of microsomal cytochrome P450 and the fatty liver phenotype. J Biol Chem 280: 31686–31698.
77. Rogers AB, Theve EJ, Feng Y, Fry RC, Taghizadeh K, et al. (2007) Hepatocellular carcinoma associated with liver-gender disruption in male mice. Cancer Res 67: 11536–11546.
78. Sternberg EM, Chrousos GP, Wilder RL, Gold PW (1992) The stress response and the regulation of inflammatory disease. Ann Intern Med 117: 854–866.

9

Modelling Size-Dependent Cannibalism in Barramundi *Lates calcarifer*: Cannibalistic Polyphenism and Its Implication to Aquaculture

Flavio F. Ribeiro, Jian G. Qin*

School of Biological Sciences, Flinders University, Adelaide, South Australia, Australia

Abstract

This study quantified size-dependent cannibalism in barramundi *Lates calcarifer* through coupling a range of prey-predator pairs in a different range of fish sizes. Predictive models were developed using morphological traits with the alterative assumption of cannibalistic polyphenism. Predictive models were validated with the data from trials where cannibals were challenged with progressing increments of prey sizes. The experimental observations showed that cannibals of 25–131 mm total length could ingest the conspecific prey of 78–72% cannibal length. In the validation test, all predictive models underestimate the maximum ingestible prey size for cannibals of a similar size range. However, the model based on the maximal mouth width at opening closely matched the empirical observations, suggesting a certain degree of phenotypic plasticity of mouth size among cannibalistic individuals. Mouth size showed allometric growth comparing with body depth, resulting in a decreasing trend on the maximum size of ingestible prey as cannibals grow larger, which in parts explains why cannibalism in barramundi is frequently observed in the early developmental stage. Any barramundi has the potential to become a cannibal when the initial prey size was <50% of the cannibal body length, but fish could never become a cannibal when prey were >58% of their size, suggesting that 50% of size difference can be the threshold to initiate intracohort cannibalism in a barramundi population. Cannibalistic polyphenism was likely to occur in barramundi that had a cannibalistic history. An experienced cannibal would have a greater ability to stretch its mouth size to capture a much larger prey than the models predict. The awareness of cannibalistic polyphenism has important application in fish farming management to reduce cannibalism.

Editor: Cédric Sueur, Institut Pluridisciplinaire Hubert Curien, France

Funding: This project was funded by CAPES Brazil for a PhD Scholarship to F Ribeiro (http://www.cambridgetrusts.org/partners/capes-brazil) The funders had no role in study design, data collection and analysis, decision to publish, or preparation of the manuscript.

Competing Interests: The authors have declared that no competing interests exist.

* E-mail: Jian.Qin@Flinders.edu.au

Introduction

Polymorphism, the occurrence of discrete intraspecific morphs, is triggered by genetic differences, phenotypic plasticity, or a combination of both [1,2]. In fish such as Arctic charr *Salvelinus alpinus* distinct intraspecific morphotypes can be a result of phenotypic plasticity associated with adaption to resources and ecological environments [3,4]. Polyphenism on the other hand refers to alternative phenotypes in a population that are originated from a single genotype in response to environmental stimuli [2,5–7]. If such phenotypic plasticity gives advantages for some individuals to ingest a larger prey and consume their conspecifics, this phenomenon is regarded as cannibalistic polyphenism.

True cannibalistic polyphenic individuals are clearly specialized in an intraspecific diet and have distinctive behaviour, morphology and life history [8], which are not common in fishes, but occur quite frequently in other taxa, such as amphibians [9]. Nonetheless, resource polymorphism has been reported in certain fish species [2]. For example, some individuals of Arctic charr exhibit a broader or larger mouth, faster growth rates and more aggressive behaviour than others [10]. In aquaculture, these traits are selected for, thereby leading to inadvertent selection of cannibalism in a farmed fish population [11], and causing frequent occurrence of intracohort cannibalism in piscivorous species. Furthermore, aquaculture conditions enhance the propensity of some individuals to become cannibals due to restriction of fish dispersing, overcrowding, and uneven food distribution, leading to size heterogeneity and cannibalism [11–13]. As a result, such conditions promote development of cannibalistic polyphenism.

The onset of intracohort cannibalism may occur shortly after hatch such as in dorada *Brycon moorei* [14], or at a later stage as in most marine fish [13] depending on the development patterns of the species. Once the cannibalistic process starts, it may persist during the juvenile phase of development as long as enough size heterogeneity enables a cannibal to prey on smaller conspecifics [13]. The current practice to control intracohort cannibalism in aquaculture is by size grading [11,15], but such procedure is labour consuming, inefficient and stressful to fish [16]. As in the prey-predator relationship of teleosts, morphological factors determine the maximum prey size that predators can ingest [17]. Assuming that a cannibal can ingest a prey if the largest body dimension of the prey cross section is equal to or smaller than the maximum mouth dimension of the cannibal, some morphological models have been used to determine the largest size variation that is acceptable so as to make the exercise of complete cannibalism impossible after size sorting [18–22]. The largest prey cross-sectional dimensions (e.g., head height, body depth or width) are reliable factors for estimating the maximum capacity of a cannibal to ingest its prey. Nevertheless, the maximum mouth dimension may be subjective by researchers' choice [22]. Gape size [23], opened mouth height [15,24], closed mouth width [22,25], and opened mouth width [19–21,26–28] have been used to predict the maximum ingestion capacity of cannibalistic fish species. However, in order to have a reliable prediction, the maximum mouth dimension must be carefully selected according to specific traits of the target species such as using mouth elasticity in snakehead *Channa striatus* [21] and orientation of the prey on cannibal mouth in orange-spotted grouper *Epinephelus coioides* [19] and giant grouper *E. lanceolatus* [20]. Furthermore, cannibalistic polyphenism has never been built into a model to predict size-dependent cannibalism in fish. As some individuals may possess larger jaws and a wider mouth [8], existing models based on the parts of a population average may underestimate the maximum prey size that a cannibal can ingest. Moreover, few models have been validated with an independent dataset, but if done, the maximum size of ingestible prey is underestimated as in snakehead [21] and largemouth bass *Micropterus salmoides* [22], or overestimated as in the giant [20] and orange-spotted [29] groupers.

The aim of this study was to determine size-dependent cannibalism in a highly cannibalistic fish, the barramundi *Lates calcarifer* (Latidae). Models were developed using the mouth width as the largest mouth dimension and the alternative assumption of polyphenism. Subsequently, the models were validated based on empirical results taken from a series of independent observations from different prey-predator pairs. Barramundi were used as the model species because it is an economically important fish for aquaculture in tropical and subtropical regions [16]. In a previous model, Parazo et al. [15] suggests that the total length (TL) of ingestible prey ranges 67–61% of the cannibal size in barramundi of 10–50 mm TL. However, Parazo's model was based on an inappropriate measurement of mouth size and the empirical validation might be prejudiced by prey size preference, as it was based on the stomach analysis of cannibals from an undisturbed population of cultured fish. Thus, the present study used a new approach to assess the maximum prey size that cannibalistic barramundi can ingest from direct observations. The new model simulates a more realistic scenario to quantify the size relationship between cannibal and victim individuals in cannibalistic fishes.

Materials and Methods

Ethics Statement

This study was carried out in strict accordance with the recommendations in the Animal Welfare Act 1985 and the Australian Code of Practice for the Care and Use of Animals for Scientific Purpose 7th Edition. The protocol, species, and number of animals used in this study were approved by the Flinders University Animal Welfare Committee (Project Number: E347). In any trial situations, each prey had an opportunity to avoid the predators in their cannibal challenge since we allocated more open space in each aquarium to facilitate prey escape. Euthanasia procedures were performed under overdose (43 mg l^{-1}) of AQUI-S® (New Zealand Ltd). All fish handling were followed by light anesthesia (15 mg l^{-1}) with AQUI-S, and all efforts were made to alleviate fish suffering.

Fish and rearing conditions

Hatchery raised barramundi *Lates calcarifer* of 34 days after hatching from the same cohort were obtained from West Beach Hatchery, West Beach, South Australia, and transported to the Animal House, Flinders University. Upon arrival, all fish were visually graded into large, medium and small sizes, and stocked into three holding tanks (300 l) filled with freshwater. Each tank was equipped with an external biofilter and kept at 27–28°C. Fish were divided into three groups and fed at different rates with dry pellets (NRD® range, 400–2,000 µm; 55% protein, 9% lipid, INVE Ltd, Thailand). Group 1: 360 large fish (1.2 fish l^{-1}) fed to satiation twice a day in order to produce large individuals to be used as cannibals; Group 2: 950 small fish (3.2 fish l^{-1}) fed once a day at a restricted ration to produce a range of small fish sizes to be used as prey on the cannibal challenge experiment; and Group 3: 650 medium fish (2.2 fish l^{-1}) fed twice a day under moderate feeding restriction in order to promote a range of fish sizes to be used for morphological measurements. Tanks were cleaned twice a day to remove unfed pellets, faeces and dead fish. Water parameters were daily checked and maintained at 27.83±0.19°C, 7.69±0.21 mg l^{-1} dissolved oxygen, 7.51±0.02 pH, and <0.5 mg l^{-1} ammonia and nitrite nitrogen. A photoperiod of 12L:12D was used at a light intensity of 350 Lux during the hours of light with abrupt transition between dark and light periods.

Morphological models construction

Periodically, 368 juveniles were sampled from fish in Group 3 for morphological measurements. Fish were collected with a hand net, euthanized with overdosed AQUI-S (43 mg l^{-1}, AQUI-S New Zealand Ltd) and immediately measured for total length (TL, mm), body depth (BD, mm) and mouth width (MW, mm) to the nearest 0.01 mm using a dissecting microscope or a digital caliper. Fish from 15 to 140 mm TL were sampled, as this comprised the size range corresponding to the time interval when intracohort cannibalism was intense in barramundi fingerling culture [30]. The selection of morphological parts for measurement was under these two assumptions: (1) cannibalistic barramundi swallow their conspecific prey in whole with head first [13]; (2) when cannibalistic barramundi ingest their conspecific prey, the maximal prey body depth was positioned laterally from side to side in the cannibal mouth. Such assumptions were used to predict the maximum prey size for barramundi cannibals from 35 to 140 mm TL. Total length (TL) was measured as the distance from the tip of the snout to the end of the caudal fin and body depth (BD) as the distance between the anterior edge of the dorsal fin and the bottom of the abdomen. Two measurements of mouth width were taken: mouth width at the close position (MWc) as the distance between the outer edges of the maxillary bones just beneath the eyes with the mouth closed; and mouth width at the open position (MWo) as the horizontal largest cross-section distance with the mouth fully stretched in an ellipse shape. With both mouth width measurements, an estimate of mouth width extension (MWE) for each fish was calculated as $MWE\ (\%MWc) = [(MWo - MWc)/MWc] \times 100$.

The morphological predictive models were developed assuming that a $TL_{cannibal}$ can swallow a TL_{prey} if the BD_{prey} is equal to or smaller than the $MW_{cannibal}$. The relationships between $MW_{cannibal}$ vs. $TL_{cannibal}$ and BD_{prey} vs. TL_{prey} were used to predict the maximum prey length (TL_{prey}) for given sizes of cannibals ($TL_{cannibal}$). Models were developed using four different estimates of mouth size: closed mouth width (MWc); maximum closed mouth width (MWcmax); opened mouth width (MWo); maximum opened mouth width (MWomax).

Cannibal challenge

A series of single pairwise trials were performed to empirically observe the maximum conspecific prey size that a cannibalistic barramundi can ingest. Cannibals from 25 to 131 mm TL were individually challenged with single conspecific prey of known sizes, starting from 45% of cannibal TL. The system consisted of 20×6 l aquaria (20×20×25 cm) connected to a freshwater recirculation system equipped with a communal 200 l biofilter and set in the same experimental room as the holding tanks. Aquaria were cleaned daily to remove faeces. Water quality and physical parameters were kept the same as those in the holding tanks.

Initially, 20 potential cannibals were sampled from fish in Group 1, anesthetized (AQUI-S, 15 mg l^{-1}), measured for TL and individually stocked into each aquarium. Then, potential prey were collected from fish in Group 2, anesthetized (AQUI-S, 15 mg l^{-1}), measured for TL, individually selected and matched their respective cannibal. No food was provided during the trials. Predation was checked twice a day (0900 and 1700 h). In case of predation, the cannibal was re-measured in order to decide the next prey size to be offered, and a new prey larger than the previous one would be selected from Group 2, anesthetized (AQUI-S, 15 mg l^{-1}), measured for TL and individually matched the same cannibal. This procedure was repeated progressively by increasing the prey size at about 5% per change according to prey size availability. As the maximum prey size approached to the maximum ingesting limit for cannibals, the incremental rate of the new prey size was reduced to about 2%. The morphological limit for cannibals was considered maximum when both cannibal and prey coexisted for over 4 days. In that occasion, both fish were measured and the cannibal was replaced by a larger one.

Successful predation events were considered completion when the prey had been fully swallowed and digested by the cannibal. Cannibals in the process of digesting prey were easily identified due to their extended belly. Such consideration avoided significant discrepancies on growth rate between cannibal and prey during the next pairing period. In some circumstances, the prey was dead on the bottom of the aquaria after having been discarded by the cannibals due to unsuccessful capture attempts. In those cases, a new prey of a similar size was paired with the same cannibal. If the cannibal would kill but not ingest the prey again, that prey size was considered the upper limit of the cannibal and the cannibal was replaced by a larger one.

Statistical analysis

All absolute estimates for body parts were regressed against TL and an analysis of covariance (ANCOVA) was used to test for homogeneity of the regression slopes of the body depth (BD, mm) and mouth width (MW, mm) estimates using total length (TL, mm) as a covariate. Linear regression analysis was used to assess the independence between mouth width extension (MWE as %MWc) and closed (MWc) and opened (MWo) mouth widths (%TL). MWE was regressed against TL to determine the capacities of mouth width extension as fish grew. Pearson's correlation analysis was used to assess the strength of correlations. All predictive models based on morphological measurements for the maximum prey to cannibal size ratio enabling the occurrence of intracohort cannibalism were developed using simple linear regression analysis. The results from the cannibal challenge experiment were used to estimate a revised model for maximum prey size for cannibals based on the empirical data. The size of the first offered prey was compared between the successful versus non-successful cannibalistic pairs with T-test to identify the criteria for the initial prey-predator size ratio that would provoke cannibalism. All statistics were considered significant at $P<0.05$.

Results

Morphological models

During the early juvenile stage (15–30 mm TL), body depth (BD) showed positive allometric growth, attaining its maximum dimension relative to body size (28% TL) when fish were around 35 mm TL (Figure 1). Thereafter, BD slightly decreased and reached 25% TL at the late juvenile stage (140 mm TL; Figure 1). Mouth width at close (MWc) or open (MWo) presented slightly negative allometric growth as fish grew larger, decreasing from 13% to 9% TL (15–135 mm TL) and from 17% to 15% TL (35–135 mm TL), respectively (Figure 1).

The relationship between absolute body depth (BD), mouth width (MW) estimates and total length (TL) fitted on linear regression equations (Table 1). ANCOVA analyses showed significant differences between the regression slopes of the body parts ($df=4$, $F=4.988$, $P<0.0001$), suggesting that absolute body depth increases faster than mouth width. The significant differences between the regression slopes of the mouth width estimates were due to the increase in the mouth width extension (MWE) as fish grew larger (Figure 2).

A marked inter-individual variability was observed for all morphological variables. Estimates for both opened and closed mouth widths presented a consistent variability during the juvenile phase (Figure 1). Inter-individual variability was also observed for MWE, varying consistently at about 30% (±15%) for the whole range of fish size (Figure 2). The positive correlation between MWc and MWo ($r=0.505$, $n=153$, $P<0.0001$; Table 1) and the non-correlation between MWo and MWE ($r=0.085$, $n=153$, $P=0.294$; Table 1) indicated that the MWo was more affected by the MWc than by the MWE. In contrast, the negative correlation between MWc and MWE ($r=-0.811$, $n=153$, $P<0.0001$; Table 1) indicated that the highest MWE (Figure 2) were associated with the smallest MWc (Table 1). Therefore, the maximum values of MWc and MWo were used to develop specific models to reflect polyphenism in mouth width.

Assuming that a $TL_{cannibal}$ could swallow a TL_{prey} if the BD_{prey} was equal to or smaller than the $MW_{cannibal}$, the maximum conspecific prey size for cannibalistic barramundi was predicted by simple linear regression (Table 2). All models predicted that the maximum prey TL increased with increasing cannibal TL (Figure 3A). However, when expressed as a proportion of cannibal TL, the models showed a slightly declining trend in the size of maximum prey as cannibal TL increased (Figure 3B). The closed mouth width (MWc) model predicted that the maximum prey size decreased from 40 to 37% or from 50 to 39% of cannibal TL considering the maximum values (MWcmax), for cannibals of 30–140 mm TL. The maximum prey size remained constant at 61% of the cannibal TL when the model was based on the opened mouth width (MWo). However, when considering maximum opened mouth width (MWomax) the model predicted a decreasing trend from 68 to 63% of cannibal TL, for a similar size range of cannibals. Such decreasing tendencies as cannibals grew larger were related to the slightly fast increase in body depth comparing with the mouth width (Table 1).

Cannibal challenge

In total, 495 prey-cannibal pairs were tested using 102 potential cannibals from 25 to 131 mm TL. There was no substantial variation of the prey size during the pairing periods. In those cases when predation did not occur, the final prey size was −50–1.40% of the initial prey size as the percent of cannibal TL ($n=55$). In all potential cannibals challenged, 75% became true cannibals ingesting at least one conspecific prey. These cannibals consumed

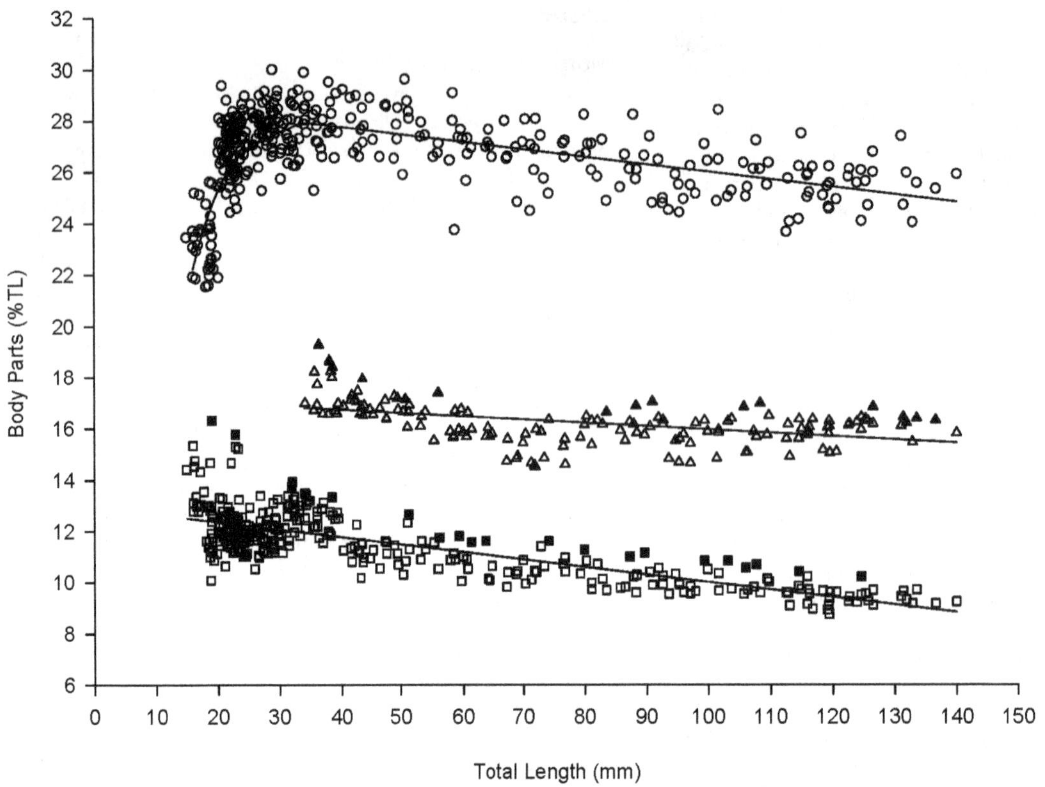

Figure 1. Morphological variation between relative body parts (%TL) and total length (TL, mm) of juvenile barramundi. Body depth (BD in circles, $n = 368$), closed mouth width (MWc in squares, $n = 360$) and opened mouth width (MWo in triangles, $n = 154$) are plotted against total length. Each symbol represents an individual estimate. Filled symbols represent maximum values of mouth width estimates.

61.6% of the total number of prey while dead prey on the bottom accounted for 20.2%. Four cases of suffocation were observed during the trials where cannibals died with the prey stuck in mouth. In addition, three half-ingestion events were observed in this study, where the cannibals predigested half of the prey and discarded the other half. Interestingly, in all these cannibalistic events, prey sizes were 65% of cannibal TL. When the prey size was firstly offered at 58.36±5.37% cannibal TL, 25% of the large fish tested did not become cannibals, but the other 75% of the large fish became cannibals when the prey size firstly offered was 50.77±2.57% cannibal TL (T-test; df = 100; $P<0.0001$).

The results of the cannibal challenged with prey showed that cannibals were able to ingest a conspecific prey larger than the size that all models could predict (Figure 3A, B). For instance, according to the models based on MWc, MWcmax, MWo or MWomax, a cannibal of 106.50 mm TL could ingest a prey of 39, 43, 65 or 68 mm TL (37, 40, 61 or 64% of cannibal TL), respectively. Results from the cannibal challenging trial showed that identical sized cannibals could ingest a conspecific prey of 77 mm TL (72% of cannibal TL). Thus, according to empirical observations, cannibals of 25–131 mm TL could ingest the prey of 78–72% of cannibal TL, respectively. Such reduction in the maximum prey size is a result of a faster growth of the body depth in relation to the mouth size (Table 1). The increase in mouth width extension as fish grew larger (Figure 2) would compensate the part of negative allometric growth of the mouth width.

Discussion

A model by Parazo et al. [15] predicted that cannibalistic barramundi of 10–50 mm TL (total length) can ingest a maximum conspecific prey size of 67–61% of cannibal TL, respectively. However, the empirical results in the present study showed that barramundi cannibals (25–131 mm TL) could ingest conspecific prey of 78–72% of cannibal TL, respectively. All predictive models using morphological traits considering the alternative assumption of cannibalistic polyphenism underestimate the maximum prey size that a cannibal can possibly ingest.

All successfully cannibalistic events in the present study were orientated by head being sucked in first and cannibals ingesting the whole prey. Moreover, cannibalistic barramundi ingested their prey horizontally, making the size of mouth width become the limiting factor for prey ingestion. Thus, using the closed mouth width (MWc) as an independent factor, the predictive model shows a maximum prey size of 40–37% cannibal TL, for the cannibals of 30–140 mm TL, respectively. Alternatively, when the model was developed with the opened mouth width (MWo), it predicts that a cannibal can ingest a maximum prey of 61% of the cannibal TL, which is almost double the size that the previous model predicted. Our model prediction is in accordance with that by Parazo et al. [15] who predicted a maximum prey size of 67–61% of cannibal TL, when cannibals were 10–50 mm TL, respectively, based on mouth size as the distance from the dorsal to the ventral boundary of the mouth opened. Whatever the case was, when predictive models were compared with the empirical results in this study, the models underestimate the maximum conspecific prey size that cannibals can ingest. Similar conclusions were drawn by Qin and

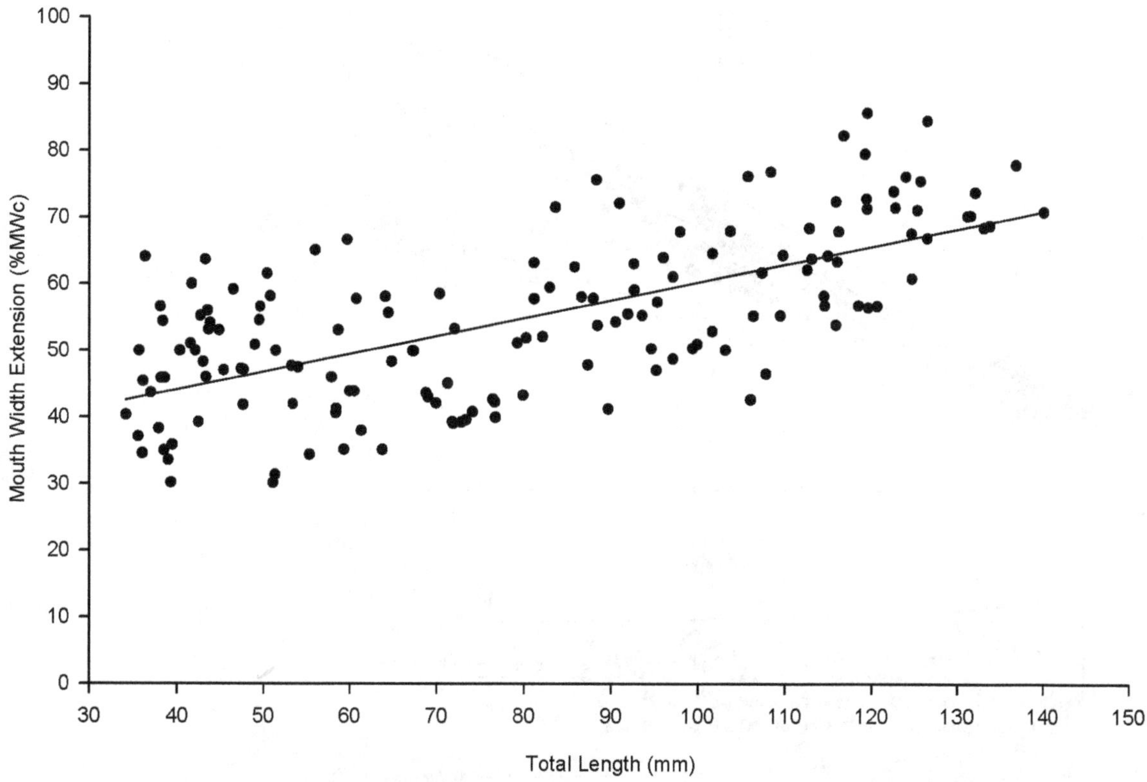

Figure 2. Relationship between mouth width extension (MWE, %MWc) and total length (TL, mm) of juvenile barramundi. MWE and TL are positive correlated ($r=0.660$, $n=154$, $P<0.0001$). The solid line represents the average MWE varying from 34 to 140 mm TL and is expressed as $MWE=0.269(0.024)TL + 33.328(2.083)$ ($r^2=0.446$, $df=152$, $F=121$, $P<0.0001$).

Table 1. Relationship of morphological parts of barramundi (15–140 mm TL).

Relationships	Equations	r^2	df	F	p intercept	p slopes
	Absolute measures (mm)					
MWc vs TL	$MWc=0.091(0.001)^{c}TL+0.850(0.034)$	0.986	359	25,181	<0.0001	<0.0001
MWcmax vs TL	$MWcmax=0.093(0.001)^{c}TL+1.508(0.071)$	0.998	21	9,620	<0.0001	<0.0001
MWo vs TL	$MWo=0.155(0.001)^{b}TL+0.462(0.117)$	0.988	153	12,873	<0.0001	<0.0001
MWomax vs TL	$MWomax=0.157(0.001)^{b}TL+1.067(0.134)$	0.999	16	11.708	<0.0001	<0.0001
BD vs TL	$BD=0.255(0.001)^{a}TL+0.483(0.063)$	0.994	367	57,299	<0.0001	<0.0001
	Relative measures (% body parts)					
MWc vs MWE	$MWE=-10.355(0.608)MWc+163.997(6.443)$	0.657	152	284.86	<0.0001	<0.0001
MWo vs MWE	$MWE=1.259(1.196)MWo+34.324(19.439)$	0.007	152	1.109	0.079	0.294
MWc vs MWo	$MWc=0.583(0.081)MWo+1.079(1.319)$	0.255	152	51.711	0.415	<0.0001

Absolute measures (mm)
MWc: closed mouth width;
MWcmax: maximum closed mouth width;
MWo: opened mouth width;
MWomax: maximum opened mouth width;
BD: body depth;
TL: total length.
Relative measures (% body parts)
MWc: closed mouth width (%TL);
MWo: opened mouth width (%TL);
MWE: mouth width extension (% MWc).

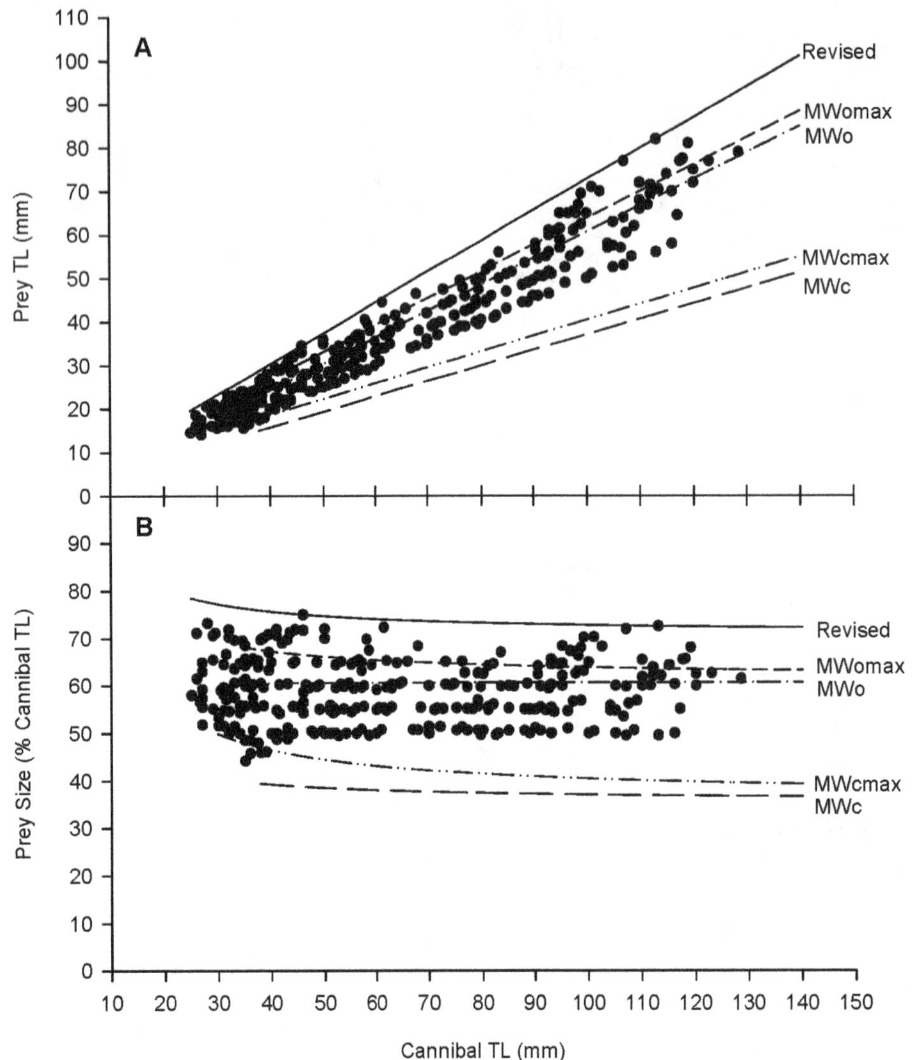

Figure 3. Maximum conspecific prey size for barramundi cannibals. Prey size in panel A is expressed as prey TL, mm, and in panel B expressed as % of cannibal TL. Regression lines include (1) the maximum size of prey ingested ("Revised" in filled circles on the top) and (2) model predictions on prey size based on closed mouth width (MWc), maximum closed mouth width (MWcmax), opened mouth width (MWo), and maximum opened mouth width (MWomax).

Table 2. Prediction of prey size (TL_{prey}, mm) from cannibal size ($TL_{cannibal}$, mm) based on different predictive model equations for cannibalistic barramundi (30–140 mm TL).

Models	Equations
MWc	$TL_{prey} = 0.3569TL_{cannibal} + 1.4392$
MWcmax	$TL_{prey} = 0.3647TL_{cannibal} + 4.0196$
MWo	$TL_{prey} = 0.6078TL_{cannibal} - 0.0824$
MWomax	$TL_{prey} = 0.6157TL_{cannibal} + 2.2902$
Revised	$TL_{prey} = 0.709TL_{cannibal} + 1.8881$

MWc: closed mouth width;
MWcmax: maximum closed mouth width;
MWo: opened mouth width;
MWomax: maximum opened mouth width;
Revised: the model based on the empirical observations on the maximum size of ingested prey.

Fast [21] on snakehead *Channa striatus* and Johnson and Post [22] on largemouth bass *Micropterus salmoides* when their predictive models were validated with empirical data.

In the present study, despite the inter-individual variability on mouth width and the capacity of mouth width extension, the models based on maximum values of the closed (MWcmax) and opened mouth widths (MWomax) underestimate the maximum ingestible capacity of cannibalistic individuals. Nevertheless, the model using MWomax predicts a slightly higher cannibalistic capacity than the model using MWo, which is closer to the empirical observation. The high inter-individual variability on MWE indicates marked polyphenism in the mouth extension capacity. However, as a negative correlation was detected between MWE and MWc, polyphenic MWE seems to be present in order to compensate the morphological disadvantages of fish with smallest MWc, but not for fish with larger MWc. As a result, the polyphenic trait of the MWo, which obviously represents the maximum predation capacity of cannibalistic barramundi, is rather a result of larger mouth width than higher mouth extension capacity. The polyphenic trait of a mouth provides not only a cannibalistic advantage, but a feeding advantage on other food. Thus, polyphenism should be considered when assessing feeding ecology of piscivorous fish species in general.

Allometric growth of the mouth is common in fish species and together with size heterogeneity it can determine the dynamic of complete cannibalism in fish [13,24]. Previous observations on barramundi feeding showed that the onset of complete cannibalism mainly occurs after metamorphosis, when fish are being weaned to inert diets [13,31]. In the present study, mouth width showed slower growth than the body depth. As both variables set the morphological boundary for complete cannibalism, both predictive and revised models show a decreasing trend on the maximum ingestible prey size as barramundi grow larger. As a result, cannibalism in barramundi is more likely to occur in early juvenile than during latter stages, which agrees with the findings on cannibalism by Otterå and Folkvord [24] for *G. morhua*, and Qin and Fast [21] for *C. striatus*.

Morphological constrains are not the only cause of a general reduction trend on cannibalism rate as fish grow larger. Cannibalistic fish usually prefer smaller prey as reported in *P. djambal* [23] and *Pseudoplatystoma punctifer* [32]. Thus, once smaller prey are succumbed to cannibalism, reducing the size heterogeneity of the population [13], cannibals are forced to move up to larger prey, which may not be energetically profitable as preying on smaller prey since such a size shift may represent an increase in pursuit and handling time and reduce energy gain per capture attempt [33–37]. In aquaculture where plenty of inert food of high energetic content is available, cannibals may choose to abandon a cannibalistic diet because such diet is not profitable anymore. In contrast, if cannibalistic individuals do enjoy growth advantages over siblings feeding on alternative diets, as observed in the Amazonian catfish *Pseudoplatystoma punctifer* [32], cannibalism will hardly become to an end as the higher growth rate of cannibals may compensate the morphological constraints as fish get larger. On the other hand, if alternative inert food is supplied accordingly, non-cannibalistic individuals may achieve more competitive growth rates [32] and growth beyond the prey spectrum of the cannibals [13]. Further studies should assess the dynamics of intracohort cannibalism in barramundi when alternative inert diet is applied at different developmental stages.

The cannibal challenge experiment was purposely designed in a small scale aiming to maximise the propensity of cannibalism. Small enclosures were used to limit escape ability of small prey and large cannibals were individually stocked, previously acclimated and deprived of alternative food, which is similar to the designs by Sogard and Olla [25] Johnson and Post [22], Hseu et al. [20] and Baras et al. [23]. In this experiment, 75% of the prey available to cannibals were ingested proving that the environment was appropriate to provoke cannibalism. Furthermore, the pairing period was defined as four days, a similar period used by Qin and Fast [21], which was assumed to be short enough to avoid significant behavioural and physiological changes in cannibals and prey, but long enough to promote maximum hunger for cannibals. In addition, cannibalistic events were orientated towards the same prey size (% cannibal TL) offered since prey and cannibals presented similar growth rates during the course of the 4-days paring period. Interestingly, barramundi could become cannibals when the first prey sizes were <50% of their predator body length, but the fish could never become a cannibal when the first prey was >58% of the cannibal size. This may indicate that, once all cannibals are removed from a barramundi population, the size difference of 50% can be a safe margin to avoid the emergence of new cannibals. Furthermore, once a fish had experienced as a cannibal, this fish would use the full morphological capacity to ingest a prey, even though the size of prey may exceed the model prediction. Challenging cannibals with an increasing prey size in the absence of alternative food may have stimulated the phenotypic plasticity in the mouth apparatus, such as hypertrophied jaw musculature [2,8] resulting in greater predation capacity when compared with the predictive models based on fish samples taken from a fish population where food was present and cannibalism was not stimulated.

Unsuccessful cannibalistic events, such as suffocation and half-ingested prey were recorded during the cannibal challenge experiment. Previous studies have used suffocation events as a reference of the maximum prey size limit for cannibals [38,39]. In the present study, unsuccessful cannibalistic events occurred when the prey size was 65% of cannibal TL, which is below the upper size limit determined by the revised model. However, when compared with the predictive model based on the opened mouth width (MWo), prey sizes were slightly larger than the model predicted. Therefore, it seems that those unsuccessful cannibalistic events were performed by hunger-motivated individuals to cannibalize a larger prey they could possibly handle. In addition, dead prey individuals were occasionally observed on the bottom of the tank. In most of these cases, cannibals resumed predation when a live prey of similar size was re-offered, suggesting that such event did not represent the maximum prey size they can ingest and they are probably associated with cannibal's difficulties to handle the prey or prey's abilities to escape from predation acts. Whatever the case, all these events can account for fingerling mortality leading to significant losses in fingerling production.

In summary, this study provides a new approach to predict cannibalistic events in fish under an aquaculture situation and offers recommendation on criteria for size grading practices. In order to reduce the incidence of intracohort cannibalism in a barramundi population, no conspecific prey smaller than 78–72% of cannibal TL should co-inhabit with cannibals from 30 to 140 mm TL, respectively. Furthermore, once all cannibals were removed from the population through size sorting, a size difference of 50% should be set as a threshold to avoid the emergence of new cannibals. Predictive models based on mouth width and body depth of a population average underestimate the maximum prey size for cannibalistic barramundi. However, when polyphenism was considered on measuring of the opened mouth width, the model became closer to the reality, suggesting that when predicting the upper prey size limit for complete cannibalism, the assumption of cannibalistic polyphenism must be considered to keep a safe margin and avoid significant losses due to cannibalistic mortality.

Some unsettled issues still exist. It is still unclear whether the fish with polyphenic mouth size are always consuming the largest prey and it is also uncertain whether these cannibals make the most of their prey's energy. Presumably, a large mouth facilitates handling and increases capture success, but energetic benefit of being a cannibal needs further investigation. In aquaculture, we recommend putting aside those fish that are cannibals because they could have broader mouth dimensions than others, and if this trait is heritable, it can complicate rearing in the future.

Acknowledgments

The authors would like to thank to Leslie Morrison and Animal House staff for their kindly technical support during the experiments of the present study and Dr. Etienne Baras for his thoughtfully comments on this manuscript.

Author Contributions

Conceived and designed the experiments: FFR JGQ. Performed the experiments: FFR. Analyzed the data: FFR JGQ. Contributed reagents/materials/analysis tools: FFR JGQ. Wrote the paper: FFR JGQ. Obtained experimental material: FFR. Formulated research concept: JGQ.

References

1. Komiya T, Fujita S, Watanabe K (2011) A Novel Resource Polymorphism in Fish, Driven by Differential Bottom Environments: An Example from an Ancient Lake in Japan. PLoS ONE 6: e17430.
2. Smith TB, Skulason S (1996) Evolutionary Significance of Resource Polymorphisms in Fishes, Amphibians, and Birds. Annual Review of Ecology and Systematics 27: 111–133.
3. Andersson J (2003) Effects of diet-induced resource polymorphism on performance in arctic charr (*Salvelinus alpinus*). Evolutionary Ecology Research 5: 213–228.
4. Hindar K, Jonsson B (1993) Ecological polymorphism in Arctic charr. Biological Journal of the Linnean Society 48: 63–74.
5. West-Eberhard MJ (1989) Phenotypic Plasticity and the Origins of Diversity. Annual Review of Ecology and Systematics 20: 249–278.
6. Hoffman EA, Pfennig DW (1999) Proximate causes of cannibalistic polyphenism in larval tiger salamander. Ecology 80: 1076–1080.
7. Closs GP, Ludgate B, Goldsmith RJ (2003) Controlling European perch (*Perca fluviatilis*): lessons from an experimental removal. Managing Invasive Freshwater Fish in New Zealand: 37–48.
8. Polis GA (1981) The evolution and dynamics of intraspecific predation. Annual Review of Ecology and Systematics 12: 225–251.
9. Crump ML (1992) Cannibalism in amphibians. In: Elgar MA, Crespi BJ, editors. Cannibalism: Ecology and evolution among diverse taxa. Oxford: Oxford University Press. pp. 256–276.
10. Jonsson B, Jonsson N (2001) Polymorphism and speciation in Arctic charr. Journal of Fish Biology 58: 605–638.
11. Hecht T, Pienaar AG (1993) A review of cannibalism and its implications in fish larviculture. Journal of the World Aquaculture Society 24: 246–261.
12. Smith C, Reay P (1991) Cannibalism in teleost fish. Reviews in Fish Biology and Fisheries 1: 41–64.
13. Baras E, Jobling M (2002) Dynamics of intracohort cannibalism in cultured fish. Aquaculture Research 33: 461–479.
14. Baras E, Ndao M, Maxi MYJ, Jeandrain D, Thomé JP, et al. (2000) Sibling cannibalism in dorada under experimental conditions. I. Ontogeny, dynamics, bioenergetics of cannibalism and prey size selective. Journal of Fish Biology 57: 1001–1020.
15. Parazo MM, Avila EM, Reyes Jr DM (1991) Size- and weight-dependent cannibalism in hatchery-bred sea bass (*Lates calcarifer* Bloch). Journal of Applied Ichthyology 7: 1–7.
16. Qin JG, Mittiga L, Ottolenghi F (2004) Cannibalism reduction in juvenile barramundi *Lates calcarifer* by providing refuges and low light. Journal of the World Aquaculture Society 35: 113–118.
17. Dabrowski K, Bardega R (1984) Mouth size and predicted food size preferences of larvae of three cyprinid fish species. Aquaculture 40: 41–46.
18. Hseu JR, Huang WB, Chu YT (2007) What causes cannibalization-associated suffocation in cultured brown-marbled grouper, *Epinephelus fuscoguttatus* (Forsskal, 1775)? Aquaculture Research 38: 1056–1060.
19. Hseu J-R, Chang H-F, Ting Y-Y (2003) Morphometric prediction of cannibalism in larviculture of orange-spotted grouper, *Epinephelus coioides*. Aquaculture 218: 203–207.
20. Hseu J-R, Hwang P-P, Ting Y-Y (2004) Morphometric model and laboratory analysis of intracohort cannibalism in giant grouper *Epinephelus lanceolatus* fry. Fisheries Science 70: 482–486.
21. Qin JG, Fast AW (1996) Size and feed dependent cannibalism with juvenile snakehead *Channa striatus*. Aquaculture 144: 313–320.
22. Johnson JM, Post DM (1996) Morphological constraints on intracohort cannibalism in age-0 largemouth bass. Transactions of the American Fisheries Society 125: 809–812.
23. Baras E, Hafsaridewi R, Slembrouck J, Priyadi A, Moreau Y, et al. (2010) Why is cannibalism so rare among cultured larvae and juveniles of *Pangasius djambal*? Morphological, behavioural and energetic answers. Aquaculture 305: 42–51.
24. Otterå H, Folkvord A (1993) Allometric growth in juvenile cod (*Gadus morhua*) and possible effects on cannibalism. Journal of Fish Biology 43: 643–645.
25. Sogard SM, Olla BL (1994) The potential for intracohort cannibalism in age-0 walleye pollock, *Theragra chalcogramma*, as determined under laboratory conditions. Environmental Biology of Fishes 39: 183–190.
26. Hecht T, Appelbaum S (1988) Observations on intraspecific aggression and coeval sibling cannibalism by larval and juvenile *Clarias gariepinus* (Clariidae: Pisces) under controlled conditions. Journal of Zoology 214: 21–44.
27. Baras E (1999) Sibling cannibalism among juvenile vundu under controlled conditions. I. Cannibalistic behaviour, prey selection and prey size selectivity. Journal of Fish Biology 54: 82–105.
28. Baras E, Hafsaridewi R, Slembrouck J, Priyadi A, Moreau Y, et al. (2012) Do cannibalistic fish possess an intrinsic higher growth capacity than others? A case study in the Asian redtail catfish *Hemibagrus nemurus* (Valenciennes, 1840). Aquaculture Research: In press.
29. Hseu J-R, Huang W-B (2012) Application of logistic regression analysis to predict cannibalism in orange-spotted grouper *Epinephelus coioides* (Hamilton) fry. Aquaculture Research: in press.
30. Schipp G, Bosmans J, Humphrey J (2007) Barramundi Farming Handbook. Darwin, NT: Department of Primary Industries, Fisheries and Mines. pp. 71.
31. Kestemont P, Jourdan S, Houbart M, Mélard C, Paspatis M, et al. (2003) Size heterogeneity, cannibalism and competition in cultured predatory fish larvae: biotic and abiotic influences. Aquaculture 227: 333–356.
32. Baras E, Silva del Aguila DV, Montalvan Naranjos GV, Dugué R, Koo FC, et al. (2011) How many meals a day to minimize cannibalism when rearing larvae of the Amazonian catfish *Pseudoplatystoma punctifer*? The cannibal's point of view. Aquatic Living Resources 24: 379–390.
33. Gill AB (2003) The dynamics of prey choice in fish: the importance of prey size and satiation. Journal of Fish Biology 63: 105–116.
34. Juanes F (1994) What determines prey size selectivity in piscivorous fishes? In: Stouder DJ, Fresh KL, Feller RJ, editors. Theory and Aplication in Fish Feeding Ecology. Columbia: University of South Carolina Press.
35. Scharf FS, Buckel JA, Juanes F, Conover DO (1998) Predation by juvenile piscivorous bluefish (*Pomatomus saltatrix*): the influence of prey to predator size ratio and prey type on predator capture success and prey profitability. Canadian Journal of Fisheries and Aquatic Sciences 55: 1695–1703.
36. Juanes F, Conover DO (1994) Piscivory and prey size selection in young-of-the-year bluefish: predator preference or size-dependent capture success? Marine Ecology Progress Series 114: 59–69.
37. Ellis T, Gibson RN (1997) Predation of 0-group flatfishes by 0-group cod: handling times and size selection. Marine Ecology Progress Series 149: 83–90.
38. Sakakura Y, Tsukamoto K (1996) Onset and development of cannibalistic behaviour in early life stages of yellowtail. Journal of Fish Biology 48: 16–29.
39. Ebisu R, Tachiara K (1993) Mortality casued by cannibalism in seed production of gold striped amberjack *Seriola lalandi*. Bulletin of Nagasaki Prefectural Institute of Fisheries 19: 1–7.

10

Molecular Phylogeny of the *Myxobolus* and *Henneguya* Genera with Several New South American Species

Mateus Maldonado Carriero[1], Edson A. Adriano[2,3]*, Márcia R. M. Silva[1], Paulo S. Ceccarelli[4], Antonio A. M. Maia[1]

1 Departamento de Medicina Veterinária, Faculdade de Zootecnia e Engenharia de Alimentos, Universidade de São Paulo, Pirassununga, São Paulo, Brazil, 2 Departamento de Ciências Biológicas, Universidade Federal de São Paulo, Diadema, São Paulo, Brazil, 3 Departamento de Biologia Animal, Instituto de Biologia, Universidade Estadual de Campinas, Campinas, São Paulo, Brazil, 4 Centro Nacional de Pesquisa e Conservação de Peixes Continentais, Instituto Chico Mendes de Conservação da Biodiversidade, Pirassununga, São Paulo, Brazil

Abstract

The present study consists of a detailed phylogenetic analysis of myxosporeans of the *Myxobolus* and *Henneguya* genera, including sequences from 12 *Myxobolus/Henneguya* species, parasites of South American pimelodids, bryconids and characids. Maximum likelihood and maximum parsimony analyses, based on 18 S rDNA gene sequences, showed that the strongest evolutionary signal is the phylogenetic affinity of the fish hosts, with clustering mainly occurring according to the order and/or family of the host. Of the 12 South American species studied here, six are newly described infecting fish from the Brazilian Pantanal wetland. *Henneguya maculosus* n. sp. and *Myxobolus flavus* n. sp. were found infecting both *Pseudoplatystoma corruscans* and *Pseudoplatystoma reticulatum*; *Myxobolus aureus* n. sp. and *Myxobolus pantanalis* n. sp. were observed parasitizing *Salminus brasiliensis* and *Myxobolus umidus* n. sp. and *Myxobolus piraputangae* n. sp. were detected infecting *Brycon hilarii*.

Editor: Donald James Colgan, Australian Museum, Australia

Funding: This study was supported by the Fundação de Amparo à Pesquisa do Estado de São Paulo - FAPESP (Proc. n° 06/59075-6) and by the Centro Nacional de Pesquisa e Conservação de Peixes Continentais of the Instituto Chico Mendes de Conservação da Biodiversidade. MMC was supported by a master's scholarship from FAPESP (Proc. n° 2009/03320-0). Research productivity grant from the Brazilian Fostering Agency CNPq (Proc. n° 477658/2010-5) to EAA. The funders had no role in study design, data collection and analysis, decision to publish, or preparation of the manuscript.

Competing Interests: The authors have declared that no competing interests exist.

* E-mail: edapadriano@gmail.com

Introduction

The phylum Myxozoa harbors a diverse group of metazoan parasites characterized by multicellular spores, with polar capsules containing an extrusible polar filament [1]. Parasites of this phylum have become increasingly important as new species are continually emerging as significant threats to the development of both farmed and natural environment fish [2].

The species have generally been described as infecting aquatic animals, especially fish, however many recent studies have reported the occurrence of myxozoans infecting mollusks [3], amphibians [4,5], reptiles [6,7], birds [8] and mammals [9]. As demonstrated firstly by Wolf and Markiw [10], it has been also proven that several other myxozoan species require an alternate invertebrate host (usually an annelid) to complete the life cycle [11–13].

Within myxozoans, 60 genera of the class Myxosporea and two of the class Malacosporea have been established. Among the myxosporeans, the genera *Myxobolus* Bütschli, 1882, and *Henneguya* Thélohan, 1892 are the most specious, and harbor species that have an important impact on their fish hosts [1,2,14,15]. These two genera are considered separated groups mainly due the presence of caudal appendages in *Henneguya* spp. [16]. However, phylogenetic studies using 18 S rDNA sequence data do not support a phylogenetic separation between *Henneguya* and *Myxobolus* [17–19].

The traditional approach for taxonomic classification of Myxozoa is based on morphological characteristics, primarily of spores, but also of the plasmodia, as well as host and organ and/or tissue specificity. Nevertheless, spore morphology is the main basis of classification and identification, as the spores are relatively rigid and exhibit low intraspecific morphological variability [20–22]. However, molecular analyses have been an important tool in the study of these parasites since the late 1990s [8,20,23]. Molecular biological methods have now been employed in myxozoan research, using a number of different approaches, such as the morphological differentiation of similar species [24–26]; study of host and tissue specificity [27]; elucidation of life cycle [11,23]; study of the phylogenetic relationships of the group [20,28–32] and determination of the phylogenetic position of the myxosporeans inside the metazoans [33].

Some studies have attempted to establish the phylogenetic position of Myxozoa within Metazoa, but the classification of these parasites still remains uncertain, with evidence of the proximity of the myxozoans to both cnidarians and bilaterians, depending on the phylogenetic analysis used [33,34]. However, in a recent study, Nesnidal et al. [35] constructed a phylogenetic tree based on a dataset including 128 genes, pointing out that these new genomic data unambiguously support the evolution of the parasitic Myxozoa from Cnidaria.

Other studies, such as those of Eszterbauer [36], Milanin et al. [30], Gleeson and Adlard [37], Hartigan et al. [5] and Adriano et al. [28], have investigated the phylogenetic relationships of the species within a particular genus. These studies have demonstrated that many factors influence species clustering, such as the phylogenetic proximity of the host, tissue tropism, geographic distribution and morphology characteristics.

The present study describes six new myxosporean species based on morphological and 18 S rDNA data and provides a detailed molecular phylogenetic analysis of the *Henneguya* and *Myxobolus* genera. This analysis includes the addition of sequences from 12 species of parasites of South American fishes and reveals the relationship of these species with parasites of fishes from other continents.

Materials and Methods

All the fish handling was approved by the ethics committee for animal welfare of the Faculdade de Zootecnia e Engenharia de Alimentos of the Universidade de São Paulo (FZEA/USP), in accordance to the Brazilian legislation (Federal Law n° 11.794 from October 8th, 2008). The field samplings were carried out with permission of SISBIO (System Authorization and Information on Biodiversity) of the Ministry of Environment of Brazil (Authorization No. 15507-1).

Specimens of six fish species were caught in two Brazilian states. In the Brazilian Pantanal wetland (Pantanal National Park - 17°50′48″S, 57°24′14″W [distance variation of 3 km], municipality of Poconé, Mato Grosso State), fishes of three species from the family Pimelodidae were examined, including 21 specimens of *Pseudoplatystoma corruscans* Spix & Agassiz, 1829 (common name *pintado*), 12 of *Pseudoplatystoma reticulatum* Eigenmann & Eigenmann, 1889 Syn.: *Pseudoplatystoma fasciatum* Linnaeus, 1766 (common name *cachara*) and 4 specimens of *Zungaro jahu* Ihering, 1898 (common name *jaú*). In the family Bryconidae, 30 specimens of *Salminus brasiliensis* Cuvier, 1816 (common name *dourado*) and 27 of *Brycon hilarii* Valenciennes, 1850 (common name *piraputanga*) were examined. The examinations were performed from October 2008 to November 2009. Other examinations were performed in March 2010 in the Mogi Guaçu River, near the Cachoeira de Emas power plant, in Pirassununga, in the state of São Paulo (21°55′37″ S, 47°22′03″ W), where 4 specimens of *S. brasiliensis* were examined, and in a fish farm in Pirassununga (Centro Nacional de Pesquisa e Conservação de Peixes Continentais-CEPTA/ICMBio), where 5 specimens of the characid *Piaractus mesopotamicus* Holmberg, 1887 (common name *pacu*) were examined.

Immediately after capture, the fish were transported alive to the field laboratory where they were euthanized by benzocaine overdose and then measured and necropsied.

Organs or tissues infected with Myxozoa plasmodia were fixed in absolute ethanol for molecular analyses and in 10% buffered formalin for morphological analyses. Each sample was composed of several plasmodia grouped together according to phenotypic characteristics such as morphologic appearance, host species and tissue tropism.

In the laboratory, the plasmodia fixed in formalin were disrupted, and a small sample of the spores was transferred to a glass slide. A total of 30 spores of each sample were measured and photographed under a Leica DM 1000 light microscope equipped with Leica Application Suite version 1.6.0 image capture software. All measurements were performed according to the patterns established by Lom and Arthur [22]. These slides were then fixed with methanol and stained with Giemsa to be deposited in the collection of the Museum of Natural History, Institute of Biology, State University of Campinas (UNICAMP), state of São Paulo, Brazil.

DNA was extracted from the samples fixed in ethanol using the DNeasy® Blood & Tissue Kit (Qiagen) following the manufacturer's instructions. The product was then quantified in a NanoDrop 2000 (Thermo Scientific, Wilmington, USA) spectrophotometer at 260 nm.

Polymerase chain reaction (PCR) was carried out, according to Adriano et al. [28] at a final volume of 25 µl using the primers MX5-MX3 (Table 1). These primers failed to amplify the samples of *Henneguya corruscans* Eiras et al., 2009 taken from *P. corruscans* and *P. reticulatum*, *Henneguya maculosus* n. sp. taken from *P. corruscans* and *Myxobolus pantanalis* n. sp. These samples were amplified with the primer pairs ERIB1-ACT1R and MYXGEN4f-ERIB10 (Table 1), which amplified two overlapping fragments of approximately 1,000 bp and 1,200 bp respectively of the 18 S rDNA gene. The other primers listed in Table 1 were used only in sequencing reactions.

The amplification reactions were conducted with 10–50 ng of genomic DNA, 2.5 µl of 1× Taq DNA Polymerase buffer (Invitrogen by Life Technologies, MD, USA), 0.75 µl of $MgCl_2$ (1.5 mM), 0.5 µl of dNTPs (0.2 mM), 0.5 µl of each primer (0.2 µM), 0.25 µl of Taq DNA polymerase (2.5 U) (Invitrogen By Life Technologies, MD, USA) and MilliQ purified water. The PCR amplifications were performed in an AG 22331 Hamburg Thermocycler (Eppendorf, Hamburg, Germany). The PCR program consisted of 35 cycles of denaturation at 95°C for 60 s, annealing at 56°C for 60 s and extension at 72°C for 90 s, preceded by an initial denaturation at 95°C for 5 min and followed by a terminal extension at 72°C for 5 min.

The amplified products were analyzed via agarose gel electrophoresis, and their sizes were estimated by comparison with the 1 kb Plus DNA Ladder (Invitrogen by Life Technologies, CA, USA). The purified products were sequenced using the BigDye® Terminator v3.1 Cycle Sequencing kit (Applied Biosystems™) in an ABI 3730 DNA Analyzer (Applied Biosystems™ Inc., CA, USA) using the primers listed in Table 1.

The sequences were visualized, assembled and edited using BioEdit 7.1.3.0 [38] and, for each sequence, a standard nucleotide–nucleotide BLAST (blastn) search was conducted [39] to verify their similarity to other myxosporean sequences in GenBank.

Representative sequences of the 18 S rDNA gene of myxosporeans of the *Henneguya* and *Myxobolus* genera infecting almost every order of fish host available in GenBank were included in the analysis. Since the regions of the 18 S rDNA gene are highly conserved, sequences shorter than 1,000 nucleotides found in the GenBank were not included in the analysis in order to maintain as high a resolution of the resulting trees as possible and to avoid loss of information due to shortening of the aligned sequences and the appearance of too many alignment gaps, as indicated by Rosenberg and Kumar [40].

The sequences of *Kudoa thyrsites* and *Kudoa alliaria* were used as an outgroup. To avoid long-branch attraction effect due to distantly related outgroup sequences, two separated analyses were performed using both maximum likelihood and maximum parsimony methods. The first analysis was conducted with a set of aligned sequences containing outgroup sequences, and the second was performed with the same dataset, but excluding the outgroup. Since there were no substantial differences for topology and bootstrap values between the trees (with and without the outgroup), the tree containing the outgroup sequences was used as the base topology of the final tree.

Table 1. List of primers used in the amplification and sequencing of the 18 S rDNA gene.

Primers	Sequences	References
MX5 (forward)	5′-CTGCGGACGGCTCAGTAAATCAGT-3′	Andree et al. [20]
MX3 (reverse)	5′-CCAGGACATCTTAGGGCATCACAGA-3′	Andree et al. [20]
MC5 (forward)	5′-CCTGAGAAACGGCTACCACATCCA-3′	Molnár et al. [27]
MC3 (reverse)	5′-GATTAGCCTGACAGATCACTCCACGA-3′	Molnár et al. [27]
ERIB1(forward)	5′-ACCTGGTTGATCCTGCCAG-3′	Barta et al. [56]
ERIB10 (reverse)	5′-CTTCCGCAGGTTCACCTACGG-3′	Barta et al. [56]
MYXGEN4f (forward)	5′-GTGCCTTGAATAAATCAGAG-3′	Diamant et al. [57]
ACT1R (reverse)	5′-AATTTCACCTCTCGCTGCCA-3′	Hallett and Diamant [58]

A total of 114 sequences were aligned using ClustalW implemented in the program BioEdit 7.1.3.0 [38] with default settings. Manual adjustments were performed by eye to correct the alignment and remove ambiguous positions from the dataset, resulting in an alignment with a total of 2,270 characters, including alignment gaps. For *Myxobolus* spp. parasites of fish from the Eurasia Palearctic region, only sequences considered valid by Molnár [41] were used.

The best evolutionary model of nucleotide substitution using the Akaike Information Criterion (AIC) was determined with the program jModeltest 0.1 [42], which identified the general time reversible model (GTR+G) as the best fit. From the data, nucleotide frequencies (A = 0.2347, C = 0.1809, G = 0.2742, T = 0.3101) and the rates of the six different types of nucleotide substitution (AC = 1.2478, AG = 3.5282, AT = 1.5099, CG = 0.7026, CT = 4.1695, GT = 1.0000) were estimated. The alpha value of the gamma distribution parameter was 0.4200. These parameters were employed in maximum likelihood (ML) analysis, which was performed using PhyML 3.0 [43] software, with bootstrap confidence values calculated using 500 replicates.

A maximum parsimony (MP) analysis was performed using PAUP* 4.0b10 [44], with a starting tree obtained via stepwise addition and a heuristic search with 320,172 replicates, with a tree bisection–reconnection (TBR) branch swapping algorithm and random sequence addition. Clade supports were assessed via a bootstrap analysis of 1,000 replicates.

The resulting trees were visualized with TreeView 1.6.6 [45] and edited and annotated in Adobe Photoshop (Adobe Systems Inc., San Jose, CA, USA).

The sequences of the South American species were aligned to produce a pairwise similarity matrix using MEGA 5.0 [46], with possible gaps and/or missing data treated via complete deletion, to determine the relationships among them and to verifying the possible occurrence of the same myxosporean infecting different tissues or organs, geographic locations or host species.

Nomenclatural Acts

The electronic edition of this article conforms to the requirements of the amended International Code of Zoological Nomenclature, and hence the new names contained herein are available under that Code from the electronic edition of this article. This published work and the nomenclatural acts it contains have been registered in ZooBank, the online registration system for the ICZN. The ZooBank LSIDs (Life Science Identifiers) can be resolved and the associated information viewed through any standard web browser by appending the LSID to the prefix "http://zoobank.org/". The LSID for this publication is: urn:lsid:zoobank.org:pub:20E2B90B-801D-43E4-B6D0-CE0561FFF7EB. The electronic edition of this work was published in a journal with an ISSN, and has been archived and is available from the following digital repositories: PubMed Central, LOCKSS.

Results

Of the fishes examined in this study, the specimens of *P. corruscans* and *P. reticulatum* were infected by *H. corruscans*, *Henneguya multiplasmodialis* Adriano et al., 2012 and two undescribed myxosporeans, one *Henneguya* sp. that infected the gill filaments (Figs. 1A and 2A) and one *Myxobolus* sp. that infected the gill arch (Figs. 1B and 2B). *Henneguya eirasi* Naldoni et al., 2011 was found only in *P. reticulatum*. The specimens of *S. brasiliensis* from the Brazilian Pantanal wetland and Mogi Guaçu River were infected by *Myxobolus macroplasmodialis* Molnár et al., 1998, and those taken from the Pantanal wetland were infected with two undescribed *Myxobolus* spp., one parasitizing the liver (Figs. 1C and 2C) and another infecting the gill filaments (Figs. 1D and 2D). The specimens of *B. hilarii* were infected by *Myxobolus oliveirai* Milanin et al., 2010 and two undescribed *Myxobolus* spp., one of which infected the spleen (Figs. 1E and 2E), while the other parasitized the kidney (Figs. 1F and 2F). *Henneguya pellucida* Adriano et al., 2005 and *Myxobolus cordeiroi* Adriano et al., 2009 were found in *P. mesopotamicus* and *Z. jahu* respectively, with both infecting the serosa of the visceral cavity of their host.

Sequencing analysis produced 15 sequences from the 18 S rDNA gene of 12 different species parasites of South American freshwater fishes. Of these species, six consisted of still undescribed myxosporeans (Table 2; Figs. 1 and 2). The other six corresponded to described species: *M. macroplasmodialis* (one sequence obtained from parasites from the Brazilian Pantanal wetland and another from the Mogi Guaçu River), *H. pellucida*, *H. eirasi*, *H. multiplasmodialis*, *H. corruscans* and *M. cordeiroi*. For *M. macroplasmodialis* and *H. pellucida*, these were the first 18 S rDNA sequences obtained. For *H. corruscans*, the sequence produced herein was obtained from specimens found to infect *P. reticulatum*, which is a different host species than that from which it was originally described, and for *H. eirasi*, *H. multiplasmodialis* and *M. cordeiroi*, the new sequences were longer than the ones available in GenBank. The undescribed species, morphometric data, spores photomicrographs and drawings of the undescribed species are presented in the description provided below (Figs. 1 and 2; Table 2).

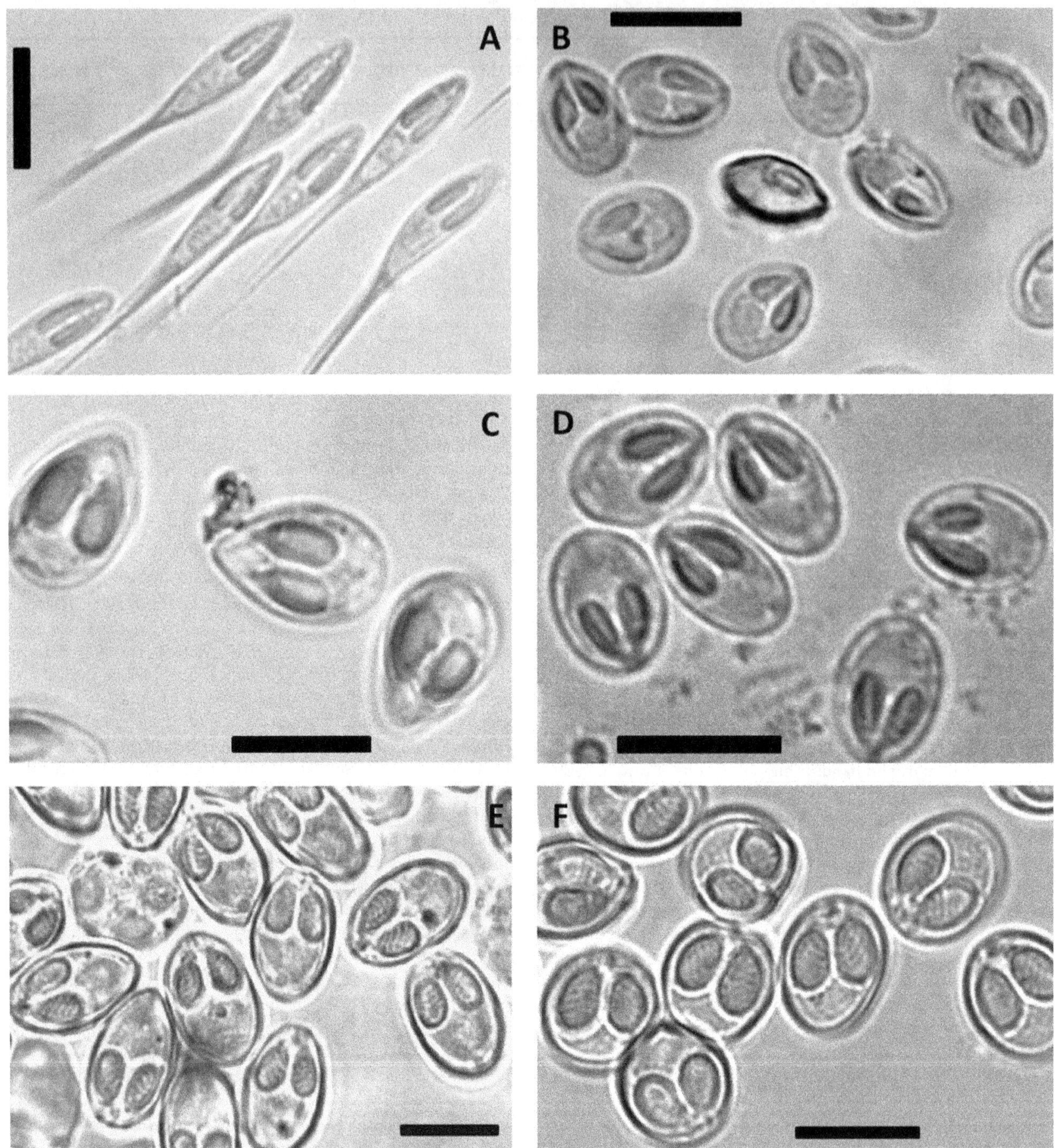

Figure 1. Photomicrographs of fresh spores of the novel myxosporean species. (A) *Henneguya maculosus* n. sp; (B) *Myxobolus flavus* n. sp.; (C) *Myxobolus aureus* n. sp.; (D) *Myxobolus pantanalis* n. sp.; (E) *Myxobolus umidus* n. sp.; (F) *Myxobolus piraputangae* n. sp. Scale bars = 10 µm.

Description of the New *Myxobolus* and *Henneguya* Species

Henneguya maculosus n. sp. urn:lsid:zoobank.org:act:68A6588A-510F-4BCD-829A-42B5FDE40236.

Description. White and elongated plasmodia, measuring 0.5 to 1.5 mm, found in the gill filaments of *P. corruscans* and *P. reticulatum*. The plasmodia exhibited spores in different stages of development. The mature spores were ellipsoidal from the frontal view, with the body measuring 13.7±0.6 µm in length, 4.1±0.2 µm in width, 3.0±0.2 µm in thickness and 17.5±1.0 µm in the caudal process. The polar capsules were elongated, filled the entire anterior half of the spore and were equal in size, measuring 5.6±0.5 µm in length and 1.6±0.2 in width. The polar filaments exhibited 6–7 turns. The measurements of the spores and polar capsules of the parasites found infecting *P. reticulatum* are presented in Table 2.

Type host. *Pseudoplatystoma corruscans* Spix & Agassiz, 1829;

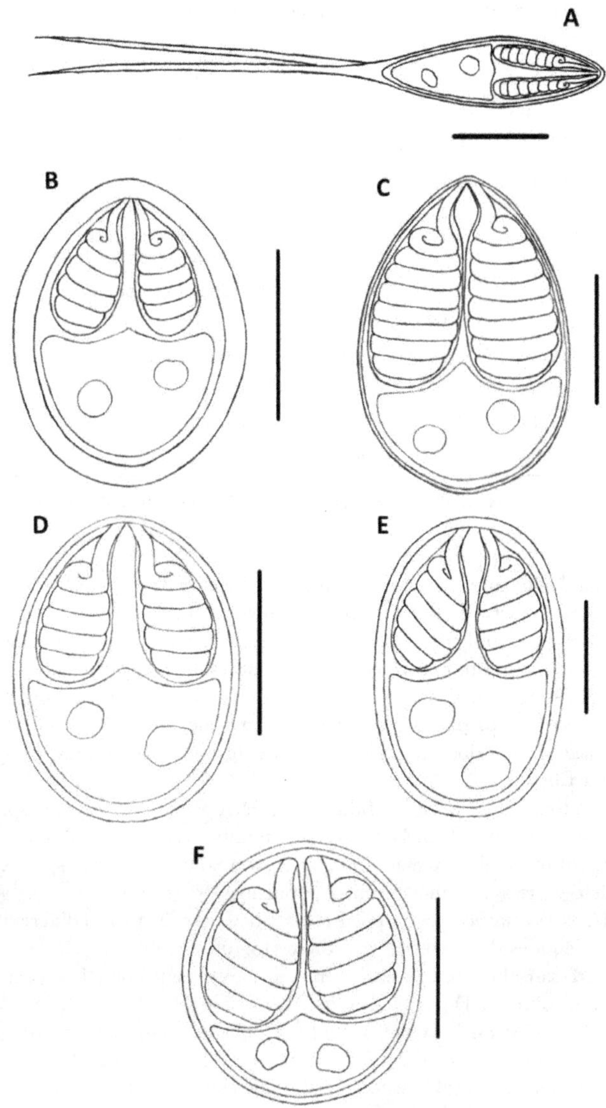

Figure 2. Schematic illustrations of the novel myxosporean species. (A) *Henneguya maculosus* n. sp.; (B) *Myxobolus flavus* n. sp.; (C) *Myxobolus aureus* n. sp.; (D) *Myxobolus pantanalis* n. sp.; (E) *Myxobolus umidus* n. sp.; (D) *Myxobolus piraputangae* n. sp. Scale bars = 5 µm.

Additional host. *Pseudoplatystoma reticulatum* Eigenmann & Eigenmann, 1889.

Prevalence. 19% (4/21) in *P. corruscans* and 16.6% (2/12) in *P. reticulatum.*

Type locality. Pantanal National Park, state of Mato Grosso, Brazil.

Site of infection. Gill filaments.

Type material. Slides with stained spores (syntype) are deposited in the collection of the Museum of Natural History, Institute of Biology, State University of Campinas (UNICAMP), State of São Paulo, Brazil (accession number ZUEC – MYX 34).

Etymology. The specific name refers to the spots in the skin of the host, which is known in Portuguese as *pintado* (in English "spotted surubim"). In Latin *maculosus* = spotted (in English) = *pintado* (in Portuguese).

Remarks. When compared with the other *Henneguya* spp. that have been described infecting *Pseudoplatystoma* fishes, *H. maculosus* n. sp. has spores with a morphology similar to that of *H. corruscans.* However, *H. maculosus* n. sp. has a longer caudal process, and the site of infection within the gills is different, as *H. maculosus* n. sp. forms plasmodia in the filaments, whereas *H. corruscans* forms round plasmodia on the lamellae. Apart from the length of the caudal process and the site of infection within the gills, the morphological features of the two species are very similar to each other, leading to inconclusive definitions regarding the separation of these species. However, combined with the differences mentioned above, molecular data (Table 3, Fig. 3) support the conclusion that these are distinct species.

Sequences consisting of 1,930 and 1,496 nucleotides were produced from the samples of *H. maculosus* n. sp. infecting *P. corruscans* (GenBank accession number KF296344) and *P. reticulatum* (GenBank accession number KF296345) respectively. Between the sequences from the samples collected in the two hosts, there were variations in 29 nucleotides, which represent 1.94% of the comparable total. The South American species that presented the most similar sequence to *H. maculosus* n. sp. found in *P. corruscans* was *H. eirasi*, which exhibited 111 different nucleotides (9.25% of the total). The comparisons with the sequence from the sample found in *P. reticulatum* are displayed in Table 3.

Myxobolus flavus n. sp. urn:lsid:zoobank.org:act:EE0C4537-6A6B-4786-9B4A-4E651A2117EE.

Description. Yellow and spherical plasmodia measuring 1 to 5 mm and containing spores in different stages of development were found in the gill arch of. *P. corruscans* and *P. reticulatum.* The mature spores were oval shaped from the frontal view. The spores obtained from plasmodia collected from *P. corruscans* measured 9.2±0.2 µm in length and 6.5±0.3 µm in width. From the lateral view, the spores measured 4.2±0.2 µm, and the valves were symmetrical. The polar capsules were similar in size and measured 4.5±0.2 µm in length and 1.6±0.1 µm in width (Table 2). The polar filaments presented 4–5 turns. The measurements of the spores and polar capsules of the parasites found infecting *P. reticulatum* are displayed in Table 2.

Type host. Pseudoplatystoma corruscans Spix & Agassiz, 1829.

Additional host. *Pseudoplatystoma reticulatum* Eigenmann & Eigenmann, 1889.

Prevalence. 38% (8/21) in *P. corruscans* and 33.3% (4/12) in *P. reticulatum.*

Type locality. Pantanal National Park, state of Mato Grosso, Brazil.

Site of infection. Gill arch.

Type material. Slides with stained spores (syntype) have been deposited in the collection of the Museum of Natural History, Institute of Biology, State University of Campinas (UNICAMP), State of São Paulo, Brazil (accession number ZUEC – MYX 39).

Etymology. The specific name refers to the yellow plasmodia formed by the parasite in the host, as in Latin, the word *flavus* = yellow.

Remarks. This is the first report of a species of the genus *Myxobolus* infecting fishes of the genus *Pseudoplatystoma.* The morphology of this parasite closely resembles that of *Myxobolus pantanalis* n. sp. (also described in this article). However, they are found in different host species and at different infection sites within the gills, with *M. pantanalis* n. sp. being found infecting the gill filaments of *S. brasiliensis* and *M. flavus* n. sp. obtained from the gill arch of *P. corruscans* and *P. reticulatum.* The molecular data show that these two myxosporeans do not share a close phylogenetic history. Compared to two other pimelodid-infecting *Myxobolus* species, *M. cordeiroi*, which infects *Z. jahu*, and *Myxobolus absonus* Cellere et al., 2002 infecting *Pimelodus maculatus* Lecépède, 1803

Table 2. Mean spore dimensions in μm ± standard deviation for the novel *Henneguya* and *Myxobolus* species from South America including their respective hosts and sites of infection.

Species	Spore length	Spore width	Spore thickness	TL	PCL	PCW	NFC	Infection site and host
H. maculosus n. sp.	13.7±0.6	4.1±0.2	3.0±0.2	17.5±1.0	5.6±0.5	1.6±0.2	6–7	Septated plasmodia in the gill filaments of *P. corruscans*
H. maculosus n. sp.	13.3±0.7	4.4±0.4	3.5±0.4	19.7±2.2	5.2±0.6	1.6±0.2	6–7	Septated plasmodia in the gill filaments of *P. reticulatum*
M. flavus n. sp.	9.2±0.2	6.5±0.3	4.2±0.2	–	4.5±0.2	1.6±0.1	4–5	Gill arch of *P. corruscans*
M. flavus n. sp.	9.3±0.3	6.6±0.3	4.0±0.2	–	4.5±0.2	1.8±0.1	4–5	Gill arch of *P. reticulatum*
M. aureus n. sp.	12.6±0.5	8.3±0.3	5.5±0.3	–	5.7±0.3	2.9±0.2	7–8	Liver of *S. brasiliensis*
M. pantanalis n. sp.	9.3±0.4	6.5±0.4	–	–	4.2±0.5	2.0±0.1	4–5	Gill filaments of *S. brasiliensis*
M. umidus n. sp.	13.5±0.7	7.8±0.4	7.7±0.1	–	5.1±0.4	2.7±0.3	4–5	Spleen of *B. hilarii*
M. piraputangae n. sp.	10.1±0.5	8.7±0.5	6.7±0.3	–	5.2±0.4	3.0±0.3	4–5	Kidney of *B. hilarii*

TL = tail length (only for *Henneguya maculosus* n. sp.); PCL = polar capsule length; PCW = polar capsule width; NFC = number of polar filament coils; dash = no data.

exhibit different morphological and morphometric characteristics in comparison to *M. flavus* n. sp.

Sequences with lengths of 1,520 and 1,530 bases of the 18 S rDNA gene were obtained from the hosts *P. corruscans* (GenBank accession number KF296346) and *P. reticulatum* (GenBank accession number KF296347) respectively. The difference between the sequences was 28 nucleotides, which represents 1.88% of the comparable total (Table 3). When compared to the other South American species, the smallest amount of interspecific differences was observed between *M. flavus* n. sp. infecting *P. corruscans* and *H. corruscans* infecting *P. reticulatum*, corresponding to 224 different nucleotides (15.17%). Comparisons with the sequence from the sample found in *P. reticulatum* are displayed in Table 3.

Myxobolus aureus n. sp. urn:lsid:zoobank.org:act:2EFA6C10-977B-4B56-B0A5-CC393833AEDB.

Description. Small white plasmodia, measuring approximately 0.4 mm in length found in the liver of *S. brasiliensis*. The mature spores were oval shaped from the frontal view and measured 12.6±0.5 μm in length and 8.3±0.3 μm in width. From the lateral view, the spores measured 5.5±0.3 μm, and the valves were smooth and symmetrical. The meeting of the two valves produced a tenuous sutural line. The two polar capsules showed the same size and measured 5.7±0.3 μm in length and 2.9±0.2 μm in width (Table 2). The polar filaments exhibited 7–8 turns.

Type host. *Salminus brasiliensis* Cuvier, 1816: Characiformes: Bryconidae.

Prevalence. 30% (9/30).

Type locality. Pantanal National Park, state of Mato Grosso, Brazil.

Site of infection. Liver.

Type material. Slides with stained spores (syntype) have been deposited in the collection of the Museum of Natural History, Institute of Biology, State University of Campinas (UNICAMP), State of São Paulo, Brazil (accession number ZUEC – MYX 35).

Etymology. The specific name refers to the Portuguese common name of the host, *dourado* (in English, "golden"), arising from the golden color of *S. brasiliensis*. In Latin, *aureus* = golden.

Remarks. When compared to the other *Myxobolus* spp. found infecting *S. brasiliensis*, *M. aureus* n. sp. parasitizes the liver, whereas the previously described *Myxobolus salminus* Adriano et al., 2009 and *M. pantanalis* n. sp. (also described in this article) are both found infecting the gills, and *M. macroplasmodialis* is found in the visceral cavity. Regarding the morphology of *M. aureus* n. sp., its spores are similar to those of *M. salminus*, exhibiting a narrow anterior portion, but *M. aureus* n. sp. spores are longer and wider, with longer and wider polar capsules. Regarding the other *Myxobolus* spp. parasitizing characiforms, the morphology of *M. aureus* n. sp. does not resemble any of the species previously described.

A total of 1,757 bases of the 18 S rDNA gene of *M. aureus* n. sp. were sequenced (GenBank accession number KF296348). According to the similarity matrix (Table 3), the South American species that carries the most similar sequence to *M. aureus* n. sp. is *Myxobolus piraputangae* n. sp. (also described here), with 199 different nucleotides detected between them (15.40%) (Table 3).

Myxobolus pantanalis n. sp. urn:lsid:zoobank.org:act:A36FA08F-ACD6-4E9C-A415-E705044734D1.

Description. White, round-to-elongated plasmodia measuring approximately 0.2 to 1.2 mm in length and found in the posterior end (distal region) of the gill filaments of *S. brasiliensis*. The mature spores were oval shaped from the frontal view and measured 9.3±0.4 μm in length and 6.5±0.4 μm in width. The two polar capsules were elongated, pear shaped and equal in size and measured 4.2±0.5 μm in length and 2.0±0.1 μm in width (Table 2). The polar filaments exhibited 4–5 turns.

Type host. *Salminus brasiliensis* Cuvier, 1816: Characiformes: Bryconidae.

Prevalence. 33.3% (10/30).

Type locality. Pantanal National Park, state of Mato Grosso, Brazil.

Site of infection. Gill filaments.

Type material. Slides with stained spores (syntype) have been deposited in the collection of the Museum of Natural History, Institute of Biology, State University of Campinas (UNICAMP), State of São Paulo, Brazil (accession number ZUEC – MYX 38).

Etymology. The specific name refers to the location in which the infected fishes were caught, the Brazilian Pantanal. In Latin, *pantanalis* = wetland.

Remarks. In *S. brasiliensis*, *M. salminus* was previously described as infecting the gills. However, the tissues infected by these two species are different, as *M. pantanalis* n. sp. is found in the surface of the filaments, and *M. salminus* is reported to infect the blood vessels of the gill filaments. These two species are also very different morphologically, as the spores of *M. pantanalis* n. sp. are

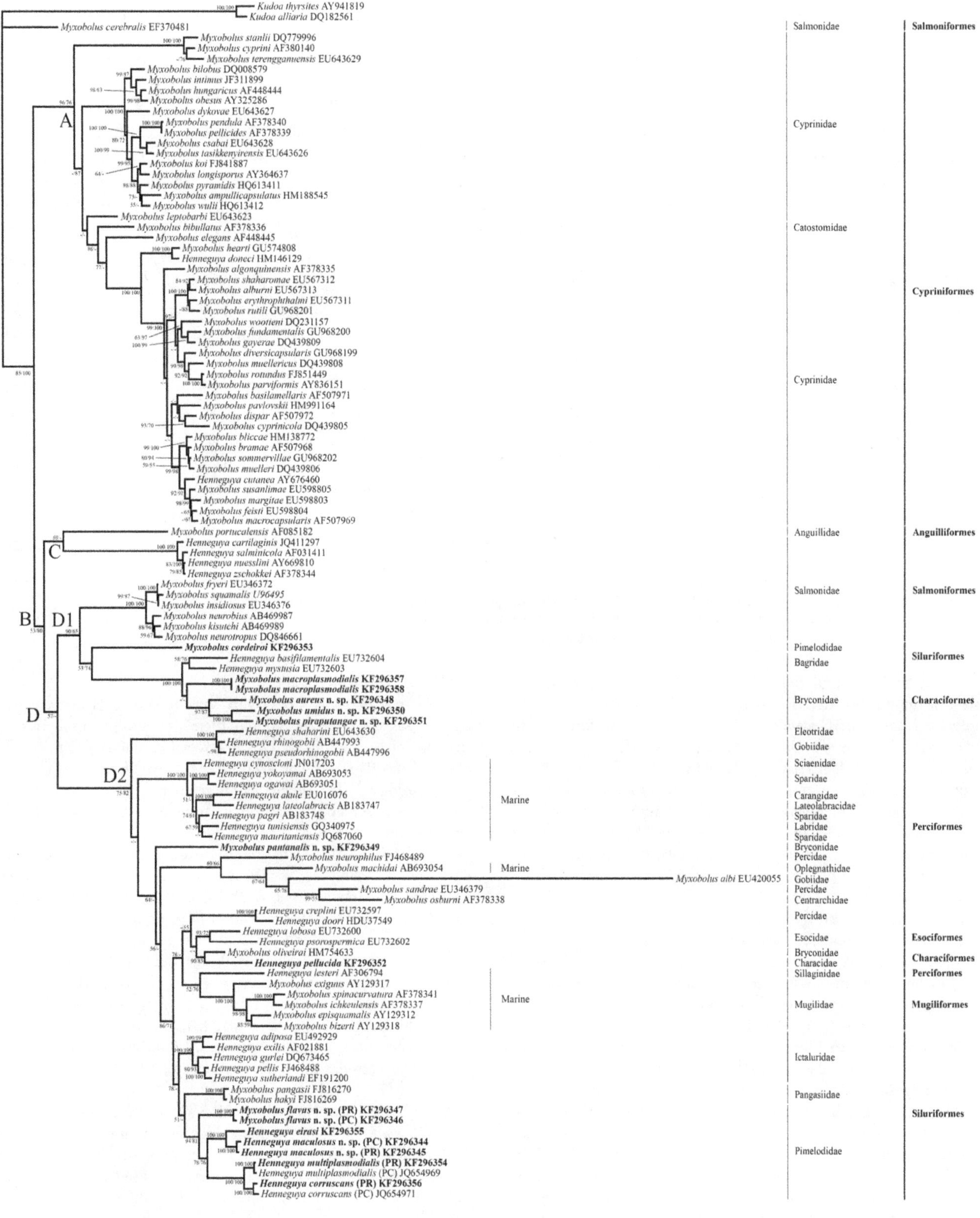

Figure 3. Maximum likelihood phylogenetic tree based on partial 18 S rDNA sequences. The bootstrap support values at branching points are listed as maximum likelihood/maximum parsimony. GenBank accession numbers are given following the name of the species. PW = Pantanal wetland; MG = Mogi Guaçu River; PC = *Pseudoplatystoma corruscans*; PR = *Pseudoplatystoma reticulatum*.

Table 3. Similarity matrix for the 18 S rDNA sequences from all available *Henneguya* and *Myxobolus* species from South America adjusted for missing data (gaps).

Species	1	2	3	4	5	6	7	8	9	10	11	12	13	14	15	16	17	18
1. *H. maculosus* n. sp.(PC)		1.94	15.48	15.88	9.25	13.56	14.04	15.53	15.82	21.04	15.45	21.73	21.73	23.76	15.80	18.13	19.03	20.38
2. *H. maculosus* n. sp. (PR)	29		17.81	16.41	9.67	18.24	18.02	15.76	16.07	23.30	17.19	24.26	24.26	23.56	15.53	17.96	24.02	25.36
3. *H. multiplasmodialis* (PC)	240	265		0.91	15.92	7.03	7.02	15.79	16.03	24.76	18.52	23.86	23.86	23.34	17.42	18.50	27.04	25.95
4. *H. multiplasmodialis* (PR)	208	213	11		15.62	6.26	6.48	15.46	15.56	24.17	18.11	24.40	24.40	23.89	17.15	17.51	26.47	25.34
5. *H. eirasi*	111	116	187	184		16.40	17.13	15.93	16.00	24.31	17.70	27.10	27.10	24.52	16.12	17.68	23.46	25.07
6. *H. corruscans* (PC)	257	271	109	82	193		2.32	15.98	15.86	20.99	17.54	22.85	22.85	26.77	17.46	18.77	21.25	21.64
7. *H. corruscans* (PR)	265	268	109	85	202	44		15.17	15.50	21.21	17.68	22.17	22.17	25.94	18.09	18.75	21.48	21.92
8. *M. flavus* n. sp. (PC)	233	232	234	203	191	236	224		1.88	23.98	16.38	23.23	23.23	23.12	16.50	18.32	26.68	26.71
9. *M. flavus* n. sp. (PR)	239	240	239	203	192	235	230	28		24.37	16.87	24.18	24.18	23.19	16.33	19.21	26.54	26.74
10. *M. aureus* n. sp.	361	344	381	314	289	360	364	359	368		22.50	16.69	16.69	25.85	23.87	26.54	15.50	15.40
11. *M. pantanalis* n. sp.	290	253	287	237	209	328	331	244	253	389		22.36	22.36	25.98	15.96	15.83	22.92	23.66
12. *M. macroplasmodialis* (PW)	321	354	352	316	320	337	327	338	359	251	332		0.00	22.91	23.05	22.63	17.66	16.86
13. *M. macroplasmodialis* (MG)	321	354	352	316	320	337	327	338	359	251	332	0		22.91	23.05	22.63	17.66	16.86
14. *M. cordeiroi*	346	336	337	313	292	393	381	341	336	381	384	329	329		21.94	23.92	25.59	26.84
15. *M. oliveirai*	236	232	262	225	192	261	271	243	246	359	241	343	343	317		15.55	25.45	25.77
16. *H. pellucida*	280	268	279	229	212	285	285	277	291	414	244	336	336	354	235		26.33	27.37
17. *M. umidus* n. sp.	244	264	301	289	259	272	276	306	297	202	297	195	195	295	283	307		10.44
18. *M. pirautangae* n. sp.	258	279	288	276	276	275	279	305	299	199	304	185	185	310	286	320	135	

The lower triangular matrix shows the actual differences, while the upper triangular matrix shows the differences in terms of percentage of nucleotides.
PC = *Pseudoplatystoma corruscans*; PR = *Pseudoplatystoma reticulatum*; PW = Pantanal wetland; MG = Mogi Guaçu River.

ellipsoids with a round anterior end, while *M. salminus* exhibits a narrow anterior portion. Regarding the other *Myxobolus* spp. that infect *S. brasiliensis*, *M. aureus* n. sp. (described above) is found in the liver, and *M. macroplasmodialis* forms large plasmodia in the visceral cavity. All other *Myxobolus* spp. infecting characiforms differed from *M. pantanalis* n. sp. regarding morphological and morphometric aspects or other phenotypic features, such as the preferential site of infection and host species. As cited above, *M. pantanalis* n. sp. exhibited a similar morphology to *M. flavus* n. sp., however, they have different sites of infection and phylogenetically distant hosts.

A sequence consisting of 1,901 bases of the 18 S rDNA gene was produced from *M. pantanalis* n. sp. (GenBank accession number KF296349). According to the similarity matrix (Table 3), the smallest amount of interspecific differences was found with *H. maculosus* n. sp. parasitizing *P. corruscans*, corresponding to 290 different nucleotides (15.45%) (Table 3).

Myxobolus umidus n. sp. urn:lsid:zoobank.org:act:926BA436-7A00-4930-8614-C91CB6D57576.

Description. White and spherical plasmodia, measuring 0.4 to 0.8 mm, found in the spleen of *B. hilarii*. The mature spores were ellipsoid in shape from the frontal view and measured 13.5±0.7 μm in length and 7.8±0.4 μm in width. From the lateral view, the spores measured 7.7±0.1 μm, and the valves were symmetrical. The two polar capsules were oval shaped and of equal size, measuring 5.1±0.4 μm in length and 2.7±0.3 μm in width, and the polar filaments exhibited 4–5 turns, arranged obliquely with respect to the axis of the capsules (Table 2).

Type host. *Brycon hilarii* (Valenciennes, 1850: Characiformes: Bryconidae).

Prevalence. 37% (10/27).

Type locality. Pantanal National Park, state of Mato Grosso, Brazil.

Site of infection. Spleen.

Type material. Slides with stained spores (syntype) have been deposited in the collection of the Museum of Natural History, Institute of Biology, State University of Campinas (UNICAMP), State of São Paulo, Brazil (accession number ZUEC – MYX 37).

Etymology. The specific name refers to the location in which the host is observed, the Brazilian Pantanal wetland, as in Latin, *umidus* = wet.

Remarks. When compared to the other *Myxobolus* spp. found infecting *B. hilarii*, *M. umidus* n. sp. differs from *M. oliveirai*, *Myxobolus brycon* Azevedo et al., 2011, and *Myxobolus piraputangae* n. sp. (also described in this article) in morphological and morphometric characteristics and other phenotypic features, such as the preferential site of infection, as this is the only species already described infecting the spleen of *B. hilarii*. Regarding species parasites of other characiform hosts, *Myxobolus serrasalmi* Walliker, 1969 exhibits spores with a similar size and is also found in the spleen of *Serrasalmus rhombeus* Linnaeus, 1766, but it has longer and wider polar capsules than *M. umidus* n. sp., as well as distinct host species.

A 1,307 base sequence of the 18 S rDNA gene was generated from *M. umidus* n. sp. (GenBank accession number KF296350). The South American species whose sequence presented the smallest amount of genetic differences compared to *M. umidus* n. sp. was *M. piraputangae* n. sp., with 135 different nucleotides (10.44%) (Table 3).

Myxobolus piraputangae n. sp. urn:lsid:zoobank.org:act:193C9F1F-4CAC-481B-A51C-06B77E8ED383.

Description. White and spherical plasmodia, measuring 0.5 to 1.0 mm, found in the kidney of *B. hilarii*. Mature spores were circular to slightly ellipsoidal in frontal view, measuring 10.1±0.5 μm in length and 8.7±0.5 μm in width. From the lateral view, the spores measured 6.7±0.3 μm, and the valves were symmetrical, with conspicuous pores for extrusion of the filaments in the anterior end of the spores. The two polar capsules were oval shaped and equal in size, measuring 5.2±0.4 μm in length and 3.0±0.3 μm in width and accounting for approximately two-thirds of the total spore length (Table 2). The polar filaments exhibited 4–5 turns arranged obliquely with respect to the axis of the capsules.

Type host. *Brycon hilarii* (Valenciennes, 1850: Characiformes: Bryconidae).

Prevalence. 7.4% (2/27).

Type locality. Pantanal National Park, state of Mato Grosso, Brazil.

Site of infection. Kidney.

Type material. Slides with stained spores (syntype) have been deposited in the collection of the Museum of Natural History, Institute of Biology, State University of Campinas (UNICAMP), State of São Paulo, Brazil (accession number ZUEC – MYX 36).

Etymology. The specific name refers to the common name of the host, which is the *piraputanga* in Brazil.

Remarks. Compared to the other *Myxobolus* species infecting *B. hilarii*, *M. piraputangae* n. sp. exhibited different spore morphology, presenting circular spores, rather than pear shaped spores, as in *M. oliveirai* and *M. brycon*, or ellipsoidal spores, as in *M. umidus* n. sp. The dimensions of these species are also very different, as *M. oliveirai* and *M. umidus* n. sp. exhibit larger spores, while those of *M. brycon* are much smaller than the spores of *M. piraputangae* n. sp. Regarding the other *Myxobolus* spp. was found infecting other characiform fishes, *M. macroplasmodialis* was described infecting *S. brasiliensis* and *Myxobolus cunhai* Penido, 1927 was described infecting *Pygocentrus piraya* Cuvier, 1819, which have spores similar in size to those of *M. piraputangae* n. sp., however, they differ in other phenotypic aspects, such as spore morphology, the site of infection and the host species.

A 1,296 nucleotide sequence of the 18 S rDNA gene from *M. piraputangae* n. sp. was generated (GenBank accession number KF296351). The South American species presenting the sequence with the smallest genetic variation from that of *M. piraputangae* n. sp. was *M. umidus* n. sp., corresponding to 135 different nucleotides (10.44%) (Table 3).

Phylogenetic Analyses

The partial 18 S rDNA gene sequences obtained from 15 samples in this study, plus three others from parasites of South American fishes, were compared with each other and with other myxozoan sequences from other geographic regions available in GenBank. BLAST searches demonstrated that these sequences exhibit a close relationship with those of other myxosporeans that have been previously sequenced, but they were not identical to any of them.

The similarity matrix of the 18 S rDNA gene sequences of the South American species showed that among the myxosporeans that infect more than one host, the intraspecific variation ranged from 0.91%, between the two isolates of *H. multiplasmodialis* found infecting *P. corruscans* and *P. reticulatum*, to 2.32%, between the samples of *H. corruscans* found infecting *P. corruscans* and *P. reticulatum* (Table 3). The *M. macroplasmodialis* sequences from *S. brasiliensis* sampled in the Brazilian Pantanal and the Mogi Guaçu River were identical. The smallest interspecific difference was observed between *H. multiplasmodialis* from *P. reticulatum* and *H. corruscans* from *P. corruscans*, which was 6.26%.

The phylogenetic analyses resulted in similar ML and MP topologies. Therefore, only the ML tree was used to represent the basic tree topology (Fig. 3), in which the clusters occurred mainly according to the phylogenetic affinities of the fish hosts. The tree was divided initially into two major clades, A and B. Clade A was formed by *Myxobolus/Henneguya* species associated exclusively with fishes belonging to the order Cypriniformes; clade B was further divided into two other clades (C and D); clade C consisted of one *Myxobolus* species parasite of anguiliforms and four *Henneguya* spp., parasites of salmoniforms; clade D was further divided to form two subclades (D1 and D2), both of *Myxobolus/Henneguya* species, which are parasites of different fish families. Subclade D1 harbored, in one branch, all the *Myxobolus* species parasites of salmoniforms, except *Myxobolus cerebralis* Hofer, 1903, which in our topology, was clustered outside the ingroup, much closer to the outgroup. In the other branch of the subclade D1, *M. cordeiroi*, a parasite of *Z. jahu*, a South American pimelodid, and *Henneguya basifilamentalis* Molnár et al., 2006 and *Henneguya mystusia* Sarkar, 1985, parasites of *Hemibagrus nemurus* Valenciennes, 1840, an Asian bagrid, were clustered. The branch D1 also harbored four *Myxobolus* species, parasites of South American characiforms of the bryconid family. The speciose subclade D2 harbored *Myxobolus/Henneguya* species parasites of marine perciforms and mugiliforms, two freshwater perciforms (*Henneguya creplini* Gurley, 1894 and *Henneguya doori* Guilford, 1963), plus *Henneguya lobosa* Cohn, 1895 and *Henneguya psorospermica* Thelohan, 1895 parasites of esociforms, and the South American *M. oliveirai* and *H. pellucida*, parasites of characiforms of the families Bryconidae and Characidae, respectively. *M. pantanalis* n. sp., another species that is parasite of a South American bryconid, appeared alone in a branch away from the other two groups of characiform parasites. Still in D2, five *Henneguya* spp., parasites of fishes of the family Ictaluridae, two *Myxobolus* spp., parasites of fishes of the pangasiid family, and five species (four *Henneguya* and one *Myxobolus*) that are parasites of fishes of the family Pimelodidae formed a large group composed of species that are parasites of siluriforms (Fig. 3).

Discussion

In this report, based on analyses of the 18 S rDNA gene sequences and morphological features, six novel species of myxozoans are described, one of which belongs to the genus *Henneguya* and five to *Myxobolus*. Of these six new species, four *Myxobolus* spp. were found infecting bryconid hosts, with *M. aureus* n. sp. and *M. pantanalis* n. sp. being parasites of *S. brasiliensis* and *M. umidus* n. sp and *M. piraputangae* n. sp. parasitizing *B. hilarii*. In *S. brasiliensis* and *B. hilarii* four myxosporean species have been described previously, with *M. macroplasmodiaslis* and *M. salminus* parasitizing *S. brasiliensis* and *M. oliveirai* and *M. brycon* parasitizing *B. hilarii*. Both *M. flavus* n. sp. and *H. maculatus* n. sp. are described here as infecting two pimelodid species, *P. corruscans* and *P. reticulatum*. The observation of these myxosporeans in these two hosts expands the number of species described as infecting these pimelodids. The parasites previously reported in these fish species include *H. corruscans*, *H. eirasi* and *H. multiplasmodialis*, which were found infecting both hosts in the natural environment in Brazilian rivers, in addition to *Henneguya pseudoplatystoma* Naldoni et al., 2009, which was recorded infecting hybrid fish, resulting from a cross between *P. corruscans* and *P. reticulatum*, on fish farms in the São Paulo and Mato Grosso do Sul states, in Brazil.

The findings presented above reveal the notable fact that *P. corruscans* and *P. reticulatum*, which are species that are phylogenetically very close, can host the same myxosporean species. These results are in accordance with the host specificity criteria noted for *Henneguya* and *Myxobolus* species by Molnár [47] and Molnár et al. [48].

The similarity matrix showed that among the myxosporean species that are able to infect both *P. corruscans* and *P. reticulatum*, the genetic divergence between the samples obtained from the different hosts was only 0.91% for *H. multiplasmodialis*, 1.88% for *M. flavus* n. sp., 1.94% for *H. maculosus* n. sp. and 2.32% for *H. corruscans*. We consider these levels to reflect intraspecific divergences. On the other hand, the interspecific differences observed among the sequences of the morphologically distinct species varied from 6.26% to 27.37% (Table 3).

Might be expected that, because of the sampling method employed, in which several cysts from the same parasite species collected from each host species were grouped together to produce a satisfactory amount of DNA, we would detect intraspecific variations among samples from the same host. This might be observable on the sequencing electropherograms, but was not in fact detected. These results suggest that, although more specific studies regarding the population structure of this species with more suitable molecular markers are required, there exist different lineages of the same species infecting these two phylogenetically similar hosts, probably on a speciation process.

There is no exact value for determining whether particular differences in 18 S rDNA genes can be used to discriminate intra- or interspecific variation in members of the phylum Myxozoa [49]. Thus, this decision must be made for each individual case, always with the aid of other aspects of the biology and/or ecology of the organisms, such as their morphology, tissue and/or organ tropism, type of host and geographic area [49]. In taxonomic analyses of Myxozoa, Molnár et al. [50] observed a variation of 3.6% among samples of *Myxobolus dujardini* Thélohan, 1892, whereas in five other species, the intraspecific differences varied from 0.1% to 2.3%. In a study addressing isolates of *M. cerebralis* from different hosts and geographic locations, Andree et al. [20] observed a variation of only 0.8%. *Myxobolus fryeri* Ferguson et al., 2008 and *Myxobolus insidiosus* Wyatt & Pratt, 1963 were considered different species by Ferguson et al. [21] even though only 0.5% variation was detected between their 18 S rDNA sequences. Because the samples presented different infection sites and spores with significantly different lengths, the authors concluded that they were separate species. The same conclusion was made by Easy et al. [51], who stated that even though *Myxobolus intramusculi* Easy et al., 2005 and *Myxobolus procerus* Kudo, 1934 were both found in the muscles of *Percopsis omiscomaycus* Walbaum, 1792 and presented differences in only 2% of their 18 S rDNA sequences, they were considered two separate species, because *M. intramusculi* was found in intramuscular tissue, whereas from *M. procerus* was found infecting the adjacent connective tissue.

This is the first study investigating the phylogenetic relationships among several *Henneguya* spp. and *Myxobolus* spp. infecting South American fishes. In the ML and MP trees produced in this study, the characteristic presenting the most apparent phylogenetic signal was the relationship of the fish hosts, with clusters occurring mainly according to the order and family. This pattern can be clearly observed in almost all clades. Clade A was composed exclusively of *Myxobolus/Henneguya* species, parasites of fishes of the order Cypriniformes, indicating that this group has a distinct evolutionary origin from the other species analyzed. Following the tendency to group according to the order/family of the fish host, *Henneguya* spp. parasites of salmonids formed a subclade in clade C, while *Myxobolus* spp. parasites of salmonids formed a subclade in D1. However, *M. cerebralis* appears separated from the *Myxobolus* spp. parasites of salmonids, revealing that this group is not monophyletic. The South American species, parasites of characiform fishes also formed polyphyletic clusters, with species occurring in subclades D1 and D2. However, in subclade D1 there is a further subclade composed exclusively of four *Myxobolus* spp. parasites of the family Bryconidae. In clade D2, the characiform parasites *M. oliveirai* (from a bryconid host) and *H. pellucida* (from a characid host) clustered together, and *M. pantanalis* n. sp. clustered away from the other characiform parasites.

Regarding the *Myxobolus/Henneguya* species parasites of siluriforms, *M. cordeiroi*, parasite of a pimelodid and *H. basifilamentalis* and *H. mystusia*, parasites of an Asian bagrid, all clustered in a subclade of D1, while in a subclade of D2, all other species parasites of siluriforms (six *Henneguya* spp. of North American ictalurids, two *Myxobolus* spp. of an Asian pangasiid and four *Henneguya* spp. and one *Myxobolus* of South American pimelodids), formed a grouping composed exclusively of parasites of this host order. Despite the polyphyletic clustering observed among parasites of siluriforms, the large majority of the species clustered in a monophyletic subclade in D2, and even *M. cordeiroi*, *H. basifilamentalis* and *H. mystusia* clustered together in D1.

In general, even though there are some inconsistencies, as reported above, the tree produced in the present study shows a strong tendency of the *Myxobolus/Henneguya* species to form clusters based on the phylogenetic relationships of the fish hosts, even if these hosts are from different biogeographic regions, as has been previously reported by other authors [28,32]. This suggests that the origins and radiations of these parasites are very ancient, perhaps as old as the hosts themselves, which, in the case of Osteichthyes, had an Early Silurian origin [52] and Mesozoic radiation [53].

The clustering of myxosporeans according to host phylogeny was also observed in a study by Gleeson et al. [54] which reported a clear separation among multivalvulid myxozoans of the genus *Kudoa*, parasites of elasmobranch and teleost hosts. However, this separation occurred at a higher taxonomic level of the hosts than that observed in this study for *Henneguya* and *Myxobolus*, since some *Kudoa* spp. have broader host specificity with infections across multiple families and even orders. Similar findings were reported by Gleeson and Adlard [37] who founded that *Chloromyxum* spp. parasites of elasmobranchs tend to cluster together despite having different geographic origins.

In clade D the occurrence of some clusters of lineages of myxozoans from a marine environment was observed, as were clusters with distribution according to the host order, with one clade composed of *Henneguya* spp. parasites of marine perciforms, and another formed by *Myxobolus* spp. parasites of mugiliforms. The only exception is the marine species *Myxobolus machidai* Li et al., 2012, which clustered with the estuarine *Myxobolus neurophilus* Guilford, 1963, *Myxobolus albi* Picon-Camacho et al., 2009 and *Myxobolus sandrae* Reuss, 1906, and the freshwater *Myxobolus osburni* Herrick, 1936.

Eszterbauer [36] suggested that the preference for a particular site of development plays an important role in determining the phylogenetic relationships among myxosporean species, and she also proposed that genetic separation based on tissue and/or organ tropism is a more ancient evolutionary characteristic than host specificity. However, the results reported by Eszterbauer [36] were based on data obtained from a study including only myxosporean species parasitizing cyprinid fishes. This does not mean that the site of infection is not a relevant evolutionary feature, but it appears that organ tropism has more evolutionary influence within clades that are defined by a stronger evolutionary signal, such as the order/family of the fish host. In accordance with Eszterbauer [36] and Fiala [18], our topology shows that tissue tropism has some phylogenetic influence in some places in the tree, especially when considering myxosporean species parasitizing hosts that are phylogenetically close. This effect can be observed in clade A, where the subclade formed by *Myxobolus diversicapsularis* Slukhai, 1984, *Myxobolus muellericus* Molnár et al., 2006, *Myxobolus rotundus*

Nemeczek, 1911 and *Myxobolus parviformis* Kallert et al., 2005 is composed exclusively of *Myxobolus* species that parasitize the gills, while the subclade formed by *Myxobolus bliccae* Donec & Tozyyakova, 1984, *Myxobolus bramae* Reuss, 1906, *Myxobolus sommervillae* Molnár et al., 2010 and *Myxobolus muelleri* Bütschli, 1882 consists only of species that infect the blood vessels of the gills. A similar situation can be observed in clade C where, in the group of *Myxobolus* spp. parasites of salmoniforms, there is a clear separation between species infecting muscle (*M. fryeri*, *Myxobolus squamalis* Egusa et al., 1990 and *M. insidiosus*) and those infecting the central nervous system (*Myxobolus neurobius* Schuberg & Schröder, 1905, *Myxobolus kisutchi* Yasutake & Wood, 1957 and *Myxobolus neurotropus* Hogge et al., 2008). In all these cases, these myxosporean species clustered primarily by host affinity and in a second step, by organ/tissue tropism.

The classic taxonomic distinction between the *Myxobolus* and *Henneguya* genera is based on a single morphological character: the presence of the caudal process in *Henneguya* spp. and its absence in *Myxobolus* spp. [16]. Our phylogenetic analyses showed that species of the *Henneguya* and *Myxobolus* genera were grouped together, and this corroborates the results reported by several other authors in studies conducted in other regions of the world [17–19,21]. This phenomenon can be observed in clade A, where *Henneguya doneci* Shulman, 1962 and *Henneguya cutanea* Dogiel & Petruschewsky, 1933 are found in a clade composed primarily of *Myxobolus* spp. The same situation occurred in clade D, where, as mentioned before, the parasites of bagrids *H. mystusia* and *H. basifilamentalis* clustered in a clade composed otherwise only of *Myxobolus* spp., as well as *M. oliveirai*, which clustered with *H. pellucida* and in the subclade composed of parasites of siluriforms, *Myxobolus hakyi* Baska et al., 2009, *Myxobolus pangasii* Molnár et al., 2006 and *M. flavus* n. sp. clustered in a predominantly *Henneguya* clade (Fig. 3). Thus, the findings of the present study of *Myxobolus*/*Henneguya* parasites of South American species, corroborate the idea that, based on analyses of ribosomal genes, there is no support for a phylogenetic separation of the *Henneguya* and *Myxobolus* genera. This absence of phylogenetic separation between these two genera led Kent et al. [17] to state that the caudal appendage of *Henneguya* spp. is not a valid feature for characterization of the genus. These authors also affirmed that the fact that species belonging the *Henneguya* and *Myxobolus* genera cluster together in groups with a monophyletic origin suggests that the genetic ability to develop spores with a caudal appendix is broadly distributed within the group, but due to still unknown reasons, only certain lineages express this character. A piece of evidence that provides support for this theory is the occurrence of some *Myxobolus* spp. that develop spores both with and without a *Henneguya*-like caudal prolongation within a single plasmodium, as reported for *Myxobolus turpisrotundus* Liu et al., 2010 [19], *Myxobolus bizerti* Bahri and Marques, 1996, *Myxobolus mulleri* Bahri, 1997 and *Myxobolus heterosporus* Baker, 1963 [55].

To date, phylogenetic studies of myxosporeans have been performed using species parasites from the Nearctic, Palearctic and Australian bioregions. In this study, for the first time, a consistent number of sequences of *Myxobolus*/*Henneguya* spp. parasites of fishes from the Neotropical region were included in a phylogenetic analysis with almost all of the valid sequences of species of these two genera available in GenBank. The results reveal that the strongest evolutionary signal for *Myxobolus*/*Henneguya* species is the phylogenetic affinity of the fish hosts, with clusters occurring mainly based on the order and/or family of the host. The origin of the Osteichthyes is estimated to the Early Silurian, around 427 million years ago [52], with Mesozoic radiation [53]. The observation that the phylogenetic affinity of the fish hosts is the strongest evolutionary signal in *Myxobolus*/*Henneguya* species suggests an ancient origin and radiation of this important and complex group of parasites.

Acknowledgments

The authors are grateful to Dr. Laerte Batista de Oliveira Alves, manager of the National Center for Research and Conservation of Continental Fish (CEPTA/ICMBio), for support during the field work, to Dr. José Augusto Ferraz de Lima, manager of Pantanal National Park, for providing the structure for the field laboratory and hosting the researchers during the fieldwork, to Ricardo Afonso Torres de Oliveira (CEPTA/ICMBio) for help in dissecting the fish and to the anonymous reviewers for their helpful suggestions that improved this article.

Author Contributions

Conceived and designed the experiments: MMC EAA PSC MRMS AAMM. Performed the experiments: MMC EAA MRMS. Analyzed the data: MMC EAA MRMS AAMM. Contributed reagents/materials/analysis tools: EAA PSC AAMM. Wrote the paper: MMC EAA.

References

1. Feist SW, Longshaw M (2006) Phylum Myxozoa. In: Woo PTK, editor. Fish diseases and disorders: Protozoan and Metazoan infections. 2ª ed. UK: CAB International. 230–296.
2. Lom J, Dyková I (2006) Myxozoan genera: definition and notes on taxonomy, life-cycle terminology and pathogenic species. Folia Parasitol 53: 1–36.
3. Yokoyama H, Masuda K (2001) *Kudoa* sp (Myxozoa) causing a post-mortem myoliquefaction of North-Pacific giant octopus *Paroctopus dofleini* (Cephalopoda : Octopodidae). B Eur Assoc Fish Pat 21: 266–268.
4. Hartigan A, Fiala I, Dyková I, Jirků M, Okimoto B, et al. (2011) A Suspected parasite spill-back of two novel *Myxidium* spp. (Myxosporea) causing disease in australian endemic frogs found in the invasive cane toad. PLoS One 6.
5. Hartigan A, Fiala I, Dyková I, Rose K, Phalen DN, et al. (2012) New species of Myxosporea from frogs and resurrection of the genus *Cystodiscus* Lutz, 1889 for species with myxospores in gallbladders of amphibians. Parasitology 139: 478–496.
6. Eiras JC (2005) An overview on the myxosporean parasites in ampphibians and reptiles. Acta Parasitol 50: 267–275.
7. Roberts JF, Whipps CM, Bartholomew JL, Schneider L, Jacobson ER (2008) *Myxidium scripta* n. sp identified in urinary and biliary tract of Louisiana-farmed red-eared slider turtles *Trachemys scripta elegans*. Dis Aquat Organ 80: 199–209.
8. Bartholomew JL, Atkinson SD, Hallett SL, Lowenstine LJ, Garner MM, et al. (2008) Myxozoan parasitism in waterfowl. Int J Parasitol 38: 1199–1207.
9. Prunescu CC, Prunescu P, Pucek Z, Lom J (2007) The first finding of myxosporean development from plasmodia to spores in terrestrial mammals: *Soricimyxum fegati* gen. et sp n. (Myxozoa) from *Sorex araneus* (Soricomorpha). Folia Parasit 54: 159–164.
10. Wolf K, Markiw ME (1984) Biology contravenes taxonomy in the Myxozoa - new discoveries show alternation of invertebrate and vertebrate hosts. Science 225: 1449–1452.
11. Atkinson SD, Bartholomew JL (2009) Alternate spore stages of *Myxobilatus gasterostei*, a myxosporean parasite of three-spined sticklebacks (*Gasterosteus aculeatus*) and oligochaetes (*Nais communis*). Parasitol Res 104: 1173–1181.
12. Marton S, Eszterbauer E (2011) The development of *Myxobolus pavlovskii* (Myxozoa: Myxobolidae) includes an echinactinomyxon-type actinospore. Folia parasit 58: 157–163.
13. Székely C, Hallett SL, Atkinson SD, Molnár K (2009) Complete life cycle of *Myxobolus rotundus* (Myxosporea: Myxobolidae), a gill myxozoan of common bream *Abramis brama*. Dis Aquat Organ 85: 147–155.
14. Eiras JC, Molnár K, Lu YS (2005) Synopsis of the species of *Myxobolus* Butschli, 1882 (Myxozoa : Myxosporea : Myxobolidae). Syst Parasitol 61: 1–46.
15. Eiras JC, Adriano EA (2012) A checklist of new species of *Henneguya* Thélohan, 1892 (Myxozoa: Myxosporea, Myxobolidae) described between 2002 and 2012. Syst Parasitol 83: 95–104.
16. Lom J, Dyková I (1992) Fine-structure of triactinomyxon early stages and sporogony - myxosporean and actinosporean features compared. J Protozool 39: 16–27.
17. Kent ML, Andree KB, Bartholomew JL, El-Matbouli M, Desser SS, et al. (2001) Recent advances in our knowledge of the Myxozoa. J Eukaryot Microbiol 48: 395–413.

18. Fiala I (2006) The phylogeny of Myxosporea (Myxozoa) based on small subunit ribosomal RNA gene analysis. Int J Parasitol 36: 1521–1534.
19. Liu Y, Whipps CM, Gu ZM, Zeng LB (2010) *Myxobolus turpisrotundus* (Myxosporea: Bivalvulida) spores with caudal appendages: investigating the validity of the genus *Henneguya* with morphological and molecular evidence. Parasitol Res 107: 699–706.
20. Andree KB, Székely C, Molnár K, Gresoviac SJ, Hedrick RP (1999) Relationships among members of the genus *Myxobolus* (Myxozoa : Bilvalvidae) based on small subunit ribosomal DNA sequences. J Parasitol 85: 68–74.
21. Ferguson JA, Atkinson SD, Whipps CM, Kent ML (2008) Molecular and morphological analysis of *Myxobolus* spp. of salmonid fishes with the description of a new *Myxobolus* species. J Parasitol 94: 1322–1334.
22. Lom J, Arthur JR (1989) A Guideline for the preparation of species descriptions in Myxosporea. J Fish Dis 12: 151–156.
23. Xiao CX, Desser SS (2000) Cladistic analysis of myxozoan species with known alternating life-cycles. Syst Parasitol 46: 81–91.
24. Bahri S, Andree KB, Hedrick RP (2003) Morphological and phylogenetic studies of marine *Myxobolus* spp. from mullet in Ichkeul Lake, Tunisia. J Eukaryot Microbiol 50: 463–470.
25. Eszterbauer E, Benko M, Dan A, Molnár K (2001) Identification of fish-parasitic *Myxobolus* (Myxosporea) species using a combined PCR-RFLP method. Dis Aquat Organ 44: 35–39.
26. Palenzuela O, Redondo MJ, Alvarez-Pellitero P (2002) Description of *Enteromyxum scophthalmi* gen. nov., sp. nov. (Myxozoa), an intestinal parasite of turbot (*Scophthalmus maximus* L.) using morphological and ribosomal RNA sequence data. Parasitology 124: 369–379.
27. Molnár K, Eszterbauer E, Székely C, Dan A, Harrach B (2002) Morphological and molecular biological studies on intramuscular *Myxobolus* spp. of cyprinid fish. J Fish Dis 25: 643–652.
28. Adriano EA, Carriero MM, Maia AAM, Silva MRM, Naldoni J, et al. (2012) Phylogenetic and host-parasite relationship analysis of *Henneguya multiplasmodialis* n. sp infecting *Pseudoplatystoma* spp. in Brazilian Pantanal wetland. Vet Parasitol 185: 110–120.
29. Fiala I, Bartošová P (2010) History of myxozoan character evolution on the basis of rDNA and EF-2 data. BMC Evol Biol 10: 228.
30. Milanin T, Eiras JC, Arana S, Maia AA, Alves AL, et al. (2010) Phylogeny, ultrastructure, histopathology and prevalence of *Myxobolus oliveirai* sp. nov., a parasite of *Brycon hilarii* (Characidae) in the Pantanal wetland, Brazil. Mem Inst Oswaldo Cruz 105: 762–769.
31. Molnár K, Cech G, Székely C (2008) Infection of the heart of the common bream, *Abramis brama* (L.), with *Myxobolus* s.l. *dogieli* (Myxozoa, Myxobolidae). J Fish Dis 31: 613–620.
32. Naldoni J, Arana S, Maia AA, Silva MR, Carriero MM, et al. (2011) Host-parasite-environment relationship, morphology and molecular analyses of *Henneguya eirasi* n. sp. parasite of two wild *Pseudoplatystoma* spp. in Pantanal Wetland, Brazil. Vet Parasitol 177: 247–255.
33. Evans NM, Holder MT, Barbeitos MS, Okamura B, Cartwright P (2010) The phylogenetic position of Myxozoa: exploring conflicting signals in phylogenomic and ribosomal data sets. Mol Biol Evol 27: 2733–2746.
34. Mallatt J, Craig CW, Yoder MJ (2012) Nearly complete rRNA genes from 371 Animalia: Updated structure-based alignment and detailed phylogenetic analysis. Mol Phylogenet Evol 64: 603–617.
35. Nesnidal MP, Helmkampf M, Bruchhaus I, El-Matbouli M, Hausdorf B (2013) Agent of whirling disease meets orphan worm: phylogenomic analyses firmly place Myxozoa in Cnidaria. PLoS One 8.
36. Eszterbauer E (2004) Genetic relationship among gill-infecting *Myxobolus* species (Myxosporea) of cyprinids: molecular evidence of importance of tissue-specificity. Dis Aquat Organ 58: 35–40.
37. Gleeson RJ, Adlard RD (2012) Phylogenetic relationships amongst *Chloromyxum* Mingazzini, 1890 (Myxozoa: Myxosporea), and the description of six novel species from Australian elasmobranchs. Parasitol Int 61: 267–274.
38. Hall TA (1999) BioEdit: a user-friendly biological sequence alignment editor and analysis program for Windows 95/98/NT. Nucleic Acids Symp Ser 41: 95–98.
39. Altschul SF, Madden TL, Schaffer AA, Zhang J, Zhang Z, et al. (1997) Gapped BLAST and PSI-BLAST: a new generation of protein database search programs. Nucleic Acids Res 25: 3389–3402.
40. Rosenberg MS, Kumar S (2001) Incomplete taxon sampling is not a problem for phylogenetic inference. Proc Natl Acad Sci USA 98: 10751–10756.
41. Molnár K (2011) Remarks to the validity of Genbank sequences of *Myxobolus* spp. (Myxozoa, Myxosporidae) infecting Eurasian fishes. Acta Parasitol 56: 263–269.
42. Posada D (2008) jModelTest: phylogenetic model averaging. Mol Biol Evol 25: 1253–1256.
43. Guindon S, Dufayard JF, Lefort V, Anisimova M, Hordijk W, et al. (2010) New algorithms and methods to estimate maximum-likelihood phylogenies: assessing the performance of PhyML 3.0. Syst Biol 59: 307–321.
44. Swofford DL (2003) PAUP*. Phylogenetic Analysis Using Parsimony (*and Other Methods). 4 ed. Sunderland, Massachusetts: Sinauer Associates.
45. Page RD (1996) TreeView: an application to display phylogenetic trees on personal computers. Comput Appl Biosci 12: 357–358.
46. Tamura K, Peterson D, Peterson N, Stecher G, Nei M, et al. (2011) MEGA5: Molecular evolutionary genetics analysis using maximum likelihood, evolutionary distance, and maximum parsimony methods. Mol Biol Evol 28: 2731–2739.
47. Molnár K (1998) Taxonomic problems, seasonality and histopathology of *Henneguya creplini* (Myxosporea) infection of the pikeperch *Stizostedion lucioperca* in Lake Balaton. Folia Parasitol 45: 261–269.
48. Molnár K, Ranzani-Paiva MJ, Eiras JC, Rodrigues EL (1998) *Myxobolus macroplasmodialis* sp. n. (Myxozoa : Myxosporea), a parasite of the abdominal cavity of the characid teleost, *Salminus maxillosus*, in Brazil. Acta Protozool 37: 241–245.
49. Gunter NL, Adlard RD (2009) Seven new species of *Ceratomyxa* Thélohan, 1892 (Myxozoa) from the gall-bladders of serranid fishes from the Great Barrier Reef, Australia. Syst Parasitol 73: 1–11.
50. Molnár K, Marton S, Eszterbauer E, Székely C (2006) Comparative morphological and molecular studies on *Myxobolus* spp. infecting chub from the river Danube, Hungary, and description of *M. muellericus* sp. n. Dis Aquat Organ 73: 49–61.
51. Easy RH, Johnson SC, Cone DK (2005) Morphological and molecular comparison of *Myxobolus procerus* (Kudo, 1934) and *M. intramusculi* n. sp. (Myxozoa) parasitising muscles of the trout-perch *Percopsis omiscomaycus*. Syst Parasitol 61: 115–122.
52. Broughton RE, Betancur RR, Li C, Arratia G, Orti G (2013) Multi-locus phylogenetic analysis reveals the pattern and tempo of bony fish evolution. PLoS Curr 5.
53. Nakatani M, Miya M, Mabuchi K, Saitoh K, Nishida M (2011) Evolutionary history of Otophysi (Teleostei), a major clade of the modern freshwater fishes: Pangaean origin and Mesozoic radiation. BMC Evol Biol 11: 177.
54. Gleeson RJ, Bennett MB, Adlard RD (2010) First taxonomic description of multivalvulidan myxosporean parasites from elasmobranchs: *Kudoa hemiscylli* n.sp. and *Kudoa carcharhini* n.sp. (Myxosporea: Multivalvulidae). Parasitology 137: 1885–1898.
55. El-Mansy A (2005) Revision of *Myxobolus heterosporus* Baker, 1963 (syn. *Myxosoma heterospora*) (Myxozoa : Myxosporea) in African records. Dis Aquat Organ 63: 205–214.
56. Barta JR, Martin DS, Liberator PA, Dashkevicz M, Anderson JW, et al. (1997) Phylogenetic relationships among eight *Eimeria* species infecting domestic fowl inferred using complete small subunit ribosomal DNA sequences. J Parasitol 83: 262–271.
57. Diamant A, Whipps CM, Kent ML (2004) A new species of *Sphaeromyxa* (Myxosporea : Sphaeromyxina : Sphaeromyxidae) in devil firefish, *Pterois miles* (Scorpaenidae), from the northern Red Sea: Morphology, ultrastructure, and phylogeny. J Parasitol 90: 1434–1442.
58. Hallett SL, Diamant A (2001) Ultrastructure and small-subunit ribosomal DNA sequence of *Henneguya lesteri* n. sp (Myxosporea), a parasite of sand whiting *Sillago analis* (Sillaginidae) from the coast of Queensland, Australia. Dis Aquat Organ 46: 197–212.

11

Outside-Host Growth of Pathogens Attenuates Epidemiological Outbreaks

Ilona Merikanto[1,2]*, Jouni Laakso[1,3], Veijo Kaitala[1]

1 Department of Biosciences, University of Helsinki, Helsinki, Finland, 2 Department of Mental Health and Substance Abuse Services, National Institute for Health and Welfare, Helsinki, Finland, 3 Centre of Excellence in Biological Interactions, Department of Biological and Environmental Science, University of Jyväskylä, Jyväskylä, Finland

Abstract

Opportunist saprotrophic pathogens differ from obligatory pathogens due to their capability in host-independent growth in environmental reservoirs. Thus, the outside-host environment potentially influences host-pathogen dynamics. Despite the socio-economical importance of these pathogens, theory on their dynamics is practically missing. We analyzed a novel epidemiological model that couples outside-host density-dependent growth to host-pathogen dynamics. Parameterization was based on columnaris disease, a major hazard in fresh water fish farms caused by saprotrophic *Flavobacterium columnare*. Stability analysis and numerical simulations revealed that the outside-host growth maintains high proportion of infected individuals, and under some conditions can drive host extinct. The model can show stable or cyclic dynamics, and the outside-host growth regulates the frequency and intensity of outbreaks. This result emerges because the density-dependence stabilizes dynamics. Our analysis demonstrates that coupling of outside-host growth and traditional host-pathogen dynamics has profound influence on disease prevalence and dynamics. This also has implications on the control of these diseases.

Editor: Sudha Chaturvedi, Wadsworth Center, United States of America

Funding: This study was funded by the Finnish Academy projects 1130724 and 125572 for Jouni Laakso and a personal grant for IM from the Research Foundation of the University of Helsinki. The funders had no role in study design, data collection and analysis, decision to publish, or preparation of the manuscript.

Competing Interests: The authors have declared that no competing interests exist.

* E-mail: ilona.merikanto@helsinki.fi

Introduction

Many pathogens are able to survive and replicate in the environment outside-host, e.g., via saprotrophism [1,2]. These kinds of pathogens can also be called opportunists as the host specificity is often low and growth within host is only an alternative reproductive strategy. The key in the transmission and survival of opportunist pathogens is that they can delay their extinction or survive indefinitely in the outside-host environment. Therefore the opportunists may thrive even though all susceptible hosts would either be treated or removed. In contrast, obligatory pathogens cannot replicate in the outside-host environment and often have higher host specificity. Thus, disease dynamics is likely to differ between opportunist and obligatory pathogens. Opportunist pathogens with the capacity of growing outside-host are also plausible ancestors in the evolution of obligatory pathogens [1]. It has also been suggested that selection favors an opportunistic strategy in general [3,4].

Although not often recognized, opportunist pathogens are very common and present a significant economical burden and health risk, yet the ecological and evolutionary dynamics of these organisms is poorly understood. Opportunist pathogens in humans include, e.g., cholera (*Vibrio cholera*) and lung infections (*Pseudomonas aeruginosa* and *Legionella pneumophila* [5–8]. Cholera outbreaks are common in countries where sanitation and drinking water quality are poor [9,10]. Lung inflammations on the other hand pose a lethal treat globally to patients with compromised immunity [11]. Other examples of opportunist pathogens are for instance bacteria from *Listeria* and *Flavobacterium* genus, such as *L. monocytogenes*, *F. psychrophilum* and *F. columnare* [12–15]. In particular, fish columnaris disease caused by *F. columnare* has become a major problem in fresh water fish farms cultivating salmonids in Finland and channel catfish (*Ictalurus punctatus*) in the United States [14–16].

F. columnare has the potential to survive through saprotrophism in the outside-host environment indefinitely and causes opportunistically infections as susceptible host are present [14–15]. Infection can result in the death of an entire fish population in a cultivation tank [17]. Also *F. psychrophilum* causes severe fish diseases, Cold-Water Disease and Rainbow Trout Fry Syndrome, in fish farms [12].

To our knowledge, only few models acknowledging outside-host growth has been reported, and the models mostly consider only short-term processes [18,19]. Theory on long-term disease dynamics of opportunist pathogens does not yet exist. Few theoretical models have been developed for environmentally transmitted pathogens that are able to survive outside the host for a certain time period [20–23], but these models are not suitable for opportunist pathogens that can interact with other species and replicate in a density-dependent manner in the outside-host environment. Many previous models regarding host-parasite interaction often assume a trade-off between virulence and transmission. High virulence would eventually increase mortality of the hosts and therefore weaken pathogen growth, as there are less susceptible hosts present. As a consequence, host dependent obligatory pathogen would die out when the host population density becomes too low [24–27]. There are host-dependent mechanisms that can enable evolution of high virulence, such as short-sighted within-host strain competition [25–33]. The concept of outside-host growth of an opportunist pathogen offers another, novel pathway to high virulence: as the fitness of an opportunist pathogen on is partially independent of the host, the trade-off between virulence and pathogen growth can be weakened or even removed altogether. The ability to survive and replicate in the outside-host environment could therefore promote high-enough virulence that leads to host extinction. Given these obvious discrepancies between the assumptions of the traditional theory of disease dynamics, and the properties of opportunist pathogens, it is essential to further the theory on the dynamics of opportunist diseases. Opportunist disease model can also be to some extent compared to predator-prey systems with more than one prey species. Theoretical work in these systems is also sparse at the moment [34].

Here we introduce a novel model that couples density-dependent growth in the environment to host-pathogen dynamics and analyze the long-term dynamics of the system. Parametrization of the model analyses are based on fish columnaris disease. We demonstrate that the ability to replicate in the outside-host environment can under some conditions lead to host extinction but not necessarily to the extinction of the opportunist pathogen. The model can also produce stable or cyclic dynamics (outbreaks), where the pathogen growth in the outside-host environment regulates the frequency and intensity of outbreaks. Especially, the outside-host growth seems to be source for unstable dynamics and increasing the strength of its density dependence has stabilizing effect on the host-pathogen dynamics in a wide range of the parameter space. Growth in the environment outside-host therefore has a profound influence on disease prevalence and dynamics that differ from the traditional theory of host-pathogen dynamics.

Methods

Model of opportunist pathogen-host interaction

We consider a deterministic continuous time model for opportunist pathogen-host interaction. The model combines SI dynamics based on model G of Anderson & May (1981) and outside-host growth of the pathogen to describe changes in time (t) in the densities of susceptible hosts (S), infected hosts (I) and pathogens in the environment outside-host (P):

$$\frac{dS}{dt} = r_s(1-S)S - \beta PS - \mu_{SI}S \quad (1)$$

$$\frac{dI}{dt} = \beta PS - (\mu_{inf} + \mu_{SI})I \quad (2)$$

$$\frac{dP}{dt} = \Lambda\mu_{inf}I + r_P(1-f_P P)P - \mu_P P \quad (3)$$

Susceptible host population (eqn. 1) grows logistically with a growth rate r_S in a density-dependent way (host carrying capacity is assumed equal to 1). Susceptible hosts die at a rate μ_{SI} and are infected at a rate β. Infected host population (eqn. 2) increases depending on transmission rate of infection ($\beta S_t P_t$). We assume that infected hosts are unable to replicate. We also assume that that infected hosts are not competing for resources with susceptible hosts. Density of I decreases by death to infection (μ_{inf}) or due to other causes (μ_{SI}). μ_{inf} is used to measure virulence. Equation (3) describes density change in the pathogen population outside-host (P) in time (t). Pathogen population outside-host (eqn. 3) increases depending on release rate (Λ) of new pathogen as infected hosts die due to infection (μ_{inf}). P increases logistically due to opportunist growth rate (r_P) in the outside-host environment, where f_P describes the strength of density dependence. Density of P decreases due to a density independent death rate μ_P. The effect of opportunist growth on disease dynamics was studied analytically by linearizing eqns. (1–3) around the equilibrium (equilibrium population densities were restricted to be positive). Linearized population dynamics are given in Appendix S1. The Jacobian eigenvalues were investigated for local stability properties [35].

Parametrization of the model and numerical simulations

Stability of *SIP* community dynamics was studied by using different combinations of parameter values between 1) pathogen growth rates (r_P) and susceptible hosts growth rates (r_S), 2) pathogen growth rates (r_P) and virulence (μ_{inf}), 3) pathogen growth rates (r_P) and pathogen death rate (μ_P) or 4) pathogen growth rates (r_P) and pathogen release rate (Λ). Parameter values were selected to cover a large range of plausible biological values for different host and opportunist pathogen organisms, such as *Flavobacterium* and *Serratia* genus, where many bacterium species are saprotrophic pathogens with multiple potential hosts [36–39]. Natural or experimental growth and mortality values due to infection regarding *Flavobacterium columnare* and *Serratia marcescens* and some of their hosts are given in Table 1. Pathogen growth rates were assumed to be lower in the analyses than those measured in experimental studies to represent more realistic situation found in natural habitats. In nature, hosts can have higher mortality to infection than to other causes [20]. Mortality of the hosts due to other reasons than to infection (μ_{SI}) was therefore standardized to a low value. Pathogen mortality (μ_P) day^{-1} was varied corresponding to realistic mortality values measured in bacteria. For example aquatic bacteria have been measured mortality rates between 0.01–0.03 h^{-1} [40]. The transmission rate (β) for the pathogen was kept low, because infectiveness is by definition lower in opportunist pathogens as compared to obligatory [41]. Parameter values used in the analyses are given in Table 2.

We carried out four different stability analyses where two parameters were varied at a time. One parameter was always the environmental growth rate of the pathogen (r_P). The parameters were given 100 different values from the value range used. For the resulting 100^2 combinations, the *SIP* community dynamics were simulated for 1700 days.

We simulated the model (1)–(3) for 3500 days to record attributes of the *SIP* community dynamics. Bifurcation diagrams were obtained by scoring the minimum and maximum values of population fluctuations after removing the initial transient. The numerical analysis of the model was performed with MATLAB v. 2011b (ODE45 solver, default tolerance settings).

Results

As a starting point it is worth of considering a variant of SI model (1)–(2) where the inflow of pathogens, P, is assumed to be constant. A straightforward analysis provides a necessary and sufficient condition for positive equilibrium S>0 and I>0:

$$\beta P + \mu_{SI} - r_S < 0 \quad (4)$$

Table 1. Reproduction (*r*) and mortality values due to infection (μ_{inf}) for saprotrophic pathogens *Flavobacterium columnare* and *Serratia marcescens* and some of their hosts based on experimental studies.

Pathogen	r_P	*Host*	μ_{inf}	r_S
F. columnare	2.4–7.2 day^{-1} [39]	Atlantic salmon, *Salmo salar*	0.2–0.3 day^{-1} [50]	0.2–1.7 day^{-1} [52]
		Rainbow trout, *Oncorhynchus mykiss*	0.2–0.4 day^{-1} [50]	0.08–0.4 day^{-1} [53,54]
		Brown trout, *Salmo trutta*	0.01–0.05 day^{-1} [50]	0.05–0.34 day^{-1} [55,56]
		Chinook salmon, *Oncorhynchus tshawytscha*	0.01–0.05 day^{-1} [50]	0.2–4.4 day^{-1} [57]
		Arctic charr, *Salvelinus alpinus*	0.2–0.3 day^{-1} [50]	0.2–9.1 day^{-1} [58,59]
		Channel catfish, *Ictalurus punctatus*	0.01–1.0 day^{-1} [38]	2.3–19.7 day^{-1} [60]
		Zebra fish, *Danio rerio*	0.2–0.4 day^{-1} [51]	15–200 day^{-1} [61]
S. marcescens	2.4–6 day^{-1} [62]	*Drosophila melanogaster*	0.002–1.0 day^{-1} [63]	11–41 day^{-1} [64]

In the fish host reproduction rates, the number of eggs produced per kg during a year, average weight range of mature fish and survival rate of eggs to fry has been taken into account.

The eigenvalues of the two-dimensional system are $\lambda_1 = \beta P + \mu_{SI} - r_S$ and $\lambda_2 = -(\mu_{SI} + \mu_{SI})$. Thus, both eigenvalues are strictly negative and the positive equilibrium is locally stable whenever it exists. In conclusion, any instability that occurs in the SIP model (1)–(3) is caused by the dynamics of the pathogen.

Consider next the pathogen dynamics (3) in the absence of density dependent growth, that is r_P = 0. A straightforward application of the results of Figures S1 and S2 provides us the first eigenvalue: $\lambda_1 = \beta P + \mu_{SI} - r_S$

which according to (4) is negative. The two other eigenvalues $\lambda_{2,3}$ satisfy the eigenvalue equation (35):

$$\lambda^2 + (\mu_{\text{inf}} + \mu_{SI} + \mu_P)\lambda = 0$$

.

Thus, λ_2 is real and negative and $\lambda_3 = 0$. Consequently, the system is now marginally stable. It follows that the possible genuine instabilities in the host-pathogen system (1)–(3) are due to the density dependent growth of the pathogen in the environment. When the density dependent growth is involved in the system then the analyses of the system becomes much more complicated as all the parameters of the system are involved in driving the behaviour of the system in a strongly nonlinear way. Moreover, as it appears our analysis below many parameters have joint consequences on the dynamics of the system. Hence, we next rely on simulation studies in characterizing the dynamics of the system with opportunistic pathogen.

Increased pathogen virulence μ_{inf} (Figure 1A), pathogen mortality μ_P (Figure 1B), and the strength of density-dependence f_P (Figure 2A) have a stabilizing effect on the *SIP* system. As f_P increases, susceptible hosts do not have enough time to reach higher maximum densities (Figure 2A). By plotting time series, this seems to be because if density dependence increases, the period of population fluctuations shortens. In contrast, increasing pathogen release rate from the host (Λ) destabilizes system dynamics (Figure 1C). Also, increased growth of the pathogen (r_P) in the outside-host environment was destabilizing, given that r_P close to μ_P (Figure 1D, 2B). This happens also when r_P is above μ_P, given that the susceptible host growth rate (r_S) is allowed increase (Figure 1). However, depending on r_S, Λ and μ_P, also a sequence of stable to periodic to stable to extinct dynamics can occur when r_P increases (Figure 1B–D). The host population (*S*+*I*) becomes often extinct when pathogen growth rate exceeds its natural mortality, i.e., $r_P > \mu_P$, depending on susceptible host growth rate r_S (Figure 1A–D, 2B). In the absence of both the pathogen's ability to have net growth outside the host (i.e., $r_P < \mu_P$) and the benefit from inside host growth (i.e., when r_S , μ_{inf} and Λ are very low), pathogen population outside host (*P*) goes extinct (Figure 1A, 1C, 1D).

The equilibrium density of pathogen *P* increases with both increasing r_S (Figure S1D) and r_P (Figures S1A, S1B, S1C). The equilibrium density of *P* is maximized when mortality to infection (μ_{inf}) or release from the host (Λ) is low (Figures S1A and S1C). In

Table 2. Parameter values that were used in stability analysis.

Parameter	*Explanation of the parameter*	*Parameter values*	*Exceptions in parameter values*
r_S	Susceptible host growth rate	0.01	0.001–0.5 (Fig. 1D, 3D, 4D)
r_P	Pathogen growth rate outside-host	0.001–0.5	0.06 (Fig. 2A) 0–0.15 (Fig. 2B)
μ_{SI}	Mortality of the susceptible and infected hosts due to other reasons than infection	10^{-3}	
μ_{inf}	Virulence (Mortality of the infected hosts due to infection)	0.1	0.001–0.5 (Fig. 1A, 3A, 4A)
μ_P	Pathogen mortality outside-host	0.1	0.1–0.5 (Fig. 1B, 3B, 4B)
β	Pathogen transmission rate to susceptible hosts from environment	10^{-5}	
Λ	Pathogen release rate from infected hosts when they die	10^5	5×10^3–1.5×10^5 (Fig. 1C, 3C, 4C)
f_P	Negative influence of pathogen population density on its growth	10^{-5}	0–10^{-6} (Fig. 2A)

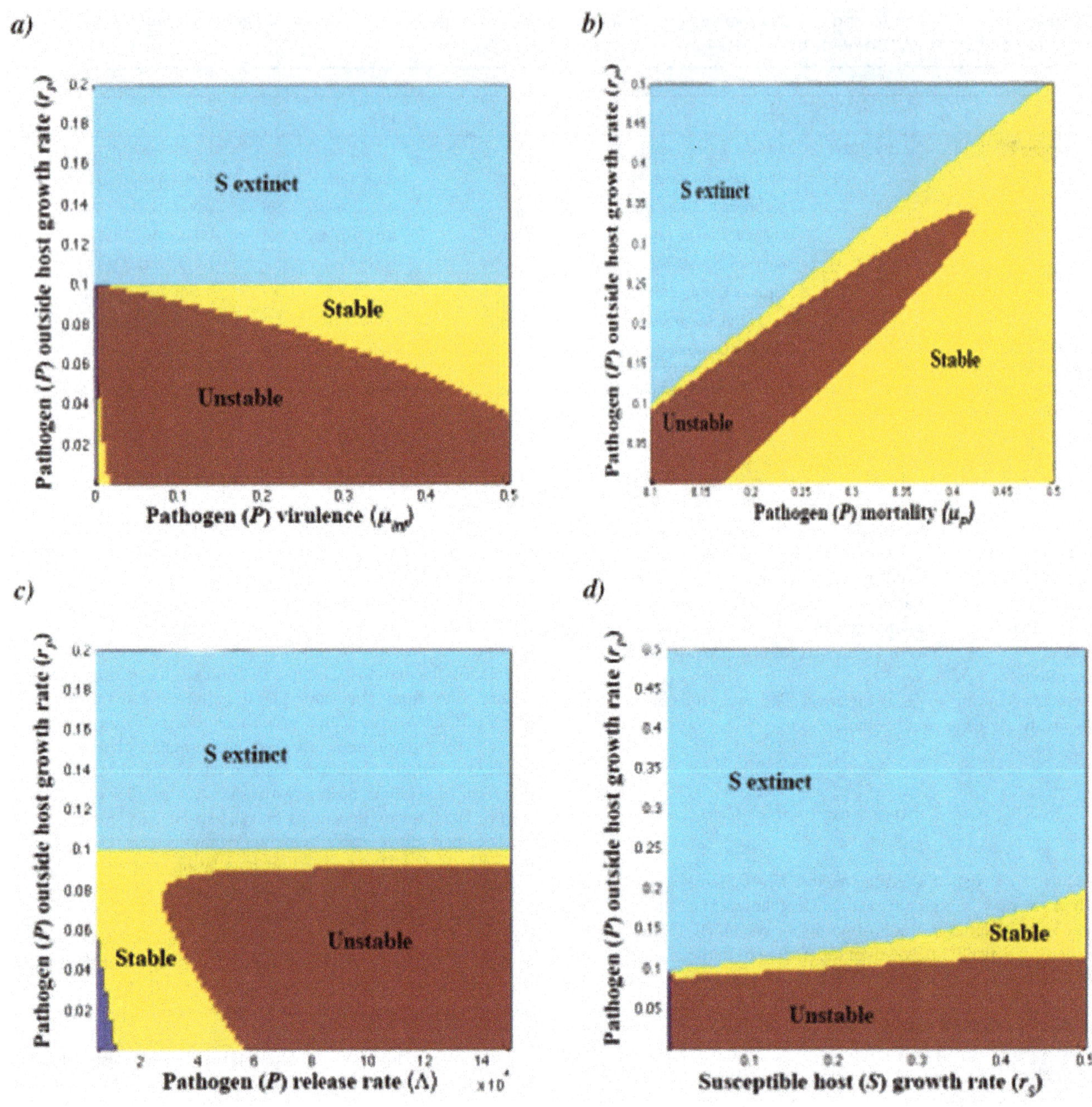

Figure 1. Stability of the *SIP* community dynamics in different combinations of outside-host growth rate of pathogen (r_P) parameter values and parameter values of a) virulence (μ_{inf}), b) pathogen mortality outside-host (μ_P), c) release rate (Λ) and d) susceptible host growth rate (r_S). Dark blue: Pathogen population outside-host (*P*) goes to extinction. Light blue: Susceptible host population (*S*) goes extinct. Yellow: *SIP* community dynamics are locally stable. Red: *SIP* community dynamics are locally unstable. Used parameter values are shown in Table 2.

situations where host (*S*) goes extinct, the pathogen (*P*) necessarily does not, due to the net environmental growth (r_P is close to zero) (Figure S1A–C, Figure S2A–C).

Discussion

We studied a new class of epidemiological models, where we assume density-dependent growth of a pathogen in the outside-host environment. Traditional epidemiological models on the other hand assume that pathogens do not actively grow in the outside-host environment, for which reason they poorly describe disease dynamics of opportunistic pathogens, such as *Vibrio cholera, Pseudomonas aeruginosa, Flavobacterium columnare, F. psychrophilum* etc. We found that density-dependent growth in the outside-host environment generates disease dynamics that can strongly differ from the traditional SI-models. Importantly, density-dependent growth in the outside-host environment has a strong stabilizing effect on the host-pathogen system. Outside-host replication also enables opportunist pathogens to remain in the environment when the density of susceptible hosts is too low for host-dependent persistence, or in the absence of the hosts. Thus, outside-host growth and the associated density depence is a potential ecological mechanism that can regulate disease outbreaks.

a)

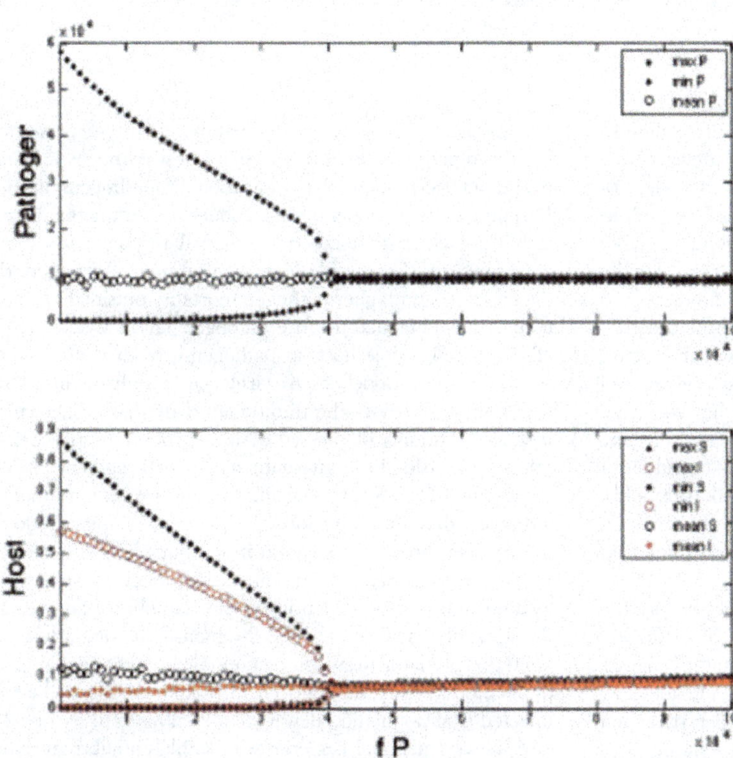
max P
min P
mean P
Pathogen
max S
max I
min S
min I
mean S
mean I
Host
f P

b)

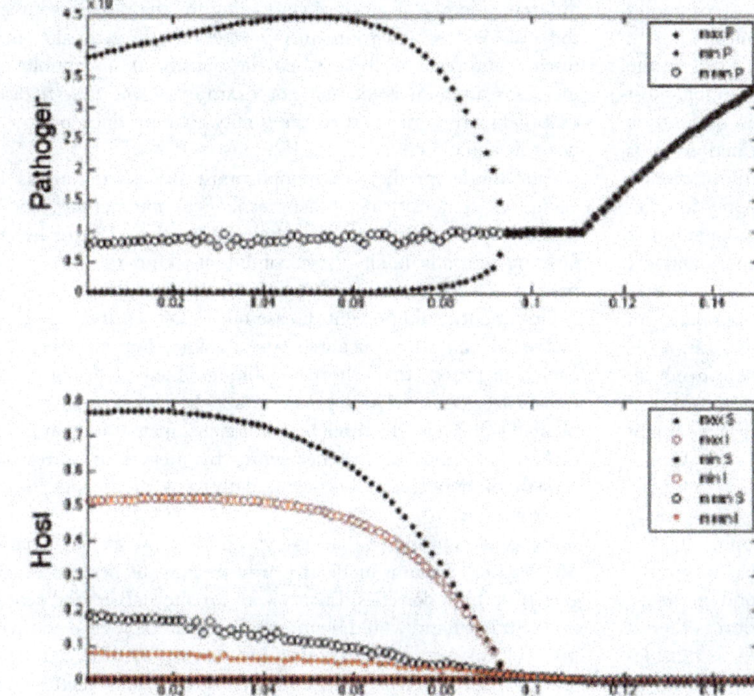
max P
min P
mean P
Pathogen
max S
max I
min S
min I
mean S
mean I
Host
r_P

Figure 2. Stabilizing effect of pathogen density-dependent growth in the outside-host environment. Bifurcation diagrammes indicate mean densities and the minimum and maximum values in the population fluctuations of the pathogen and host after the initial transients in the simulations have been removed. a) Increasing the strength of density-dependence stabilizes the population dynamics at $f_P = 4.2 \times 10^{-5}$ ($r_P = 0.06$). b) For increasing pathogen growth rate population dynamics stabilize at $r_P = 0.096$ while host extinction occurs at $r_P = 0.112$. Used parameter values are shown in Table 2.

The idea that pathogens are able to survive in the outside-host environment is not new. For example, Day (2002) developed a model where pathogens were able to survive in the outside-host environment by producing resting spores. This model, as well as models by Roche and co-authors (2011) and Boldin & Kisdi (2011), however do not assume that a pathogen can replicate in the outside-host environment. The model developed by Godfray and co-authors (1999) acknowledged outside-host growth of a pathogen and competition in the outside-host environment. However, this model assumes density-independent outside-host growth, and thus, it applies only to processes with limited time horizon, e.g., short-term biological control [18]. To our knowledge, only one previous study considers density-dependent growth outside-host [19]; Joh and co-authors however focused on outbreak thresholds. Their model also lacks background mortality of the pathogen in an environmental reservoir, and thus is not realistic for modelling long-term dynamics of opportunist disease.

Our model can produce cyclic dynamics (outbreaks) where the pathogen growth in the outside host environment regulates the frequency and intensity of outbreaks. In the traditional SI-models destabilizing factors of host-pathogen dynamics include low reproduction rate of susceptible hosts, time lag between pathogen replication inside host, and transmission to new hosts [42]. However, in our model increase in the outside-host pathogen growth rate, and release rate of new pathogens promote cyclic dynamics (outbreaks). The size of the free-living pathogen population (directly related to the increase of new infections) is increased either by higher pathogen release rates, higher growth rate of susceptible hosts, or trivially by higher pathogen growth rates in the outside-host environment.

Also increasing the growth rate of susceptible host promotes cyclic dynamics, when the pathogen growth in the outside host is near pathogen mortality. However, as the pathogen growth in the outside host increases, higher growth rate of susceptible host promotes stable dynamics. As the pathogen growth in the outside host increases even further, the result is host extinction. It is possible to make some comparisons between opportunist disease dynamics and predator-prey dynamics. In predator-prey interactions, unstable dynamics are possible when prey population is abundant and the growth rate of predator is on the other hand low [34,43].

Opportunist pathogenicity functions as a density amplifier of otherwise saprotrophic bacteria growth as opportunists can often replicate more efficiently via pathogenicity than via saprotrophism due to host being richer resource as compered to saprotrophic resource. Abrams and Matsuda (1996) predicted that there might not be apparent competition between prey species if efficient predation of other prey species increases death rate of both prey species similarly [44]. We assume that opportunist pathogenicity leads to trade-off with the efficiency to grow saprotrophically, but as opportunists are able to increase their abundance in greater quantities by pathogenicity in the outside host environment, leads this also to greater consumption of saprotrophic resources than if bacteria would only grow saprotrophically. However, as outside-host growth might lead to host extinction, this situation could be compared to apparent competition found in predator-prey systems [45].

It is notable that the outside-host density-dependence and increased pathogen mortality stabilize host-pathogen dynamics. In traditional SI-models, the replication of pathogens inside hosts [46] functions as a density-dependent constrain that also has a stabilizing effect that could be equivalent to density-dependent constrains in the outside-host environment. Increased density-dependency and higher pathogen mortality outside hosts are likely to be stabilizing because the pathogen has a weaker regulatory effect on susceptible hosts as pathogen growth is more restricted. Also, in our model high virulence stabilizes host-pathogen dynamics similarly to the traditional SI-models. Higher virulence increases mortality of infected hosts and therefore speeds pathogen release to outside host environment. This strengthens intraspecific competition for resources in the outside-host environment. High virulence therefore functions indirectly as a density-dependent mechanism through replication in the outside-host environment.

If we would disregard density dependency in our model and assume that pathogen density would equilibrate faster than host densities, the model could be 2-dimensional similarly as Lotka-Volterra predator-prey interaction, where change in pathogen population density in time (eq. 3) could be solved in terms of infected host population density (eq. 2). This kind of model would give linear functional response with stable equilibrium dynamics.

The fact that the opportunist pathogen does not necessarily need the host for long-term survival could lead to development of high virulence, which is in contrast with predictions made for obligatory pathogens where virulence is expected to decline in over time [24–27]. High host infection rate coupled to high pathogen release rate increases the pathogen fitness. This could offer a competitive benefit to opportunist pathogens, e.g., when different microbe strains are competing for the same resources in the outside-host environment. Outside-host growth also makes disease outbreaks possible when the density of susceptible hosts increases after an epidemic. For example, *Vibrio cholerae* disease outbreaks are connected to the ability to reproduce outside the hosts in aquatic environment [8].

Our model predicts that opportunist disease outbreaks may occur when density dependent ecological interactions, such as outside-host competition, are relaxed. Intensive plant or livestock farming typically fulfills these conditions. Thus the model could predict how decreasing pathogen growth rate in the outside-host environment could prevent disease outbreaks. Furthermore, other biological control applications are possible. For instance, it has been suggested that thorn-inhabiting bacteria, such as the *Clostridium* genus, have a potential anti-herbivory role in thorny plants [47]. Thus, the model could for instance also be applied to biological control against herbivores by increasing saprotrophic growth of opportunist herbivore pathogens in the outside host environment.

It has been proposed that host-specific enemy, such as a parasite or predator, is often ineffective way to prevent pest populations growth to high densities. Enemy with an alternative food resource on the other hand would be more efficient in biological control, as it is able to sustain high population density by using alternative food resource even if pest population sizes would fluctuate [48]. Opportunist pathogen would in this sense be ideal in biological control as it is able to replicate in the outside host environment independently of the host but prevents efficiently host population growth. For example, saprotrophic *Serratia entomophila* bacteria has been used with success in the biological control of New Zealand grass grub (*Costelytra zealandica*) [18].

The model could also be applied to biological control of the saprotrophically transmitting *Flavobacterium columnare* bacterium. Biological control of *F. columnare*, as well as for some other fish pathogen bacteria from the *Flavobacterium* genus, is needed due to the negative side effects of increasing use of antibiotics. It has been suggested that conditions in fish tanks are ideal environment for the evolution of high virulence opportunist bacteria and therefore opportunist disease outbreaks. These conditions include high susceptible host density, lack of natural bacterial predators or competitors of *F. columnare*, and high availability of dead fish material, fish food [15]. The antibiotic treatment of infected fish is unable to remove *F. columnare* from the fish tanks, as they are able to survive and replicate outside fish [14], and are constantly reintroduced to the tanks from inflow water from natural fresh waters [49]. We suggest that the more effective way to treat columnaris disease in fish farm would be more efficient removal of saprotrophic resources, such as dead fish material and faeces from the tanks and thus decrease pathogens' ability to grow outside the host and by increasing the diversity of natural bacterial enemies in fish tanks.

Summary

We modeled the dynamics of opportunist disease capable of density-dependent environmental growth. The disease cycles (outbreaks) crucially depend on the outside-host density-dependent growth. Interestingly, the density dependent outside-host pathogen growth strongly stabilizes disease dynamics. Saprotrophic opportunism is an efficient life-history strategy because the ability to replicate in the outside-host environment potentially gives large fitness benefit as compared to non-pathogenic strains or obligatory pathogen. That the opportunist pathogens are also able to survive in the outside-host environment even when there are no susceptible hosts available, could promote the evolution of higher virulence regardless of the virulence-transmission tradeoff that the obligatory pathogens have to face. Capability to the outside-host growth is also a novel ecological mechanism for disease outbreaks.

Supporting Information

Appendix S1 Local Stability Analysis.

Figure S1 Equilibrium densities of pathogen population outside-host (P) with different combinations of outside-host growth rate of pathogen (r_P) values and parameter values of a) virulence (μ_{inf}), b) pathogen mortality outside-host (μ_P), c) release rate (Λ) and d) susceptible host growth rate (r_S). Used parameter values are shown in Table 2.

Figure S2 Equilibrium densities of susceptible host population (S) with different combinations of outside-host growth rate of pathogen (r_P) values and parameter values of a) virulence (μ_{inf}), b) pathogen mortality outside-host (μ_P), c) release rate (Λ) and d) susceptible host growth rate (r_S). Used parameter values are shown in Table 2.

Acknowledgments

The authors thank K. Viipale for conceptual help.

Author Contributions

Conceived and designed the experiments: IM JL VK. Performed the experiments: IM. Analyzed the data: IM. Contributed reagents/materials/analysis tools: IM JL VK. Wrote the paper: IM.

References

1. Casadevall A (2008) Evolution of intracellular pathogens. Ann Rev Microbiol 62: 19–33.
2. Veneault-Fourrey C, Martin F (2011) Mutualistic interactions on a knife-edge between saprotrophy and pathogenesis. Curr Opin Plant Biol 14: 444–450.
3. Hudson PJ, Rizzoli A, Grenfell BT, Heesterbeek H, Dobson AP (2002) The Ecology of Wildlife Diseases. Oxford: Oxford Univ Press.
4. Walther BA, Ewald PW (2004) Pathogen survival in the external environment and the evolution of virulence. Biol Rev Camb Philos Soc 79: 849–869.
5. Friedman H, Yamamoto Y, Klein TW (2002) Legionella pneumophila pathogenesis and immunity. Semin Pediatr Infect Dis 13: 273–279.
6. Leclerc H, Schwartzbrod L, Dei-Cas E (2002) Microbial agents associated with waterborne diseases. Crit Rev Microbiol 28: 371–409.
7. Hall-Stoodley L, Stoodley P (2005) Biofilm formation and dispersal and the transmission of human pathogens. Trends Microbiol 13: 7–10.
8. Rahman MH, Biswas K, Hossain MA, Sack RB, Mekalanos JJ, et al. (2008) Distribution of genes for virulence and ecological fitness among diverse Vibrio cholerae population in a cholera endemic area: tracking the evolution of pathogenic strains. DNA Cell Biol 27: 347–355.
9. Mandal S, Mandal MD, Pal NK (2011) Cholera: a great global concern. Asian Pac J Trop Med 4: 573–580.
10. Murugaiah C (2011) The burden of cholera. Crit Rev Microbiol 37: 337–348.
11. Parker D, Prince A (2011) Innate immunity in the respiratory epithelium. Am J Respir Cell Mol Biol 45: 189–201.
12. Madetoja J, Nyman P, Wiklund T (2000) Flavobacterium psychrophilum, invasion into and shedding by rainbow trout Oncorhynchus mykiss. Dis Aquat Organ 43: 27–38.
13. Freitag NE, Port GC, Miner MD (2009) Listeria monocytogenes – from saprophyte to intracellular pathogen. Nat Rev Microbiol 7: 623–628.
14. Kunttu HM, Valtonen ET, Jokinen EI, Suomalainen LR (2009) Saprophytism of a fish pathogen as a transmission strategy. Epidemics 1: 96–100.
15. Pulkkinen K, Suomalainen LR, Read AF, Ebert D, Rintamäki P, et al. (2010) Intensive fish farming and the evolution of pathogen virulence: the case of columnaris disease in Finland. Proc Roy Soc Lond B 277: 593–600.
16. Wagner BA, Wise DJ, Khoo LH, Terhune JS (2002) The epidemiology of bacterial diseases in food-size channel catfish. J Aquat Anim Health 14: 263–272.
17. Suomalainen LR, Tiirola MA, Valtonen ET (2005) Effect of Pseudomonas sp. MT5 baths on Flavobacterium columnare infection of rainbow trout and on microbial diversity on fish skin and gills. Dis Aquat Organ 63: 61–68.
18. Godfray HCJ, Briggs CJ, Barlow ND, O'Callaghan M, Glare TR, et al. (1999) A model of insect-pathogen dynamics in which a pathogenic bacterium can also reproduce saprophytically. Proc Biol Sci 266: 233–240.
19. Joh RI, Wang H, Weiss H, Weitz JS (2009) Dynamics of indirectly transmitted infectious diseases with immunological threshold. Bull Math Biol 71: 845–862.
20. Anderson RM, May RM (1981) The population dynamics of microparasites and their invertebrate hosts. Phil Trans R Soc Lond B 291: 451–524.
21. Day T (2002) Virulence evolution via host exploitation and toxin production in spore-producing pathogens. Ecol Lett 5: 471–476.
22. Roche B, Drake JM, Rohani P (2011) The curse of the Pharaoh revisited: evolutionary bi-stability in environmentally transmitted pathogens. Ecol Lett 14: 569–575.
23. Boldin B, Kisdi E (2011) On the evolutionary dynamics of pathogens with direct and environmental transmission. Ecol Lett 14: 569–575.
24. Read AF (1994) The evolution of virulence. Trends Microbiol 2: 73–81.
25. Frank SA (1996) Models of parasite virulence. Q Rev Biol 71: 37–78.
26. Levin BR (1996) The evolution and maintenance of virulence in microparasites. Emerg Infect Dis 2: 93–102.
27. Lipsitch M, Moxon ER (1997) Virulence and transmissibility of pathogens: what is the relationship? Trends Microbiol 5: 31–37.
28. Kaltz O, Shykoff JA (2002) Within- and among-population variation in infectivity, latency and spore production in a host-pathogen system. J Evol Biol 15: 850–860.
29. Bell AS, de Roode JC, Sim D, Read AF (2006) Within-host competition in genetically diverse malaria infections: parasite virulence and competitive success. Evolution Int J Org Evolution 60: 1358–1371.

30. Montarry J, Corbiere R, Lesueur S, Glais I, Andrivon D (2006) Does selection by resistant host trigger local adaptation in plant-pathogen systems? J Evol Biol 19: 522–531.
31. Wang IN (2006) Lysis timing and bacteriophages fitness. Genetics 172: 17–26.
32. Balmer O, Stearns SC, Schötzau A, Brun R (2009) Intraspecific competition between co-infecting parasite strains enhances host survival in African trypanosomes. Ecology 90: 3367–3378.
33. Lafforgue G, Sardanyés J, Elena SF (2011) Differences in accumulation and virulence determine the outcome of competition during Tobacco etch virus coinfection. PLoS One 6: e17917.
34. Abrams PA (2000) The evolution of predator–prey interactions: theory and evidence. Annu. Rev Ecol Syst. 31: 79–105.
35. Edelstein-Keshet L (1988) Mathematical models in biology. New York: Random House.
36. Grimont PAD, Grimont F (1978) The genus *Serratia*. Annu Rev Microbiol 32: 221–248.
37. Mahlen SD (2011) *Serratia* Infections - from military experiments to current practice. Clin Microb Rev 24: 755–791.
38. Soto E, Mauel MJ, Karsi A, Lawrence ML (2008) Genetic and virulence characterization of Flavobacterium columnare from channel catfish (*Ictalurus punctatus*). J Appl Microbiol 104: 1302–1310.
39. Suomalainen LR, Kunttu H, Valtonen ET, Hirvelä-Koski V, Tiirola M (2006) Molecular diversity and growth features of Flavobacterium columnare strains isolated in Finland. Dis Aquat Organ 70: 55–61.
40. Servais P, Billen G, Rego JV (1985) Rate of bacterial mortality in aquatic environments. Appl Environ Microbiol 49: 1448–1454.
41. Tirri R, Lehtonen J, Lemmetyinen R, Pihakaski S, Portin P (2003) Biologian sanakirja. Helsinki: Otava.
42. Anderson RM, May RM (1978) Regulation and stability of host-parasite population interactions.: II. Destabilizing processes. J Anim Ecol 47: 249–267.
43. Abrams PA, Matsuda H (1997) Prey Adaptation as a Cause of Predator-Prey Cycles. Evolution 51: 1742–1750.
44. Abrams PA, Matsuda H (1996) Positive indirect effects between prey species that share predators. Ecology 77: 610–616.
45. Holt RD (1977) Predation, apparent competition, and the structure of prey communities. Theor. Popul. Biol. 12:197–229.
46. Anderson RM, May RM (1978) Regulation and stability of host-parasite population interactions: I. Regulatory processes. J Anim Ecol 47: 219–247.
47. Halpern M, Raats D, Lev-Yadun S (2007) Plant biological warfare: Thorns inject pathogenic bacteria into herbivores. Env Microbiol 9: 584–592.
48. Murdoch WW (1973) Functional response of predators. J Appl Ecol 10: 335–342.
49. Kunttu HM, Sundberg LR, Pulkkinen K, Valtonen ET (2012) Environment may be the source of *Flavobacterium columnare* outbreaks at fish farms. Environ Microbiol Rep In press.
50. Suomalainen LR, Tiirola M, Valtonen ET (2006) Chondroitin AC lyase activity is related to virulence of fish pathogenic Flavobacterium columnare. J Fish Dis 29: 757–763.
51. Chang MX, Nie P (2008) RNAi suppression of zebrafish peptidoglycan recognition protein 6 (zfPGRP6) mediated differentially expressed genes involved in Toll-like receptor signaling pathway and caused increased susceptibility to *Flavobacterium columnare*. Vet Immunol Immunopathol 124: 295–301.
52. Danie DS, Trial JG, Stanley JC (1984) Species profiles: lifehistories and environmental requirements of coastal fish and invertebrates (Yorth Atlantic) – Atlantic salmon. U.S. Fish Wildl. Serv. FWS/OBS-82/11.22. U.S. Army Corps of Engineers, TR EL-82-4. 19 pp.
53. Tyler CR, Sumpter JP, Witthames PR (1990) The dynamics of oocyte growth during vitellogenesis in the rainbow trout (*Oncorhynchus mykiss*). Biol Reprod 43: 202–209.
54. Winckler-Sosinski LT, Schwarzbold A, Schulz U (2005) Survival of rainbow trout *Oncorhynchus mykiss* Walbaum, 1792 (*Salmoniformes - Salmonidae*) eggs in an altitude stream in southern Brazil. Acta Limnol Bras 17: 465–472.
55. Arslan M, Aras NM (2007) Structure and Reproductive Characteristics of Two Brown Trout (*Salmo trutta*) Populations in the Coruh River Basin, North-eastern Anatolia, Turkey. Turk J Zool 31: 185–192.
56. Harshbarger TJ, Porter PE (1979) Survival of brown trout eggs: two planting techniques compared. Prog Fish-Cult 41: 206–209.
57. Quinn TP, Bloomberg S (1992) Fecundity of Chinook salmon (*Oncorhynchus tshawytscha*) from the Waitaki and Rakaia Rivers, New Zealand. New Zealand Journal of Marine and Freshwater Research 26: 429–434.
58. Gillet C (1991) Egg production in an Arctic charr (*Salvelinus alpinus* L.) brood stock: effects of temperature on the timing of spawning and the quality of eggs. Aquar Living Resour 4: 109–116.
59. Atse CB, Audet C, de la Noue J (2002) Effects of temperature and salinity on the reproductive success of Arctic charr, *Salvelinus alpinus* (L.): egg composition, milt characteristics and fry survival. Aquacult Res 33: 299–309.
60. Sink TD, Lochmann RT (2008) Effects of dietary lipid source and concentration on channel catfish (*Ictalurus punctatus*) egg biochemical composition, egg and fry production and egg and fry quality. Aquaculture 283: 68–76.
61. Gioacchini G, Maradonna F, Lombardo F, Bizzaro D, Olivotto I, et al. (2010) Increase of fecundity by probiotic administration in zebrafish (Danio rerio). Reproduction 140: 953–959.
62. Hiltunen T, Friman VP, Kaitala V, Mappes J, Laakso J (2011) Predation and resource fluctuations drive eco-evolutionary dynamics of a bacterial community. Acta Oecologica (doi:10.1016/j.actao.2011.09.010.)
63. Flyg C, Kenne K, Boman HG (1980) Insect pathogenic properties of *Serratia marcescens*: phageresistant mutants with a decreased resistance to *Cecropia* immunity and a decreased virulence to *drosophila*. J Gen Microbiol 120: 173–181.
64. Gowen JW, Johnson LE (1946) On the mechanism of heterosis. I. Metabolic capacity of different races of *Drosophila melanogaster* for egg production. Am Naturalist 80: 149–179.

12

Sulphonamide and Trimethoprim Resistance Genes Persist in Sediments at Baltic Sea Aquaculture Farms but Are Not Detected in the Surrounding Environment

Windi Indra Muziasari[1], Satoshi Managaki[2], Katariina Pärnänen[1], Antti Karkman[1], Christina Lyra[1], Manu Tamminen[1], Satoru Suzuki[3], Marko Virta[1]*

1 Department of Food and Environmental Sciences, University of Helsinki, Helsinki, Finland, **2** Department of Environmental Sciences, Musashino University, Tokyo, Japan, **3** Centre for Marine Environmental Studies (CMES), Ehime University, Matsuyama, Ehime, Japan

Abstract

Persistence and dispersal of antibiotic resistance genes (ARGs) are important factors for assessing ARG risk in aquaculture environments. Here, we quantitatively detected ARGs for sulphonamides (*sul1* and *sul2*) and trimethoprim (*dfrA1*) and an integrase gene for a class 1 integron (*intI1*) at aquaculture facilities in the northern Baltic Sea, Finland. The ARGs persisted in sediments below fish farms at very low antibiotic concentrations during the 6-year observation period from 2006 to 2012. Although the ARGs persisted in the farm sediments, they were less prevalent in the surrounding sediments. The copy numbers between the *sul1* and *intI1* genes were significantly correlated suggesting that class 1 integrons may play a role in the prevalence of *sul1* in the farm sediments through horizontal gene transfer. In conclusion, the presence of ARGs may limit the effectiveness of antibiotics in treating fish illnesses, thereby causing a potential risk to the aquaculture industry. However, the restricted presence of ARGs at the farms is unlikely to cause serious effects in the northern Baltic Sea sediment environments around the farms.

Editor: Axel Cloeckaert, Institut National de la Recherche Agronomique, France

Funding: The research was funded by the Maj and Tor Nessling Foundation, Academy of Finland and Japan Society for the Promotion of Science (JSPS) grant. The funders had no role in study design, data collection and analysis, decision to publish, or preparation of the manuscript.

Competing Interests: The authors have declared that no competing interests exist.

* E-mail: windi.muziasari@helsinki.fi

Introduction

Aquaculture production is increasing worldwide as a source of fish for human consumption. Aquaculture introduces land-derived microbes, nutrients, metals and other chemicals such as antibiotics to the water environment. The prophylactic and therapeutic use of antibiotics results in the occurrence of antibiotic-resistant bacteria and antibiotic resistance genes (ARGs) in the aquaculture environment [1,2]. This may lead to seawater and the sediment becoming reservoirs for ARGs [3]. The ARGs in aquaculture environments can be transferred horizontally among microbes and ultimately be transferred to fish pathogens [4]. Thus, the presence of ARGs in aquaculture environments may lead to inefficiency in treating fish diseases using antibiotics [5]. To avoid production losses in the fish-farming industry, it is important to control the occurrence and spread of ARGs in aquaculture facilities.

The spreading of ARGs in the environment is mediated by horizontal gene transfer (HGT) [6]. Therefore, genes associated with HGT should also be examined when determining the prevalence of ARGs in the environment. Integrons can contribute to the occurrence of HGT of ARGs in bacterial populations [7,8] as a consequence of possessing a site-specific recombination system capable of capturing gene cassettes containing ARGs [9]. Integrons can be carried by mobile genetic elements such as transposons and plasmids that promote their wide distribution within bacterial communities [8]. Class 1 integrons, which contain an *intI1* gene encoding an integrase of the tyrosine recombinase family, are known to carry gene cassettes containing ARGs [10]. Class 1 integrons have been found in cultured fish pathogens [11] and from cultured bacteria in the aquaculture environments [4,12].

Sulphonamides potentiated with trimethoprim or ormethoprim and florfenicol are some of the antibiotics commonly used in aquaculture [5]. Consequently, the presence of several antibiotic-resistant bacteria in aquaculture environments has been reported previously using culture-dependent methods. Bacteria of the *Actinobacter* [13] and *Bacillus* genera [14] have been observed to carry the resistance genes to sulphonamides and *Aeromonas* and *Pseudomonas* to florfenicol [15]. Also aquatic bacteria of the genera *Proteus* and *Pseudomonas* can carry the resistance genes to both sulphonamides and trimethoprim [4]. However, culture-dependent methods may introduce bias when determining the prevalence of ARGs due to the inability to cultivate the majority of bacteria from environmental samples [16].

Culture-independent methods, including the measurement of gene copy numbers by quantitative polymerase chain reaction (qPCR), give a less biased estimation of the ARG amounts in the environment. qPCR has been widely used to study ARGs in environmental samples [17–19] and aquaculture environments [14,20]. However, little is known about the long-term persistence of ARGs at aquaculture sites and their dispersal to the surrounding environment [20]. To investigate these aspects, we collected sediment samples below two open-cage fish farms located in the Turku Archipelago, Finland in the northern Baltic Sea during the summers of 2006 to 2012. Sediment samples were also collected 200-m and 1000-m from fish farms, as well as from transect sites at 200-m intervals up to 1000 m from one farm to observe the dispersal of ARGs to the surrounding sediment environment. We quantified ARGs for sulphonamides, trimethoprim, and florfenicol and an integrase gene of class 1 integrons using qPCR. The antibiotic concentrations (sulphamethoxazole, sulphadiazine and trimethoprim) were measured using liquid chromatography-mass spectrometry (LC-MS). Our findings show that the ARGs were abundant and persistent in the sediments below the fish farm cages during the 6-year observation period, but were not detected in the sediments even at the closest 200-m distance from the cages. Moreover, a correlation was found between the amount of class 1 integrons and the *sul1* gene.

Results

Antibiotic resistance genes and class 1 integrons in sediments

The detection of trimethoprim resistance genes (*dfrA1, dfrA2, dfrA5, dfrA12, dfrA15, dfrA16, dfrA17,* and *dfrA19),* sulphonamide resistance genes (*sul1, sul2* and *sul3*), florfenicol resistance gene (*floR*) and an integrase gene (*intI1*) of class 1 integrons was done for six sediment samples chosen from northern Baltic Sea fish farm sites, using standard PCR. From the targeted ARGs, the *dfrA1, sul1, sul2* and *intI1* genes were found at two fish farms (FIN1 and FIN2). The amounts of the genes detected were measured, using qPCR in all 51 sediment samples. The copy numbers of the four genes were under the detection limit in every sediment sample taken 200 m to 1000 m from the farms as shown in Figure 1. The dispersal of the ARGs was not considerable, since the genes were not detected even at the closest 200-m distance from each farm.

The *sul1, sul2* and *intI1* genes were present in every sample and the *dfrA1* gene in most samples from the FIN1 and FIN2 farms throughout the 10 sampling times from 2006 to 2012 (Figure 1). The *sul1, sul2* and *intI1* gene copy numbers at the two farms were similar, with about 10^{-3}–10^{-2} copies in proportion to the 16S ribosomal RNA (rRNA) gene copies. The *dfrA1* gene copy numbers varied approximately 10^{-5}–10^{-2} copies in proportion to the 16S rRNA gene copies. The abundance of the *dfrA1* gene was lower than and significantly different from that of the *sul1* ($P<0.001$), *sul2* ($P<0.001$) and *intI1* ($P<0.001$) genes.

Correlation between ARGs and class 1 integrons in the sediments

Linear regression analysis was performed to test whether the copy number of the class 1 integron was correlated with any of the three ARGs detected (*dfrA1, sul1* and *sul2*) and thus could have played a role in the prevalence of the ARGs. Significant correlation ($F_{1,22}$ = 19.39; P = 0.000225; R^2 = 0.47) was found only between the average copy numbers of the *intI1* and *sul1* genes (Figure 2). The prevalence of the *sul1* gene in the Baltic Sea farm sediment may therefore be associated with class 1 integrons.

Antibiotic concentrations in the sediments

The sulphamethoxazole, sulphadiazine, and trimethoprim concentrations were measured, using LC-MS analysis to estimate the presence of selection pressure in the sediments. The antibiotic concentrations in the sediments are shown in Table 1. All the antibiotic concentrations measured in the farm sediments were very low (1.5–101 ng g^{-1} of dry sediment) and some even below the detection limit (<1 ng g^{-1} of dry sediment). The antibiotic concentrations were also below the detection limit in the sediments taken 200 m to 1000 m from the farms. Hence, there was no clear selection pressure in the sediments.

Discussion

Our results showed that the sulphonamide resistance genes (*sul1* and *sul2*) and trimethoprim resistance gene (*dfrA1*) were persistent in the Baltic Sea farm sediments during the 6-year observation period. We assume that the Baltic Sea farms have relatively small impact from human habitats and agriculture and therefore municipal and agricultural ARG sources can be excluded. The antibiotics and organic matter from uneaten fish feeds and fish excrement can enter the sediments directly from the water since there is no purification process in the open-cage farming system. Thus, the selection for the resistant bacteria and ARGs may have occurred in the medicated fish feeds [20] or inside the fish intestines and fish faeces that entered the sediments [21,22], or selection through the antibiotics present in the sediments [3,22]. However, the LC-MS results indicated low concentrations of sulphonamides and trimethoprim, suggesting that there was no clear selection pressure in the farm sediments. Furthermore, sulphonamides are decomposed by chemical and biological factors with half-lives of 7–85 days [23]. The persistence of ARGs in the farm sediments may, therefore, have been due to a constant introduction of ARGs from external sources, such as uneaten medicated fish feeds and fish faeces [16].

On the other hand, very low concentrations of the antibiotics may also play a role in maintaining the ARGs in the farm sediments by selection and enrichment of resistant bacteria by a subinhibitory antibiotic concentration, which was shown in a previous study [24]. Moreover, the presence of antibiotics in a subinhibitory concentration may induce the HGT system in bacterial communities [25], which further increases the prevalence of ARGs. Therefore, the potential for low antibiotic concentrations in maintaining resistant bacteria also needs to be examined to further understand the persistence of ARGs in northern Baltic Sea farm sediments.

In this study, the dispersal of ARGs (*sul1, sul2 and dfrA1*) from the Baltic Sea aquaculture farms to the surrounding sediments was not detected. Previous work from the same fish farms locations shows similar results; the four tetracycline resistance genes studied were not detected in the sediments even at the closest 200-m distance from the farm cages [20]. These results suggest that the resistance genes are potentially a problem for the fish-farming industry, but impact less the surrounding sediment environment in the northern Baltic Sea.

The copy number of the class 1 integron *intI1* was significantly correlated with the *sul1* gene copies in the farm sediments. This was expected, since the *sul1* gene is one of the backbone genes of the 3′-conserved segments in class 1 integrons [10]. Class 1 integrons are, therefore, likely involved in the prevalence of the *sul1* gene in Baltic Sea fish farm sediments. Similar correlations between the copy numbers of the *intI1* and *sul1* genes have been observed in riverine sediments in Haihe, China [18] and in Colorado, USA [19]. Only *dfrA1* of the eight trimethoprim resistance genes analysed (*dfrA1, dfrA2, dfrA5, dfrA12, dfrA15, dfrA16, dfrA17 and dfrA19*), which are commonly associated with class 1 integrons [6], was detected in the Baltic Sea farm sediments. There was no significant correlation between the amounts of the *intI1* gene of class 1 integrons and the *dfrA1* gene copies, suggesting that the *dfrA1* gene was not associated with class 1 integrons in the farm sediments. The prevalence of the *dfrA1*

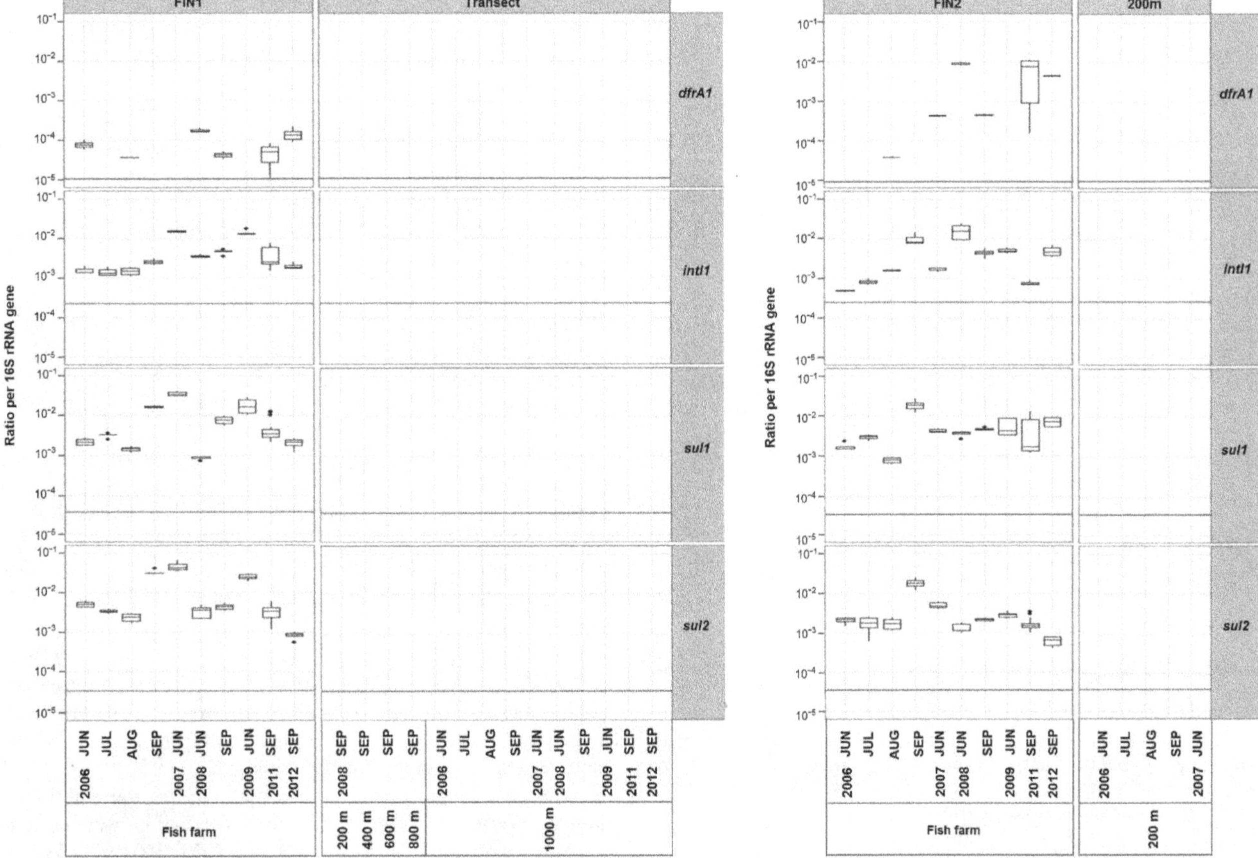

Figure 1. Gene copy numbers in the sediments. The ratios of the trimethoprim resistance gene (*dfrA1*), sulphonamide resistance gene (*sul1* and *sul2*) and an integrase gene of a class 1 integron (*intI1*) to the 16S rRNA gene copies were elevated in the farm sediments. None of these genes were detected in sediment samples at the closest 200-m distance from the two farms during the 6-year observation period. The missing data values in the plot mean that the respective gene copy numbers were below the limit of detection in the qPCR assays. The gene quantification limit normalized to the average numbers of the 16S rRNA gene copies is indicated by a grey line.

gene may have been mediated by other mobile elements or even independently of them [9].

The copy number of *dfrA1* was significantly lower than that of *sul*. Although the amount of *dfrA1* genes was lower than the amount of *sul* genes, the prevalence of both genes in the aquaculture environment deserves equal focus, because sulphonamides and trimethoprim are used in combination as so-called potentiated sulphonamides [5,26]. While many studies have demonstrated sulphonamide resistance in aquaculture environments [17], to our knowledge, this is the first study reporting qPCR measurements of the trimethoprim resistance gene (*dfrA1*) in aquaculture-impacted sediment samples.

In conclusion, the persistence of ARGs in farm sediments may lead to problems in the efficiency of antibiotics used to treat fish diseases and eventually to production losses at the fish farms. It is important for the fish-farming management to control the use of antibiotics to avoid the emergence of ARGs at the farms. Since the ARGs were not detected 200 m and 1000 m from the farms, their presence at the farms is unlikely to cause serious effects in the aquatic environment surrounding fish farms in the northern Baltic Sea in the current environmental conditions. However, a change in environmental conditions or an extended exposure to nearby fish farming activity could conceivably lead to an emergence of antibiotic resistance genes in the future. The sources and the spread of ARGs in aquaculture process chains should also be studied, as well as their potential risk to human health.

Materials and Methods

Study site and sampling

The sediment samples were collected from two fish farms (FIN1 and FIN2) and nearby areas located in the Turku Archipelago, Finland in the northern Baltic Sea from 2006 to 2012. The northern Baltic Sea is a unique brackish water marine environment (mean salinity: 6.7 parts per thousand) and no tide [27]. The sampling locations are described in Table 2. Both the FIN1 and FIN2 farms use open-cage systems in which the fish are kept in net cages that allow free transfer of uneaten fish feeds and fish excrement from the cages to the surrounding waters and eventually to the sediments. The farms raise European whitefish (*Coregonus lavaretus* (L.)) and rainbow trout (*Oncorhynchus mykiss* (Walbaum)). Each farm produces approximately 50 tons of fish annually. The record amount of antibiotics used at the fish farms was not available.

Northern Baltic Sea aquaculture farms operate only in summer. Sampling was done 10 times during the 6-year observation: June, July, August and September 2006, June 2007, June and September 2008, June 2009, September 2011 and September 2012. In addition, transect interval samples were collected at sites

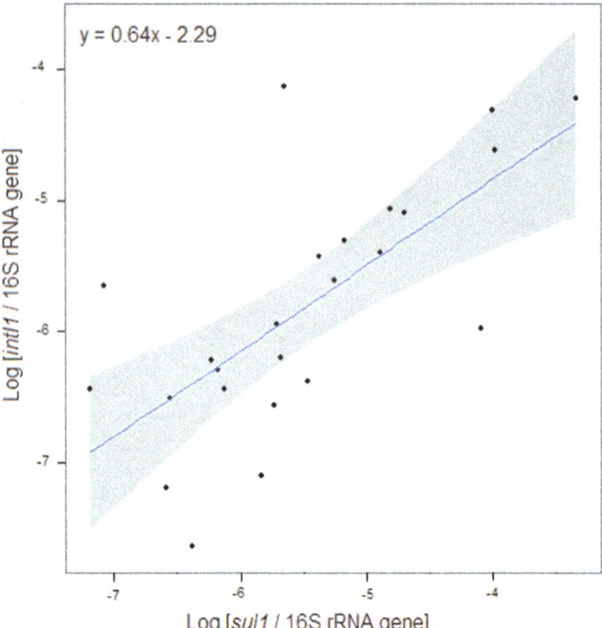

Figure 2. Correlation analysis. Linear regression model with log-transformed variables between the *intI1* and *sul1* gene copy numbers in the sediments below the northern Baltic Sea farms ($F_{1,22}$=19.39; P=0.000225; R^2=0.47). Each point represents the average ratio of a gene copy number normalized to the 16S rRNA gene copy number in every sediment sample. The blue line indicates the regression model and the grey area the 95% confidence intervals.

200-m up to 1000-m distance from the FIN1 farm on September 2008. Three replicate samples were collected in each year from 2006 to 2009 and the replicates from each year were pooled. In 2011 and 2012, three biological replicates were individually collected. In all, 51 samples from the FIN1 farm and FIN2 farm and surrounding areas were collected using a Limnos sediment probe (Limnos Ltd., Turku, Finland). Each sample was homogenized manually inside a zipper storage plastic bag and immediately frozen on dry ice. The sediments were stored at −80°C until DNA extraction.

DNA extraction

The environmental total DNA was extracted from 0.5 g wet weight sediment, using the FastDNA® SPIN kit for soil (MP Biomedicals, Illkrich, France). The standard protocol was modified by adding an extra washing step with 5.5 M of guanidine thiocyanate (Sigma Life Science, Steinheim, Germany), according to the manufacturer's instructions for removing humic acids. The DNA quality and concentration were analysed with a Nanodrop 1000 spectrophotometer (Thermo Scientific, Wilmington, DE, USA). The extracted DNA was stored at −20°C.

Standard PCR

The PCR primers and conditions for detecting the presence of targeted genes in the sediments were described previously: sulphonamide resistance genes (*sul1, sul2* and *sul3*) [17], trimethoprim resistance genes associated with mobile elements (*dhfr1, dfrII, dfrV, dhfrIX, dfrXII, dhfrXV, dfr16, dfr17* and *dfrA19*) [28], currently known as *dfrA1, dfrA2, dfrA5, dfrA9, dfrA12, dfrA15, dfrA16, dfrA17* and *dfrA19* [10], florfenicol resistance gene (*floR*) [28] as well as the integrase gene of the class 1 integron (*intI1*) [29]. Six sediment samples from the FIN1 farm, FIN2 farm and 1000-m distance from the FIN1 farm, taken during years 2007 and 2011, were chosen for gene detection in PCR.

The 25-µl PCR reactions consisted of 1× Taq buffer with $(NH_4)2SO_4$, 2 mM $MgCl_2$, 0.2 mM of each deoxyribonucleotide triphosphate (dNTP), 50 U Taq DNA polymerase, recombinant (Finnzymes, ThermoFisher Scientific, Espoo, Finland), 0.2 µM of each primer (Oligomer Oy, Helsinki, Finland) and the DNA template. The negative controls had nuclease-free water and were done for every PCR reaction. All the PCR reactions were done in triplicate using a PTC-200 thermal cycler (MJ Research, Watertown, MA, USA). The PCR products were purified, using QIAquick PCR purification kit (Qiagen, Hilden, Germany) and sequenced by the DNA sequencing service at the Institute of Biotechnology (University of Helsinki, Finland) to confirm the sequences of the PCR product.

Quantitative PCR measurement

The primers for the *dfrA1* gene [28] and the *intI1* gene [29] used in standard PCR were not optimal for the qPCR assays in this study. Thus, new primer sets were designed. The *dfrA1* and *intI1* gene sequences (accession numbers indicated in Table 3) were submitted to Primer3 v.2.3.4 to produce primer sets with melting temperatures above 60°C and amplifying 150–250 base-pair (bp)-

Table 1. Antibiotic concentrations in sediments.

Sites	Sampling times	Sulfamethoxazole (ng g^{-1})	Sulfadiazine (ng g^{-1})	Trimethoprim (ng g^{-1})
FIN1-Farm	JUN 2007	<1	<1	<1
	JUN 2009	3.6	2.3	<1
	SEP 2011	7.1	3.2	1.5
1000-m distance from FIN1 farm	JUN 2009	<1	<1	<1
	SEP 2011	<1	<1	<1
FIN2-Farm	JUN 2007	41	23	5.7
	SEP 2011	101	47	8.9
200-m distance from FIN2 farm	JUN 2007	<1	<1	<1

The detection limit = 1 ng g^{-1} of dry sediment. The relative standard deviations of the target antibiotics ranged between 8% and 13%.

Table 2. Sampling sites and their descriptions.

Sites	Mean value at sampling times[a]			Locations
	depth (m)	T (°C)	pH	
FIN1Farm	6 (+/−SD 1)	15.3 (+/−SD 3.2)	7.6 (+/−SD 0.5)	Located in the middle of a 400-m-wide strait
Distance from FIN1 farm	8 (+/−SD 0.4)	15.1 (+/−SD 2.7)	4.2 (+/−SD 1.6)	A site 1000-m distance from the FIN1 farm. In addition, a transect was sampled along the strait of the FIN1 farm at 200-m intervals up to 1000 m
FIN2-Farm	7.4 (+/SD 1.9)	16 (+/−SD 2.7)	7.9 (+/−SD 0.5)	Located next to the seashore in an 800-m-wide strait
Distance from FIN2 farm	5.1 (+/SD 2.6)	16.5 (+/−SD 2)	8.1 (+/−SD 0.3)	A site 200-m distance from the FIN2 farm

[a]Mean values of depth, temperature (T) and pH were measured from bottom seawater at sampling sites located in the archipelago area in the northern Baltic Sea.

long fragments. The primers designed were compared with known gene sequences that were retrieved from GenBank and aligned with mafft [30] to choose primer sets that are located in conserved regions. Basic Local Alignment Search Tool (BLAST) analysis of the primer sequences against the National Center for Biotechnology Information (NCBI) database was performed to avoid nonspecific amplification.

Plasmids R388, RSF1010 and pUV441 were used as the *sul1*, *sul2*, and *sul3* gene standards. Presynthesized pUC57 vectors containing either the entire 459-bp sequence of the *dfrA1* gene or a 250-bp fragment of the *floR* gene were ordered from GenScript (Piscataway, NJ, USA). The chromosomal DNA of *Escherichia coli* K12 was used as the 16S rRNA gene standard. The plasmid standards were linearized and purified. The standard copy numbers per µl were calculated using the estimated molar mass for the DNA bp (650 Da/1 bp). Determination of the qPCR efficiency was done, using a five-point 10-fold dilution series. In addition, inhibition tests were performed as previously described [31] to observe whether the sediment samples had the same amplification efficiency as the standard. The average inhibition was 2.5% (+/− SD 1.9%).

The qPCR was performed using a 7300 real-time PCR system (Applied Biosystems, Foster City, CA, USA). The 20-µl qPCR reactions contained 1× DyNAmo Flash SYBR Green Master Mix (Thermo Scientific) with 1× of ROX passive reference dye, 0.625-µM primers for the targeted gene listed in Table 3 and freshly diluted DNA sample. Nuclease-free water was added instead of DNA samples for the no template control (NTC) sample. The qPCR programs consisted of 7 min at 95°C, 40 cycles of 10 s at 95°C, 30 s at the annealing temperatures (Ta) listed in Table 3 and melting curve analysis. Three technical replicates of each sample, NTC and standard dilution were performed in each measurement and the measurements were repeated once to determine the reproducibility of the qPCR assays. Based on previous studies of the validation of qPCR assays [32,33], each assay was performed until the qPCR assay characteristics were achieved as described in Text S1. The limit of quantification (LOQ) of the qPCR assay was 5.92×10^{-5} copies in proportion to the average number of 16S rRNA gene copies for *sul1*, *sul2*, *sul3* and *floR*, 1.18×10^{-5} copies in proportion to the average of 16S rRNA gene copies for *dfrA1* and 2.96×10^{-4} copies in proportion to the average number of 16S rRNA gene copies for *intI1*. Data analysis was performed manually, using the 7300 System SDS v.1.2 Software (Applied Biosystems).

Statistical analysis

Student's t-test was performed to determine whether the copy numbers of the ARGs from all farm samples varied significantly. The correlation between the average gene copy numbers of the integrase gene of the class 1 integron and the ARGs detected in the

Table 3. Primers used for qPCR assays.

Target gene	Primers	Sequence 5′ - 3′	Ta (°C)	Product (bp)	References
sul1	sulI-FW	CGC ACC GGA AAC ATC GCT GCA C	64	163	[13]
	sulI-RV	TGA AGT TCC GCC GCA AGG CTC G			
sul2	sulII-FW	TCC GAT GGA GGC CGG TAT CTG G	60	191	[13]
	sulII-RV	CGG GAA TGC CAT CTG CCT TGA G			
sul3	sulIII-FW	TCC GTT CAG CGA ATT GGT GCA G	60	128	[13]
	sulIII-RV	TTC GTT CAC GCT TTA CAC CAG C			
dfrA1	dfrA-q-FW	TTC AGG TGG TGG GGA GAT ATA C	60	150	This Study
	dfrA-q-RV	TTA GAG GCG AAG TCT TGG GTA A			[NC_006385.1]
floR	floR-right	TCG TCA TCT ACG GCC TTT TC	60	188	[28]
	floR-left	CTT GAC TTG ATC CAG AGG GC			
intI1	intI1-a-FW	CGA AGT CGA GGC ATT TCT GTC	60	217	This study
	intI1-a-RV	GCC TTC CAG AAA ACC GAG GA			[NC_017659.1]
16S rRNA	pA	AGA GTT TGA TCC TGG CTC AG	60	350	[34]
	358R	CTG CTG CCT CCC GTA GG			[35]

farm sediments was analysed, using linear regression. The t-test and the linear regression with log-transformed variables were performed using RStudio v.0.97.168 (RStudio, Boston, MA, 2012). All models were considered to be significant at p-values less than 0.05.

Analytical methods for antibiotic quantification

The liquid chromatography analyses were performed with a Hewlett-Packard 1100 (Hewlett-Packard Co., Palo Alto, CA, USA). The reagents used and extraction methods are described in Text S2. The antibiotics were separated on a reverse-phase column (YMC Pro C18, 3 μm, 150 mm×2 mm; YMC America Inc., Allentown, PA, USA) operated at 30°C at a flow rate of 0.15 ml min^{-1}. The mobile-phase solvents were water-acidified with 1% (v/v) formic acid (eluent A) and methanol-acidified with 1% (v/v) formic acid (eluent B) to a pH of 2.5. The HPLC gradient programmes are contained in Table S1 in Text S2. The antibiotics were detected with a Micromass Quattro Ultima triple-quadrupole MS (Micromass, Milford, MA, USA), equipped with electrospray ionization. The analyses were performed in the positive ion mode. The protonated molecular ion ([M+H]+) of the compounds was selected as the parent ion. Detection was performed in the multiple reaction-monitoring mode, using the two most intense and specific fragment ions. Table S2 in Text S2 lists the optimized conditions of the individual analytes. To identify the antibiotics, we compared the retention times and the area ratios of the two product ions in each sample with the average retention time and peak ratios of the standards in all measurements. The criteria difference between the samples and the standard was within 0.3 min for the retention time and 20% for the area ratio of the two product ions. The concentrations of the samples were calculated by an external standard method, based on the peak area of the sum of the two product ions monitored. The calibration lines of five concentration points (10, 20, 30, 40, and 50 μg l^{-1} in water-methanol [1:1]) of the individual antibiotics were used for quantification. The linearity of the calibration curve in this range was confirmed (R^2>0.99). The final concentrations of most of the samples in the vials were within the range of the calibration lines.The LOQs were defined as 10 times the noise level of the baseline in the chromatograms signal to noise (S/N) ratio and were 1 ng g^{-1}.

Ethics statement

No specific permits were required for the field study described. The sampling did not affect any protected or endangered organisms.

Acknowledgments

We would like to thank Kornelia Smalla (Julius Kühn-Institut Federal Research Centre for Cultivated Crops, Braunschweig, Germany) for providing the plasmids R388, RSF1010 and pUV441. We thank Johanna Muurinen and Leena Pitkänen for technical assistance during sampling.

Author Contributions

Conceived and designed the experiments: WIM MT MV. Performed the experiments: WIM SM KP. Analyzed the data: WIM SM AK CL MT SS MV. Contributed reagents/materials/analysis tools: SM MV. Wrote the paper: WIM SM KP.

References

1. Cabello FC (2006) Heavy use of prophylactic antibiotics in aquaculture: A growing problem for human and animal health and for the environment. Environ Microbiol 8: 1137–1144.
2. Cabello FC, Godfrey HP, Tomova A, Ivanova L, Dölz H, et al. (2013) Antimicrobial use in aquaculture re-examined: Its relevance to antimicrobial resistance and to animal and human health. Environ Microbiol 15: 1917–1942.
3. Nonaka L, Ikeno K, Suzuki S (2007) Distribution of tetracycline resistance gene, tet(M), in gram-positive and gram-negative bacteria isolated from sediment and seawater at a coastal aquaculture site in japan. Microbes Environ 22: 355–364.
4. Shah SQA, Colquhoun DJ, Nikuli HL, Sørum H (2012) Prevalence of antibiotic resistance genes in the bacterial flora of integrated fish farming environments of Pakistan and Tanzania. Environ Sci Technol 46: 8672–8679.
5. Serrano PH (2005) Responsible use of antibiotics in aquaculture. FAO Fisheries Technical Paper 469.
6. Blahna MT, Zalewski CA, Reuer J, Kahlmeter G, Foxman B, et al. (2006) The role of horizontal gene transfer in the spread of trimethoprim-sulfamethoxazole resistance among uropathogenic escherichia coli in Europe and Canada. J Antimicrob Chemother 57: 666–672.
7. Stokes HW, Nesbø CL, Holley M, Bahl MI, Gillings MR, et al. (2006) Class 1 integrons potentially predating the association with Tn402-like transposition genes are present in a sediment microbial community. J Bacteriol 188: 5722–5730.
8. Stalder T, Barraud O, Casellas M, Dagot C, Ploy MC (2012) Integron involvement in environmental spread of antibiotic resistance. Front Microbiol 3(119): 1–14.
9. Hall RM, Collis CM (1995) Mobile gene cassettes and integrons: capture and spread of genes by site-specific recombination. Mol Microbio 15: 593–600.
10. Partridge SR, Tsafnat G, Coiera E, Iredell JR (2009) Gene cassettes and cassette arrays inmobile resistance integrons. FEMS Microbiol Rev 33: 757–784.
11. L'abée-Lund TM, Sørum H (2001) Class 1 integrons mediate antibiotic resistance in the fish pathogen aeromonas salmonicida worldwide. Microbial Drug Resistance 7: 263–272.
12. Buschmann AH, Tomova A, López A, Maldonado MA, Henriquez MA, et al. (2012) Salmon aquaculture and antimicrobial resistance in the marine environment. PLoS ONE 7: 1–11.
13. Hoa PTP, Nonaka L, Viet PH, Suzuki S (2008) Detection of the sul1, sul2, and sul3 genes in sulfonamide-resistant bacteria from wastewater and shrimp ponds of North Vietnam. Sci of the Total Environ 405: 377–384.
14. Gao P, Mao D, Luo Y, Wang L, Xu B, et al. (2012) Occurrence of sulfonamide and tetracycline-resistant bacteria and resistance genes in aquaculture environment. Water Res 46: 2355–2364.
15. Akinbowale OL, Peng H, Barton MD (2006) Antimicrobial resistance in bacteria isolated from aquaculture sources in Australia. J Appl Microbiol 100: 1103–1113.
16. Suzuki S, Hoa PTP (2012) Distribution of quinolones, sulfonamides, tetracyclines in aquatic environment and antibiotic resistance in indochina. Front Microbiol 3(67): 1–8.
17. Pei R, Kim SC, Carlson KH, Pruden A (2006) Effect of river landscape on the sediment concentrations of antibiotics and corresponding antibiotic resistance genes (ARG). Water Res 40: 2427–2435.
18. Luo Y, Mao D, Rysz M, Zhou Q, Zhang H, et al. (2011) Occurrence and transport of tetracycline, sulfonamide, quinolone, and macrolide antibiotics in the Haihe river basin, China. Environ Sci Technol 45: 1827–1833.
19. Pruden A, Arabi M, Storteboom HN (2012) Correlation between upstream human activities and riverine antibiotic resistance genes. Environ Sci Technol 46: 11541–11549.
20. Tamminen M, Karkman A, Lohmus A, Muziasari WI, Takasu H, et al. (2011) Tetracycline resistance genes persist at aquaculture farms in the absence of selection pressure. Environ Sci Technol 45: 386–391.
21. Giraud E, Douet DG, Le Bris H, Bouju-Albert A, Donnay-Moreno C, et al. (2006) Survey of antibiotic resistance in an integrated marine aquaculture system under oxolinic acid treatment. FEMS Microbiol Ecol 55: 439–448.
22. KÜmmerer K (2009) Antibiotics in the aquatic environment - A review - part I. Chemosphere 75: 417–434.
23. Lai H, Wang TS, Chou CC (2011) Implication of light sources and microbial activities on degradation of sulfonamides in water and sediment from a marine shrimp pond. Bioresour Technol 102: 5017–5023.
24. Gullberg E, Cao S, Berg OG, Ilbäck C, Sandegren L, et al. (2011) Selection of resistant bacteria at very low antibiotic concentrations. PLoS Pathogens 7: 1–9.
25. Hastings PJ, Rosenberg SM, Slack A (2004) Antibiotic-induced lateral transfer of antibiotic resistance. Trends Microbiol 12: 401–404.

26. Capone DG, Weston DP, Miller V, Shoemaker C (1996) Antibacterial residues in marine sediments and invertebrates following chemotherapy in aquaculture. Aquaculture 145: 55–75.
27. Ojaveer H, Jaanus A, Mackenzie BR, Martin G, Olenin S, et al. (2010) Status of biodiversity in the baltic sea (biodiversity of the baltic sea). PLoS ONE 5(9): 1–19.
28. Szczepanowski R, Linke B, Krahn I, Gartemann KH, Gützkow KH, et al. (2009) Detection of 140 clinically relevant antibiotic-resistance genes in the plasmid metagenome of wastewater treatment plant bacteria showing reduced susceptibility to selected antibiotics. Microbiol 155(7): 2306–2320.
29. Hardwick SA, Stokes HW, Findlay S, Taylor M, Gillings MR (2008) Quantification of class 1 integron abundance in natural environments using real-time quantitative PCR. FEMS Microbiol Lett 278(2): 207–212.
30. Katoh K, Asimenos G, Toh H (2009) Multiple alignment of DNA sequences with MAFFT. Methods Mol Biol 537: 39–64.
31. Goebel NL, Turk KA, Achilles KM, Paerl R, Hewson I, et al. (2010) Abundance and distribution of major groups of diazotrophic cyanobacteria and their potential contribution to N. Environ Microbiol 12: 3272–3289.
32. Bustin SA, Benes V, Garson JA, Hellemans J, Huggett J, et al. (2009) The MIQE guidelines: Minimum information for publication of quantitative real-time PCR experiments. Clin Chem 55:4: 611–622.
33. Smith CJ, Osborn AM (2009) Advantages and limitations of quantitative PCR (Q-PCR)-based approaches in microbial ecology. FEMS Microbiol Ecol 67: 6–20.
34. Edwards U, Rogall T, Blöcker H, Emde M, Böttger EC (1989) Isolation and direct complete nucleotide determination of entire genes. Characterization of a gene coding for 16S ribosomal RNA. Nucl Acids Res 17(19): 7843–7853.
35. Lopez-Gutierrez J, Henry S, Hallet S, Martin-Laurent F, Catroux G, et al. (2004) Quantification of a novel group of nitrate-reducing bacteria in the environment by real-time PCR. J Microbiol Methods 57: 399–407.

13

A Preliminary Study on *Oxya fuscovittata* (Marschall) as an Alternative Nutrient Supplement in the Diets of *Poecillia sphenops* (Valenciennes)

Arijit Ganguly[1], Ranita Chakravorty[1], Angshuman Sarkar[2], Dipak K. Mandal[1], Parimalendu Haldar[1], Julieta Ramos-Elorduy[3], Jose Manuel Pino Moreno[3]*

1 Department of Zoology, Visva-Bharati University, Santiniketan, West Bengal, India, **2** Department of Statistics, Visva-Bharati University, Santiniketan, West Bengal, India, **3** Departamento de Zoología, Universidad Nacional Autonoma de Mexico, Mexico City, Distrito Federal, México

Abstract

Growth of the ornamental fish industry is being hindered by the scarcity of low cost feed; hence alternative protein supplements should be explored. In this context the present study aims to evaluate whether the grasshopper *Oxya fuscovittata* could be used as a supplement for fish meal in the diets of *Poecillia sphenops*, which is one of the most common ornamental fishes worldwide. The present work is divided into three phases: In the first phase proximate composition of the grasshopper is obtained and five diets are prepared where fish meal is gradually replaced by *Oxya* meal and named as control, D1, D2, D3 and D4. All the diets are formulated on iso-nitrogenous basis where the protein percentage is fixed at 400 g/kg. The second phase deals with feeding trial and in the third phase all the data of the feeding trial are subjected to a linear model. The feeding trial shows that the control, D1 and D2 fed fishes have almost similar results. The linear model proves that the variation in the indices are mainly due to replacement of fish meal by *Oxya* meal, not due to the variations of rice husk and mustard oil cake that are also used to formulate the diets of the present study. From the results two *Oxya* supplemented diets, i.e. D1 and D2 are proved to be almost equivalent to the control diet. Hence it is concluded that *Oxya* meal is able to replace 25% to 50% of fish meal from the diets of *P. sphenops*.

Editor: Alexander Chong Shu-Chien, Universiti Sains Malaysia, Malaysia

Funding: These authors have no support or funding to report.

Competing Interests: The authors have declared that no competing interests exist.

* Email: jpino@ib.unam.mx

Introduction

Fish farmers in general are facing difficulty in raising their fishes because of an alarming increase in the cost of feed which represents 60–70% of the total production cost [1]; and this is also true for ornamental fish industry. In concert with the high rising production cost, the market price of the ornamental fishes is also gaining height [2]. Moreover the conventional protein sources for fish industry i.e. fish meal, soy meal and other grains are being competed for by the human population as well [3]. Recently, with a global population boom the increasing demand and uncertain availability of these traditional protein sources resulted in the quest of other protein rich alternatives for inclusion into the fish diets [4,5]. In this context, insects being nutritionally rich could be a possible alternative [1]. Among the edible insects, acridids (i.e. short-horn grasshoppers of family Acrididae of the order Orthoptera) might have a great future to be established as a sustainable mini-livestock. Acridids are suitable insects for consumption because according to the report of Xiaoming et al. acridids have a high nutritional value and can be used to formulate good quality feed for livestock [6]. But for a constant supply to the feed developing companies there is a need to build acridid farms, for which a suitable species should be found out. In this context Anand et al. reports an estimation of probable annual biomass that could be produced in acridid farms [7]. According to their conclusions *Oxya fuscovittata* (Marschall) is most suitable because being tetra-voltine, they can complete four life cycles annually and could produce a huge biomass. Keeping this in mind the present study aims to utilize *O. fuscovittata* as a supplement of fish meal in the diets of *Poecilia sphenops* (Valenciennes).

Materials and Methods

The present work is divided into three phases; in the first phase analysis of nutrient composition of the ingredients are carried out, which is followed by feed formulation and diet quality assessment; in the second phase the feeding trial is conducted with the selected ornamental fish. During diet formulation fish meal is gradually replaced by *Oxya* meal; but to formulate diets on iso-nitrogenous basis, two other ingredients are needed to vary. Consequently the amounts of those ingredients differ from diet to diet. In this context it is hard to infer that the significant differences in the results of feeding trial is due to replacement of fish meal by *Oxya* meal, or due to the variations in the amounts of the other major ingredients. Hence, in the third phase all the data of the feeding trial are fitted to a linear model, to conclude which ingredient has greater effect on the variations of the observed indices during the feeding trial.

Table 1. Proximate composition of main ingredients.

Ingredients	CP(g/Kg±SD)	EE(g/Kg±SD)	C(g/Kg±SD)	TA(g/Kg±SD)
Fish Meal	483.1±07.2c	92.8±19.0b	381.7±10.1b	42.4±17.5a
Oxya Meal	640.7±02.3d	64.8±03.1a	241.8±04.7a	52.7±0.9a
Rice Husk	54.42±01.5a	49.0±08.0a	742.3±11.8d	154.3±11.1b
Mustard Oil Cake	306.3±12.8b	52.7±12.5a	606.5±12.2c	34.5±11.9a

Note: CP = crude protein, EE = ether extract (crude fat), C = carbohydrate, TA = total ash. Values with different letters within a column are significantly different (P<0.001) using Tukey's range test.

The present work has been conducted according to the rules of "Institutional Animal Ethics Committee" of Visva-Bharati University, and approved by the committee according to the Indian law.

Selection of model ornamental fish

To accomplish the present work we have chosen *P. sphenops* (more commonly "black molly") which could be easily reared, could complete life cycle within a short period and easy to breed. Moreover, it is one of the most common ornamental fishes to be seen in the aquariums.

Diet formulation

For the formulation of various supplementary diets *Oxya* meal, fish meal (dried *Harpadon nehereus* F. Hamilton 1822), mustard oil cake (MOC) and rice husk are used as major ingredients. Carboxymethyl cellulose (CMC) is used as a binder for the diets and calcium, table salt (NaCl) and vitamin-mineral mixture are used as additive. First of all proximate composition of the major ingredients are carried out according to AOAC methods [8]. Nitrogen is estimated using Kjeldahl method and crude protein is calculated by multiplying the amount of nitrogen with the factor 6.25 (988.05). Crude fat is estimated with soxhlet apparatus (920.39). Ash contents are determined by subjecting the samples to muffle furnace at 550°C for 6 hrs (942.05). The amount of carbohydrate content is calculated by difference method (100– sum of protein, fat and ash).

Five diets are prepared gradually replacing fish meal by *Oxya* meal with the ratio of 100:0, 75:25, 50:50, 25:75 and 0:100. The diet having "fish meal: *Oxya* meal" with the ratio of 100:0 is considered as control because there is no supplement of *Oxya* meal. The other diets are named D1, D2, D3 and D4 respectively. The amount of ingredients of all of the formulated diets are mixed in such a way so that each of them contain about 400 g/kg crude protein, because according to literature mollies need this much protein in their diets for optimal growth [9]. Maintaining the

Table 2. Ingredient composition and quality of formulated diets.

Ingredients	Control	D1	D2	D3	D4
Fish meal	600.00 g	450.00 g	300.00 g	150.00 g	0.00 g
Acridid meal	0.00 g	150.0 g	300.0 g	450.0 g	600.0 g
Mustard oil cake	339.00 g	267.80 g	173.70 g	80.10 g	12.70 g
Rice husk	11.00 g	82.20 g	176.30 g	269.90 g	337.30 g
Carboxy methyl cellulose	30.00 g	30.00 g	30.00 g	30.00 g	30.00 g
Vitamin-Mineral mixture[a]	7.00 g	7.00 g	7.00 g	7.00 g	7.00 g
NaCl	8.00 g	8.00 g	8.00 g	8.00 g	8.00 g
Calcium additive	5.00 g	5.00 g	5.00 g	5.00 g	5.00 g
Total	1 Kg	1 Kg	1 Kg	1 Kg	1 Kg
Formulated diet quality					
CP (g/Kg ± SD)	399.4±6.2a	400.1±5.3a	401.1±6.7a	399.9±8.1a	400.6±7.6a
EE (g/Kg ± SD)	63.1±11.3a	57.3±9.8a	52.7±12.3a	61.2±11.8a	52.3±8.7a
C (g/Kg ± SD)	498.4±7.7a	517.1±4.8b	522.1±10.1b	523.1±8.6b	523.3±4.4b
TA (g/Kg ± SD)	44.1±4.6b	27.5±3.1a	25.2±4.8a	21.8±3.6a	23.8±2.7a
Energy (kJ/g ± SD)	18.94±0.06a	19.15±0.04b	19.15±0.07b	19.39±0.06c	19.19±0.05b
P/E(mg protein per kJ ± SD)	20.82±0.19a	20.89±0.13a	21.14±0.11b	20.61±0.08a	21.18±0.19b
Striking time (in seconds)	6.50±1.22a	6.72±0.68a	7.06±1.57a	7.24±0.28a	6.36±0.16a

Note: CP = crude protein, EE = ether extract (crude fat), C = carbohydrate, TA = total ash, P/E = protein to energy ratio. Values with different letters within a column are significantly different (P<0.001) using Tukey's range test.

[a]Provided *per* Kg of diet: Vitamin A (as Acetate) 69000 IU, Vitamin D3 6900 IU, Vitamin E Acetate 172.5 mg, Ascorbic Acid 1035 mg, Thiamine Mononitrate 69 mg, Riboflavine 69 mg, Pyridoxine HCl 20.7 mg, Cyanocobalamine 103.5 mcg, Niacinamide 690 mg, Vitamin K 1.7 mg, Calcium Pentothenate 112.4 mg, Calcium Phosphate 890.1 mg, Magnesium Oxide 414 mg, Ferrous Fumerate 221.1 mg, Zinc (from zinc sulphate) 15.2 mg, Copper (from copper sulphate) 23.4 mg, Manganese (from manganese sulphate) 14 mg, Sodium Molybdate 1.7 mg, Sodium Borate 6.1 mg.

Table 3. Water quality of the experimental tanks.

Parameters	Range	Mean ± SD
Temperature (°C)	21.40–30.50	27.18±1.25
pH	7.34–7.71	7.52±0.48
Dissolve Oxygen (mg/L)	3.14–3.82	3.61±0.52
Hardness (mg/L)	162.40–167.60	165.67±2.20

amount of protein by adjusting the proportion of ingredients is done by the Pearson's square method. After formulation, diets are first made to dough forms that are subjected to a pelletizer (RJK-FFF-PM-52, Rajkumar Agro Engineers Pvt. Ltd, India), to make floating type pellets.

Assessment of diet quality

Crude protein, crude fat (ether extract), carbohydrate and total ash (TA) are estimated according to AOAC as already mentioned [8]. Energy contents of the diets are estimated by oxygen bomb calorimeter and expressed in kJ/g. Protein to energy (P/E) ratio is an important measure for the quality assessment of the formulated diets and it is expressed as "mg of protein/kJ of energy". Feed acceptability test is assessed as the time taken by the first fish to strike the first particle after the diet dropped into the tanks [10]. The time is recorded in seconds using a stop watch.

Water quality

Analysis of water of the experimental tanks has been done fortnightly in terms of temperature, pH, dissolved oxygen (DO) and hardness. Temperature and pH is measured using a combined pH meter and thermometer (ZPHI-9100), Zico India Ltd., but DO and hardness is measured by titration according to APHA [11].

Collection of experimental species

Same variety of twenty females and ten males of *P. sphenops* are bought from the local market of Bhubandanga, Santiniketan, West Bengal, India. They are reared in the laboratory using a *Spirulina* based diet till the females give birth to sufficient amount of hatchlings. One hundred and fifty (150) fingerlings of similar initial weight (about 0.01 g) are selected to start the feeding experiment, and are kept for 7 days of acclimation period prior to feeding.

Experimental setup

150 fingerlings are divided into 5 sets with 3 replications having 10 individuals in each group. They are placed in 15 experimental tanks of 26 L capacities. Half of the total volume of water is changed daily from each tank. The tanks are completely drained and thoroughly scrubbed once a week to prevent algal growth. Adult body weight and length are measured at the end of experiment after 176 days.

Feeding the fishes

The experimental fishes are fed with known amount of oven dried feed twice daily at 6.30 and 18.30 hrs for five minutes. The uneaten diets are siphoned off and oven dried. The amount of dried feed consumed is calculated by subtracting the left over dried feed from the known amount of dried feed offered.

Estimation of consumption, utilization, survival, growth and reproductive potential of *P. sphenops* when fed with the formulated diets

Growth is measured in terms of specific growth rate (SGR) and condition factor (K) according to De Silva & Anderson, and Moyle & Cech using the following equations [12,13].

$$SGR = [(\log W_{t2} - \log W_{t1}) \times 100]/T$$

$$K = 100 \times \left(\text{weight}/\text{length}^3\right)$$

Food consumption and utilization is measured in terms of protein efficiency ratio (PER), feed conversion ratio (FCR), and feed conversion efficiency (FCE) as suggested by De Silva and Anderson using the following equations [12].

Table 4. Results of Survival, SGR, PER, FCR, FCE, K and reproductive potential of *P. sphenops* fed with various formulated diets.

Response indices	Control	D1	D2	D3	D4
% Survival	84.19±2.36a	83.78±2.11a	83.93±1.82a	82.97±2.32a	83.31±1.97a
SGR	1.52±0.10c	1.52±0.08c	1.5±0.17c	1.26±0.06b	1.04±0.14a
PER	14.97±0.28d	15.97±0.24e	12.18±0.40c	3.16±0.26b	1.13±0.05a
FCR	1.16±0.03a	1.14±0.02a	1.18±0.05a	1.86±0.08b	2.21±0.11c
FCE	3.40±0.23d	3.62±0.14d	2.76±0.14c	0.97±0.13b	0.45±0.09a
K	1.66±0.0187c	1.55±0.192c	1.94±0.206c	1.01±0.097b	0.55±0.016a
Reproductive potential/female	14.33±1.53a	14.67±1.15a	15.67±0.58a	14.33±1.15a	15.33±0.58a

Note: SGR = Specific growth rate, PER = Protein efficiency ratio, FCR = Feed conversion ratio, FCE = Feed conversion efficiency, K = Condition factor. Values are means ± SD. Values with different letters within a row are significantly different ($P<0.001$) using Tukey's range test.

Table 5. Estimate of the regression coefficients of the effects of different factors used in the linear model on each response variable.

Response indices	Estimate of the regression coefficients			
	Intercept [P value]	Replacement [P value]	Rice husk [P value]	Mustard oil cake [P value]
SGR	0.0003 [0.4999]	0.0080 [0.4970]	0.0055 [0.4979]	0.0047 [0.4982]
PER	0.0004 [0.4998]	0.0074 [0.4972]	0.0107 [0.4960]	0.0044 [0.4983]
FCR	0.0002 [0.4999]	0.0089 [0.4967]	0.0016 [0.4994]	0.0051 [0.4981]
FCE	0.0004 [0.4999]	0.0070 [0.4974]	0.0090 [0.4966]	0.0042 [0.4984]
K	0.0003 [0.4999]	0.0077 [0.4971]	0.0067 [0.4975]	0.0045 [0.4983]

Note: SGR = Specific growth rate, PER = Protein efficiency ratio, FCR = Feed conversion ratio, FCE = Feed conversion efficiency, K = Condition factor, Intercept = constant effect, Replacement = *Oxya* meal that replaced fish meal, P value = Area under the t distribution above the observed t value of the student t-test that was conducted to know whether their corresponding factors had significant effect on the response.

$$PER = WWt/DWp$$

$$FCR = F/WWt$$

$$FCE = WWt/F$$

[Where W_{t1} = initial weight, W_{t2} = final weight, T = number of days, F = average dry weight of food ingested per individual, WWt = wet weight gained and DWp = dry weight of protein in feed].

Survival is measured in terms of percentage of individuals alive up to sexual maturity, whereas reproductive potential is measured in terms of total number of offspring born per female in the experimental sets.

Statistical analysis phase 1 and 2

All the data are presented as means ± SD. Each variable is first subjected to one way analysis of variance (ANOVA). Thereafter separation of mean values according to significance is done within each variable by Tuky's range test. All the statistical analyses are carried out using S Plus version 4.0 and PAST version 3.02.

Statistical analysis phase 3

To make sure whether the responses obtained in the feeding trial are due to the replacement of fish meal by *Oxya* meal or due to the variation of rice husk and mustard oil cake, the data of the response indices are fitted to the following linear model using MATLAB program (MATLAB 7.1)–

$$y = \beta_0 + \beta_1 x + \beta_2 (60 - x) + \beta_3 Z_1 + \beta_4 Z_2 + e$$

[Where, y = response indices, β_0 = effect of some constant factors that are identical for all the formulated diet fed sets, β_1 = effect due to x, x = *Oxya* meal, β_2 = effect due to (60-x), (60-x) = fish meal, β_3 = effect due to Z_1, Z_1 = rice husk, β_4 = effect due to Z_2, Z_2 = mustard oil cake, e = error].

$$\text{Or, } y = \beta_0 + 60\beta_2 + (\beta_1 - \beta_2)x + \beta_3 Z_1 + \beta_4 Z_2 + e$$

This equation could be simplified as–

$$y = \beta_0^* + \beta_1^* x + \beta_3 Z_1 + \beta_4 Z_2 + e$$

[Where, $\beta_0^* = (\beta_0 + 60\beta_2)$, and $\beta_1^* = (\beta_1 - \beta_2)$. In other words according to this model, β_0^*, β_1^*, β_3, and β_4 are the "estimate of the regression coefficients" of the effects due to their corresponding factors (i.e. intercept or the constant effect, *Oxya* meal that replaced fish meal, rice husk and mustard oil cake respectively) on the response indices].

In the next step, "estimate of the regression coefficients" are individually subjected to student t-test to know whether their corresponding factors have significant effect on the response indices. If the null hypothesis is accepted, one could conclude that the corresponding factor has no significant effect and if the null hypothesis is rejected, one could conclude that the corresponding factor has significant effect on the response.

Table 6. Comparison between the sum of squares of the effects due to Group1 and Group2 on the response indices.

Response Variables	Sum of squares due to Group1 [$\times 10^3$]	Sum of squares due to Group2 [$\times 10^3$]	P-value of the ratio of sum of squares
SGR	215.9000000	0.7788000	<0.001
PER	0.9156000	0.2053000	0.0883
FCR	543.9200000	26.7680000	<0.01
FCE	21.7300000	3.3322000	<0.01
K	135.9900000	3.4697000	<0.01

Note: SGR = Specific growth rate, PER = Protein efficiency ratio, FCR = Feed conversion ratio, FCE = Feed conversion efficiency, K = Condition factor, Group1 = replacement of fish meal by *Oxya* meal, Group2 = rice husk and mustard oil cake in combination.

After this test the four major factors of the formulated diets (i.e. *Oxya* meal, fish meal, rice husk and mustard oil cake) are divided into two groups–

replacement of fish meal by *Oxya* meal = Group1

rice husk + mustard oil cake = Group2

Then the effect of Group1 on the response indices are statistically compared with the same of Group2, to conclude which group has greater effect on the variation of the response indices. This is done by comparing the sum squares (SS) of the effect of Group1 and Group 2 because SS represents the sum of squared difference from the mean, and this difference gives rise to variance. This part is carried out using the same model already programmed in MATLAB.

Results

Diet formulation

Proximate compositions of the major ingredients i.e. rice husk, *Oxya* meal, MOC and fish meal are tabulated in table 1. It reveals that crude protein level is highest in *Oxya* meal (more than 640 g/kg) and crude fat is highest in fish meal (more than 90 g/kg). Contents of carbohydrate and total ash show higher values for rice husk. After getting the proximate composition of the major ingredients, their required proportions in the formulated diets are calculated. The final compositions of the diets are depicted in table 2.

Assessment of diet quality

Quality of the formulated diets is presented in table 2. Almost 400 g/kg of crude protein level is obtained in each of the diets with no significant difference ($p>0.05$, Tuky's range test). Similarly the values of crude fat (ether extract) again show statistically insignificant variation. The values of carbohydrate contents are also very similar; however, in the control diet it is slightly lower than the others ($p<0.05$, Tuky's range test). Total ash has a result which is reciprocal to the values of carbohydrate where the control diet shows statistically higher ($p<0.001$, Tuky's range test) results. Energy is quite similar, but the control and D3 has significantly lowest and highest mean values respectively ($p<0.001$, Tuky's range test). Protein to energy ratios (P/E) are almost the same (nearly 21 mg/kJ) for all the diets. Results of the acceptability test reveals that the time taken by the first fish to strike the first particle of the formulated diets is within about 6–7 seconds for each case. Hence no significant variation is obtained by Tuky's range test.

Assessment of water quality of the experimental tanks

The temperature, pH, dissolved oxygen and hardness of water during the feeding period are tabulated in table 3. Temperature of all the experimental tanks range from 21.4 to 30.5°C with a mean value of 27.18°C. The water is found to be slightly alkaline as the pH range falls between 7.34–7.71 with a mean value of 7.52. Dissolved oxygen has an average value of 3.61 mg/L and it ranges between 3.14–3.82 mg/L in all the experimental tanks. Assessment of hardness reveals that the water of the experimental tanks is hard with a mean value of 165.67 mg/L that ranges between 162.4–167.6 mg/L.

Consumption, utilization, survival, growth and reproductive potential of *P. sphenops* when fed with the formulated diets

The results of survival, SGR, PER, FCR, FCE, K and reproductive potential of *P. sphenops* are summarized in table 4. More than 80% survival is obtained for all the diet fed sets and they do not vary significantly. But for the other indices as a whole there is a tendency of the control diet, along with D1 and D2 fed *P. sphenops* to have similar results that are significantly better than D3 and D4 fed groups. In case of specific growth rate (%SGR) the values vary insignificantly ($p>0.05$, Tuky's range test) between the control, D1 and D2 fed sets and these values are higher than D3 and D4 fed ones. The values of feed conversion ratio (FCR) also show better results for the control, D1 and D2 diets as the results are significantly lower ($p<0.05$, Tuky's range test) than the groups fed with D3 and D4. On the contrary unlike SGR and FCR, the results of protein efficiency ratio (PER) and feed conversion efficiency (FCE) statistically vary between the control, D1 and D2 fed sets. For PER among these three diets D1 fed groups show highest values ($p<0.05$, Tuky's range test) followed by control and D2 fed groups. For FCE, though the values are insignificant between control and D1 fed groups, the sets fed with D2 shows a slightly lower value ($p<0.05$, Tuky's range test). Condition factor (K) also shows a similar trend, values vary insignificantly between control, D1 and D2 fed mollies and then it gradually decreases in D3 and D4 fed sets. Each female of all the experimental tanks are found to give birth to 13–16 offspring. On an average the values are insignificant ($p>0.05$, ANOVA, Tuky's range test) when data are compared between the sets fed on different formulated diets.

Linear model of response variables

The estimate of the regression coefficients of the effects of different factors used in the linear model on each response index are depicted in table 5. Survival and reproductive potential are excluded from this analysis as they vary insignificantly from diet to diet ($p>0.05$). Values of the student t-test that are conducted to know whether their corresponding factors have significant effect on the response indices are also presented in the same table. The insignificant p values ($p>0.05$, Tuky's range test) show that the null hypothesis is rejected for all. This means all the factors in question (i.e. replacement of fish meal by *Oxya* meal, rice husk and mustard oil cake) has certain amount of effect on the variation of the response data that are obtained during the feeding trial. Now, as it has been evident that all the factors in question are playing some role on the variation of the response data, in the next step it is essential to find out which factor is playing greater role. For this purpose the sum of squares (SS) of the effects due to replacement of fish meal by *Oxya* meal (Group1) is compared with the sum of squares (SS) of the effects due to rice husk and mustard oil cake in combination (Group 2). Results of the sum of squares along with their corresponding p values have been tabulated as table 6. This table has revealed that only excluding PER, for rest of the cases, Group1 has much greater effect on the variations in response data that are obtained during the feeding trial. This proves that the effect of replacement of fish meal by *Oxya* meal has greater influence on the variation of response than the effect of rice husk and mustard oil cake in combination.

Discussion

It is already known that mollies prefer hard water with a pH range of 7.5–8.2, and 18°C−28°C is the suitable temperature for them [14]. In the present study the result of water quality reveals that the water in the experimental tanks is hard with a temperature and pH within the optimum range. However, dissolved oxygen is a little lower than the optimum need (i.e. above 5 mg/L) for most of the fishes [15,16]. The main cause behind this lower DO may be because no aerator is used in the tanks. However, mollies have been reported to get adapted to hypoxic conditions as they could tolerate a level of DO as low as 1 mg/L [17]. In this context it could be stated that the water condition of the experimental tanks are ideal for rearing mollies.

Protein is one of the most important constituents in feed for fish growth. Before designing any experiment on the effect of formulated diets on growth of fishes, the dietary percentage of protein needed for their optimal growth should be obtained first [18]. Shim and Chua worked on poeciliids like *Poecilia reticulata* Peters 1859 and according to them 300–400 g/kg of protein in feed is optimum for their growth [19]. Parker also opines that 400 g/kg dietary protein is optimum for mollies [9]. Hence 400 g/kg of crude protein is fixed for all the formulated diets in the present study. According to some workers, 100–200 g/kg of lipid in fish diets gives optimum growth rates without producing an excessive fatty carcass [20]. In this context it may seem that the diets of the present study contain a little lower amount of fat (near about 50–60 g/kg). However, report of Fernando et al. states that poeciliids are fed with diets having 30–70 g/kg of fat in Singapore [21]. This report supports the view that the formulated diets of the present study contain sufficient amount of fat. It is well known that fish apparently do not have carbohydrate requirement, but it could be metabolized by many fishes and usually herbivorous fishes can metabolize carbohydrates better than carnivorous species [22,23]. According to the report of National Research Council warm water fish (such as mollies) are able to use high carbohydrate levels as an energy source [24]. This phenomenon is advantageous because carbohydrate sources are inexpensive as seen in the present case where all of the formulated diets contain a high amount of carbohydrate. Protein to energy (P/E) ratio is another important measure that gives us a clear picture whether the balance between protein and energy contents is maintained in the diets in proper amount. According to some authors for achieving maximum growth, the P/E ratio in fish diet should range from 19–22 mg/kJ [25–27]. These reports are encouraging because in the present case all of the formulated diets have P/E value of nearly 21 mg/kJ.

Before feeding trial it is necessary to be sure whether the experimental fish recognizes the diets as a palatable one. The results of low feed striking time with insignificant variation positively indicate about the palatability of all the formulated diets. The feeding trial reveals that the results of growth, food consumption and utilization are very similar when fed on control, D1 and D2. On the contrary the values when fed on D3 and D4 are of a lower grade. Among the very little information in literature on using insects as alternative protein supplement for fish, satisfactory results for *Heterobranchus longifilis* Valenciennes 1840 has been reported, where 50% fish meal is replaced by termite meal [28]. Martinez et al. also report satisfactory results when 30% cockroach meal is added to formulate diets for Japanese carp [29]. Along with food consumption, utilization and growth, the measurement of condition factor (K) is also important as it could express the final health of the adult individuals in terms of length and weight at the end of the feeding trial. Results of the condition factor follow a similar trend as observed in case of the other indices already mentioned, because here also the results of control, D1 and D2 diets show ideal results (more than 1.5 [30,31]). Unsatisfactory results are obtained in case of D3 and D4 fed groups might be because of mollies cannot tolerate a composition where more than 50% of fish meal is replaced by the *Oxya* meal. Or this could be an effect of variations in the amount of rice husk and mustard oil cake.

To clear the doubt whether the difference in response variables of feeding trial are due to replacement of fish meal by *Oxya* meal, or due to the effect of change in the amount of rice husk and mustard oil cake, a statistical explanation is important. In this regard results of the linear model quite clearly reveal that only excluding PER, variations of all the indices under consideration occur mainly due to the replacement of fish meal by *Oxya* meal. Here it should be kept in mind that PER is actually a measure of weight gain with respect to the amount of protein consumed through food. The statistical model shows that *Oxya* meal, fish meal, rice husk and mustard oil cake, all have certain effect on the response variables; it is quite clear that all these four ingredients contribute in the protein part of the formulated diets. In this context it could be explained that may be a significant amount of protein contribution by all the four major ingredients leads to an insignificant result when the effect of replacement of fish meal by *Oxya* meal is compared with the combined effect of rice husk and mustard oil cake in case of PER. Results of the present work is encouraging because it is quite clear from this study that at least up to 50% of fish meal could be effectively replaced by *Oxya* meal. This amount of replacement will have no negative effect on the growth and reproduction of *P. sphenops*. Moreover, it is also clear that the variations on the response indices are mainly due to this replacement and not due to the varied amount of rice husk and mustard oil cake.

The results of the present study as a whole support the view of rearing *O. fuscovittata* in "acridid farms" that will easily and constantly provide nutritionally rich animal protein supplement to the aqua-feed developers to formulate supplementary diets for ornamental fishes making the industry more viable. Moreover, if the concept of "acridids as alternative food source" is popularized it could lower the rate of overexploitation of fish meal. Consequently the "demand: supply" ratio of the fish meal could come down, resulting a lower market price of the fish meal also. However, similar studies with various proportions of *Oxya* supplement in the diets of other ornamental fishes are necessary in future. These works also should emphasize on obtaining the digestibility of the formulated diets, which is lacking in the present work. Then only one could be more assured of the suitability of *Oxya* meal as an effective alternative protein source.

Acknowledgments

The authors are thankful to the head of the Department of Zoology, Visva-Bharati University for providing necessary laboratory facilities. Mrs. Mousumi Das is specially acknowledged for her kind help regarding crude protein estimation.

Author Contributions

Conceived and designed the experiments: AG RC DM PH. Performed the experiments: AG RC AS DM PH JR JP. Analyzed the data: AS. Contributed reagents/materials/analysis tools: AG RC DM PH. Contributed to the writing of the manuscript: AG RC As DM PH JR JP.

References

1. Van Huis A, Itterbeeck JV, Klunder H, Mertens E, Halloran A, et al. (2013) Edible insects: Future prospects for food and feed security. FAO Forestry paper 171. Rome: Food and Agriculture Organization of the United Nations. 201 p.
2. Mohanta KN, Subramanian S (2011) Nutrient of common fresh water ornamental fishes. Technical Bulletin No. 27. Goa: ICAR Research Complex for Goa. 55 p.
3. Hossain MH, Ahammad MU, Howlider MAR (2003) Replacement of fish meal by broiler offal in broiler diet. Int J Poult Sci 2: 159–163.
4. Naylor RL, Goldburg RJ, Primavera JH, Kautsky N, Beveridge MCM, et al. (2000) Effect of aquaculture on world fish supplies. Nature 405: 1017–1024.

5. Nyirenda J, Mwabumba M, Kaunda E, Sales J (2000) Effect of substituting animal protein source with soybean meal in diets of *Oreochromis karongae* (Trewavas 1941). Naga, ICLARM q 23: 13–15.
6. Xiaoming C, Ying F, Hong Z, Zhiyong C (2010) Review of the nuritive value of edible insects. In: Durst PB, Johnson DV, Leslie RL, Shono K, editors. Forest insects as food: humans bite back. Proceedings of a workshop on Asia-Pacific resources and their potential for development. Bangkok: FAO Regional Office for Asia and the Pacific. 85–92.
7. Anand H, Das S, Ganguly A, Haldar P (2008) Biomass production of acridids as possible animal feed supplement. **J Environ &** Sociobiol 5: 181–190.
8. AOAC (Association of official analytical chemists) (1990) Official methods of analysis, 15 th edition, volume 1. Virginia: Association of Official Analytical Chemists. 737 p.
9. Parker R (2011) Aquaculture Science. Stamford, CT: Cengage Learning. 672 p.
10. Sogbesan OA, Ugwumba AAA (2006) Bionomics evaluation of garden snail (*Limicolaria aurora*, Jay, 1937; Gastropoda: Limicolaria) meat meal in the diet of *Clarias gariepinus* fingerlings (Burchell, 1822). Nig J Fisheries 2: 358–371.
11. APHA (American Public Health Association) (1985) Standard methods for the examination of water and waste water, 16 th edition. Washington DC: American Public Health Association. 1268 p.
12. De Silva SS, Anderson A (1998) Fish nutrition in aquaculture. London: Chapman and Hall. 319 p.
13. Moyle PB, Cech JJ (1996) Fishes: An introduction to ichthyology. Englewood Cliffs, NJ: Prentice Hall Inc. 590 p.
14. Bailey M, Sandford G (2001) The ultimate aquarium. London: Anness publishing Ltd. 256 p.
15. Lawrence C (2007) The husbandry of zebrafish (*Danio rerio*): A review. Aquaculture 269: 1–20.
16. Bahnasawy MH, El-Ghobashy AE, Abdel-Hakim NF (2009) Culture of the Nile tilapia (*Oreochromis niloticus*) in a recirculating water system using different protein levels. Egypt Aquat Biol Fish 13: 1–15.
17. Timmerman CM, Chapman LJ (2004) Hypoxia and interdemic variation in *Poecilia latipinna*. J Fish Biol 65: 635–650.
18. Guillaume J (1997) Protein and amino acids. In: D'Abramo LR, Conclin DE, Akiyama DM, editors. Crustacean nutrition, Advances in world aquaculture. Louisiana: World Aquaculture Society. 26–41.
19. Shim KF, Chua YL (1986) Some studies on the protein requirement of the guppy *Poecilia reticulata* (Peters). J Aquariculture Aquat Sci 4: 79–84.
20. Cowey CB, Sargent JR (1979) Nutrition. In: Hoar WS, Randall J, editors. Fish physiology. New York: Academic Press. 1–69.
21. Fernando AA, Phang VPE, Chan SY (1991) Diets and feeding regimes of poeciliid fishes in Singapore. Asian Fish Sci 4: 99–107.
22. Furuichi M, Yone Y (1981) Availability of carbohydrate in nutrition of carp and red sea breem. B Japan Soc Sci Fish 48: 945–948.
23. Kalita P, Mukhopadhyay PK, Mukherjee AK (2007) Evaluation of the nutritional quality of four unexplored aquatic weeds from North East India for the formulation of cost-effective fish feeds. Food Chem 103: 204–209.
24. National Research Council (NRC) (1983) Nutrient requirements of warm water fishes and shellfishes. Washington: National Academy Press.102 p.
25. Hasan MR, Moniruzzaman M, Farooque AMO (1990) Evaluation of leucaena and water hyacinth leaf meal as dietary protein sources for the fry of Indian major carp, Labeo rohita (Hamilton). In: Hirano R, Hanyu I, editors. Second Asian Fisheries Forum. Manila: Asian Fisheries Society. 275–278.
26. Akand AM, Hassan MR, Habib MAB (1991) Utilization of carbohydrate and lipid as dietary energy sources by stinging catfish H. fossilis (Bloch). In: DeSilva S, editor. Fish nutrition research in Asia. Proceedings of the fourth Asian fish nutrition workshop. Manila: Asian Fisheries Society. 93–100.
27. Arockiaraj AJ, Muruganandam M, Marimuthu K, Haniffa MA (1999) Utilization of carbohydrates as a dietary energy source by stripped murrel *Channa striatus* (Bloch) fingerlings. Acta Zool Taiwanica 10: 103–111.
28. Sogbesan O, Ugwumba AAA (2008) Nutritional evaluation of termite (*Macrotermes subhyalinus*) meal as animal protein supplements in the diets of *Heterobranchus longifilis* (Valenciennes, 1840) Fingerlings. Turk J Fish Aquat Sc 8: 149–157.
29. Martínez HH, Ramos-Elorduy J, Pino-Moreno JM, Acosta-Castaneda C (2008) Evaluation of diets including nymphs of *Periplaneta americana*. Cienc Pesq 16: 25–30. (In Spanish with English abstract).
30. Lagler KF (1956) Freshwater Fishery Biology. Iowa: W.C. Brown Company. 421 p.
31. Wade JW (1992) The relationship between temperature, food intake and growth of brown trout, *Salmon trutta* (L.) fed natural and artificial pelleted diet in earth ponds. J Aquat Sci 7: 59–71.

14

Schooling Increases Risk Exposure for Fish Navigating Past Artificial Barriers

Bertrand H. Lemasson[1*¤a], James W. Haefner[1], Mark D. Bowen[2¤b]

1 Department of Biology and Ecology Center, Utah State University (USU), Logan, Utah, United States of America, **2** Fisheries and Wildlife Resources Group, United States Bureau of Reclamation, Denver, Colorado, United States of America

Abstract

Artificial barriers have become ubiquitous features in freshwater ecosystems and they can significantly impact a region's biodiversity. Assessing the risk faced by fish forced to navigate their way around artificial barriers is largely based on assays of individual swimming behavior. However, social interactions can significantly influence fish movement patterns and alter their risk exposure. Using an experimental flume, we assessed the effects of social interactions on the amount of time required for juvenile palmetto bass (*Morone chrysops* × *M. saxatilis*) to navigate downstream past an artificial barrier. Fish were released either individually or in groups into the flume using flow conditions that approached the limit of their expected swimming stamina. We compared fish swimming behaviors under solitary and schooling conditions and measured risk as the time individuals spent exposed to the barrier. Solitary fish generally turned with the current and moved quickly downstream past the barrier, while fish in groups swam against the current and displayed a 23-fold increase in exposure time. Solitary individuals also showed greater signs of skittish behavior than those released in groups, which was reflected by larger changes in their accelerations and turning profiles. While groups displayed fission-fusion dynamics, inter-individual positions were highly structured and remained steady over time. These spatial patterns align with theoretical positions necessary to reduce swimming exertion through either wake capturing or velocity sheltering, but diverge from any potential gains from channeling effects between adjacent neighbors. We conclude that isolated performance trials and projections based on individual behaviors can lead to erroneous predictions of risk exposure along engineered structures. Our results also suggest that risk perception and behavior may be more important than a fish's swimming stamina in artificially modified systems.

Editor: Carlos Garcia de Leaniz, Swansea University, United Kingdom

Funding: Funding for this work was provided by the Bureau of Reclamation's Tracy Fish Improvement Program (No. FC810748) and the U.S. Army Engineer Research & Development Center's basic research program in network sciences (BT25/NS/14-63). Dr. Mark Bowen, a co-author on this study, is a former employee of this institution and contributed to the design of the experiment. Mark did not participate in data collection and analyses, nor was he involved in the decision to publish the findings. While Mark did not write the manuscript, he has commented on the final product.

Competing Interests: The authors have declared that no competing interests exist.

* Email: brilraven@gmail.com

¤a Current address: Environmental Laboratory, United States Army Engineer Research and Development Center, Santa Barbara, California, United States of America
¤b Current address: Turnpenny Horsfield Associates, Ashurst, Southampton, Hampshire, United Kingdom

Introduction

Nearly 65% of surveyed rivers and their aquatic habitats are threatened by human activity and climate change, with extinction rates among fresh water fish species rivaling those of past geological events [1]. A wide variety of artificial structures pepper the earth's river systems and are essential for diverting water to meet the needs of human societies, such as providing municipal drinking water, creating hydropower, and supporting agriculture [2,3]. Unfortunately, a single facility alone can impede or divert millions of fish per year [4] and the cumulative impacts of multiple facilities across the landscape have lead to ecosystem fragmentation and isolation in many systems [5]. Human manufactured disturbances can also exacerbate an animal's risk exposure if its life history strategy results in maladaptive behavior in novel settings [6], which can result in both immediate and long-lasting impacts on fitness [7,8]. A prominent risk minimizing strategy undertaken by most fish species is to form into social groups or schools of varying coherence [9], yet we know little of this behavior's impacts on individual risk as fish are forced to navigate past artificial barriers in their environment.

Human barriers, such as dams, water diversion facilities and pumping stations can expose fish to a variety of dangers, including physical harm from collisions, exhaustion from swimming exertion [10–12], increased stress levels [13], and elevated predation [2,14]. Juvenile fish are particularly susceptible to the risks posed by combinations of these stressors and suffer the highest mortality rates when navigating through artificial bottlenecks [3]. Even if individuals escape harm, many migratory species are on a tight physiological schedule (e.g., the 'smolt window' in Salmonids). Migratory fish species respond to environmental cues, such as photoperiod, temperature, and flow, and undergo physiological changes to prepare for their osmotic transition between freshwater and saltwater regimes [15]. The optimal navigation strategy is therefore to pass artificial barriers quickly in order to minimize the risk of any immediate threats [10], which would also reduce any migration delays that can have negative consequences on future stock success [15,16].

Engineering and operational efforts predominantly rely upon estimates of average individual swimming stamina and behavior to expedite the safe passage of a region's fish. The design objective is to enable a fish's capacity to avoid flows or structural designs that would otherwise either impede its progress or increase its mortality risk [17,18]. Traditionally, most assays on barrier exposure or individual swimming stamina are conducted with large juveniles or adults that are tested in isolation [17,19–21]. Recent efforts have questioned the transferability of these empirical estimates to field conditions, pointing out that adaptive behaviors are not simply a function of individual physiological and biomechanical performance metrics [2,22]. Experiments with groups of fish challenged with bypassing a barrier have shown large differences in their navigational performance relative to estimates reported from prior trials on individuals [22]. However, we know little of how social interactions affect individual swimming patterns under such conditions. This information would be particularly useful in defining movement parameters in agent-based modeling efforts that couple individual behaviors and environmental conditions to assess how animals navigate past engineered structures in their environment [23,24].

Most juvenile fish form into groups of varying social and physical organization [9]. While this strategy has primarily evolved as a means to reduce the risk of predation, it can also effectively mitigate travel costs. Social cues can improve migration success by serving to average out individual directional uncertainties along a gradient, or migration route, [25,26] and can enhance directional decision-making [27]. Organized formations may also convey net energetic benefits, such as reducing the drag that individuals experience [9,28–30]. Despite all of these potential benefits, environmental context can alter the adaptive value of an ingrained behavior. For instance, if environmental conditions generate severe disorientation then solitary navigation is hypothesized to become preferable to a social navigation strategy, such as schooling in fish [31]. Similarly, deviations from optimal positions in schooling fish can be energetically costly and definitive empirical evidence to support any hydrodynamic advantage to schooling remains elusive [29]. While context may reshape the costs and benefits of social movement strategies, like schooling, this strategy should none-the-less directly impact the basic movement parameters related to the speed and orientation of animals on the move.

In this study we tested the hypothesis that social interactions alter the swimming behaviors of fish when they are forced to navigate past an artificial barrier. Palmetto bass, a *Morone chrysops* × *M. saxatilis* hybrid, were selected for their availability and propensity to display polarized schooling behavior when young. We began by determining individual swimming performance using ramped velocity tests to establish the water velocity necessary to challenge the average individual's swimming stamina (experiment I). Subjects were then released upstream of a behavioral barrier under solitary and social conditions to evaluate how neighbors altered individual swimming behaviors and, subsequently, impacted risk exposure (experiment II). We found that risk exposure varied dramatically between fish swimming by themselves or within schools, where risk was measured as the time taken to navigate successfully downstream past the barrier. A fine-scale kinematic analysis further revealed how individual behaviors varied under each treatment. We conclude by discussing how the observed patterns relate to existing theory and empirical evidence concerning fish swimming behavior.

Materials and Methods

Ethics Statement

All work was conducted within the U.S. Bureau of Reclamation's water lab in Denver, CO, and was included within BHL's dissertation work at USU. The BOR facility did not have an Institutional Animal Care and Use Committee (IACUC) in place and so all handling procedures followed the ethical standards outlined by the National Research Council [32], which also aligns with those of the American Fisheries Society [33].

Fish husbandry

Juvenile palmetto bass were obtained from Keo Fish Farms, AR, and maintained in flow-through cylindrical tanks (1.2 m in diameter, 1.4 m high), fed daily to satiation and experienced light: dark cycles typical of summer months in the northern hemisphere. Experiments were conducted in a 16 m flume (Figure 1). Water temperatures in the rearing pens and experimental flume varied between 19–20°C. A total of 283 fish were used in our experiments and all fish were used only once. Fish were drawn at random from the main population tank and transferred to the flume within an aerated bucket. After their trial each fish was allowed to rest in the aerated bucket until they resumed normal swimming behavior. Fish were then moved to a recuperation tank with the same dimensions and environmental conditions as that of the main population. If any fish was unresponsive for 24 h after their trial, due to either exhaustion or impingement against a retention screen, they were euthanized using MS-222 (Tricaine Methanesulfonate) at a prescribed dose of 400 mg/L.

Experiment I. Individual swimming stamina

We determined the expected swimming stamina of our subjects using the critical swimming speed paradigm. Critical swimming speed, υ_c, is measured by incrementally exposing individuals to increased water speeds ($\Delta\upsilon$) for fixed time intervals (Δt) until the subject is exhausted [34].

The speed observed during the final successfully completed interval (υ_f) is then adjusted by the proportion of time spent in the subsequent interval in which exhaustion occurred ($t_{max}/\Delta t$):

$$\upsilon_c = \upsilon_f + \Delta\upsilon \cdot \left(\frac{t_{max}}{\Delta t}\right), \quad (1)$$

The performance trials estimate when an individual approaches its energetic limits, thereby providing a more conservative measure of its stamina. While fish are capable of swimming faster than this limit it generally requires them to transition to an anaerobically dominated phase and rapidly exhaust their oxygen supply [34,35]. Individuals varied in body size so we standardized all υ_c measurements to body length (i.e., fork length, FL) and exhaustion time (t_{max}) was recorded after a subject had collapsed against the rear screen for a second time. Fish ranged in size from $4-6.5$ cm with a mean of 5.1 ± 0.6 cm and weighed between $1-4.9$ g with a mean wet weight of 2.3 ± 0.9 g ($\pm$SD). Pilot trials indicated that fish would occasionally exploit velocity refuges within the flume, so we adopted a novel approach and conducted our swimming trials in a wire cage that kept subjects suspended above the bottom and in the center of the flume (Figure 1a, b). The test cage was a half-cylinder (30 × 46 cm, 15 cm radius) composed of 0.6 cm wide wire mesh that was covered with a Plexiglas lid. The lid had a 10 cm baffle along its edges that caused the cage to rise with the water level during the velocity trials. An acoustic Doppler velocimeter was used to confirm that flow values within the cage

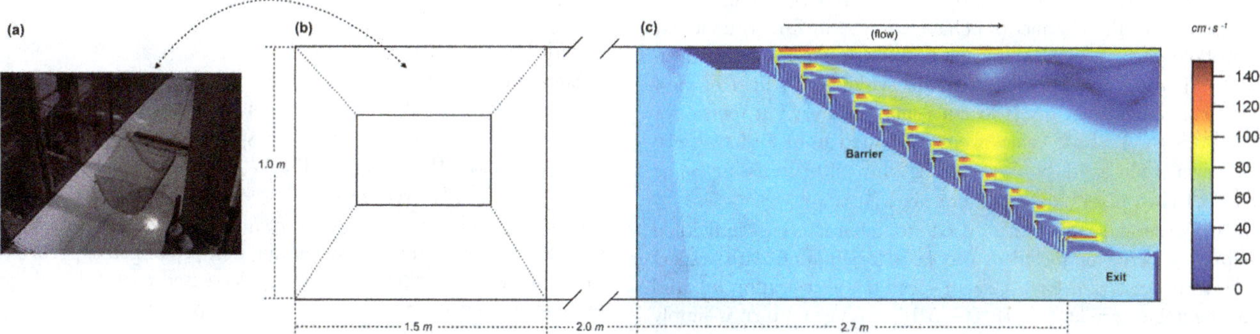

Figure 1. Top-down view of the experimental flume. The area included the swimming apparatus used in the stamina trials (a), the holding pen (b) and the louver-style hydraulic barrier (c), which leads to a 23 cm wide exit. Flume dimensions are in meters and flows within the flume were recreated using a Computational Fluid Dynamics model (CFD), with color profiles representing the speed of the water in cm·s^{-1} in section (c). The floating wire mesh cage (a) was suspended within area (b) with a series of cords. All stamina trials were done before the downstream barrier was installed. The upstream and down-stream boundaries of the holding area were enclosed with 1 × 1 cm^2 wire screens.

did not differ from those in the main channel. A 10-minute adjustment period at 10 cm·s^{-1} ($\approx$2 FL·s^{-1}) preceded each trial, after which the flow was incremented by 5 cm·s^{-1} using 20 or 7 min intervals. Constrained time intervals were deemed necessary for filming during the subsequent barrier trials to preserve film, yet υ_c measurements based on short speed increments can inflate performance estimates [34]. We therefore compared both time intervals to determine how exposure time affected the expected swimming stamina of our subjects, $\bar{\upsilon}_c$. A total of 38 fish were used with $N = \{23,15\}$ for the 20 and 7 min treatments, respectively, and performance data were recorded by an observer (see File S1, section SI-1 for further details).

Experiment II. Solitary *vs.* schooling behavior

To determine if social interactions affect the time fish spend exposed to an artificial barrier we released fish upstream of a barrier either alone or in groups and filmed their swimming behavior from above. A louver was installed in the flume and angled to the oncoming water to guide fish to a bypass exit (Figure 1c). The barrier consisted of a series of vertical slats, each 2.5 cm apart and set perpendicular to the flow. A louver is designed to generate turbulence patterns meant to elicit avoidance maneuvers in fish, thereby passively guiding them towards an exit route. We focused on this type of behavioral barrier as part of an extension of earlier studies aimed at exploring how modeling fish behavior can inform management decisions at fish passage facilities (for further details on barrier history, design, and application see [19,23,36,37]). Fish exposure time (T_e) was measured as the time subjects spent within the barrier area, which equates to their risk of entrainment or impact in such artificial systems [10,13]. We filmed our subjects using three Panasonic PV DV51 digital cameras that were installed above the barrier area so as to have overlapping fields of view. Each camera recorded at 29.97 frames per second and we stored our film segments on mini digital video tapes (60 min storage capacity). A Plexiglas sheet with a 15 cm baffle covered the test area to reduce any distortion from the water's surface.

Subjects were initially placed in a holding pen upstream of the diversion barrier either alone or in groups of 14. Fish ranged in lengths from 4.9 – 6.9 cm ($\overline{\mathrm{FL}} = 5.8 \pm 0.4$ cm) and weighed between 1.4–5.0 g with a mean of 1.8±0.4 g (wet weight, ± SD). The number of subjects used in the social treatment conforms to the number needed to significantly influence movement decisions [27]. A 10-minute acclimation period followed with an initial flow of 10 cm·s^{-1}, which seemed to be sufficient time for fish to cease any erratic swimming movements and display either station holding or minor movements. After this settlement period flows were then increased in 5 cm·s^{-1} increments at 3–5 min intervals, which allowed the system to increase steadily until flows reached a maximum downstream speed of 48.6±1.7 cm·s^{-1} ($\approx$8 FL·s^{-1}), with an average cross-channel speed of 0.53±0.35 cm·s^{-1} (±*SD*). Maximum water velocity was selected based on our findings from the stamina trials and falls within the lower limit of flows recorded at a full-scale diversion facility that employs the same diversion design [37]. Flow increases here reflect a trade-off. Raising the water velocity too slowly risked premature exhaustion in our subjects from prolonged exposures, while raising the velocity too quickly risked creating a very turbulent system and disorientating the fish. Once the final water velocity had stabilized, the downstream screen of the holding pen was raised and individuals were allowed to drift passively towards the barrier area. We analyzed 14 of 21 solitary trials and 11 of 16 social trials. While a fixed number of fish were released in each social treatment ($N = 14$), the fish would self-assemble into groups that varied in size over time. The size of a given group or school at any point in time was empirically defined from the observed dynamics (see File S1, section SI-2 and Analysis below for details). The unbalanced number of replicates stems from the removal of trials in which the subjects either displayed errant behaviors or remained upstream of the barrier beyond our designated recording limit (15 min, See File S1, section SI-3.1). Trials began when subjects entered the barrier area and ended when they either reached the exit or passed through the barrier's slats.

Analysis

Our primary response variables were the critical swimming speeds (υ_c) in experiment I and exposure times (T_e) in experiment II. These data were skewed and therefore compared using the non-parametric Wilcox Rank Sum test. Secondary variables of interest described the swimming behavior of our subjects in experiment II and were extracted from our digital video recordings. Fish head and tail positions were manually tracked at 10 frames per second and each fish's centroid, $\mathbf{x}$, was estimated from these points to create a path from each field of view. Fish paths were then smoothed using a 5-point running median to improve velocity and acceleration estimates. Although individuals occasionally drifted vertically, movements were largely confined to the bottom of the channel (a behavior found in fish with similar morphologies and swimming gaits under comparable experimental conditions [12,37]). Measurements within each field of view were converted from pixels to cm using a conversion metric for each axis ($c_{x,y}$) whose values were calculated from a virtual grid laid over each field of view during the post-processing stage ($c_x=0.1$ cm · pixel^{-1}; $c_y=0.2$ cm pixel^{-1}).

The secondary variables used to characterize individual swimming behavior included: rheotaxis (orientation with respect to flow; downstream $\theta=0°$), swim speed (v), acceleration (a), and turning angle, $\varphi=||\mathrm{acos}(\hat{\mathbf{v}}(t)\cdot\hat{\mathbf{v}}(t-\Delta t))||$. Any given turn is simply the difference between an individual's current heading and its past one, or $\hat{\mathbf{v}}$ at (t) and $(t-\Delta t)$, respectively. These turn angles fall between 0° and 180° as fish turn left or right in any given time step, so we report them as an absolute deviations in headings. Swim speeds were calculated by differencing fish centroids over time, correcting for water velocity and standardizing to average body length. Accelerations were then calculated by differencing the adjusted velocities. We controlled for spatial correlations among individuals in the social condition by randomly selecting a proxy fish from each replicate to represent socially influenced movement. Within each trial, fish were considered to be members of the same group when they were within 5 body lengths of one another; an interaction threshold empirically determined to balance the growth and decay rates in group membership observed in the data (Figure S1 in File S1). Fish released in groups often moved in and out of camera range, which prevented us from tracking the fate of any given individual for the entire duration of a trial. To avoid misidentifications and ensure path continuity, we randomly selected a representative sequence from each replicate for analysis, ensuring that a group of fish had distinct entry and exit points in the video sequence. In addition, groups in the social treatment were highly dynamic, continuously fragmenting and coalescing as they moved in and out of the filming area. We therefore investigated how these fission-fusion dynamics impacted overall risk exposure. Group cohesion or fusion was characterized by the mean observed size of our randomly selected groups within each trial i ($\overline{G_i}$) and we used the variability in these group sizes over time ($\sigma^2(G_i)$) as a metric of the group's instability, or tendency to fragment, during a trial. These group-level metrics were recorded within each trial and related to the overall pattern in exposure times across trials. At the individual-level we investigated how nearest neighbor positions varied over space and time within groups. Nearest neighbor positions were based on the distance between a proxy fish i and its j neighbors over time. Distance was calculated as the magnitude of the directional vector extending between position vectors, from fish $\mathbf{x}_i$ to fish $\mathbf{x}_j$, where $d_{j,i}(t)=|\mathbf{d}_{j,i}(t)|=|\mathbf{x}_j(t)-\mathbf{x}_i(t)|$. The bearing to a proxy fish's neighbor, $\beta_{j,i}$, was the angular difference between the proxy's current heading, $\hat{\mathbf{v}}_i$, and its neighbor's bearing $\mathbf{d}_{j,i}$, given as $\beta_{j,i}(t)=\mathrm{acos}\left(\hat{\mathbf{v}}_i\cdot\hat{\mathbf{d}}_{j,i}\right)(t)$. Time-series analyses were used to assess the reliability of all global averages reported in order to avoid spurious estimates from unsteady or biased trends in the data, as well as accounting for varying track lengths across trials (see SI-3). All analyses were conducted in R version 3.0.2. Laboratory experiments were originally conducted in late summer of 2003 and all analyses were repeated during the summer of 2013. Data supporting the table and figures are stored in the Knowledge Network for Biocomplexity repository, data package knb.480.1 (https://knb.ecoinformatics.org/).

Results

Individual swimming stamina

Standardized critical swimming speeds were statistically equivalent between the 20 min and 7 min intervals, being 9.2±2.0 FL·s^{-1} and $\bar{v}_c=9.6\pm1.5$ FL·s^{-1}, respectively (± SD; Wilcoxon rank sum test, $W_{0.05(2),15,23}=192$, $P=0.574$; global mean = 9.3±1.8 FL·s^{-1}). We recorded a dramatic decrease in post exercise mortality from 43% to 0% when trial increments declined from 20 to 7 min intervals, suggesting that prolonged exposure to elevated water velocities substantially decreased each fish's ability to recover from their trials.

Solitary *vs.* schooling behavior

Solitary individuals tended to turn downstream and swim with the current (negative rheotaxis), while grouped fish predominantly faced upstream (positive rheotaxis; Watson's test, $U^2_{0.05,14,11}=0.4094$, $P<0.001$, Figure 2, Table 1). These differences in preferred orientation occurred early on and remained steady over time (Figure S2 in File S1). Differences in turning angles between treatments were marginal, but significantly different from one another (Watson's test, $U^2_{0.05,14,11}=0.4161$, $P<0.001$; Table 1). Solitary individuals also swam at significantly slower swimming speeds than their schooling counterparts (Wilcoxon rank sum $W_{0.05(2),14,11}=0$; $P<0.001$), yet displayed stronger shifts in acceleration ($W_{0.05(2),14,11}=128$, $P=0.01$, Table 1). The swim speeds of these solitary fish were unsteady and increased over time, as opposed to their accelerations, which remained steady as they moved downstream. In contrast, the proxy fish swimming in groups displayed steady speed and acceleration profiles throughout their tests (Figure S3 in File S1). Taken together these behavioral discrepancies led to a 23-fold increase in exposure times when fish were released in groups rather than alone (Table 1, Wilcoxon rank sum, $W_{0.05(2),14,11}=0$, $P<0.001$). While exposure times under the social treatment were skewed and contained an outlier, omitting this trial had little impact on the median exposure times or any of the remaining kinematic data reported in Table 1. Interestingly, fish in either treatment were equally capable of safely reaching the exit (94% solitary; 98% social) and both treatment groups showed similar post exposure mortalities from either impacting the barrier or from impingement against a downstream retention screen (10% vs. 12%). In summary, a comparison of Table 1 and Figure 2 demonstrates that solitary fish turned with the current and so left the system quickly without having to exert themselves (low swim speeds). In contrast, fish traveling in groups faced into the current and effectively worked harder to hold their station (greater swim speeds), thereby substantially increasing their exposure times.

Schooling dynamics

Fish released in groups displayed varying degrees of cohesion within and across trials, splitting and remerging as they moved in and out of camera range. Exposure time was not significantly correlated with mean group size (Spearman rank-correlation; $T_e \times \overline{G_i}$, $S_{0.05(2)(11)}=155$; $P=0.87$), but was strongly correlated with group instability ($T_e \times \sigma^2(G_i)$, $\rho_s=-0.76$, $S_{0.05(2)(11)}=290$; $P=0.01$). Figure 3 shows how overall exposure times were negatively correlated with increasing group instability, while no distinctive pattern arose with respect to the average size of the groups (where mean group size is shown by the diameter of the points). Retaining the outlier in exposure times (Figure 3) changes neither the direction nor the significance of either of these group-level associations. At the individual-level we found no correlations

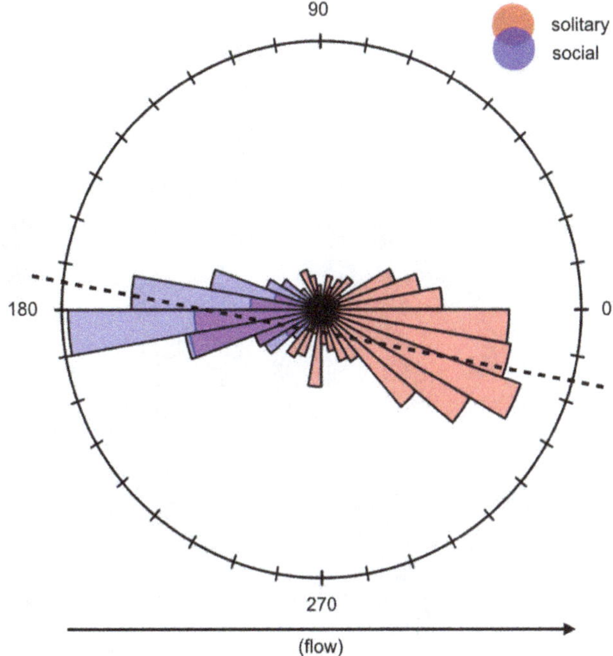

Figure 2. Circular histograms for the orientation patterns recorded under both solitary (red) and schooling (purple) conditions. Color modes are semi-transparent to show regions of overlap in the data from each condition. Schooling data represent only the proxy member's orientations from each replicate. The figure shows θ values pooled across all subjects within each category while in our statistical analyses we used the mean orientations from each replicate for both conditions. Upstream and downstream orientations are 180° and 0°, respectively. The dashed line represents the relative angle of the barrier with respect to the flow of water.

between either the nearest neighbor distances or bearings with regards to overall exposure time. Nearly all of the proxy fish chosen from each group (10/11) maintained a steady and close association with their nearest neighbors (median $\tilde{d}_1 = 0.72 \pm 0.40$ L; ± MAD) and predominantly trailed these neighbors at bearings between 30° and 330° (Figure 4). We found no evidence of any temporal trends in nearest neighbor positions or bearings, so while average trends in $\tilde{d}_1$ and $\tilde{\beta}_1$ fluctuated over time these processes remained relatively steady (see Figures S4 and S5 in File S1). There was, however, greater variability in $\tilde{\beta}_1$ values over time, which is not unexpected. As fish speed up or slow down even minor shifts in the relative distances between individuals can result in larger changes in bearings, as a fish's closest neighbor can suddenly shift orthogonally, or from front to back.

Discussion

Fish released in groups up-stream of a diversion barrier formed schools and took longer to navigate downstream past the barrier than fish that were released alone. Schooling has evolved as an adaptive behavior largely because it decreases the immediate risks that fish face, whether it be by diluting the threat of predation, or reducing directional uncertainty [9]. Anthropogenic structures in rivers and streams are dangerous places for juvenile fish [2,11,12,14,15,17] and their risk of mortality increases with their exposure to these environments [10]. Additionally, the immediate threats that fish face when navigating past diversion barriers (e.g., collision, exhaustion, stress) can also be compounded by indirect costs, such as detracting from more profitable opportunities like searching for resources or mates in other areas (e.g., the risk allocation hypothesis) [38]. Our results demonstrate that the long held practice of managing water diversion facilities based on individual fish swimming stamina [17,19–21] can significantly underestimate how long fish will linger along a diversion barrier.

Results from our stamina trials indicated that a current of 48.6 $cm \cdot s^{-1}$ approached the upper limits of our subjects' swimming stamina ($46 - 49$ $cm \cdot s^{-1}$). Under these circumstances our subjects would generally not be expected to hold their station for extended periods. Yet, water velocity alone is a poor predictor of fish residency times within and across species [23,24], suggesting that individual behavior may either reduce or exacerbate exposure times. As fish move downstream with a current even minor trajectory deviations brought about by obstacle avoidance maneuvers can change the time they take to traverse an area. For instance, a biased-random walk model, parameterized to reflect fish swimming speeds, demonstrates that the expected exposure time of a particle can increase by a factor of 5 simply by accounting for self-propulsion and obstacle avoidance behavior [37]. Within the relatively uniform flows found in our flume a passive particle released from the holding pen would take approximately 12 s to drift downstream past the barrier if it traveled in a straight line towards the exit. Although the shorter exposure times observed with the solitary individuals may fall within some, albeit large, margin of error of this expectation, the 23 fold increase in exposure time by those fish released in groups represents a substantial deviation from a purely hydraulic prediction. Group members swam at twice the speed of their solitary counterparts, showed less erratic swimming behavior, and were predominantly moving upstream. To understand how such pronounced differences could have arisen requires a brief review of

Table 1. Swimming characteristics of fish traveling either alone or within schools.

Metric	Solitary	Schooling	Units
Orientation, θ	317.0 ± 48.9	187.3 ± 17.0	degrees
Turn angles, φ	2.2 ± 6.9	0.5 ±1.4	degrees
Speed, υ	4.2 ± 1.2	7.6 ±0.4	FL s^{-1}
Acceleration, a	13.2 ± 5.2	6.8 ± 2.2	FL s^{-2}
Exposure time, T_e	3.5 ± 0.6	82.0 ± 40.0	s

Data from schools are pooled from each replicate schools' proxy fish and do not represent group averages. We report the mean ± 1 SD for each circular metric (θ and φ). The linear metrics showed varying degrees of skewness, so we provide median ± median absolute deviation (MAD) for υ, a, and T_e. All metrics were significantly different between treatments.

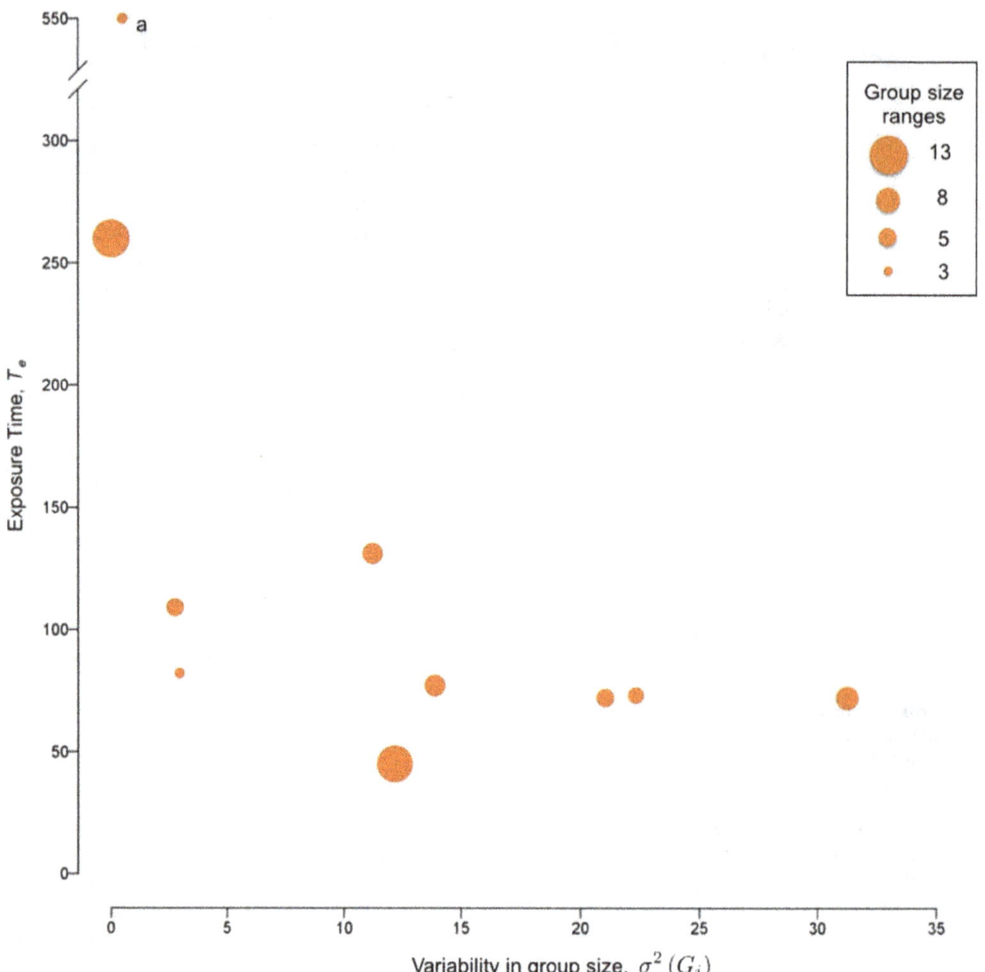

Figure 3. Non-linear correlation in exposure time with increasing variability in group size. Points represent the degree of variance observed across group sizes within each trial, while their diameters are scaled by the mean group size from those trials. While fish in the social treatment were released in groups of 14 individuals, these groups often fragmented and randomly selected representative groups varied in size both within and across trials. Overall, fish in the social treatment formed groups that ranged in size from 2–14 individuals with mean group sizes across trials ranging from 3.3 to 13. Solitary stragglers were also not uncommon (see SI-2). While extreme event (a) may represent a biologically plausible scenario, its absence does not detract from the relationship between variance in group size and exposure time.

the physical, energetic, and behavioral factors that may influence the observed patterns and our ability to draw broader conclusions from them.

Palmetto bass share diet and habitat preferences with their maternal species in the wild and integrate themselves into striped bass spawning migrations [39,40], which suggests at least parallels in their movement behaviors under natural conditions. Palmetto bass also display an innate ability to form polarized schools when young and so they provide a pragmatic means to study the physical implications of coordinated motion. Hatchery-bred hybrids are likely to suffer the same biases displayed by hatchery fish, such as ecological naïveté with respect to foraging or predator avoidance. Yet, despite such limitations hatchery reared fish have proven useful in studying the fundamental mechanics of collective motion, yielding insights into topics ranging from gradient detection [41] to social learning and decision-making [42,43]. These emergent behaviors stem from the basic physical mechanics of coordinated motion, which can have ecologically relevant impacts in wild populations. For instance, the speed and distance travelled by Chum (*Oncorhynchus keta*) and Pink (*O. nerka*) salmon generally increases with the size of the schools they form [44]. As a group grows in size, or density, the frequency of local interactions increases and can result in directional feedbacks that reduce overall turning variability among individuals [25,44]. Similar behavior has been documented in locusts [45], demonstrating that the underlying physical mechanics can have an impact that transcends taxa and context. Care should none-the-less be taken in interpreting our results beyond the physical ramification of displaying polarized schooling, such as the potential ecological reasons that influence when or why a fish species will school.

Data also suggests that the size of the group may play a role. For example, the ability of Atlantic salmon (*Salmo salar*) and American Shad (*Alosa sapidissima*) to navigate through bypass weirs decreases with increases in the size of the schools they form, with the majority of fish breaking off from larger groups and passing their respective barriers in pairs or alone [46]. Fish exposure time in our study was negatively correlated with group instability, suggesting that fish in groups may have been passing through the system more quickly as social interactions degraded.

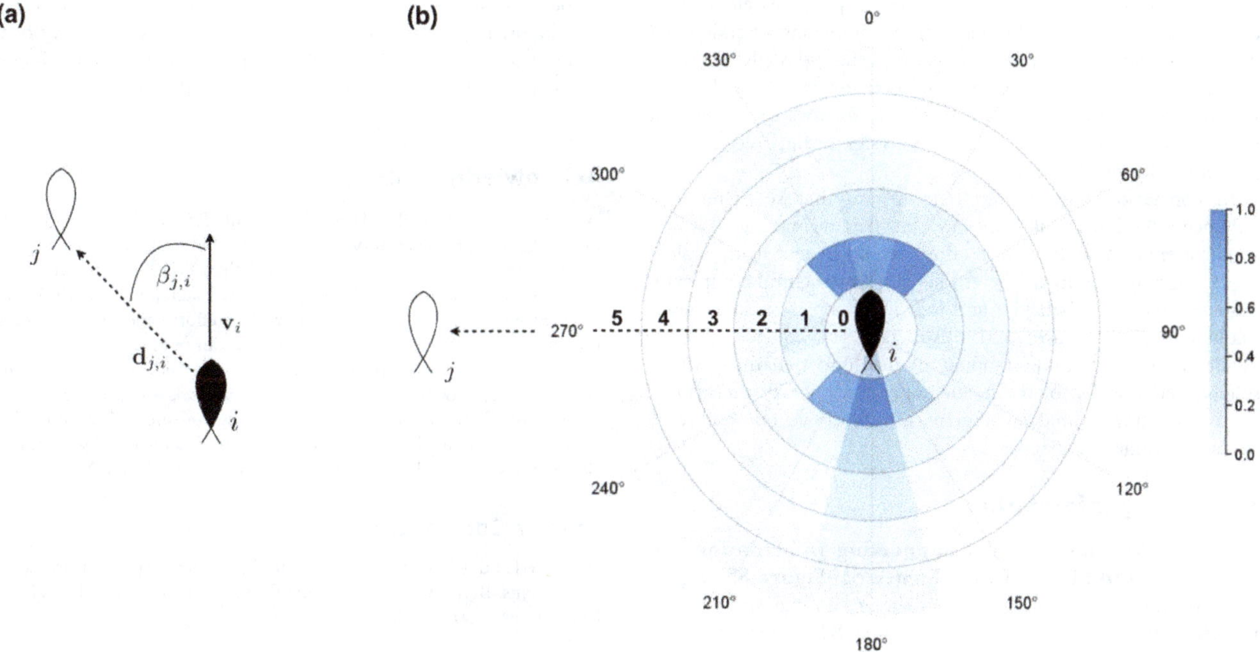

Figure 4. Nearest neighbor distributions within schools based on data pooled across all observations (replicates and time). Neighbor positions are taken from the perspective of each group's proxy fish. The distance, d_1, and bearing, β_1, between a subject and its nearest neighbor (a) are presented in circular sections of 1 body length increments (FL) and binned by 30° intervals (b). Color values represent the normalized frequency of observations per bin and range from 0 (white) to 1 (dark blue). The highest concentration of nearest neighbors fall within 1 FL of the proxy fish, which tended to trail their leading neighbors at angles ranging from ±30° from their heading.

Taken together these associations highlight the need to test for a group size effect to determine if there is a causal relationship and, if so, determine how this factor can influence management strategies.

Fish traveling in groups in our study followed leading neighbors more closely than those swimming alongside them (Figure 4b). This pattern can arise from two alternative, yet not mutually exclusive, ecological mechanisms. In the early 1970's, Daniel Weihs theorized that hydrodynamic advantages can arise predominantly from three means: channeling effects between parallel neighbors, flow sheltering behind neighbors, or by capitalizing on the energy generated by trailing vortices [28,29]. Channeling effects are enhanced when individuals are separated by <1 body length and these effects decay rapidly with distance [28,29]. Our subjects tended to position themselves further than this optimum, yet predominantly followed or led neighbors at bearings that closely corresponded with those predicted by Weihs' theoretical diamond configuration (±30°). Moreover, these patterns were not only spatially structured, but also remained relatively steady over time. Taken together these patterns suggest that potential hydrodynamic gains were theoretically plausible and, if present, more likely to stem from either flow avoidance or vortex capturing than from channeling effects between parallel swimmers.

Recent evidence has shown that leader positions are significantly correlated with an individual's metabolic scope (MS), with spatial positioning from a school's front to rear being negatively correlated with the constituents' MS [47]. Individual fish have also been shown to display less muscle activity when drafting behind the vortices shed from a stationary object [29], so it is certainly possible for individuals to take advantage of similar physical mechanisms that may arise within a school. However, we continue to find opposing predictions from theoretical efforts exploring the hydrodynamic benefits of schooling [48,49]. The leader-follower dynamics seen in our subjects are found outside of advection dominated systems [50–52] and may therefore arise for ecological reasons beyond any hydrodynamic advantage. Leadership positions are typically ephemeral in fish groups as they generally lack a social hierarchy [52] and are more commonly attributed to foraging information [50–52], aerobic capacity [47] and risk taking [9]. All of our individuals were fed to satiation, were naïve as to the location of the exit route, and either rarely or never exceeded their expected critical swimming speed. So, while the positional patterns in Figure 4b provide evidence to support some of Weihs' assumptions, another plausible explanation is that risk perception can play an equal, if not more important, role in such artificial environments.

Consider that 'leaders' within groups can either possess information, or merely display bold tendencies [50,53]. Bold animals typically commit quickly to reactionary behaviors, such as fleeing or attacking, and show less variability in their movements than the typical individual [54]. We might therefore expect bold individuals to move quickly and decidedly downstream past the barrier, as displayed by the solitary travelers. However, our solitary subjects showed greater variability in their turning angles and accelerations than those fish released in groups. Such elevated variability is better associated with shy or risk-averse individuals than with bold ones [54], which suggests that solitary individuals were simply more skittish than those released in groups and moved through the system quickly due to their orientation in the water. While social interactions affect an individual's reaction to stressful scenarios (e.g., predators), such interactions also significantly impact stress responses at the behavioral and metabolic level [55,56]. We can simplify this metabolic argument by considering that in the simplest models of collective behavior, individuals average the motion of their neighbors in order for any degree of coordinated or cohesive motion to emerge [9]. This elementary assumption alone invariably leads to a dampening of individual movement variability, which may, in turn, indirectly decrease metabolic rates caused by erratic movements. Regardless of whether fish traveling in groups were able to work less to remain upstream of the barrier, their interactions significantly reduced any propensity for erratic movements.

In conclusion our findings demonstrate that schooling can enhance individual risk exposure for fish swimming through artificial environments. Future efforts would benefit from exploring how group size, hydraulic gradients, and structural complexity influence schooling behavior and individual exposure times along manufactured structures and obstructions in aquatic systems. While a thorough understanding of the fluid dynamics within schools remains out of reach, including basic social interactions when modeling animal movements may improve our ability to provide reliable risk assessments.

Supporting Information

File S1 This file contains supporting information for the article and also contains Figure S1–Figure S5. Figure S1, The influence of interaction range, r, on the size (a) and number (b) of groups observed. Figure S2, Time series of rheotactic patterns observed in the solitary and social travel conditions. Figure S3, Time series of observed swimming velocity and acceleration in both the solitary and social conditions. Figure S4, Time series of nearest neighbor values, d_1. Figure S5.

Acknowledgments

We are grateful to the U.S. Bureau of Reclamation's Hydraulic Investigations and Laboratory Services group for hydraulic modeling and flume arrangements. We are indebted to Nathan Welch for assistance with laboratory work and useful discussions, Vinay Patel and Jason Derico for their patient video processing and fish tracking efforts, and Eva Enders and Jordan Watson for helpful comments on an earlier version of the manuscript. We also thank three anonymous reviewers for their time and constructive comments. Funding for this work was provided by the Bureau of Reclamation's Tracy Fish Improvement Program (No. FC810748) and the U.S. Army Engineering Research & Development Center's basic research program in network sciences (BT25/NS/14–63).

Author Contributions

Conceived and designed the experiments: BHL JWH MDB. Performed the experiments: BHL. Analyzed the data: BHL. Wrote the paper: BHL JWH. Revised the paper: MDB.

References

1. Vörösmarty CJ, McIntyre PB, Gessner MO, Dudgeon D, Prusevich A, et al. (2010) Global threats to human water security and river biodiversity. Nature 467: 555–561.
2. Schilt C (2007) Developing fish passage and protection at hydropower dams. Appl Anim Behav Sci 104: 295–325.
3. Coutant C, Whitney R (2000) Fish behavior in relation to passage through hydropower turbines: A review. Trans Am Fish Soc 129: 351–380.
4. Aasen G (2010) 2009 fish salvage at the tracy fish collection facility. Technical Report R0785504, CA Dept Fish Game.
5. Poff N, Hart D (2002) How dams vary and why it matters for the emerging science of dam removal. Bioscience 52: 659–668.
6. Tuomainen U, Candolin U (2011) Behavioural responses to human-induced environmental change. Biol Rev 86: 640–657.
7. Frid A, Dill L (2002) Human-caused disturbance stimuli as a form of predation risk. Cons Ecol 6: 11.
8. Creel S, Christianson D (2008) Relationships between direct predation and risk effects. Trends Ecol Evol 23: 194–201.
9. Krause J, Ruxton G (2002) Living in Groups. London, UK: Oxford University Press.
10. Fletcher R (1985) Risk analysis for fish diversion experiments: pumped intake systems. Trans Am Fish Soc 114: 652–694.
11. Swanson C, Young PS, Cech Jr JJ (2004) Swimming in two-vector flows: performance and behavior of juvenile Chinook salmon near a simulated screened water diversion. Trans Am Fish Soc 133: 265–278.
12. Swanson C, Young PS, Cech JJ (2005) Close encounters with a fish screen: Integrating physiological and behavioral results to protect endangered species in exploited ecosystems. Trans Am Fish Soc 134: 1111–1123.
13. Young PS, Swanson C, Cech JJ (2010) Close encounters with a fish screen iii: Behavior, performance, physiological stress responses, and recovery of adult delta smelt exposed to two-vector flows near a fish screen. Trans Am Fish Soc 139: 713–726.
14. Naughton GP, Keefer ML, Clabough TS, Jepson MA, Lee SR, et al. (2011) Influence of pinniped-caused injuries on the survival of adult Chinook salmon (*Oncorhynchus tshawytscha*) and steelhead trout (*Oncorhynchus mykiss*) in the Columbia River basin. Can J Fish Aquat Sci 68: 1615–1624.
15. McCormick SD, Hansen LP, Quinn TP (1998) Movement, migration, and smolting of Atlantic salmon (*Salmo salar*). Can J Fish Aquat Sci 55: 77–92.
16. Castro-Santos T, Haro A (2003) Quantifying migratory delay: a new application of survival analysis methods. Can J Fish Aquat Sci 60: 986–996.
17. Castro-Santos T, Haro A (2006) Biomechanics and fisheries conservation. In: Randall D, Farrell A, editors, Fish Physiology, Fish Biomechanics. San Diego, CA: Academic Press, volume 23, pp. 469–523.
18. Tudorache C, Viaene P, Blust R, Vereecken H, De Boeck G (2008) A comparison of swimming capacity and energy use in seven european freshwater fish species. Ecol Freshw Fish 17: 284–291.
19. Bates D, Vinsonhaler R (1956) Use of louvers for guiding fish. Trans Am Fish Soc 86: 38–57.
20. Beamish F (1978) Swimming capacity. In: Hoar W, Randall D, editors, Fish Physiology. New York: Academic Press, pp. 101–187.
21. Beach M (1984) Fish pass design - criteria for the design and approval of fish passes and other structures to facilitate the passage of migratory fish in rivers. Technical Report 78, Minist Agric, Fish and Food, Directorate of Fisheries Research, Lowestoft, UK.
22. Haro A, Castro-Santos T, Noreika J, Odeh M (2004) Swimming performance of upstream migrant fishes in open-channel flow: a new approach to predicting passage through velocity barriers. Can J Fish Aquat Sci 61: 1590–1601.
23. Haefner J, Bowen M (2002) Physical-based model of fish movement in fish extraction facilities. Ecol Model 152: 227–245.
24. Goodwin R, Nestler J, Anderson J, Weber L, Loucks D (2006) Forecasting 3-d fish movement behavior using a eulerian-lagrangian-agent method (ELAM). Ecol Model 192: 197–223.
25. Larkin PA, Walten A (1969) Fish school size and migration. J Fish Res Bd Can 26: 1372–1374.
26. Simons A (2004) Many wrongs: the advantage of group navigation. Trends Ecol Evol 19: 453–455.
27. Ward AJW, Herbert-Read JE, Sumpter DJT, Krause J (2011) Fast and accurate decisions through collective vigilance in fish shoals. Proc Nat Acad Sci 108: 2312–2315.
28. Weihs D (1975) Some hydrodynamical aspects of fish schooling. In: Wu T, Brokaw C, Brennen C, editors, Symposium on swimming and flying in nature. New York: Plenum. pp. 203–218.
29. Liao J (2007) A review of fish swimming mechanics and behaviour in altered flows. Phil Trans R Soc B 362: 1973–1993.
30. Svendsen JC, Skov J, Bildsoe M, Steffensen JF (2003) Intra-school positional preference and reduced tail beat frequency in trailing positions in schooling roach under experimental conditions. J Fish Biol 62: 834–846.
31. Codling E, Pitchford J, Simpson S (2007) Group navigation and the "many-wrongs principle" in models of animal movements. Ecology 88: 1864–1870.
32. Institute of Laboratory Animal Research NRC Commission on Life Sciences (1996) Guide for the Care and Use of Laboratory Animals. The National Academies Press. URL http://www.nap.edu/openbook.php?record_id=5140.
33. AFS (2004) Guidelines for the use of fishes in research. Technical report, American Fisheries Society.
34. Hammer C (1995) Fatigue and exercise tests with fish. Comp Biochem Physiol 112A: 1–20.
35. Plaut I (2001) Critical swimming speed: its ecological relevance. Comp Biochem Physiol A 131: 41–50.
36. Bowen M, Baskerville-Bridges B, Hess L, Karp C, Siegfried S, et al. (2004) Empirical and experimental analyses of secondary louver efficiency at the tracy fish collection facility: March 1996 to november 1997. Technical report, Tracy fish collection facility studies, California, vol. 11 U.S. Bureau of Reclamation, Denver Technical Service Center, CO, USA.
37. Lemasson B, Haefner J, Bowen M (2008) The effect of avoidance behavior on predicting fish passage rates through water diversion structures. Ecol Model 219: 178–188.
38. Lima SL, Bednekoff PA (1999) Temporal Variation in Danger Drives Antipredator Behavior: The Predation Risk Allocation Hypothesis. Am Nat 153: 649–659.

39. Patrick W, Moser M (2001) Potential competition between hybrid striped bass (*morone saxatilis* × *m. americana*) and striped bass (*m. saxatilis*) in the cape fear river estuary, north carolina. Estuaries 24: 425–429.
40. Rash J, Ney J (2012) Comparative ecology of juvenile striped bass and juvenile hybrid striped bass in claytor lake, virginia. Am Fish Soc Sym 79: 1–19.
41. Berdahl A, Torney CJ, Ioannou CC, Faria JJ, Couzin ID (2013) Emergent Sensing of Complex Environments by Mobile Animal Groups. Science 339: 574–576.
42. Brown C, Laland K (2002) Social enhancement and social inhibition of foraging behaviour in hatchery-reared Atlantic salmon. J Fish Biol 61: 987–998.
43. Miller N, Garnier S, Hartnett AT, Couzin ID (2013) Both information and social cohesion determine collective decisions in animal groups. Proc Nat Acad Sci 110: 5263–5268.
44. Hoar WS (1958) Rapid learning of a constant course by travelling schools of juvenile Pacific salmon. J Fish Res Bd Can 15: 251–274.
45. Buhl J, Sumpter D, Couzin I, Hale J, Despland E, et al. (2006) From disorder to order in marching locusts. Science 312: 1402–1406.
46. Haro A, Odeh M, Noreika J, Castro-Santos T (1998) Effect of Water Acceleration on Downstream Migratory Behavior and Passage of Atlantic Salmon Smolts and Juvenile American Shad at Surface Bypasses. Trans Am Fish Soc 127: 118–127.
47. Killen SS, Marras S, Steffensen JF, McKenzie DJ (2012) Aerobic capacity influences the spatial position of individuals within fish schools. Proc Roy Soc B 279: 357–364.
48. Gazzola M, Chatelain P, van Rees WM, Koumoutsakos P (2011) Simulations of single and multiple swimmers with non-divergence free deforming geometries. J Comput Phys 230: 7093–7114.
49. Hemelrijk CK, Reid D, Hildenbrandt H, Padding JT (2014) The increased efficiency of fish swimming in a school. Fish Fish: 1467–2979.
50. Reebs S (2000) Can a minority of informed leaders determine the foraging movements of a fish shoal? Anim Behav 59: 403–409.
51. Reebs S (2010) Temporal complementarity of information-based leadership. Behav Process 84: 685–686.
52. Krause J, Hoare D, Krause S, Hemelrijk C, Rubenstein D (2000) Leadership in fish shoals. Fish Fish 1: 82–89.
53. Kurvers R, Eijkelenkamp B, van Oers K, van Lith B, van Wieren S, et al. (2009) Personality differences explain leadership positions in barnacle geese. Ani Behav 78: 447–453.
54. Dahlbom S, Lagman D, Lundstedt-Enkel K, Sundström L, Winberg S (2011) Boldness predicts social status in zebrafish (*danio rerio*). PloS One 6: e23565.
55. Stankowich T, Blumstein D (2005) Fear in animals: a meta-analysis and review of risk assessment. Proc Roy Soc B 272: 2627–2634.
56. Fox H, White S, Kao M, Fernald R (1997) Stress and dominance in a social fish. J Neurosci 17: 6463–6469.

15

Counter-Regulatory Response to a Fall in Circulating Fatty Acid Levels in Rainbow Trout. Possible Involvement of the Hypothalamus-Pituitary-Interrenal Axis

Marta Librán-Pérez, Cristina Velasco, Marcos A. López-Patiño, Jesús M. Míguez, José L. Soengas*

Laboratorio de Fisioloxía Animal, Departamento de Bioloxía Funcional e Ciencias da Saúde, Facultade de Bioloxía, Universidade de Vigo, Vigo, Spain

Abstract

We hypothesize that a decrease in circulating levels of fatty acid (FA) in rainbow trout *Oncorhynchus mykiss* would result in the inhibition of putative hypothalamic FA sensing systems with concomitant changes in the expression of orexigenic and anorexigenic factors ultimately leading to a stimulation of food intake. To assess this hypothesis, we lowered circulating FA levels treating fish with SDZ WAG 994 (SDZ), a selective A1 adenosine receptor agonist that inhibits lipolysis. In additional groups, we also evaluated if the presence of intralipid was able to counteract changes induced by SDZ treatment, and the possible involvement of the hypothalamus-pituitary-interrenal (HPI) axis by treating fish with SDZ in the presence of metyrapone, which decreases cortisol synthesis in fish. The decrease in circulating levels of FA in rainbow trout induced a clear increase in food intake that was associated with the decrease of the anorexigenic potential in hypothalamus (decreased POMC-A1 and CART mRNA abundance), and with changes in several parameters related to putative FA-sensing mechanisms in hypothalamus. Intralipid treatment counteracted these changes. SDZ treatment also induced increased cortisol levels and the activation of different components of the HPI axis whereas these changes disappeared in the presence of intralipid or metyrapone. These results suggest that the HPI axis is involved in a counter-regulatory response in rainbow trout to restore FA levels in plasma.

Editor: Patrick Prunet, Institut National de la Recherche Agronomique (INRA), France

Funding: Ministerio de Ciencia e Innovación and European Fund for Regional Development (AGL2013-46448-3-1-R and FEDER) to JLS. Xunta de Galicia (Consolidación e estruturación de unidades de investigación competitivas do sistema universitario de Galicia, CN 2012/004) to JLS. Ministerio de Ciencia e Innovación (Predoctoral Fellowship, Program FPI) to MLP. The funders had no role in study design, data collection and analysis, decision to publish, or preparation of the manuscript.

Competing Interests: The authors have declared that no competing interests exist.

* Email: jsoengas@uvigo.es

Introduction

Specialized neurons within mammalian hypothalamus have been suggested to detect increases in plasma levels of long-chain fatty acid (LCFA), but not short-chain (SCFA) or medium-chain (MCFA) FA through several mechanisms [1,2,3], such as i) FA metabolism through inhibition of carnitine palmitoyltransferase 1 (CPT-1) to import FA-CoA into the mitochondria for oxidation; ii) binding to FA translocase (FAT/CD36), and further modulation of transcription factors like peroxisome proliferator-activated receptor type α (PPARα), and sterol regulatory element-binding protein type 1c (SREBP1c); iii) activation of protein kinase C-θ; and iv) mitochondrial production of reactive oxygen species (ROS) by electron leakage resulting in an inhibition of ATP-dependent inward rectifier potassium channel (K_{ATP}) activity. Changes in these systems have been associated [4] with the modulation of hypothalamic homeobox domain transcription factor (BSX), forkhead box 01 (Fox01), and phosphorylated cAMP response-element binding protein (pCREB). The action of these factors would result in the inhibition of the orexigenic factors agouti-related protein (AgRP) and neuropeptide Y (NPY), and the enhancement of the anorexigenic factors pro-opio melanocortin (POMC) and cocaine and amphetamine-related transcript (CART) ultimately leading to decreased food intake [1,4].

In fish, a reduced food intake has been observed after feeding fish with lipid-enriched diets or in fish containing high fat stores [5,6,7,8,9,10,11] raising the question whether lipid sensing mechanisms regulating food intake may be also present in fish [12,13]. Accordingly, we observed in rainbow trout *Oncorhynchus mykiss* that intraperitoneal [14] or intracerebroventricular [15] administration of oleate (LCFA) or octanoate (MCFA) elicited an inhibition in food intake. Furthermore, the treatment induced a response in the hypothalamus compatible with FA sensing including reduced potential of lipogenesis and FA oxidation, decreased potential of K_{ATP}, and modulation of FAT/CD36 with subsequent changes in the expression of transcription factors [14,15,16]. This response is comparable in general with that reported in mammals with the main difference of the capacity of fish to respond to increased levels of an MCFA like octanoate [13]. Changes in these hypothalamic pathways can be also related to the control of food intake, since changes in mRNA levels of neuropeptides such as NPY and POMC-A1 were also noted [14,15,16]. In the hypothalamus of another fish species, the orange-spotted grouper (*Epinephelus coioides*), the involvement of

FA metabolism and mitochondrial activity in the orexigenic effects of NPY has been also suggested [17].

The fall of blood glucose levels is sensed in central glucosensor areas eliciting counter-regulatory responses to restore glucose levels as demonstrated in mammals [18] and fish [19,20]. This mechanism has been evidenced in rat for metabolites other than glucose, such as FA. Thus, the counter-regulatory response to decreased circulating levels of FA has been associated with the activation of the hypothalamus-pituitary-adrenal (HPA) axis, and, therefore, to enhanced circulating levels of glucocorticoids whose lipolytic action would restore plasma FA levels [21,22,23]. In fish fed with diets containing low lipid levels, an increase in food intake has been described [6,8,9,10,24,25] but to date there is no evidence available regarding the existence of counter-regulatory responses to decreased FA levels. We hypothesize that the decrease in circulating levels of FA in rainbow trout would result in the down-regulation of putative hypothalamic FA sensing systems [13] with concomitant changes in the expression of orexigenic and anorexigenic factors ultimately leading to a stimulation of food intake. To assess this hypothesis, we lowered circulating FA levels in rainbow trout treating fish with SDZ WAG 994 (SDZ), a selective A1 adenosine receptor agonist that inhibits lipolysis [26]. We also evaluated if the presence of intralipid (a lipid emulsion of phospholipid-stabilized soybean oil for intravenous administration, Sigma Chemical Co) was able to counteract changes induced by SDZ treatment. Furthermore, we also evaluated the possible involvement of the hypothalamus-pituitary-interrenal (HPI) axis (fish equivalent to mammalian HPA) in the counter-regulatory response. Thus, we modified an accessible part of the HPI axis such as cortisol levels by treating fish with SDZ in the presence of metyrapone, which decreases cortisol synthesis in fish [27,28].

Materials and Methods

Ethics statement

The experiments described comply with the Guidelines of the European Union Council (2010/63/UE), and of the Spanish Government (RD 55/2013) for the use of animals in research. The Ethics Committee of the Universidade de Vigo approved the procedures.

Fish

Rainbow trout obtained from a local fish farm (A Estrada, Spain) were maintained for 1 month in 100 litre tanks under laboratory conditions and 12L:12D photoperiod in dechlorinated tap water at 15°C. Fish weight was 99±3 g. Fish were fed once daily (09.00 h) to satiety with commercial dry fish pellets (Dibaq-Diproteg SA, Spain).

Experimental design

Following acclimation, fish were fasted for 24 h before treatment to ensure fish had basal levels of metabolic hormones including cortisol. On the day of experiment, a first set of fish were anaesthetized in tanks with 2-phenoxyethanol (Sigma, 0.2% v/v), and weighed. Then, 15 fish per group received intraperitoneally (IP) 10 $mL.Kg^{-1}$ injection of saline solution alone (control, C), or containing SDZ (SDZ; Tocris, 60 $\mu g.Kg^{-1}$), metyrapone (M; Sigma, 1 $mg.Kg^{-1}$), both SDZ and metyrapone (SDZ+M), or both SDZ and intralipid (SDZ+IL; Sigma I-141, 3 $mL.Kg^{-1}$). Blood, hypothalamus, and head kidney samples were taken 6 h after treatment, which was chosen on the basis of previous studies in which such time period was necessary to achieve changes in the FA sensing mechanisms when levels of FA were increased [14,16]. Initial concentrations of SDZ were selected based on studies carried out previously in mammals [22,29], and then in preliminary studies (data not shown) we evaluated different SDZ doses. Since SDZ is known to reduce mean arterial pressure and heart rate at high doses [29] we selected a dose (60 $\mu g.Kg^{-1}$) able to lower levels of circulating FA without inducing any other apparent alteration. The concentrations of metyrapone and intralipid were selected based on previous studies carried out in rainbow trout [27,28,30] and mammals [22], respectively. In each group, 10 fish were used to assess enzyme activities and metabolite levels whereas the remaining 5 fish were used for the assessment of mRNA levels by qRT-PCR. In each sampling, fish were anesthesized as above, and blood was collected from the caudal vein with a heparinised syringe. Fish were then sacrificed by decapitation, and hypothalamus and head kidney (area containing interrenal cells, i.e. those involved in glucocorticoid synthesis in fish) were taken and stored as previously described [31,32,33].

In a second set of fish, we evaluated changes in food intake after IP administration of SDZ or SDZ+intralipid. Fish were randomly assigned to experimental groups in different tanks and fasted for 24 h before injection. Then, 8 fish per group were anesthesized and IP injected as above. Food intake was assessed 3 days before treatment (to define baseline data), and then 6 and 24 h after treatment. After feeding, the food uneaten remaining at the bottom (conical tanks) and feed waste were withdrawn, dried and weighed. The amount of food consumed by all fish in each tank was calculated as previously described as the difference from the feed offered [31,32,34]. Results are shown as the mean ± SEM of the data obtained in three different tanks per treatment.

Assessment of metabolite levels and enzyme activities

Levels of FA, total lipid, triglyceride, glucose, and lactate in plasma were determined enzymatically using commercial kits (Wako for FA; Spinreact for total lipid, triglyceride, and lactate; Biomérieux for glucose) adapted to a microplate format. Plasma cortisol levels were assessed by ELISA using a commercially available kit (Cayman).

Samples used to assess hypothalamic metabolite levels were homogenized immediately by ultrasonic disruption in 7.5 vols of ice-cooled 0.6 M perchloric acid, and neutralized (using 1 M potassium bicarbonate). The homogenate was centrifuged (10,000 g), and the supernatant used to assay tissue metabolites. Tissue FA, total lipid, and triglyceride levels were determined enzymatically using commercial kits as described above for plasma samples.

Samples for enzyme activities were homogenized by ultrasonic disruption with 9 vols of ice-cold-buffer consisting of 50 mM Tris (pH 7.6), 5 mM EDTA, 2 mM 1,4-dithiothreitol, and a protease inhibitor cocktail (Sigma). The homogenate was centrifuged (10,000 g) and the supernatant used immediately for enzyme assays. Enzyme activities were determined using a microplate reader INFINITE 200 Pro (Tecan) and microplates. Reaction rates of enzymes were determined by the increase or decrease in absorbance of NAD(P)H at 340 nm or, in the case of CPT-1 activity, of 5,5′-Dithiobis(2-nitrobenzoic acid)-CoA (DTNB) complex at 412 nm. The reactions were started by the addition of supernatant (15 μl) at a pre-established protein concentration, omitting the substrate in control wells (final volume 265–295 μl), and allowing the reactions to proceed at 20°C for pre-established times (3–10 min). Enzyme activities are expressed per protein level, which was assayed according to the bicinchoninic acid method with bovine serum albumin (Sigma) as standard. Enzyme activities were assessed at maximum rates determined by preliminary tests to determine optimal substrate concentrations. ATP-citrate lyase (ACLY, *EC* 4.1.3.8) activity was assessed in a

Table 1. Nucleotide sequences of the PCR primers used to evaluate mRNA abundance by RT-PCR (qPCR).

	Forward primer	Reverse primer	Data base	Accession Number
3β-HSD	TCACAGGGTCAACGTCAAAG	CCTCCTTCTTGGTCTTGCTG	GenBank	S72665.1
11βH	ATTTGCCCTGTACGAGTTGG	GGATGATGATGTCTCTGACTG	GenBank	AF179894
ACC	TGAGGGCGTTTTCACTATCC	CTCGATCTCCCTCTCCACT	Sigenae	tcbk0010c.b.21_5.1.om.4
ACLY	CTGAAGCCCAGACAAGGAAG	CAGATTGGAGGCCAAGATGT	GenBank	CA349411.1
CART	ACCATGGAGAGCTCCAG	GCGCACTGCTCTCCAA	GenBank	NM_001124627
CPT-1c	CGCTTCAAGAATGGGGTGAT	CAACCACCTGCTGTTTCTCA	GenBank	AJ619768
CPT-1d	CCGTTCCTAACAGAGGTGCT	ACACTCCGTAGCCATCGTCT	GenBank	AJ620356
CRF	ACAACGACTCAACTGAAGATCTCG	AGGAAATTGAGCTTCATGTCAGG	GenBank	AF296672
CRFBP	GGAGGAGACTTCATCAAGGTGTT	CTTCTCTCCCTTCATCACCCAG	GenBank	AY363677
CS	GGCCAAGTACTGGGAGTTCA	CTCATGGTCACTGTGGATGG	Tigr	TC89195
EF-1α	TCCTCTTGGTCGTTTCGCTG	ACCCGAGGGACATCCTGTG	GenBank	AF498320
FAS	GAGACCTAGTGGAGGCTGTC	TCTTGTTGATGGTGAGCTGT	Sigenae	tcab0001c.e.06 5.1.s.om.8
FAT/CD36	CAAGTCAGCGACAAACCAGA	ACTTCTGAGCCTCCACAGGA	DFCI	AY606034.1
Kir6.x-like	TTGGCTCCTCTTCGCCATGT	AAAGCCGATGGTCACCTGGA	Sigenae	CA346261.1.s.om.8:1:773:1
LXRα	TGCAGCAGCCGTATGTGGA	GCGGCGGGAGCTTCTTGTC	GenBank	FJ470291
MCD	TCAGCCAGTACGAAGCTGTG	CTCACATCCTCCTCCGAGTC	Sigenae	BX869708.s.om.10
NPY	CTCGTCTGGACCTTTATATGC	GTTCATCATATCTGGACTGTG	GenBank	NM_001124266
P450scc	ATGCGTCAGGACACTAACAC	CAGCGGTATCATCTTCAGCA	GenBank	S57305.1
POMC-A1	CTCGCTGTCAAGACCTCAACTCT	GAGTTGGGTTGGAGATGGACCTC	Tigr	TC86162
PPARα	CTGGAGCTGGATGACAGTGA	GGCAAGTTTTTGCAGCAGAT	GenBank	AY494835
SREBP1c	GACAAGGTGGTCCAGTTGCT	CACACGTTAGTCCGCATCAC	GenBank	CA048941.1
StAR	CTCCTACAGACATATGAGGAAC	GCCTCCTCTCCCTGCTTCAC	GenBank	AB047032
SUR-like	CGAGGACTGGCCCCAGCA	GACTTTCCACTTCCTGTGCGTCC	Sigenae	tcce0019d.e.20_3.1.s.om.8
UCP2a	TCCGGCTACAGATCCAGG	CTCTCCACAGACCACGCA	GenBank	DQ295324

3βHSD, 3β-hydroxysteroid dehydrogenase; 11βH, 11β-hydroxylase; ACC, Acetyl-CoA carboxylase; ACLY, ATP-citrate lyase; CART, cocaine- and amphetamine-related transcript; CPT-1, carnitine palmitoyl transferase type 1; CRF, corticotrophin releasing factor; CRFBP, corticotrophin releasing factor binding protein; CS, citrate synthetase; EF-1α, elongation factor 1α; FAS, fatty acid synthetase; FAT/CD36, fatty acid translocase; Kir6.x-like, inward rectifier K^+ channel pore type 6.x-like; LXRα, liver X receptor α; MCD, malonyl CoA dehydrogenase; NPY, neuropeptide Y; P450scc, cytochrome P450 cholesterol side chain cleavage; POMC-A1, pro-opio melanocortin A1; PPARα, peroxisome proliferator-activated receptor type α; SREBP1c, sterol regulatory element-binding protein type 1c; StAR, steroidogenic acute regulatory protein; SUR-like, sulfonylurea receptor-like; UCP2a, mitochondrial uncoupling protein 2a.

tris-HCl buffer (50 mM, pH 7.8) containing 100 mM KCl, 10 mM $MgCl_2$, 20 mM citrate, 10 mM β-mercaptoethanol, 5 mM ATP, 0.3 mM NADH, 7 $U.ml^{-1}$ malate dehydrogenase, and 50 μM Coenzyme A (omitted for controls). Fatty acid synthase (FAS, *EC* 2.3.1.85) activity was assessed in a phosphate buffer (100 mM, pH 7.6) containing 0.1 mM NADPH, 25 μM Acetyl-CoA, and 30 μM Malonyl-CoA (omitted for controls). Hydroxyacil-CoA dehydrogenase (HOAD, *EC* 1.1.1.35) activity was assessed in a imidazole buffer (50 mM, pH 7.6) containing 0.15 mM NADH and 3.5 mM Acetoacetyl-CoA (omitted for controls). CPT-1 (*EC* 2.3.1.21) activity was assessed in a tris-HCl buffer (75 mM, pH 8.0) containing 1.5 mM EDTA, 0.25 mM DTNB, 35 μM palmitoyl CoA, and 0.7 mM L-carnitine (omitted for controls).

mRNA abundance analysis by quantitative RT-PCR

Total RNA extracted from tissues using Trizol reagent (Life Technologies) was treated with RQ1-DNAse (Promega). 4 μg total RNA were reverse transcribed into cDNA using Superscript II reverse transcriptase (Promega) and random hexaprimers (Promega). Gene expression levels were determined by real-time quantitative RT-PCR (q-PCR) using the iCycler iQ (BIO-RAD). Analyses were performed on 1 μl cDNA using the MAXIMA SYBRGreen qPCR Mastermix (Thermo Fisher Scientific), in a total PCR reaction volume of 25 μl, containing 50–500 nM of each primer. mRNA abundance of transcripts 3β-hydroxysteroid dehydrogenase (3βHSD), 11β-hydroxylase (11βH), acetyl-CoA carboxylase (ACC), ACLY, CART, corticotrophin releasing factor (CRF), corticotrophin releasing factor binding protein (CRFBP), FAT/CD36, CPT-1, citrate synthetase (CS), FAS, inward rectifier K^+ channel pore type 6.x-like (Kir6.x-like), liver X receptor α (LXRα), malonyl CoA dehydrogenase (MCD), NPY, cytochrome P450 cholesterol side chain cleavage (P450scc), POMC-A1, PPARα, SREBP1c, steroidogenic acute regulatory protein (StAR), sulfonylurea receptor-like (SUR-like), and mitochondrial uncoupling protein 2a (UCP2) was determined as previously described in the same species [35,36,37,38,39,40,41,42,43,44,45]. Sequences of the forward and reverse primers used for each gene expression are shown in Table 1. Relative quantification of the target gene transcripts was done using elongation factor 1α (EF-1α) gene expression as reference, which was stably expressed in this experiment.

Thermal cycling was initiated with incubation at 95°C for 15 min using hot-start iTaq DNA polymerase activation; 40 steps of PCR were performed, each one consisting of heating at 95°C for 15 s for denaturing, annealing at specific temperatures for 30 s,

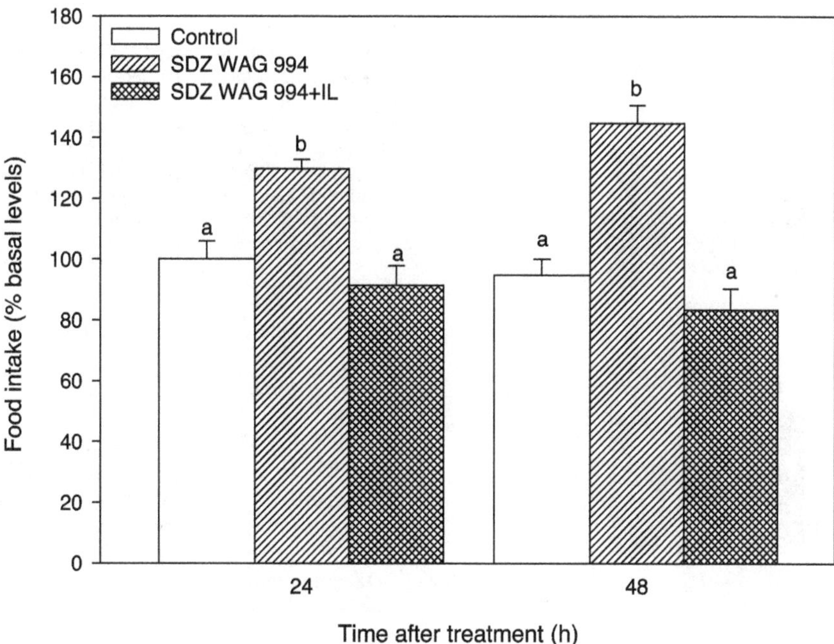

Figure 1. Changes in food intake after intraperitoneal treatment. Food intake of rainbow trout after 6 h or 24 h of intraperitoneal administration of 10 mL.Kg^{-1} of saline solution alone (control) or containing SDZ WAG 994 alone (60 µg.Kg^{-1}) or containing SDZ WAG 994 (60 µg.Kg^{-1}) together with intralipid (3 mL.Kg^{-1}) solution. Different letters indicate significant differences ($P<0.05$) among treatments at the same time. Food intake is shown as mean+S.E.M. of the percentage of food ingested with respect to basal levels (calculated as the average of food intake the three days previous to experiment). The results are shown as mean+S.E.M. of the results obtained in three different tanks in which 8 fish were used per group in each tank. Different letters indicate significant differences ($P<0.05$) among treatments at each time.

and extension at 72°C for 30 s. Following the final PCR cycle, melting curves were systematically monitored (55°C temperature gradient from 55 to 95°C) to ensure amplification of only one fragment. Each sample was assessed in triplicate. Samples without reverse transcriptase and samples without RNA were run for each reaction as negative controls. Only efficiency values between 85–100% were accepted (the R^2 for all the genes assessed was always higher than 0.985). Relative quantification of the target gene transcript with the EF-1α reference gene transcript was made following the Pfaffl method [46].

Statistics

Comparisons among groups were carried out with two-way ANOVA with treatment and time as main factors for food intake data whereas the remaining parameters were compared with one-way ANOVA. Post-hoc comparisons were carried out with a Student-Newman-Keuls test, and differences were considered statistically significant at $P<0.05$.

Results

Food intake increased after 6 and 24 h of treatment with SDZ, and the presence of intralipid counteracted the increase (Fig. 1).

Levels of metabolites assessed in plasma are shown in Fig. 2. FA (Fig. 2A) and triglyceride (Fig. 2B) levels decreased after SDZ treatment compared with all other groups. Total lipid levels (Fig. 2C) decreased after treatment with SDZ or SDZ +IL compared with the remaining groups, and the decrease was weaker in the SDZ +IL group. Glucose levels increased after treatment with SDZ +M compared with control and SDZ groups (Fig. 2D). Lactate levels increased after SDZ +IL treatment compared with SDZ +M (Fig. 2E). Cortisol levels were higher in the SDZ group than in all other groups (Fig. 2F).

In hypothalamus (Fig. 3) SDZ treatment decreased levels of FA (vs. all other groups) and triglycerides (vs. control, SDZ +M and SDZ +IL groups) whereas no significant changes were noted for total lipid levels.

Enzyme activities assessed in hypothalamus are shown in Fig. 4. FAS activity was lower in the SDZ and M groups compared with the remaining groups (Fig. 4B). HOAD activity was lower in the SDZ group than in control, SDZ +M, and SDZ +IL groups (Fig. 4D). We did not observe significant differences for CPT-1 (Fig. 4A) and ACLY (Fig. 4C) activities.

RNA abundance of transcripts related to fatty acid sensing in hypothalamus is shown in Table 2. FAT/CD36 level was lower in the SDZ group compared with control, M, and SDZ +IL groups. ACC level increased in the M group compared with control, SDZ +M, and SDZ +IL groups. ACLY level increased after M treatment compared with control, SDZ, and SDZ +IL treatments. CPT-1c level decreased in the SDZ group compared with control and SDZ +IL groups. Levels of CPT-1d in the SDZ group were lower than in control, M, and SDZ +IL groups. FAS level increased in the group treated with SDZ compared with all other groups. MCD values were lower in the SDZ group than in control and SDZ +IL groups. UCP2a values were higher in the SDZ group than in control, SDZ +M, and SDZ +IL groups whereas those of SDZ +IL group were lower than in all other groups. The level of SUR-like was lower in the M and SDZ +M groups than in all other groups. PPARα level increased after treatment with M or SDZ +M compared with all other groups. CART level decreased after treatment with SDZ or SDZ +M compared with control and SDZ +IL groups. POMC-A1 level decreased after treatment with SDZ or SDZ +M compared with control and SDZ +IL groups with the decrease being more important for SDZ +M; the value of SDZ +IL was higher than in all other groups. Finally, no

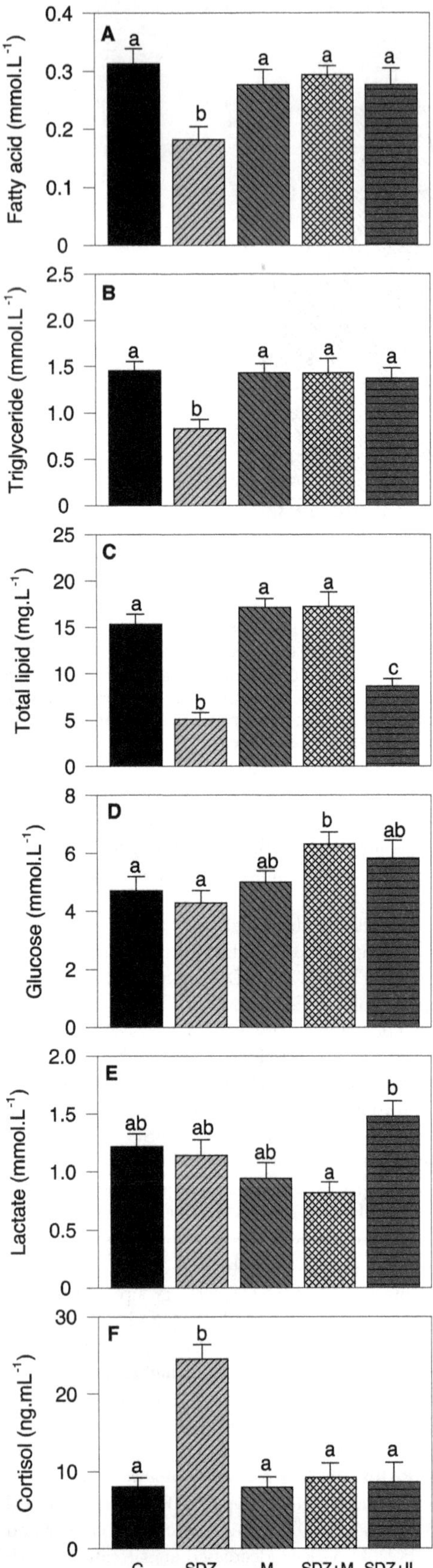

Figure 2. Changes in the levels of plasma metabolites after intraperitoneal treatment. Levels of fatty acid (A), triglyceride (B), total lipid (C), glucose (D), lactate (E), and cortisol (F) in plasma of rainbow trout after 6 h of intraperitoneal administration of 10 mL.Kg^{-1} of saline solution alone (control, C) or containing SDZ WAG 994 (SDZ, 60 μg.Kg^{-1}), metyrapone (M, 1 mg.Kg^{-1}), both SDZ WAG 994 and metyrapone (SDZ+M), or both SDZ WAG 994 and intralipid (3 mL.Kg^{-1}) solution (SDZ+IL). The results are shown as mean+S.E.M. of 10 fish per treatment. Different letters indicate significant differences ($P<0.05$) among treatments.

significant changes were apparent for CS, Kir6.x-like, LXRα, and NPY.

RNA abundance of transcripts related to HPI axis in hypothalamus and head kidney is shown in Table 3. In hypothalamus, CRF value was higher in the SDZ group compared with all other groups whereas value of SDZ +IL group was higher than in M and SDZ +M groups. CRFBP levels increased after treatment with SDZ or M compared with control, SDZ +M, and SDZ +IL groups. 3βHSD level increased after treatment with SDZ compared with all other groups; values in the SDZ +M group were lower than in all other groups. 11βH level increased after treatment with SDZ or M compared with all other groups with the increase being higher for M group, and decreased after treatment with SDZ +M compared with all other groups. P450scc level increased in the group treated with SDZ compared with the remaining groups; the level in the M and SDZ +M groups was lower than in all other groups with the decrease being more important for SDZ +M. Finally, StAR value increased after treatment with SDZ compared with all other groups.

Discussion

Lowering circulating FA levels in rainbow trout decreased anorexigenic potential in hypothalamus and increased food intake

Treatment with SDZ was effective in reducing not only circulating FA levels but also those of triglyceride and total lipid, which validates the experimental design. These changes are similar to those reported in rat after similar treatment [22,29,47]. There are no comparable references available in fish, though in zebrafish bezafibrate treatment was also able to decrease plasma triglyceride and cholesterol levels [48]. Moreover, the presence of intralipid was able to counteract the action of SDZ in rainbow trout resulting in levels of FA and triglyceride similar to those of controls, again in agreement with that reported in rat [22].

Food intake increased in fish treated with SDZ. This allows us to suggest, for the first time in fish, that hypothalamus sense the decreased levels of FA resulting in an orexigenic response stimulating food intake. We cannot exclude the possibility that SDZ could have a direct effect on food intake independent of that elicited by the inhibition of lipolysis. However, the finding that the presence of intralipid counteracted the increased food intake suggest that the effect of reduced FA levels is more likely. This increased food intake is in agreement with that reported in different fish species, including rainbow trout, after feeding diets with low lipid content [6,8,9,10,24,25]. Furthermore, the food intake response is the opposite of that observed when the same species was treated with specific FA such as oleate or octanoate [14]. Changes elicited in the levels of circulating lipids relate to changes observed in hypothalamus since SDZ treatment decreased levels of FA and triglyceride, and levels recovered those of controls after additional treatment with intralipid. Therefore, changes in circulating metabolites in plasma are translated into changes in metabolite levels in hypothalamus thus supporting the FA sensing capacity of that tissue not only to increased levels of FA as previously described [14,15,16] but also to decreased levels of circulating FA as herein reported. Accordingly, we evaluated changes in parameters related to putative FA sensing systems in hypothalamus in order to relate them to changes observed in food intake.

SDZ treatment affected the FA sensing system related to FA metabolism since it decreased the activity of FAS and HOAD as well as mRNA abundance of CPT1c, CPT1d, and MCD and increased mRNA abundance of FAS. Furthermore, the presence of intralipid counteracted these changes. We have previously observed that this system responds to increased levels of FA with decreased lipogenic potential and decreased capacity of FA oxidation [14,15,16]. Therefore, we expected that changes observed in parameters related to FA metabolism would be different than those observed after increasing levels of specific FA such as oleate or octanoate [14]. Accordingly, we observed opposite changes for FAS activity and mRNA abundance as well as for mRNA abundance of CPT1c and MCD. However, other parameters that displayed changes in situations of increased circulating levels of oleate or octanoate [14,15,16] such as CPT-1 and ACLY activities and mRNA abundance of ACC, ACLY, and CS did not display any change in the present study. The responses observed suggest that the FA sensing system related to FA metabolism only respond partially to decreased circulating levels of FA. There are no similar studies available in fish hypothalamus [13] whereas in peripheral tissues like liver lipogenic potential is up-regulated when fish were fed with diets with low lipid content [49].

The putative FA sensing system related to FA transport through FAT/CD36 and subsequent modulation of transcription factors partially responded to decreased circulating levels of FA since several parameters did not show changes compared with controls (mRNA abundance of PPARα, SREBP1c, and LXRα). Only mRNA abundance of FAT/CD36 decreased after SDZ treatment. This change is the opposite of that previously observed when FA levels increased [14] suggesting a decreased potential for binding capacity of FAT/CD36 in parallel with the decreased circulating FA levels.

As for the putative FA sensing system related to K_{ATP} channel and mitochondrial activity, Kir6.x-like and SUR-like mRNA abundance displayed no changes when circulating FA levels decreased. However, levels of UCP2a mRNA were higher in the group treated with SDZ than in controls, a change opposite of that described when FA levels increased [14,15].

Despite the relatively few changes observed in the FA sensing systems evaluated, mRNA abundance of anorexigenic peptides (POMC-A1 and CART) clearly decreased in hypothalamus of fish treated with SDZ. This response is the opposite of that observed under conditions of increased circulating FA levels where the abundance of these transcripts increased [13]. Moreover, the increase observed in the mRNA abundance of POMC-A1 in the SDZ+IL group could be associated with the returning of food intake in that group to values similar of those in control group. Furthermore, we did not observe significant changes in NPY mRNA abundance when FA levels decreased, in contrast with the decrease observed when FA levels increased [14,15]. These changes suggest a decreased anorexigenic potential in hypothalamus of SDZ-treated fish, which is in agreement with changes noted in food intake. However, CRF mRNA levels increased under the same conditions what considering the anorexigenic nature of this peptide suggest a more complex relationship between changes in mRNA levels of neuropeptides and food intake.

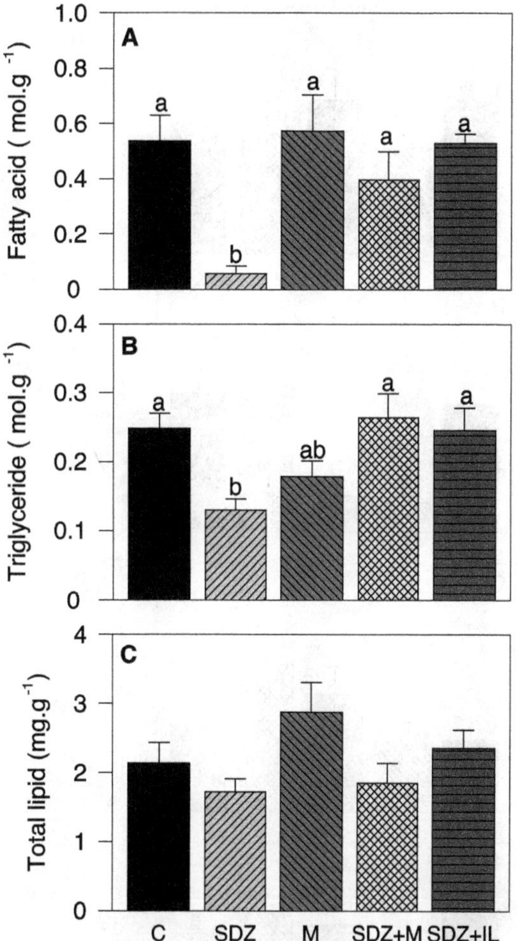

Figure 3. Changes in the levels of metabolites in hypothalamus after intraperitoneal treatment. Levels of fatty acid (A), triglyceride (B), and total lipid (C) in hypothalamus of rainbow trout after 6 h of intraperitoneal administration of 10 mL.Kg^{-1} of saline solution alone (control, C) or containing SDZ WAG 994 (SDZ, 60 μg.Kg^{-1}), metyrapone (M, 1 mg.Kg^{-1}), both SDZ WAG 994 and metyrapone (SDZ+M), or both SDZ WAG 994 and intralipid (3 mL.Kg^{-1}) solution (SDZ+IL). The results are shown as mean+S.E.M. of 10 fish per treatment. Different letters indicate significant differences ($P<0.05$) among treatments.

Therefore, it seems that the decrease in circulating FA levels induced by SDZ treatment resulted in decreased levels of FA in hypothalamus but only few of the parameters involved in putative FA sensing systems addressed in hypothalamus changed accordingly. However, since the expression of several neuropeptides and, more important, food intake changed under those conditions, we can suggest, for the first time in non-mammalian vertebrates, that central capacity for sensing decreased FA levels exists in fish. However, this modulation of FA sensors is not exactly the opposite of that observed when FA levels increase [14,15,16]. The difference in the response of these systems to increased or decreased FA levels could be due to different factors. Any of these factors could be: i) the involvement of another putative FA sensing mechanism, such as that mediated by PKC-θ [2], or ii) the fact that we evaluated the decrease of FA in general whereas in the other studies carried out in rainbow trout [14,15,16] we evaluated the increase of specific FA, such as oleate and octanoate. In fact, in

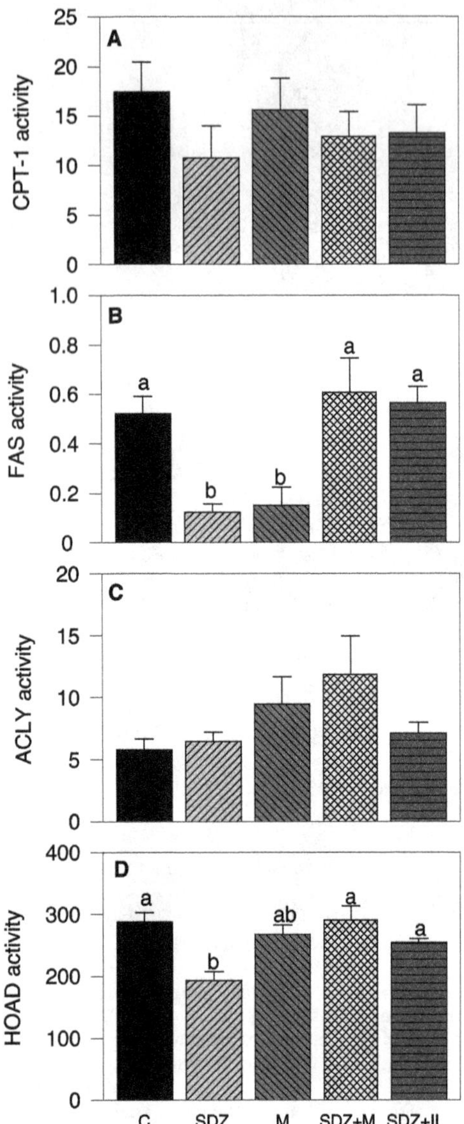

Figure 4. Changes in the activity of enzymes in hypothalamus after intraperitoneal treatment. Activities (mU.mg^{-1} protein) of CPT-1 (A), FAS (B), ACLY (C), and HOAD (D) in hypothalamus of rainbow trout after 6 h of intraperitoneal administration of 10 mL.Kg^{-1} of saline solution alone (control, C) or containing SDZ WAG 994 (SDZ, 60 μg.Kg^{-1}), metyrapone (M, 1 mg.Kg^{-1}), both SDZ WAG 994 and metyrapone (SDZ+M), or both SDZ WAG 994 and intralipid (3 mL.Kg^{-1}) solution (SDZ+IL). The results are shown as mean+S.E.M. of 10 fish per treatment. Different letters indicate significant differences ($P<0.05$) among treatment.

a recent study in rat, the counter-regulatory response to decreased FA levels depends on the type of FA [23].

The activation of the HPI axis is probably involved in the FA counter-regulatory response to decreased levels of FA

In mammals, different hormones of lipolytic action such as growth hormone [50], epinephrine [51], glucagon [52] or glucocorticoids [21,22,23] were associated with the counter-regulatory response to decreased levels of FA. In the present study, we have evaluated the possibility that cortisol (main glucocorticoid in fish) could be involved in the counter-regulatory response observed in rainbow trout.

SDZ treatment induced an increase in circulating levels of cortisol, which was not observed in the SDZ+IL group. This allows us to suggest that the activation of the HPI axis is involved in the response to decreased levels of FA. This response is comparable to that observed in rat where the presence of intralipid also counteracted the decrease in circulating FA levels induced by SDZ treatment [22]. We have also monitored in head kidney the mRNA abundance of proteins involved in cortisol synthesis such as 11βH, 3βHSD, P450scc and StAR. Abundance of these transcripts increased after treatment with SDZ, and values returned to normality when fish were co-treated with SDZ and intralipid. In rainbow trout up-regulation of these transcripts corresponds with enhanced cortisol levels in plasma [53,54]. Therefore, changes observed in these transcripts agree well with those of cortisol levels. Furthermore, we have also monitored hypothalamic CRF and CRFBP mRNA abundance that in the same species are normally changing in parallel with the levels of cortisol in plasma under short-term periods [55]. SDZ treatment increased levels of both transcripts and, once more, the presence of intralipid counteracted such elevation. Interestingly, an inverse relationship between plasma glucose and FA was observed in plasma in agreement with that seen in many fish species under stress situations in which the HPI axis is activated [56]. Altogether, these results suggest that decreased circulating levels of FA activate the HPI axis. However, we cannot discard a direct action of SDZ on cortisol synthesis.

Further support of the involvement of the HPI axis comes from results obtained in the groups treated with metyrapone. Metyrapone is an inhibitor of 11βH, and its treatment resulted in decreased circulating levels of cortisol in previously stressed fish [27,28,30] or no changes if fish were not previously stressed [57,58], as observed in the present study. A glucocorticoid feedback response (i.e., increased CRF mRNA levels in response to decreased cortisol levels) has been demonstrated in rainbow trout when plasma cortisol levels changed after metyrapone treatment [27,59]. Thus, the lack of changes in cortisol levels in the M group is also reflected by the absence of changes in hypothalamic CRF mRNA levels. Changes observed in hypothalamus of metyrapone-treated fish (up-regulation of ACC, ACLY and PPARα mRNA abundance) generally fit with those previously observed in catfish brain [58] where increased lipogenic capacity occurred after metyrapone treatment. However, since metyrapone treatment alone induced changes in several parameters assessed in hypothalamus and head kidney, we cannot discard that several of the effects could be attributable to an interactions between SDZ and metyrapone.

Based on the comparison between SDZ and SDZ+M groups, metyrapone treatment was effective in counteracting the increased cortisol levels elicited by SDZ treatment. This effect of metyrapone is in agreement with that previously observed in the same species under different stressful conditions [27,28,30]. Furthermore, the treatment with both metyrapone and SDZ induced in plasma changes in parameters in a way that values were comparable to those of the control group and different than those of the SDZ group, such as for plasma FA, triglyceride or total lipid levels. This return to normality is also evident in parameters assessed in hypothalamus where FA and triglyceride levels that decreased in the SDZ group turned back to normal values in the group treated with SDZ and metyrapone. In parameters related to the HPI axis, an effective blockade of its functioning was apparent as demonstrated by the strong decrease observed in transcript abundance of the four proteins related to cortisol metabolism

Table 2. Relative mRNA abundance of transcripts related to fatty acid sensing in hypothalamus of rainbow trout after 6 h of intraperitoneal administration of 10 mL.Kg^{-1} of saline solution alone (control, C) or containing SDZ WAG 994 (SDZ, 60 μg.Kg^{-1}), metyrapone (M, 1 mg.Kg^{-1}), both SDZ WAG 994 and metyrapone (SDZ+M), or both SDZ WAG 994 and intralipid (3 mL.Kg^{-1}) solution (SDZ+IL).

	C	SDZ	M	SDZ+M	SDZ+IL
Fatty acid transport					
FAT/CD36	1±0.05a	0.63±0.04b	1.13±0.09a	0.89±0.12ab	1.23±0.08a
Fatty acid metabolism					
ACC	1±0.16a	1.15±0.12ab	1.80±0.25b	0.99±0.08a	1.06±0.16a
ACLY	1±0.11a	1.07±0.11a	1.70±0.14b	1.30±0.20ab	0.99±0.14a
CPT-1c	1±0.08a	0.62±0.05b	0.90±0.09ab	0.69±0.10ab	1.15±0.11a
CPT-1d	1±0.06ab	0.57±0.04c	0.96±0.12ab	0.64±0.08bc	1.24±0.11a
CS	1±0.16	0.88±0.12	1.46±0.20	0.93±0.16	1.15±0.12
FAS	1±0.08a	1.58±0.09b	1.17±0.12a	1.16±0.09a	1.14±0.13a
MCD	1±0.08a	0.57±0.06b	0.75±0.13ab	0.72±0.05ab	1.43±0.23a
Mitochondrial uncoupling					
UCP2a	1±0.12a	1.51±0.10b	1.34±0.14a	0.82±0.14a	0.47±0.07c
K_{ATP} channel					
Kir6.x-like	1±0.13	0.98±0.08	0.92±0.11	0.86±0.09	1.40±0.21
SUR-like	1±0.12a	0.81±0.07a	0.58±0.07b	0.49±0.03b	1.03±0.10a
Transcription factors					
LXRα	1±0.13	0.73±0.06	1.22±0.16	0.86±0.09	1.03±0.12
PPARα	1±0.15a	1.02±0.13a	2.39±0.18b	1.88±0.14b	1.26±0.16a
SREBP1c	1±0.11	0.96±0.11	1.31±0.15	0.90±0.09	1.32±0.12
Neuropeptides					
CART	1±0.11a	0.53±0.13b	0.75±0.05ab	0.57±0.08b	1.13±0.11a
NPY	1±0.07	1.09±0.09	1.28±0.15	1.06±0.09	1.03±0.12
POMC-A1	1±0.06a	0.54±0.05b	0.75±0.08ab	0.37±0.06b	2.50±0.22c

Data represent mean of 5 measurements. Data is expressed with respect to the control group (expression results were normalized by EF-1α mRNA levels, which did not show changes among groups). Different letters indicate significant differences ($P<0.05$) among treatments.

assessed in head kidney in the SDZ+metyrapone group. These changes are associated with those observed in the levels of cortisol in plasma, which were similar to those of the control group and lower than in the SDZ-treated group. Further support comes from changes observed in CRF and CRFBP mRNA abundance in the hypothalamus where a significant inhibition of the response

Table 3. Relative mRNA abundance of transcripts related to hypothalamus-pituitary-interrenal axis in hypothalamus and head kidney of rainbow trout after 6 h of intraperitoneal administration of 10 mL.Kg^{-1} of saline solution alone (control, C) or containing SDZ WAG 994 (SDZ, 60 μg.Kg^{-1}), metyrapone (M, 1 mg.Kg^{-1}), both SDZ WAG 994 and metyrapone (SDZ+M), or both SDZ WAG 994 and intralipid (3 mL.Kg^{-1}) solution (SDZ+IL).

	C	SDZ	M	SDZ+M	SDZ+IL
Hypothalamus					
CRF	1±0.11a	1.74±0.10b	0.69±0.09a	0.63±0.07a	1.19±0.11a
CRFBP	1±0.09a	1.49±0.07b	1.88±0.17b	1.35±0.10a	1.01±0.12a
Head kidney					
3βHSD	1±0.17a	1.77±0.12b	1.30±0.15a	0.32±0.03c	0.86±0.14a
11βH	1±0.13a	1.91±0.14b	3.17±0.39c	0.31±0.09d	1.28±0.20a
P450scc	1±0.15a	1.68±0.07b	0.52±0.11c	0.08±0.01d	1.08±0.15a
StAR	1±0.18a	1.92±0.13	1.23±0.10a	0.77±0.09a	1.26±0.11a

Data represent mean of 5 measurements. Data is expressed with respect to the control group (expression results were normalized by EF-1α mRNA levels, which did not show changes among groups). Different letters indicate significant differences ($P<0.05$) among treatments.

observed with the SDZ treatment was evident. Altogether, these results suggest us that the HPI axis is involved in the response induced by decreased circulating FA levels produced by SDZ treatment.

The remaining parameters assessed in hypothalamus related to FA sensing displayed in general in the group treated with SDZ and metyrapone values similar to those of controls and different than those elicited by SDZ treatment alone. Thus, metyrapone treatment effectively counteracted changes elicited by SDZ treatment in a way similar to those elicited by the presence of intralipid. Therefore, we suggest that the changes observed in FA sensing mechanisms involved in the counter-regulatory orexigenic response in hypothalamus could relate to the activation of the HPI axis. Since increased CRF production in hypothalamus activates this axis, we could speculate that CRF is modulating the activity of hypothalamic neurons that integrate metabolic information resulting in altered production of anorexigenic and orexigenic neuropeptides and finally in the food intake response, in a way comparable to that previously suggested for glucosensing in the same species [60]. However, blocking the HPI axis with metyrapone did not alter the decreased anorectic potential deduced by the down regulation observed in POMC-A1 and CART transcripts by SDZ treatment in a way similar to the lack of effect of metyrapone treatment on mRNA abundance of NPY in hypothalamus of stressed rainbow trout [59]. Moreover, stress situations and factors related to the HPI axis such as CRF have been demonstrated to be clearly anorexigenic in several fish species including rainbow trout [61,62] and a clear increase in mRNA levels of CRF was also observed in hypothalamus of SDZ-treated fish. Therefore, the modulation of food intake control by HPI activation is complex and other factors besides those herein assessed, are likely involved.

Conclusions

In summary, we have obtained evidence, for the first time in fish (and in a non-mammalian vertebrate), about the existence of a counter-regulatory response to a fall in circulating FA levels. The response is apparently associated with food intake control and the activation of HPI axis. Thus, the decrease in circulating levels of FA in rainbow trout induces an increase in food intake that is associated with the decrease of the anorexigenic potential in hypothalamus and with changes in several parameters related to putative FA-sensing mechanisms in hypothalamus. The treatment with intralipid counteracted these changes. The decrease in FA levels apparently induces a counter-regulatory response in rainbow trout in which the activation of the HPI axis is likely involved. This activation probably not related to the control of food intake through FA sensor systems [13] but to the modulation of lipolysis in peripheral tissues to restore FA levels in plasma. This counter-regulatory response initiated in the hypothalamus, probably through changes in CRF, and this activation would arrive to peripheral tissues such as liver where metabolic changes would occur accordingly. However, we cannot exclude the possibility that i) the fall in plasma FA was sensed outside the brain, and the information be transmitted via an afferent neural pathway or a humoral factor, or ii) that other lipolytic hormones in fish such as GH, glucagon or catecholamines [63,64,65,66,67] could be involved in the FA counter-regulatory response. All these possibilities clearly deserve further studies.

Author Contributions

Conceived and designed the experiments: JMM JLS. Performed the experiments: MLP CV MALP. Analyzed the data: MLP CV MALP. Contributed reagents/materials/analysis tools: MALP JMM JLS. Contributed to the writing of the manuscript: MLP CV MALP JMM JLS.

References

1. López M, Lelliott CJ, Vidal-Puig A (2007) Hypothalamic fatty acid metabolism: a housekeeping pathway that regulates food intake. BioEssays 29: 248–261.
2. Benoit SC, Kemp CJ, Elias CF, Abplanalp W, Herman JP, et al. (2009) Palmitic acid mediates hypothalamic insulin resistance by altering PKC-θ subcellular localization in rodents. J Clin Invest 119: 2577–2589.
3. Blouet C, Schwartz GJ (2010) Hypothalamic nutrient sensing in the control of energy homeostasis. Behav Brain Res 209: 1–12.
4. Diéguez C, Vazquez MJ, Romero A, López M, Nogueiras R (2011) Hypothalamic control of lipid metabolism: focus on leptin, ghrelin and melanocortins. Neuroendocrinology 94: 1–11.
5. Shearer KD, Silverstein J, Plisetskaya EM (1997) Role of adiposity in food intake control of juvenile chinook salmon (*Oncorhynchus tshawytscha*). Comp Biochem Physiol A 118: 1209–1215.
6. Silverstein JT, Shearer KD, Dickhoff WW, Plisetskaya EM (1999) Regulation of nutrient intake and energy balance in salmon. Aquaculture 177: 161–169.
7. Rasmussen RS, Ostenfeld TH, Roensholdt B, McLean E (2000) Manipulation of end-product quality of rainbow trout with finishing diets. Aquaculture Nutr 6: 17–23.
8. Gélineau A, Corraze G, Boujard T, Larroquet L, Kaushik S (2001) Relation between dietary lipid level and voluntary feed intake, growth, nutrient gain, lipid deposition and hepatic lipogenesis in rainbow trout. Reprod Nutr Dev 41: 487–503.
9. Johansen SJS, Ekli M, Jobling M (2002) Is there lipostatic regulation of feed intake in Atlantic salmon *Salmo salar* L.? Aquaculture Res 33: 515–524.
10. Johansen SJS, Sveier H, Jobling M (2003) Lipostatic regulation of feed intake in Atlantic salmon *Salmo salar* L. defending adiposity at the expense of growth? Aquaculture Res 34: 317–331.
11. Figueiredo-Silva AC, Kaushik S, Terrier F, Schrama JW, Médale F, et al. (2012) Link between lipid metabolism and voluntary food intake in rainbow trout fed coconut oil rich in medium-chain TAG. Brit J Nutr 107: 1714–1725.
12. FrØiland E, Jobling M, Björnsson BT, Kling P, Ravuri CS, et al. (2012) Seasonal appetite regulation in the anadromous Arctic charr: evidence for a role of adiposity in the regulation of appetite but not for leptin in signalling adiposity. Gen Comp Endocrinol 178: 330–337.
13. Soengas JL (2014) Contribution of glucose- and fatty acid sensing systems to the regulation of food intake in fish. A review. Gen Comp Endocrinol In press. http://dx.doi.org/10.1016/j.ygcen.2014.01.015.
14. Librán-Pérez M, Polakof S, López-Patiño MA, Míguez JM, Soengas JL (2012) Evidence of a metabolic fatty-acid sensing system in the hypothalamus and Brockmann bodies of rainbow trout: implications in food intake regulation. Am J Physiol Regul Integr Comp Physiol 302: R1340–R1350.
15. Librán-Pérez M, Otero-Rodiño C, López-Patiño MA, Míguez JM, Soengas JL (2014) Central administration of oleate or octanoate activates hypothalamic fatty acid sensing and inhibits food intake in rainbow trout. Physiol Behav 129: 272–279.
16. Librán-Pérez M, López-Patiño MA, Míguez JM, Soengas JL (2013) Oleic acid and octanoic acid sensing capacity in rainbow trout *Oncorhynchus mykiss* is direct in hypothalamus and Brockmann bodies. PLoS ONE 8: e59507.
17. Tang Z, Sun C, Yan A, Wu S, Qin C, et al. (2013) Genes involved in fatty acid metabolism: molecular characterization and hypothalamic mRNA response to energy status and neuropeptide Y treatment in the orange-spotted grouper *Epinephelus coioides*. Mol Cell Endocrinol 376: 114–124.
18. Marty N, Dallaporta M, Thorens B (2007) Brain glucose sensing, counteregulation, and energy homeostasis. Physiology 22: 241–251.
19. Polakof S, Mommsen TP, Soengas JL (2011). Glucosensing and glucose homeostasis: from fish to mammals. Comp Biochem Physiol B 160: 123–149.
20. Polakof S, Panserat S, Soengas JL, Moon TW (2012) Glucose metabolism in fish: a review. J Comp Physiol B 182: 1015–1045.
21. Clément L, Cruciani-Guglielmacci C, Magnan C, Vincent M, Douared L, et al. (2002) Intracerebroventricular infusion of a triglyceride emulsion leads to both altered insulin secretion and hepatic glucose production in rats. Pflügers Arch Eur J Physiol 445: 375–380.
22. Oh YT, Oh K-S, Kang I, Youn JH (2012) A fall in plasma free fatty acid (FFA) level activates the hypothalamic-pituitary-adrenal axis independent of plasma glucose: evidence for brain sensing of circulating FFA. Endocrinology 153: 3587–3592.
23. Oh YT, Kim J, Kang I, Youn JH (2014). Regulation of hypothalamic-pituitary-adrenal axis by circulating free fatty acids in male wistar rats: role of individual free fatty acids. Endocrinology 155: 923–931.
24. Schrama JW, Saravanan S, Geurden I, Heinsbroek LTN, Kaushik S, et al. (2012) Dietary nutrient composition affects digestible energy utilisation for growth: a study of Nile tilapia (*Oreochromis niloticus*) and a literature comparison across fish species. Brit J Nutr 108: 277–289.

25. Saravanan S, Geurden I, Figueiredo-Silva AC, Kaushik S, Verreth JAJ, et al. (2013) Voluntary feed intake in rainbow trout is regulated by diet-induced diferences in oxygen use. J. Nutr. 143, 781–787.
26. Jacobson KA, Gao Z-G (2006) Adenosine receptors as therapeutic targets. Nat Rev Drug Discov 5: 247–264.
27. Bernier NJ, Peter RE (2001) Appetite-suppressing effects of urotensin I and corticotropin-releasing hormone in goldfish (*Carassius auratus*). Neuroendocrinology 73: 248–260.
28. Dindia L, Faught E, Leonenko Z, Thomas R, Vijayan MM (2013) Rapid cortisol signaling in response to acute stress involves changes in plasma membrane order in rainbow trout liver. Am J Physiol Endocrinol Metab 304: E1157–E1166.
29. Cox BF, Perrone MH, Welzel GE, Greenland BD, Colussi DJ, et al. 1997. Cardiovascular and metabolic effects of adenosine A1-receptor agonists in streptozotocin-treated rats. J Cardiovasc Pharmacol 29: 417–426.
30. Milligan CL (2003) A regulatory role for cortisol in muscle glycogen metabolism in rainbow trout *Oncorhynchus mykiss* Walbaum. J Exp Biol 206: 3167–3173.
31. Polakof S, Míguez JM, Soengas JL (2008) Changes in food intake and glucosensing function of hypothalamus and hindbrain in rainbow trout subjected to hyperglycemic or hypoglycemic conditions. J Comp Physiol A 194: 829–839.
32. Polakof S, Míguez JM, Soengas JL (2008) Dietary carbohydrates induce changes in glucosensing capacity and food intake in rainbow trout. Am J Physiol Regul Integr Comp Physiol 295: R478–R489.
33. Conde-Sieira M, Alvarez R., López-Patiño MA, Míguez JM, Flik G, et al. (2013) ACTH-stimulated cortisol release from head kidney of rainbow trout is modulated by glucose concentration. J Exp Biol 216: 554–567.
34. De Pedro N, Pinillos ML, Valenciano AI, Alonso-Bedate M, Delgado MJ (1998) Inhibitory effect of serotonin on feeding behavior in goldfish: involvement of CRF. Peptides 19: 505–511.
35. Geslin M, Auperin B (2004) Relationship between changes in mRNAs of the genes encoding steroidogenic acute regulatory protein and P450 cholesterol side chain cleavage in head kidney and plasma levels of cortisol in response to different kinds of acute stress in the rainbow trout (*Oncorhynchus mykiss*). Gen. Comp. Endocrinol. 135, 70–80.
36. Ducasse-Cabanot S, Zambonino-Infante J, Richard N, Medale F, Corraze G, et al. (2007) Reduced lipid intake leads to changes in digestive enzymes in the intestine but has minor effects on key enzymes of hepatic intermediary metabolism in rainbow trout (*Oncorhynchus mykiss*). Animal 1: 1272–1282.
37. Geurden I, Aramendi M, Zambonino-Infante J, Panserat S (2007) Early feeding of carnivorous rainbow trout (*Oncorhynchus mykiss*) with a hyperglucidic diet during a short period: effect on dietary glucose utilization in juveniles. Am J Physiol Regul Integr Comp Physiol 292: R2275–R2283.
38. Kolditz C, Borthaire M, Richard N, Corraze G, Panserat S, et al. (2008) Liver and muscle metabolic changes induced by dietary energy content and genetic selection in rainbow trout (*Oncorhynchus mykiss*). Am J Physiol Regul Integr Comp Physiol 294: R1154–R1164.
39. Lansard M, Panserat S, Seiliez I, Polakof S, Plagnes-Juan E, et al. (2009) Hepatic protein kinase B (Akt)-target of rapamycin (TOR)-signalling pathways and intermediary metabolism in rainbow trout (*Oncorhynchus mykiss*) are not significantly affected by feeding plant-based diets. Brit J Nutr 102: 1564–1573.
40. Cruz-Garcia L, Minghetti M, Navarro I, Tocher DR (2009) Molecular cloning, tissue expression and regulation of liver X receptor (LXR) transcription factors of Atlantic salmon (*Salmo salar*) and rainbow trout (*Oncorhynchus mykiss*). Comp Biochem Physiol B 153: 81–88.
41. Conde-Sieira M, Agulleiro MJ, Aguilar AJ, Míguez JM, Cerdá-Reverter JM, et al. (2010) Effect of different glycaemic conditions on gene expression of neuropeptides involved in control of food intake in rainbow trout; interaction with stress. J Exp Biol 213: 3858–3865.
42. Polakof S, Médale F, Skiba-Cassy S, Corraze G, Panserat S (2010) Molecular regulation of lipid metabolism in liver and muscle of rainbow trout subjected to acute and chronic insulin treatments. Domestic Anim Endocrinol 39: 26–33.
43. Polakof S, Médale F, Larroquet L, Vachot C, Corraze G, et al. (2011) Insulin stimulates lipogenesis and attenuates beta-oxidation in white adipose tissue of fed rainbow trout. Lipids 46: 189–199.
44. Figueiredo-Silva AC, Saravanan S, Schrama JW, Kaushik S, Geurden I (2012) Macronutrient-induced differences in food intake relate with hepatic oxidative metabolism and hypothalamic regulatory neuropeptides in rainbow trout (*Oncorhynchus mykiss*). Physiol Behav 106: 499–505.
45. Librán-Pérez M, Figueiredo-Silva AC, Panserat S, Geurden I, Míguez JM, et al. (2013). Response of hepatic lipid and glucose metabolism to a mixture or single fatty acids: possible presence of fatty acid-sensing mechanisms. Comp Biochem Physiol A 164: 241–248.
46. Pfaffl MW (2001) A new mathematical model for relative quantification in real-time RT-PCR. Nucleic Acids Res. 29: e45.
47. Kashiwagi A (1995) Rationale and hurdles of inhibitors of hepatic gluconeogenesis in treatment of diabetes mellitus. Diabetes Res Clin Pract 28: S195–S200.
48. Velasco-Santamaría YM, Korsgaard B, Madsen SS, Bjerregaard P (2011) Bezafibrate, a lipid-lowering pharmaceutical, a potential endocrine disruptor in male zebrafish (*Danio rerio*). Aquat Toxicol 105: 107–118.
49. Martinez-Rubio L, Wadsworth S, González Vecino JL, Bell JG, Tocher DR (2013) Effect of dietary digestible energy content on expression of genes of lipid metabolism and LC-PUFA biosynthesis in liver of Atlantic salmon (*Salmo salar* L.). Aquaculture 384–387: 94–103.
50. Kreitschmann-Andermahr I, Suarez P, Jennings R, Evers N, Braban G (2010) GH/IGF-I regulation in obesity-mechanisms and practical consequences in children and adults. Horm Res Paediatr 73: 153–160.
51. Watt MJ, Holmes AG, Steinberg GR, Mesa JL, Kemp BE, et al. (2004) Reduced plasma FFA availability increases net triacyl-glycerol degradation, but not GPAT or HSL activity, in human skeletal muscle. Am J Physiol Endocrinol Metab 287: E120–E127.
52. Quabbe HJ, Lyucks AS, Lage M, Schwartz C (1983) Growth hormone, cortisol, and glucagon concentrations during plasma free fatty acid depression: different effects of nicotininc acid and an adenosine derivative (BM 11.189). J Clin Endocrinol Metab 57: 410–414.
53. Aluru N, Vijayan MM (2006) Aryl hydrocarbon receptor activation impairs cortisol response to stress in rainbow trout by disrupting the rate-limiting steps in steroidogenesis. Endocrinology 147: 1895–1903.
54. Alderman SL, McGuire A, Bernier NJ, Vijayan MM (2012) Central and peripheral glucocorticoid receptors are involved in the plasma cortisol response to an acute stressor in rainbow trout. Gen Comp Endocrinol 176: 79–85.
55. Jeffrey JD, Gollock MJ, Gilmour KM (2014) Social stress modulates the cortisol response to an acute stressor in rainbow trout (*Oncorhynchus mykiss*). Gen Comp Endocrinol 196: 8–16.
56. Wendelaar Bonga SE (1997) The stress response in fish. Physiol Rev 77: 591–625.
57. Leach GJ, Taylor MH (1980) The role of cortisol in stress-induced metabolic changes in *Fundulus heteroclitus*. Gen Comp Endocrinol 42: 219–227.
58. Tripathi G, Verma P (2003) Pathway-specific response to cortisol in the metabolism of catfish. Comp Biochem Physiol B 136: 463–471.
59. Doyon C, Leclair J, Trudeau VL, Moon TW (2006) Corticotropin-releasing factor and neuropeptide Y mRNA levels are modified by glucocorticoids in rainbow trout, *Oncorhynchus mykiss*. Gen Comp Endocrinol 146: 126–135.
60. Conde-Sieira M, Librán-Pérez M, López-Patiño MA, Míguez JM, Soengas JL (2011) CRF treatment induces a readjustment in glucosensing capacity in the hypothalamus and hindbrain of rainbow trout. J. Exp. Biol. 214, 3887–3894.
61. Bernier NJ, Craig PM (2005) CRF-related peptides contribute to stress response and regulation of appetite in hypoxic rainbow trout. Am J Physiol Regul Integr Comp Physiol 289: R982–R990.
62. Bernier NJ, Alderman SL, Bristow EN (2008) Heads or tails? Stressor-specific expression of corticotropin-releasing factor and urotensin I in the preoptic area and caudal neurosecretory system of rainbow trout. J Endocrinol 196: 637–648.
63. Harmon JS, Rieniets LM, Sheridan MA (1993) Glucagon and insulin regulate lipolysis in trout liver by altering phosphorylation in triacylglycerol lipase. Am J Physiol Regul Integr Comp Physiol 265: R255–R260.
64. O'Connor PK, Reich B, Sheridan MA (1993) Growth hormone stimulates hepatic lipid mobilization in rainbow trout, *Oncorhynchus mykiss*. J Comp Physiol B 163: 427–431.
65. Fabbri E, Capuzzo A, Moon TW (1998) The role of circulating catecholamines in the regulation of fish metabolism: An overview. Comp Biochem Physiol C 120: 177–192.
66. Albalat A, Gómez-Requeni P, Rojas P, Médale F, Kaushik S, et al. (2005) Nutritional and hormonal control of lipolysis in isolated gilthead seabream (*Sparus aurata*) adipocytes. Am J Physiol Regul Integr Comp Physiol 289: R259–R265.
67. Sangiao-Alvarellos S, Arjona FJ, Míguez JM, Martin del Rio MP, Soengas JL, et al. (2006) Growth hormone and prolactin actions on osmoregulation and energy metabolism of gilthead sea bream (*Sparus auratus*). Comp Biochem Physiol A 144: 491–500.

16

Proxy Measures of Fitness Suggest Coastal Fish Farms Can Act as Population Sources and Not Ecological Traps for Wild Gadoid Fish

Tim Dempster[1,2]*, Pablo Sanchez-Jerez[3], Damian Fernandez-Jover[3], Just Bayle-Sempere[3], Rune Nilsen[4], Pal-Arne Bjørn[4], Ingebrigt Uglem[5]

1 Department of Zoology, University of Melbourne, Parkville, Victoria, Australia, **2** SINTEF Fisheries and Aquaculture, Trondheim, Norway, **3** Department of Marine Sciences and Applied Biology, University of Alicante, Alicante, Spain, **4** NOFIMA, Tromsø, Norway, **5** Norwegian Institute for Nature Research, Trondheim, Norway

Abstract

Background: Ecological traps form when artificial structures are added to natural habitats and induce mismatches between habitat preferences and fitness consequences. Their existence in terrestrial systems has been documented, yet little evidence suggests they occur in marine environments. Coastal fish farms are widespread artificial structures in coastal ecosystems and are highly attractive to wild fish.

Methodology/Principal Findings: To investigate if coastal salmon farms act as ecological traps for wild Atlantic cod (*Gadus morhua*) and saithe (*Pollachius virens*), we compared proxy measures of fitness between farm-associated fish and control fish caught distant from farms in nine locations throughout coastal Norway, the largest coastal fish farming industry in the world. Farms modified wild fish diets in both quality and quantity, thereby providing farm-associated wild fish with a strong trophic subsidy. This translated to greater somatic (saithe: 1.06–1.12 times; cod: 1.06–1.11 times) and liver condition indices (saithe: 1.4–1.8 times; cod: 2.0–2.8 times) than control fish caught distant from farms. Parasite loads of farm-associated wild fish were modified from control fish, with increased external and decreased internal parasites, however the strong effect of the trophic subsidy overrode any effects of altered loads upon condition.

Conclusions and Significance: Proxy measures of fitness provided no evidence that salmon farms function as ecological traps for wild fish. We suggest fish farms may act as population sources for wild fish, provided they are protected from fishing while resident at farms to allow their increased condition to manifest as greater reproductive output.

Editor: Yan Ropert-Coudert, Institut Pluridisciplinaire Hubert Curien, France

Funding: Funding was provided by the Norwegian Research Council Havet og kysten program to the CoastACE project (no: 173384). The funders had no role in study design, data collection and analysis, decision to publish, or preparation of the manuscript.

Competing Interests: SINTEF Fisheries and Aquaculture is a private, not-for-profit research institute. It is not a commercial funder of this research. The authors have declared that no competing interests exist.

* E-mail: tim.dempster@sintef.no

Introduction

An ecological trap arises when an artificial habitat is introduced into a natural environment, attracts animals to its vicinity and the subsequent association leads to negative ecological consequences for the animal [1]. Animals may prefer an artificial habitat over natural habitats if it mimics the set of ecological cues which signify a good quality habitat, despite other ecological processes rendering the habitat of low quality and leading to poorer reproduction or survival. Robertson and Hutto [2] suggest that ecological traps derive from habitat alteration that operates in one of three ways; (1) increasing the attractiveness of an environment by enhancing the set of cues that animals recognise as attractive; (2) decreasing the suitability of a habitat; or (3) doing both (1) and (2) simultaneously. Alternatively, artificial habitats of high quality, where individuals increase in condition, reproduce better or have improved survival, all of which may ultimately lead to positive population growth rates, act as population sources.

Objective testing of whether ecological traps exist is well embedded in the literature concerning terrestrial systems [2], yet few studies have investigated whether they exist in marine environments. Artificial structures that aggregate fish (fish aggregation devices; FADs) have been previously suggested to act as ecological traps by acting as a super-stimulus and misleading fish to make inappropriate habitat selections [3]. Coastal sea-cage fish farms are widespread artificial structures in coastal waters, producing over 2.5 million tons of fish each year [4]. They have previously been described as analogous to FADs, attracting and aggregating large assemblages of wild fish in their immediate vicinity [5]. Attraction and aggregation of tons of wild fish to the immediate surrounds of Norway's coastal salmon farms [5,6] meets Robertson and Hutto's [2] first condition for the formation of an ecological trap. However, whether the fish farm area is poorer in habitat quality for wild fish than natural adjacent habitats, thus meeting Robertson and Hutto's [2] second condition, remains unknown. Relative habitat quality is a key component in determining the extent to which fish farms may act as population sources or ecological traps for wild fish.

Along the Norwegian coastline, 1198 coastal sea-cage salmonid farm concessions used 1.2 million tons of fish food to produce 829 000 t in 2008 [7]. Farming is concentrated in particular fjords, with farms spaced several kilometres apart. Wild saithe are the most abundant species associated with salmon farms within fjord systems [6,8]. Saithe use farms as a loose network of preferred habitats, moving repeatedly among farms and remaining resident at specific farms for weeks to months [9]. Atlantic cod are also attracted to fish farms in number [6] and may reside in their vicinity for months at a time [10,11]. Attraction of wild fish to salmon farms is likely to have a range of fitness consequences due to the modified environment fish farms induce, both in the altered trophic network around farms and the close proximity of hundreds of thousands to millions of farmed salmonids. Diet, body fat content, fatty acid composition and parasite loads may all be altered when wild fish closely associate with farms [12,13,14]. Simultaneous analysis of this suite of factors at an extensive number of locations is required to resolve whether farms function as population sources or ecological traps [15].

Here, we tested the hypotheses that the diets, indices of condition and parasite loads of cod and saithe associated with salmon farms differed from those of fish present at locations distant from salmon farms. To ensure broad generality of the results, we sampled fish in three intensive fish farming areas along the latitudinal extent of salmon farming in Norway (59°N to 70°N).

Materials and Methods

Study locations and experimental design

Saithe and cod were sampled from the three salmon farming areas (Ryfylke 59°N, Hitra 63°N and Øksfjord 70°N) from the same Atlantic salmon (*Salmo salar*) farms and during the same season (summer) as aggregation sizes were determined [6]. Within each salmon farming area, fish were sampled at three farms and two to six non-farm control locations (Fig. 1). Farm-associated fish were captured within 5 m of cages containing salmon. The number of non-farm locations varied from two to six depending on the area and species of wild fish sampled (Saithe: Ryfylke 2, Hitra 4, Øksfjord 3; Cod: Ryfylke 3, Hitra 6, Øksfjord 3). Control fish were sampled from locations 4 to 20 km distant from the nearest farm (Fig. 1) to limit the possibility of sampling fish at non-farm locations that had interacted recently with a farm. The 4 km minimum limit was based on telemetry-derived observations of the predominant movements of wild cod and wild saithe [9,10,11] in the vicinity of fish farms.

All fish were sampled with standardised hook and line fishing gear. Collections by hook and line select for feeding fish, but are more suitable for accurate counts of the number of external parasites than other catch methods such as trawling or gill nets which may remove external parasites through abrasion. Moreover, capture by any other method beside the cages at fish farms is impractical due to possible negative interactions of fishing gear with fish farming structures. Collections were made at each location from June to September 2007 during the period where feed input to salmon farms is high [7].

Size, diet and condition indices

Upon capture, fish were immediately examined for the presence of external parasites (see parasite sampling section below) and then placed on ice. Fish were weighed and measured to the nearest 0.5 cm (fork length; FL). Each fish was dissected and liver and gonad weights were obtained. Sex for each fish was determined by macroscopic examination of the gonads. In gadoid species, such as Atlantic cod, lipids are stored primarily in the liver [16] making liver weight a measure of spawner quality [17]. Therefore, we calculated three condition indices: body condition, the hepatosomatic index and the gonadosomatic index. Fulton's condition index (FCI) was calculated with the formula: $FCI = (W/FL^3) \times 100$, where W = wet weight–stomach content weight and FL = fork length (cm). The hepatosomatic index (HSI) was calculated using the formula: $HSI = (LW/W) \times 100$, where LW = liver weight and W = wet weight–stomach content weight. The gonadosomatic index (GSI) was calculated using the formula: $GSI = (GW/W) \times 100$, where GW = gonad weight and W = wet weight–stomach content weight.

Stomach contents from the foregut were examined and prey species were identified to the lowest taxonomic level possible and weighed. Prey categories were later reduced to 11 for saithe (waste salmon feed, Brachyura, Osteichthyes, Polychaeta, Caridea, zooplankton, Phaeophyceae, Bivalvia (principally *Mytilus* sp.), Ophiuridae, Hydroida (principally *Ectopleura larynx*), and other organic matter) and 13 for cod (waste salmon feed, Brachyura, Osteichthyes, Polychaeta, Caridea, Phaeophyceae, Bivalvia (*Mytilus* sp.), Holothuria, Ophiuridae, Echinoidea, Octopoda, Amphipoda and other organic matter).

Parasite sampling

Fish were examined to estimate the incidence of parasites that may have occurred in increased incidence around fish farms through direct transfer from the farmed salmonids (e.g. mobile sea lice) or through indirect means, such as the modified farm environment increasing the density of con-specific fish or the pool of intermediate hosts available to these parasites, thus increasing their incidence. Immediately upon capture, saithe and cod were examined for the incidence of mobile sea lice (*Caligus* spp.) and attached parasitic copepods (*Clavella* sp.) on all external surfaces, and inside the mouth and gills. In August, 100 mobile sea lice from un-associated (hereafter UA) and farm-associated (hereafter FA) fish were collected in all salmon farming areas to identify the species composition of mobile sea lice. We hypothesised that FA cod and saithe would have elevated levels of *Caligus* compared to UA fish either through direct transfer of adult *Caligus* from caged salmon or elevated levels of *Caligus* larvae in the waters surrounding farms.

Gills of cod were examined for the presence and abundance of *Lernaeocera branchialis*, a copepod parasite of cod which invasively attaches to the gills and feeds on blood [18]. For *Clavella* sp. and *L. branchialis*, we hypothesised that no differences in infestation levels would be detected between FA and UA fish, as no direct transfer route between salmon farms and wild fish has been established for these parasites.

Livers were dissected from both species of fish and inspected for the third stage (L3) larvae of the parasitic nematode *Anisakis simplex* [19]. Infection intensity was scored on a semi-quantitative scale form 0 to 3: 0 = *A. simplex* absent; 1 = mild infestation; 2 = moderate infestation; and 3 = heavy infestation. We hypothesised that L3 larvae of *A. simplex* would be less abundant in FA than UA fish as high consumption of lost feed at farms would mean lower consumption of natural prey items such as crustaceans, squid and fish, which may contain L3 larvae.

Statistical analyses

As gonadal development was minimal during the non-spawning season sampling period and diets in the non-spawning season are not known to vary among male and female cod and saithe, we pooled the sexes for dietary analyses and analyses of condition.

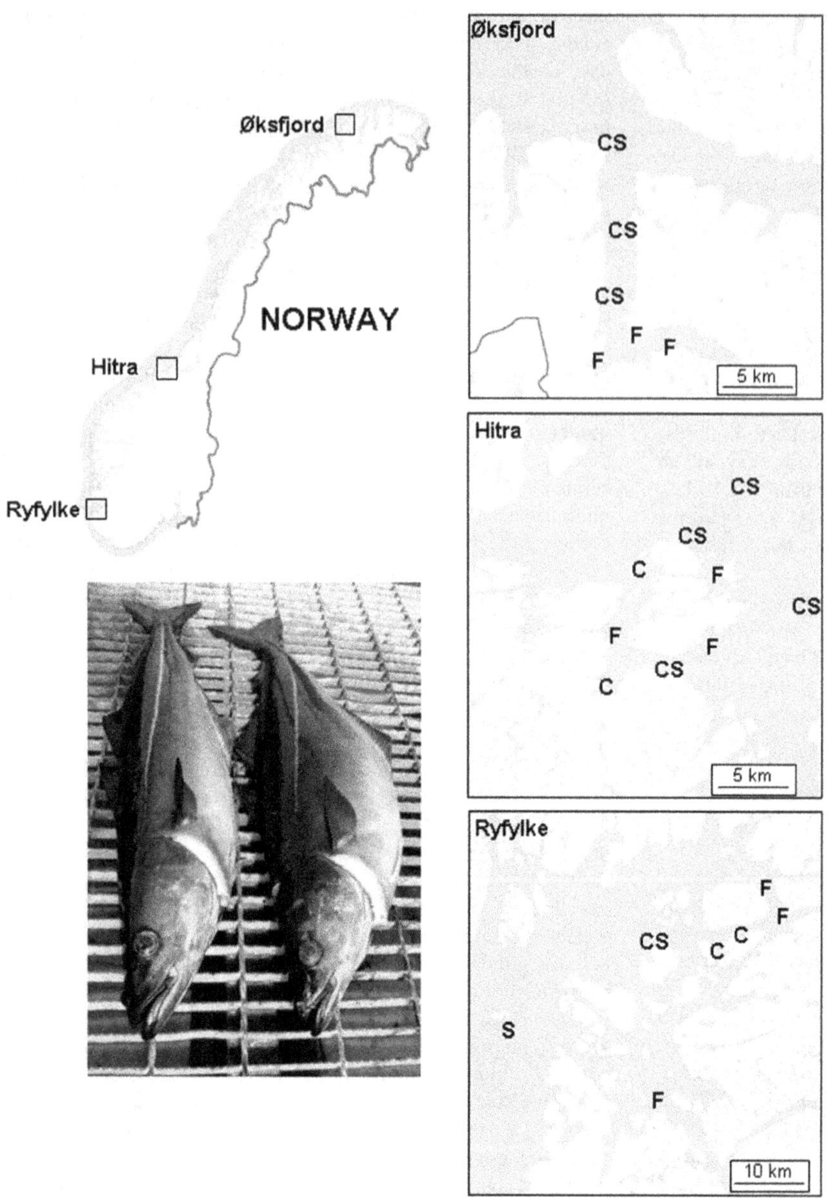

Figure 1. Map of the study locations in the three Norwegian salmon farming areas of Ryfylke, Hitra and Øksfjord. (F) = salmon farm sampling location for both saithe *Pollachius virens* and Atlantic cod *Gadus morhua*; (S) = non-farm sampling location for saithe; (C) = non-farm sampling location for Atlantic cod. The picture shows an un-associated (left) and farm-associated (right) saithe of similar length but distinctly different morphology sampled from Hitra.

Further, as differences in the incidence of parasites among male and female gadoids have rarely been found [20], and no differences are known for the parasite species investigated here, we pooled the sexes for parasite analyses.

Non-parametric multivariate techniques were used to compare dietary compositions among farm and non-farm locations. All multivariate analyses were performed using the PRIMER statistical package. Prior to calculating the Bray-Curtis similarity matrices, the dietary data were pooled across all individuals sampled within each location and month by summing the total weights of prey items within each prey category to reduce the stress of MDS representation. Fourth root transformations were made to weigh the contributions of common and rare dietary categories in the similarity coefficient. Non-metric multidimensional scaling (nMDS) was used as the ordination method. Variables that had more influence on similarities within groups and dissimilarities among groups of locations or depths, determined by ANOSIM (analysis of similarity), were calculated using the SIMPER (similarity percentages) procedure. The ANOSIM permutation test was used to assess the significance of differences among farm and non-farm locations. As diets of both saithe and cod at farms contained feed pellets, we repeated all analyses with this prey category removed to determine if differences in diet among farm and non-farm locations remained significant.

To test for differences in fish size (fork length; FL), stomach content weight, FCI, HSI, GSI and the incidence of the various parasites among farm and non-farm locations in each of the three fish farming areas, we used Generalized Linear Models (GLMs).

Prior to the GLMs, heterogeneity of variance was tested with Cochran's *C*-test. Data were ln(*x*+1) transformed if variances were significantly different at $p = 0.05$. Comparisons across fish in all size classes were made in each of the three farming areas for cod. To ensure that any differences detected in comparisons were not related to the different sizes of fish in the FA and UA treatments, we used FL as a co-variate in analyses of stomach content weight, condition and parasite loads. For saithe, as HSIs>10% are indicative of a waste feed dominated diet for several months and wild saithe fed solely on natural diets do not have HSIs>10% [21], we tested if the incidence of the various parasites differed among FA fish with HSIs>10%, FA fish with HSIs<10%, and UA fish with HSIs<10%. To detect if the parasite loads we detected were related to the body condition (FCI) of wild fish, we applied multiple regression analysis for both cod and saithe.

Results

Size structures of farm-associated and un-associated fish

In total, 355 FA and 215 UN saithe were captured at sizes ranging from 21.5–108.5 cm fork length (FL) and weights from 0.1–12.5 kg. 171 FA and 178 UA cod were collected at sizes ranging from 28.5–121.0 cm FL and weights from 0.23–18.0 kg. Saithe were captured at all farms, while cod were only available at 8 of the 9 farms (all except one farm at Hitra). Significant differences were detected in mean fork lengths among UA and FA groups for both species in all three farming areas (Table 1). FA saithe were larger than UA saithe at two of the three farming areas (Hitra and Øksfjord), but significantly smaller at Ryfylke. FA cod were significantly larger than UA cod in Ryfylke and Hitra but not Øksfjord, and the magnitude of the difference varied greatly among the areas.

Diets of farm-associated and un-associated fish

Saithe captured from non-farm locations had a higher proportion of empty stomachs (31%) and lower average stomach content weight (8.6 g) compared to FA saithe (16%, 20.2 g). For cod, both FA (18%) and UA fish (19%) had similar proportions of empty stomachs, although stomach content weight was higher in FA (32.9 g) than UA fish (23.2 g). 44.3% of saithe and 20% of cod captured around farms had waste feed in their stomachs. Overall, waste feed accounted for 71% (14.2 g) and 25% (8.3 g) of the diet by weight of FA saithe and cod, respectively.

Table 1. Mean sizes of samples of saithe (*Pollachius virens*) and Atlantic cod (*Gadus morhua*) used to compare diet, condition and parasite loads across farm-associated (FA) and farm unassociated (UA) locations in each of the three Norwegian salmon farming areas.

	FA/UA	Ryfylke		Hitra		Øksfjord	
		n	FL (cm)	n	FL (cm)	n	FL (cm)
P. virens	FA	97	50.1±0.7[b]	148	40.2±1.2[a]	110	46.2±0.8[a]
	UA	30	54.2±0.9[a]	88	34.3±1.1[b]	97	43.4±0.8[b]
G. morhua	FA	13	63.3±4.5[a]	89	52.8±1.6[a]	65	62.3±2.5
	UA	12	46.7±4.4[b]	75	45.7±1.9[b]	91	58.5±1.5

Superscripts ([a,b]) indicate a significant difference was detected between the FA and UA groups at $p<0.05$.

The 2-dimensional nMDS plot based on weights of prey groups by location and month revealed clear separation of the diets of FA and UA fish for both saithe (Fig. 2a) and cod (Fig. 2b). ANOSIM indicated that differences in diets between FA and UA fish were significant (saithe: $R_{global} = 0.69$, $p = 0.001$; cod: $R_{global} = 0.45$, $p = 0.003$). When pellets were removed from the analysis, differences in diets between FA and UA fish remained significant (saithe: $R_{global} = 0.52$, $p = 0.01$; cod: $R_{global} = 0.38$, $p = 0.02$).

Diets of FA saithe clustered together, regardless of sampling location and month, while diets of UA saithe were more variable (Fig. 2a). UA saithe diets were characterised by similar weights of relatively few dietary items. Over 70% of group similarity was accounted for by fish (41.5%), zooplankton (16.8%), crustaceans (8.0%) and ophiuroids (4.5%). Over 80% of similarity in FA saithe diets was due to waste feed (45.7%), fish (14.8%), mussels (10.5%) and zooplankton (10.2%). Dissimilarities in diets between UA and FA saithe were due to large differences in the abundance of a few of the major items (waste feed 32.3% F>C, fish 14.3% C>F, zooplankton 10.3% F>C and mussels 9.9% F>C).

Similarities in UA cod diets were predominantly due to similar weights of fish (39.7%), crabs (24.3%), ophiuroids (9.7%) and crustaceans (6.5%) while similarities in FA cod diets were predominantly due to fish (37.6%), polychaetes (19.6%), pellets (14.6%) and crabs (9.6%). FA cod consumed more waste feed, polychaetes and fish (dissimilarities of 18.9%, 12.1% and 7.8%, respectively) while UA cod consumed more Ophiuridae, crabs and mussels (dissimilarities of 11.5%, 9.9% and 7.8%, respectively).

Body, liver and gonad condition of farm-associated and un-associated fish

FA saithe had significantly higher average FCIs (1.06-1-12 times) than UA fish in all three farming areas (Fig. 1, Fig. 3a). Average HSIs were significantly higher in saithe (1.4–1.8 times) collected around farms compared to UA fish at Hitra and Øksfjord (Fig. 3b). No difference was detected for FA and UA fish sampled from Ryfylke. As the June-September sampling period occurred after the main spawning period for saithe and many of the individuals sampled were less than 2 kg in size and thus likely to be immature, no difference was detected in average GSIs among farm and UA fish in any of the three areas (Fig. 3c).

FCIs, HSIs and GSIs of cod were clearly affected by association with salmon farms. FCIs were consistently 1.06–1.11 times greater in FA than UA cod in all three areas (Fig. 3d). Similarly, average HSIs varied among the three areas, but were consistently 2.0–2.8 times greater in cod collected around farms compared to UA fish (Fig. 3e). In contrast to saithe, where average GSIs in FA and UA fish were similar, average GSIs in cod were significantly greater (1.7–4.8 times) in FA than UA cod in all three areas, despite the timing of sampling in the post-spawning period (Fig. 3f).

Parasite loads of farm-associated and un-associated fish

Significant differences in the abundances of parasites were detected in both directions, with FA or UA fish having greater levels of particular parasites in certain fish farming areas. From the collections in August, two species of mobile sea-lice were identified on both saithe and cod in all three areas: *Caligus elongatus* and *C. curtus*. Significantly higher numbers of sea lice occurred on FA saithe with HSIs>10 or <10 compared to UA fish with HSIs<10 at Ryfylke (2.5 to 3.5 times) and Hitra (3.1 to 3.7 times), but not at Øksfjord (Fig. 4). *Clavella* sp. abundances were significantly higher in FA saithe with HSIs>10 or <10 compared to UA fish at Hitra (1.8 to 2.1 times). FA saithe with

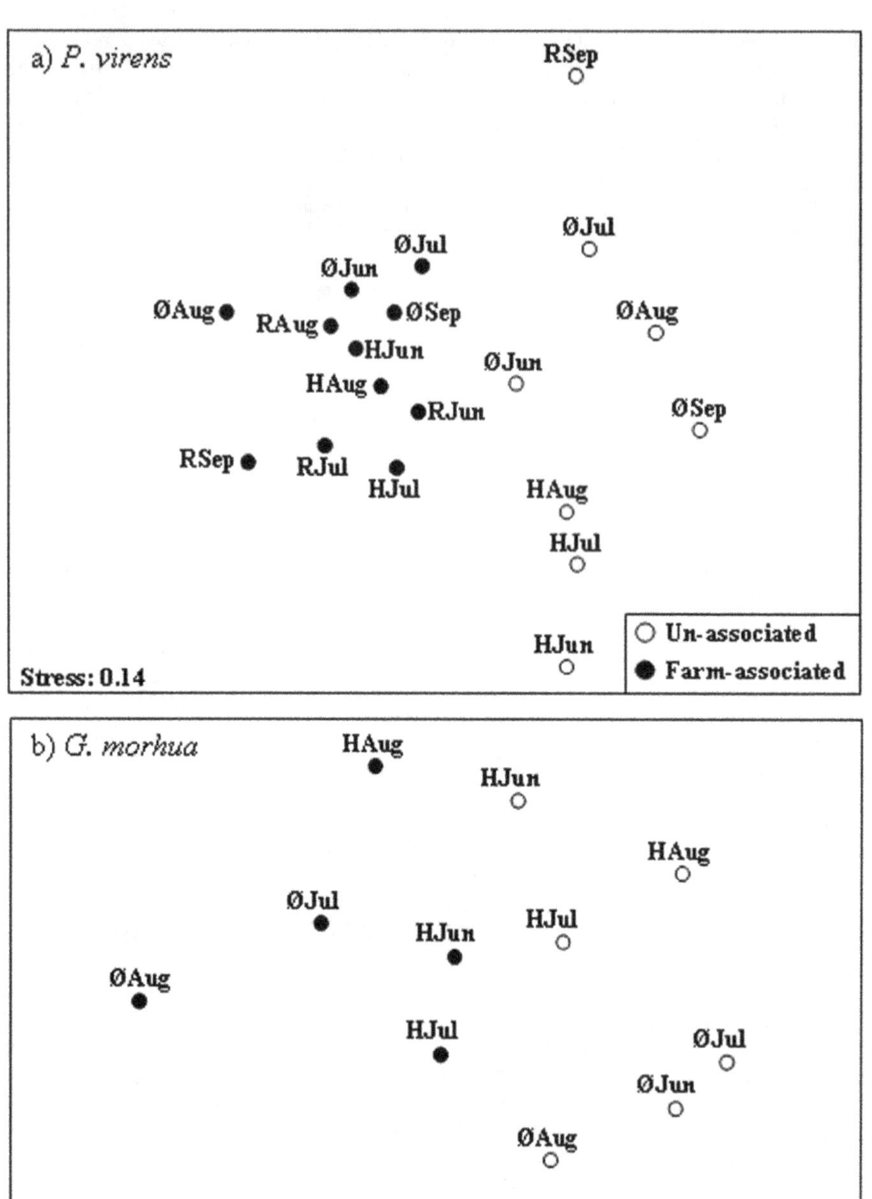

Figure 2. Non-metric multi-dimensional scaling plots of dietary items of saithe *Pollachius virens* and Atlantic cod *Gadus morhua* sampled from farm and non-farm locations throughout Norway from June to September. a: saithe; b: Atlantic cod. Each point is based on mean weights of prey categories for the specific month. Jun = June; Jul = July; Aug = August; Sep = September. R = Ryfylke; H = Hitra; Ø = Øksfjord.

HSIs>10 had 2.6 to 3.6 greater abundances of *Clavella* sp. than both FA saithe with HSIs<10 and UA fish in Øksfjord. No differences in *Clavella* sp. abundance among the three groups were detected at Ryfylke. For the *Anisakis simplex* index, FA saithe had consistently lower values that UA saithe across the three locations. FA saithe with HSIs>10 had 1.6 to 2.1 times lower *A. simplex* infestations than FA saithe with HSIs<10 and UA fish.

Caligus spp. occurred in abundances 2.4 times higher on FA cod at Øksfjord compared to UA cod, whereas no significant differences between FA and UA cod were detected at Ryfylke and Hitra (Fig. 5). No significant differences were detected for *Clavella* sp. or *Anisakis simplex* L3 larvae between farm-associated and UA fish in any of the three areas. The gill parasite *Lernaeocera branchialis* occurred in significantly higher abundance (2.8 times) in UA cod than FA cod in Øksfjord, with no difference detected in the other two areas.

Multiple regression analysis of parasite loads versus body condition revealed that none on the four species of parasites investigated for cod were significantly related to FCI ($F = 1.12$, $p = 0.35$; $R^2 = 0.02$; *Caligus* spp.: $p = 0.11$; *Clavella* sp.: $p = 0.17$; *L. branchialis*: $p = 0.86$; A. *simplex*: $p = 0.49$). For saithe, the multiple regression was significant ($F = 9.7$, $p < 0.001$; $R^2 = 0.05$), with *Clavella* spp. positively related to FCI ($p = 0.003$), *Anisakis* sp.

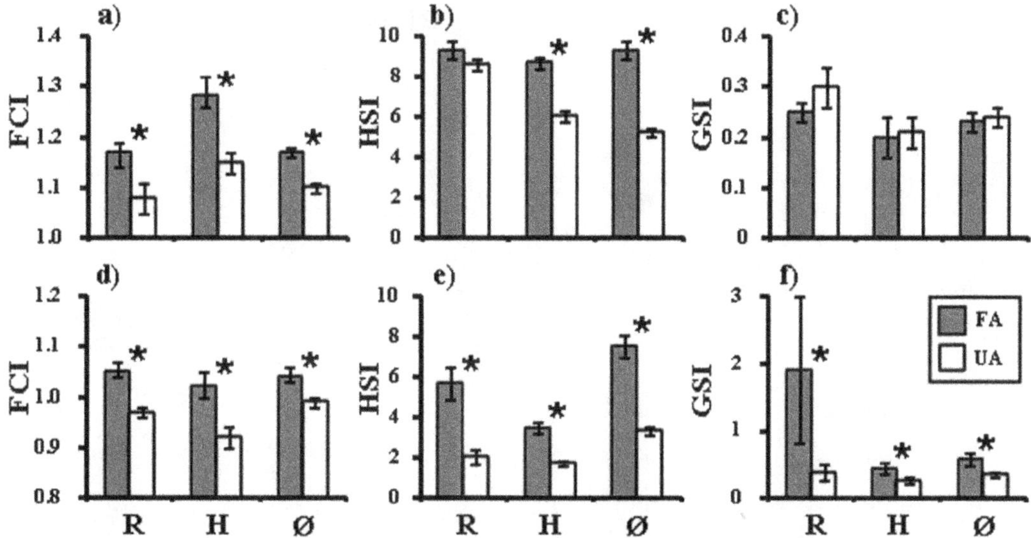

Figure 3. Condition indices of farm-associated (FA) and un-associated (UA) saithe *Pollachius virens* and Atlantic cod *Gadus morhua* in each of the three intensive fish farming areas. a, b, c: saithe; d, e, f: Atlantic cod. R = Ryfylke; H = Hitra; Ø = Øksfjord. FCI = Fulton's condition index; HSI = Hepatosomatic index; GSI = Gonadosomatic index. * indicates a significant difference at $p<0.05$ was detected among the groups.

strongly negatively related to FCI ($p<0.001$), and no relationship evident for mobile sea lice (*Caligus* spp.: $p = 0.59$).

Discussion

Proxy measures of fitness of farm-associated and un-associated fish

We have demonstrated that proxy measures of fitness (FCI, HSI, abundances of specific parasites and diet) of wild saithe and cod caught in close association with salmon farms differ significantly from their counterparts captured distant from farms. These effects are likely to be general across the spatial extent of salmon farming in Norway (59°N–70N°) and apply to a substantial pool of fish aggregated around farms. Dempster et al. [6] conservatively estimated that over 12000 tons of wild fish, principally saithe and cod, were aggregated at Norway's 1198 salmon farms on any given day in summer based on video-derived estimates of aggregations at the same 9 farms investigated here. Conclusions derived from this study are therefore based upon these abundance estimates, but are limited to the summer months during which samples were taken.

Salmon farms clearly increased the amount of food consumed by closely associated saithe and cod, indicating a strong trophic link between farms and wild fish. Stomachs of FA saithe contained more than twice the amount of food by weight than UA fish with stomach content weight similarly elevated in FA cod (1.4 times). Food pellets are high in fish proteins and oils and thus provide a high energy source of feed [22], although with distinctly different fatty acid distributions from natural diets [13]. While waste feed dominated diets of FA saithe and cod, the composition of dietary items still differed among FA and UA fish when waste pellets were removed from analyses, indicating that the availability of other types of prey differed between farm and non-farm locations. Salmon farms are known to have modified meio- and macro-fauna communities [23] and modified fish assemblages [6] compared to control locations, which likely contributed to the dietary differences.

The increased body and liver condition observed in FA saithe and cod is likely linked to the trophic subsidy that farms provide. Livers are the principal lipid and thus energy stores in gadoids [16]. High HSIs are indicative of high total lipid energy, which is known as a direct proxy to egg production in gadoid fish [17]. Moreover, lipid energy reserves 3–4 months prior to spawning are the best proxy for fecundity [24]. In this context, association with fish farms throughout summer and autumn could increase the fecundity of saithe and cod, which spawn in early spring, even if these fish migrate away from farms months prior to spawning.

While fecundity, in terms of egg numbers or size, may increase through FA fish having high energy reserves, the composition of stored lipids in FA saithe and cod may differ from those of UA fish which consume a natural diet (Fernandez-Jover et al. unpubl. data). This may effect egg quality as farm-feeds contain low proportions of highly unsaturated fatty acids (HUFAs) and arachidonic acids, which are key to fertilization rates and egg quality [25]. If the waste-feed dominated diet alters the fatty acid composition of saithe and cod livers and has a negative effect upon egg quality during vitellogenesis, the increased condition evident in FA fish may not translate to a proportional increase in spawning success. Experimental manipulations of wild saithe and cod fed diets containing different proportions of waste feed for various durations and the subsequent evaluation of the effect this has on egg and larval quality are required to determine the extent of this potentially negative effect.

Some parasites were found in elevated abundances in FA fish. We hypothesised that mobile sea-lice would occur in higher abundances on FA fish due to direct transfer or greater infestation levels as larvae occurred in greater abundance. This was the case for saithe at Hitra and Ryfylke and cod in Øksfjord. Similarly the attached copepod *Clavella* sp. was detected in elevated abundances in FA saithe at Hitra and Øksfjord. In contrast to the mobile sea-lice and *Clavella* sp. loads, the gill parasite *Lernaeocera branchialis* and the internal parasite *Anisakis* simplex were only ever detected in lower levels in FA fish. For *L. branchialis*, significant differences between FA and UA fish were only detected in Øksfjord, where UA cod had higher levels. Significantly lower *A. simplex* infections occurred in FA saithe with HSIs>10 in all three farming areas, suggesting that the longer-term residence at salmon farms required to generate an HSI>10 [21] plays an important role in reducing

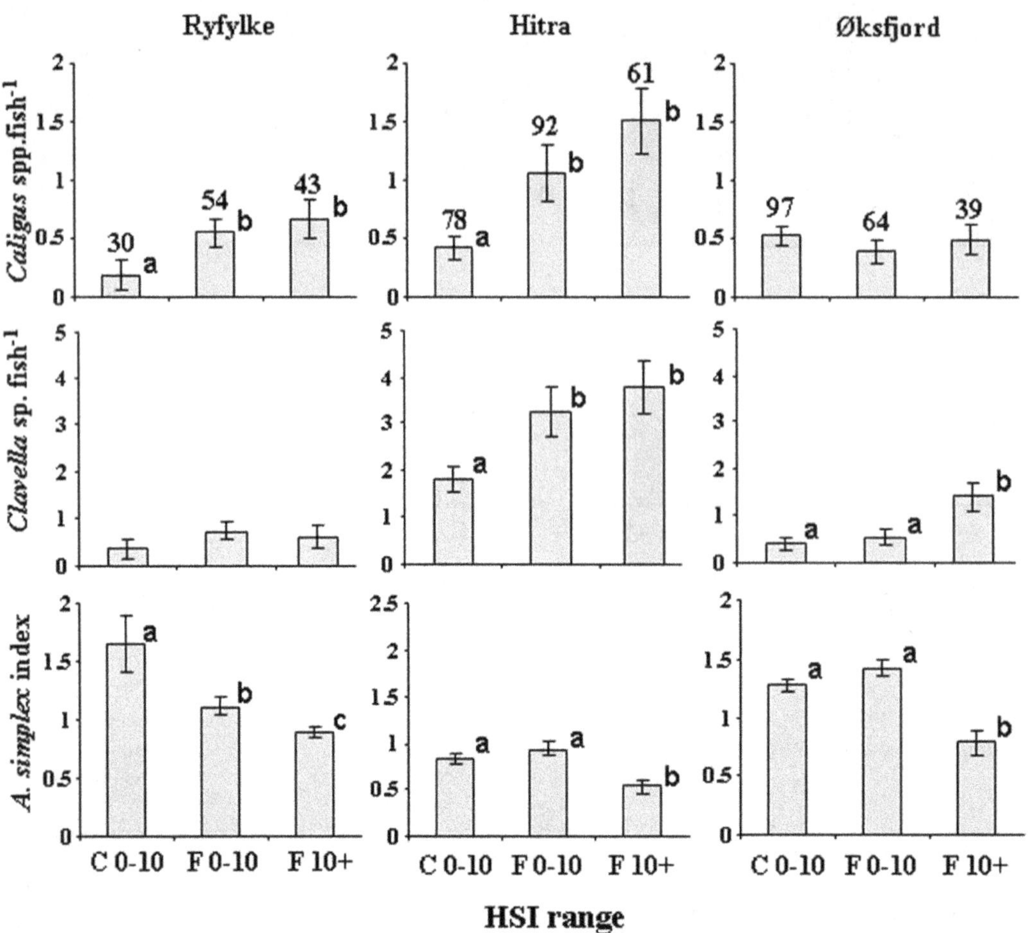

Figure 4. Mean abundances (± SE) of common parasites of saithe *Pollachius virens* in fish with a hepatosomatic index (HSI)<10 taken from non-farm locations (UA), and fish with HSI<10 and HSI>10 captured in association with Atlantic salmon farms (FA) in the three intensive fish farming areas. HSI = Hepatosomatic index. Superscripts ([a,b,c]) indicate a significant difference was detected among the groups at the $p<0.05$ level. Numbers above bars give the number of fish sampled for each comparison.

the level of *A. simplex* infection. The strong trophic link between saithe and fish farms, with saithe diets containing >70% by weight of lost feed pellets which are free of *A. simplex* A3 larvae, reduces the amount of potential hosts of A3 larvae such as small fish and crustaceans that saithe consume [19].

Elevated levels of *Caligus* spp. and *Clavella* sp. detected in FA fish may have had detrimental effects upon condition. Limited information exists to assess the threshold levels at which *Caligus* spp. and *Clavella* sp. infestations cause reductions in condition in cod and saithe, although heavy infestation of *Clavella adunca* can produce a moderate reduction in cod condition [20]. However, mobile sea-lice infestations of gadoids were generally close to the range of those typically recorded in Norwegian fjord and coastal waters (1 to 2 *C. elongatus* gadoid^{-1}; [26]). *L. branchialis* is considered the most serious metazoan parasite of wild cod [18,20] and can cause mortality, loss of condition and affect reproductive output. Similarly, heavy *Anisakis simplex* infestation has the capacity to reduce the condition of wild gadoids [20]. The reduction of both of these parasites in FA fish at some locations was therefore likely to have led to increased average condition compared to control fish. However, multiple regression analyses revealed that farm-modified parasite loads did not have major effects on the somatic condition of cod. For saithe, *Clavella* sp. abundance was positively correlated with condition, while the *A. simplex* infestation index was strongly negatively correlated with condition. Regardless of these relationships, body condition was significantly higher for FA fish than UA fish for both cod and saithe across all farming locations. As the body condition index integrates all factors that influence the condition of a fish over its recent life history, including the effects of parasites upon condition, our data suggests that the trophic subsidy that farms provide elevates body condition such that any effects on condition related to modified parasite loads were negligible in comparison.

In addition to the parasite species investigated here, gadoid fish such as *Gadus morhua* and *Pollachius virens* are infected by over 100 pathogens and parasites, at least 20 of which may be directly transferred among salmonids and gadoids [27]. These include some of the most significant diseases prevalent in salmon aquaculture, including *Vibrio anguillarum*, salmonid alphavirus and infectious pancreatic necrosis virus [28,29]. If these pathogens are enhanced in wild fish aggregated at fish farms they could negatively affect condition and survival; further research is required in this field.

Fish farms: ecological traps or population sources for wild fish?

In contrast to the detrimental effects of salmon farming detected at the population-level for wild salmonids (sea lice: [30,31,32];

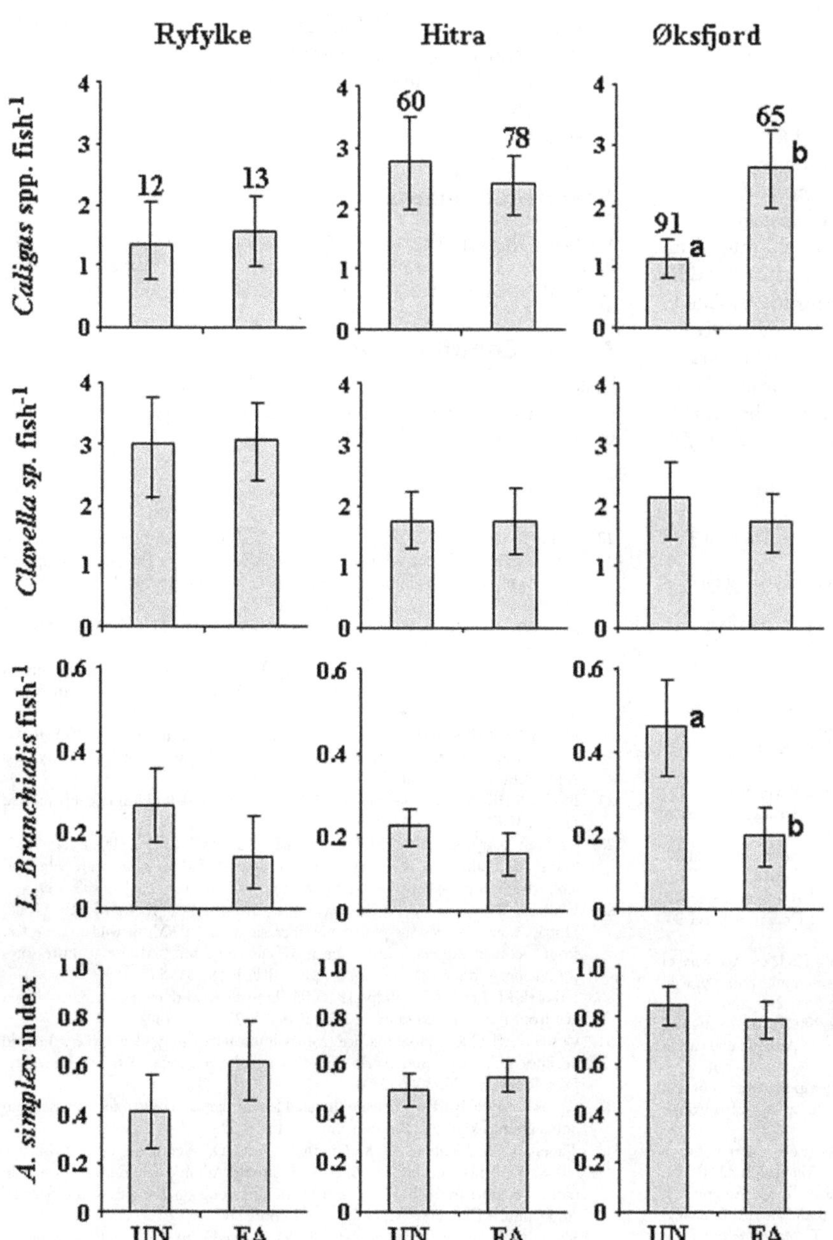

Figure 5. Mean abundances (± SE) of common parasites of farm-associated (FA) and un-associated (UA) Atlantic cod *Gadus morhua* in each of the three intensive salmon farming areas. HSI = Hepatosomatic index. Superscripts ([a,b]) indicate a significant difference was detected between the two groups at the $p<0.05$ level. Numbers above bars give the number of fish sampled for each comparison.

escapes: [33,34]), we did not detect significant negative effects of the co-occurrence of wild saithe and cod with salmon farms. The diet and condition data indicate that wild saithe and cod benefited from their associations with salmon farms through access to greater amounts of food which translated to enhanced condition. While *Caligus* spp. and *Clavella* sp. loads were elevated at some farming locations compared to controls and *Anisakis* sp. and *Lernaeocera branchialis* loads were lowered at some farming locations compared to controls, it appears that any effects these modified parasite loads may have had on the condition of wild cod and saithe were overridden by the trophic subsidy that farms supply. The results provide no evidence that salmon farms act as ecological traps for wild cod and saithe that aggregate in their vicinity, provided that: 1) the modified fatty acid distributions and elevated organohalogen levels in fat stores in livers that results from a fish farm modified diet [35,36] does not negatively affect physiological processes, vitellogenesis or egg and larval quality; 2) salmon farms do not amplify any of the numerous pathogens not investigated here that salmonids and gadoids share [27]; and 3) that attraction to farms does not disrupt natural spawning migrations or behavior. Future research should seek to discern the effects of both salmon and cod farms during the spawning season for cod resident in fjords containing farms, as a range of different effects are possible during this period, including mass spawning of farmed cod in cod farms [37] and possible avoidance of fjords containing salmon farms by spawning cod [38].

As saithe and cod condition is enhanced by farms, an opportunity exists to protect wild fish around salmon farms where they are aggregated and vulnerable to fishing, thus enabling farms to act as a population source. Presently, fishing adjacent to salmon farms occurs [39], although the importance of this activity to overall catches is unknown. Stocks of fjord cod in southern Norwegian waters, in particular, are depressed due to chronic overfishing [40]. Therefore, to ensure farms do not act as ecological traps for cod via increased fishing mortality alone, restrictions on the fishing of cod in the vicinity of farms could be introduced. Spatial protection from fishing would allow an opportunity for the enhanced condition that cod and saithe generate due to their association with salmon farms to translate to enhanced spawning success. Fish farms have recently become targets of significant fishing pressure [41] in other coastal ecosystems, thus the principle of restricting fishing around farms to ensure they do not function as ecological traps may be broadly applicable. As large, multi-specific aggregations of wild fish aggregate around coastal fish farms wherever they occur [5,6], we predict that significant conservation benefit would be derived through the protection of tens of thousands of tons of wild coastal fish in high spawning condition if this measure were implemented worldwide.

Acknowledgments

We thank Dr Bengt Finstad (NINA) who assisted in identifying mobile sea lice. We thank the salmon farming companies Marine Harvest, Greig and Lerøy Midnor for access to farm locations throughout the study.

Author Contributions

Conceived and designed the experiments: TD PSJ IU. Performed the experiments: TD PSJ DFJ JBS RN PAB IU. Analyzed the data: TD. Contributed reagents/materials/analysis tools: TD. Wrote the paper: TD.

References

1. Battin J (2004) When good animals love bad habitats: ecological traps and the conservation of animal populations. Cons Biol 18: 1482–1491.
2. Robertson BA, Hutto RL (2006) A framework for understanding ecological traps and an evaluation of existing evidence. Ecology 87: 1075–1085.
3. Hallier JP, Gaertner D (2008) Drifting fish aggregation devices could act as an ecological trap for tropical tuna species. Mar Ecol Prog Ser 353: 255–264.
4. FAO (2009) Fishstat Plus. Aquaculture production: quantities 1950–2005. Rome: FAO.
5. Dempster T, Sanchez-Jerez P, Bayle-Sempere JT, Giminez-Casualdero F, Valle C (2002) Attraction of wild fish to sea-cage fish farms in the south-western Mediterranean Sea: spatial and short-term variability. Mar Ecol Prog Ser 242: 237–252.
6. Dempster T, Uglem I, Sanchez-Jerez P, Fernandez-Jover D, Bayle-Sempere JT, et al. (2009) Coastal salmon farms attract large and persistent aggregations of wild fish: an ecosystem effect. Mar Ecol Prog Ser 385: 1–14.
7. Norwegian Directorate of Fisheries (2009) Statistics for Aquaculture 2008. http://www.fiskeridir.no/fiskeridir/kystsone_og_havbruk/statistikk.
8. Dempster T, Sanchez-Jerez P, Uglem I, Bjørn PA (2010) Species-specific patterns of aggregation of wild fish around fish farms. Est Coast Shelf Sci 86: 271–275.
9. Uglem I, Dempster T, Bjorn PA, Sanchez-Jerez P (2009) High connectivity of salmon farms revealed by aggregation, residence and repeated migrations of wild saithe (*Pollachius virens*) among farms. Mar Ecol Prog Ser 384: 251–260.
10. Uglem I, Bjørn PA, Dale T, Kerwath S, Økland F, et al. (2008) Movements and spatiotemporal distribution of escaped farmed and local wild Atlantic cod (*Gadus morhua* L.). Aqua Res 39: 158–170.
11. Uglem I, Bjørn PA, Mitamura H, Nilsen R (2010) Spatiotemporal distribution of coastal and oceanic Atlantic cod (*Gadus morhua* L.) sub-groups after escape from a farm. Aquacult Environ Interact 1: 11–20.
12. Diamant A, Colonri A, Ucko M (2007) Parasite and disease transfer between cultured and wild coastal marine fish. CIESM Workshop Monograph 32: 49–54.
13. Fernandez-Jover D, Lopez-Jimenez JA, Sanchez-Jerez P, Bayle-Sempere J, Gimenez-Casalduero F, et al. (2007) Changes in body condition and fatty acid composition of wild Mediterranean horse mackerel (*Trachurus mediterraneus*, Steindachner, 1868) associated to sea-cage fish farms. Mar Environ Res 63: 1–18.
14. Fernandez-Jover D, Sanchez-Jerez P, Bayle-Sempere JT, Valle C, Dempster T (2008) Seasonal patterns and diets of wild fish assemblages associated with Mediterranean coastal fish farms. ICES J Mar Sci 65: 1153–1160.
15. Chalfoun AD, Martin TE (2007) Assessments of habitat preferences and quality depend on spatial scale and metrics of fitness. J Appl Ecol 44: 983–992.
16. Lambert Y, Dutil J-D (1997) Can simple condition indices be used to monitor and quantify seasonal changes in the energy reserves of Atlantic cod (*Gadus morhua*)? Can J Fish Aquat Sci 54(Suppl. 1): 104–112.
17. Marshall CT, Yaragina NA, Lambert Y, Kjesbu OS (1999) Total lipid energy as a proxy for total egg production by fish stocks. Nature 402: 288–290.
18. Khan RA, Ryan K, Barber DE, Lee EM (1993) Effect of a single *Lernaeocera branchialis* (Copepoda) on growth of Atlantic cod. J Parasitol 79(6): 954.
19. Klimpel S, Palm HW, Ruckert S, Piatkowski U (2004) The life cycle of *Anisakis simplex* in the Norwegian Deep (northern North Sea). Parasitol Res 94: 1–9.
20. Hemmingsen W, Mackenzie K (2001) The parasite fauna of the Atlantic cod, *Gadus morhua* L. Adv Mar Biol 40: 1–80.
21. Gjøsæter J, Otterå H, Slinde E, Nedreaas K, Ervik A (2008) Release of nutrients and excess feed from aquaculture. Kyst og Havbruk 2008: 52–55 (In Norwegian).
22. Tacon AGJ, Metian M (2008) Global overview on the use of fish meal and fish oil in industrially compounded aquafeeds: Trends and future prospects. Aquaculture 285: 146–158.
23. Kutti T, Hansen PK, Ervik A, Høisæter T, Johannessen P (2007) Effects of organic effluents from a salmon farm on a fjord system. II. Temporal and spatial patterns in infauna community composition. Aquaculture 262(2–4): 355–366.
24. Skjæraasen JE, Nilsen T, Kjesbu OS (2006) Timing and determination of potential fecundity in Atlantic cod (*Gadus morhua*). Can J Fish Aquat Sci 63: 310–320.
25. Salze G, Tocher DR, Roy WJ, Robertson DA (2005) Egg quality determinants in cod (*Gadus morhua* L.): egg performance and lipids in eggs from farmed and wild broodstock. Aqua Res 36: 1488–1499.
26. Heuch PA, Øines Ø, Knutsen JA, Schram TA (2007) Infection of wild fishes by the parasitic copepod *Caligus elongatus* on the south east coast of Norway. Dis Aquat Org 77: 149–158.
27. Bricknell IR, Bron JE, Bowden TJ (2006) Diseases of gadoid fish in cultivation: a review. ICES J Mar Sci 63: 253–266.
28. Graham DA, Jewhurst H, McLoughlin MF, Sourd P, Rowley HM, et al. (2006) Sub-clinical infection of farmed Atlantic salmon Salmo salar with salmonid alphavirus - a prospective longitudinal study. Dis Aquat Org 72: 193–199.
29. Wallace IS, Gregory A, Murray AG, Munro ES, Raynard RS (2008) Distribution of infectious pancreatic necrosis virus (IPNV) in wild marine fish from Scottish waters with respect to clinically infected aquaculture sites producing Atlantic salmon, *Salmo salar* L. J Fish Dis 31: 177–186.
30. Krkošek M, Lewis MA, Volpe JP (2005) Transmission dynamics of parasitic sea lice from farm to wild salmon. Proc R Soc B 272: 689–696.
31. Costello MJ (2009) How sea lice from salmon farms may cause wild salmonid declines in Europe and North America and be a threat to fishes elsewhere. Proc R Soc B 276: 3385–3394.
32. Krkošek M (2010) Host density thresholds and disease control for fisheries and aquaculture. Aquacult Environ Interact 1: 21–32.
33. Thorstad EB, Fleming IA, McGinnity P, Soto D, Wennevik V, Whoriskey F (2008) Incidence and impacts of escaped farmed Atlantic salmon Salmo salar in nature. Report from the Technical Working Group on Escapes of the Salmon Aquaculture Dialogue. NINA Special Report 36: 1–112.
34. Jensen Ø, Dempster T, Thorstad E, Uglem I, Fredheim A (2010) Escapes of fish from Norwegian sea-cage aquaculture: causes, consequences and methods to prevent escape. Aquacult Environ Interact 1: 71–83.
35. Bustnes JO, Lie E, Herske D, Dempster I, Bjørn PA, et al. (2010) Salmon farms as a source of organohalogenated contaminants in wild fish. Environ Sci Technol 44: 8736–8743.
36. Fernandez-Jover D, Martinez-Rubio L, Sanchez-Jerez P, Bayle-Sempere JT, Lopez-Jiminez JA, et al. (2011) Waste feed from coastal fish farms: a trophic subsidy with compositional side effects for wild gadoids. Est Coast Shelf Sci; (in press).
37. Jørstad KE, van der Meeren T, Paulsen OI, Thomsen T, Thorsen A, et al. (2008) "Escapes" of eggs from farmed cod spawning in net pens: recruitment to wild stocks. Rev Fish Sci 16: 1–11.
38. Bjørn PA, Uglem I, Kerwath S, Sæther BS, Nilsen R (2008) Spatiotemporal distribution of Atlantic cod (*Gadus morhua* L.) with intact and blocked olfactory sense during spawning season in a Norwegian fjord with intensive salmon farming. Aquaculture 286: 36–44.
39. Maurstad A, Dale T, Bjørn PA (2007) You wouldn't spawn in a septic tank, would you? Human Ecology 35(5): 601–610.
40. Berg E (2008) Norwegian coastal cod still in decline. Kyst og Havbruk 2008(2): 79—81 (In Norwegian).
41. Akyol O, Ertosluk O (2010) Fishing near sea-cage farms along the coast of the Turkish Aegean Sea. J Appl Ichthyol 26: 11–15.

17

Fluorescent *In Situ* Hybridization: A New Tool for the Direct Identification and Detection of *F. psychrophilum*

Nicole Strepparava[1]*, Thomas Wahli[2], Helmut Segner[2], Bruno Polli[3], Orlando Petrini[1]

1 Cantonal Institute of Microbiology, Bellinzona, Switzerland, **2** Centre for Fish and Wildlife Health, University of Bern, Bern, Switzerland, **3** Cantonal Office of Hunting and Fisheries, Bellinzona, Switzerland

Abstract

F. psychrophilum is the causative agent of Bacterial Cold Water Disease (BCW) and Rainbow Trout Fry Syndrome (RTFS). To date, diagnosis relies mainly on direct microscopy or cultural methods. Direct microscopy is fast but not very reliable, whereas cultural methods are reliable but time-consuming and labor-intensive. So far fluorescent *in situ* hybridization (FISH) has not been used in the diagnosis of flavobacteriosis but it has the potential to rapidly and specifically detect *F. psychrophilum* in infected tissues. Outbreaks in fish farms, caused by pathogenic strains of *Flavobacterium* species, are increasingly frequent and there is a need for reliable and cost-effective techniques to rapidly diagnose flavobacterioses. This study is aimed at developing a FISH that could be used for the diagnosis of *F. psychrophilum* infections in fish. We constructed a generic probe for the genus *Flavobacterium* ("Pan-Flavo") and two specific probes targeting *F. psychrophilum* based on 16S rRNA gene sequences. We tested their specificity and sensitivity on pure cultures of different *Flavobacterium* and other aquatic bacterial species. After assessing their sensitivity and specificity, we established their limit of detection and tested the probes on infected fresh tissues (spleen and skin) and on paraffin-embedded tissues. The results showed high sensitivity and specificity of the probes (100% and 91% for the Pan-Flavo probe and 100% and 97% for the *F. psychrophilum* probe, respectively). FISH was able to detect *F. psychrophilum* in infected fish tissues, thus the findings from this study indicate this technique is suitable as a fast and reliable method for the detection of *Flavobacterium* spp. and *F. psychrophilum*.

Editor: Yung-Fu Chang, Cornell University, United States of America

Funding: This work was supported by the Intramural fund from the Cantonal Institute of Microbiology. The funders had no role in study design, data collection and analysis, decision to publish, or preparation of the manuscript.

Competing Interests: The authors have declared that no competing interests exist.

* E-mail: nicole.strepparava@bluewin.ch

Introduction

Bacteria belonging to the genus *Flavobacterium* are non-fermentative, catalase- and oxidase-positive, gram-negative bacteria that occur in abiotic and biotic compartments of many ecosystems (e.g. soil, fresh and marine water, fish). Some species, in particular *F. brevis, F. columnare, F. johnsoniae, F. branchiophilum* and *F. psychrophilum*, are ubiquitous, opportunistic pathogens that may cause disease symptoms in injured or immunologically weak animals and sometimes also in humans [1,2,3,4]).

F. psychrophilum is a pathogenic agent causing both external and systemic infections in fish. One of the diseases caused by *F. psychrophilum* is the so-called Bacterial Cold Water (BCW) Disease, which is geographically widespread and affects a variety of fish species [5,6]. BCW is characterized by epidermal necrosis leading to saddle-like skin lesions, usually near the dorsal fin, but also the mouth or gills may be affected, particularly in juvenile fish. The Rainbow Trout Fry Syndrome (RTFS) is a severe systemic infection that occurs in general when bacteria accumulate in the liver or spleen of salmonids. It causes high mortalities in cultured fish stocks, primarily when the infection occurs in small rainbow trout [7]. It is not yet clear, however, whether RTFS is the result of a systemic infection or an advanced form of a superficial infection.

Diagnosis of *F. psychrophilum* infection is lengthy and time-consuming, being mainly based on macroscopic and microscopic examination of fresh spleen samples and culture methods. *F. psychrophilum* is a fastidious, slow-growing, opportunistic pathogen, the growth of which is inhibited by the presence of other microorganisms; selective plates are not available and the colonies are often overgrown by other fast-growing bacteria. In addition, *F. psychrophilum* grows optimally at 15°C, an incubation temperature not routinely used in diagnostic labs [8]. As a result, *F. psychrophilum* is easily overseen during sample processing and the number of incorrect diagnoses can therefore be quite high. A rapid, sensitive and specific detection method enabling diagnosis of *F. psychrophilum* at an early stage of infection would help to prevent further spread of the disease.

Fluorescent *in situ* Hybridization (FISH) is frequently used to detect bacterial species in environmental and clinical samples and species-specific probes have been developed for the rapid identification of pathogenic species [9,10]; FISH has already been employed to identify flavobacteria, using probes designed on 16S rRNA gene sequences targeting the *Cytophaga-Flavobacterium-Bacteroides* (CFB) group [11,12]. So far, however, specific probes for the genus *Flavobacterium* in general or for the fish pathogenic species *F. psychrophilum* in particular are not available.

This study is aimed to develop genus- and species-specific probes that can be used to detect and identify *Flavobacterium* spp. and *F. psychrophilum* in particular, and to test the usefulness of this technique in the early diagnosis *in situ* of infections caused by *F. psychrophilum* in salmonids.

Methods

Strains Used

We used pure cultures of *Flavobacteria* and other bacterial species isolated from soil, water and fish, as well as clinical isolates of related and unrelated bacterial species. A list of all tested bacteria and their origin is presented in Table S1.

Water was collected in fish farms (inlets, water from fish tanks and water at the outlets of the fish farms). Swabs from immersed soil or tank surfaces were suspended in 1 ml of sterile water. For each sample 100 µl of suspension and a 1:10 dilution thereof were plated onto CY-Agar (medium 67 DSMZ for *F. psychrophilum*: 0.3% casitone, 0.136% $CaCl_2 \cdot H_2O$, 0.1% yeast extract, 1.5% agar) as well as on Enriched Cytophaga Agar Medium (CYAM) (medium 1133 DSMZ for *F. columnare*: 0.2% tryptone, 0.05% beef extract, 0.05% yeast extract, 0.02% sodium acetate, 1.5% agar).

Fish suspected to be infected were sent by fish farmers in a container with water. A sample of their external mucus was taken and the fish were then killed by immersion in 0.01% benzocaine followed by cutting along the vertebral column allowing for the removal of the spleen. The external mucus, gills and spleens of rainbow trout (*Oncorhynchus mykiss*) and brown trout (*Salmo trutta fario* and *Salmo trutta lacustris)* were collected and homogenized separately in 200 µl of sterile water. The homogenates were plated on both CY and CYAM.

All samples were incubated at 15°C for 5 to 10 days. Growing colonies were transferred onto fresh plates and pure cultures were conserved in 1 ml skimmed milk [7% Skim Milk (Becton Dickinson, Switzerland), 10% bovine serum and 20% glycerol] at −80°C.

For FISH, symptomless fish (brown trout fario and rainbow trout) from a fish farm in Rodi (Cantonal Fish Farm, Ticino, Switzerland) in which no signs of infection were present were treated as described above. The body surfaces were swabbed using 70% ethanol to prevent contamination of the spleens by normal external bacterial flora. The spleens were removed and stored at −20°C until the time of the experiments and were then homogenized by grinding them in 200 µl of sterile water.

Approval for the animal experiments and the water collection was obtained from the Federal Veterinary Office (FVO, Switzerland) and the Ticino Cantonal Veterinary Office (Authorization 03/2010 and 04/2010).

Identification of the Isolates

Based on growth characteristics, colonies suspected to be *Flavobacterium* spp. were transferred onto fresh plates of CY-Agar and CYAM and grown for 5 days. DNA of all samples was extracted using the Instagene kit (Bio-Rad, Hercules (CA).

Putative *Flavobacterium* strains were identified by 16S rRNA gene sequencing [13]. All other clinical and environmental isolates were identified by MALDI-TOF mass spectrometry [14,15]. When identification by MALDI-TOF MS was not possible, identification was carried out by 16S rRNA gene sequencing.

16S rRNA Gene Sequencing

16S rRNA gene PCR was carried out using the universal primers uniL 26f (5′-ATTCTAGAGTTTGATCATGGCTCA-3′) and uniR 1392r (5′-ATGGTACCGTGT-GACGGGCGGTGTGTA-3′) [16] PCR amplifications were carried out in a total volume of 50 µl. 25 µl of Taq PCR Master Mix (QIAGEN, Switzerland), 1.5 µl of each primer, 17 µl of water and 5 µl of DNA were mixed and the PCR was performed at the following conditions: 5 min. at 94°C, 35 cycles of 30 s at 94°C, 30s at 52°C and 1 min. at 72°C with a final elongation of 7 min. at 72°C. Purification of PCR products was carried out with PCR clean-up NucleoSpin® ExtractII (Macherey-Nagel, Germany). Sequencing was performed using the BigDye Terminator v1.1 Cycle Sequencing kit (Applied Biosystems, Switzerland) according to the manufacturer's instructions. Reactions were carried out in a total volume of 15 µl containing 3 µl of BigDye®, 1.5 µl of BigDye® Buffer, and 2.4 µl of a 1 µM primer solution. The same primers used for PCR were also used for the 16S rRNA gene sequencing. Thermal cycling conditions were 1 min at 96°C, followed by 25 cycles of 10s at 96°C, 5s at 50°C and 4 min at 60°C. The sequencing products were purified on a 0.025 mm membrane filter in a Tris-EDTA buffer solution (pH 8) before sequencing with Hi-Di™ Formamide (Applied Biosystems) on a AB Prism 310 Genetic Analyzer (Applied Biosystems). The obtained sequences were compared with data included in GenBank (http://blast.ncbi.nlm.nih.gov/).

Probes Used

Oligonucleotide FISH probes were manually designed by aligning all 16S rDNA of *Flavobacterium* strains of interest. Sequences were downloaded from GenBank for the following species: *Flavobacterium psychrophilum* (AY662493, AB297676, AB297673), *F. branchiophilum* (D14017), *F. columnare* (AM230485, AB015481, AB010951, AB180738), *F. granuli* (AB180738), *F. johnsoniae* (AM921621), *F. degerlachei* (AJ441005), *F. flevense* (AJ440988), *F. frigidarium* (EU000241), *F. frigoris* (AJ440988), *F. hibernum* (L39067), *F. hydatis* (M58764), *F. limicola* (AB075230), *F. pectinovorum* (AM230490), *F. succinicans* (AM230492), and *F. omnivorum* (AF433174). Sequence alignment was performed using MEGA4. Probes were named by their position after alignment of all *Flavobacterium* sequences using *Escherichia coli* HM371196 as the outgroup.

The possible target regions were chosen by evaluating which region within the 16S rRNA secondary structure of *E. coli* would be most suitable for probe design [17]. This led to the construction of the generic *Flavobacterium* probe ("Pan-Flavo": Flavo285; Table 1). Pan-Flavo was labeled with Cyanine dye (CY3) at the 5′ end.

Two probes (FlavoP77 and FlavoP477, Table 1) were designed for the specific detection of *F. psychrophilum* using the same 16S rDNA alignment of sequences as described above, using, however, seven additional *F. psychrophilum* strains (AB297675, AB297484, AB297483, AB297674, AB297671, AB297672, AB297494) to cater for internal *F. psychrophilum* variability. These two oligonucleotide probes were then labeled with Carboxyfluorescein (FAM) at the 5′ end.

To test for a possible cumulative effect of different fluorochromes applied on the same slide, a Pan-Flavo probe was constructed with the same primer as above, but without labeling.

The sensitivity and specificity of the Pan-Flavo and *F. psychrophilum* probes were tested on several *Flavobacterium* species (*F. psychrophilum, F. columnare, F. branchiophilum, F. johnsoniae, F. fryxellicola, F. frigidimaris, F. aquatile, F. psychrolimnae, F. succinicans, F. aquidurense, F. hercynium, F. hydatis, F. limicola, F. pectinovorum) and Chryseobacterium* spp. strains isolated from our samples as well as on non-*Flavobacteriaceae* isolates (Table S1).

The specificity of the probes was also tested *in silico* using the Ribosomal Database Project (RDP) [18,19], thus providing evidence that the designed probes match the sequences present in the database and therefore making them suitable for the *in vivo* assays.

Table 1. Probes used, target microorganisms and DNA target regions. [Cyanine dye (CY3); Carboxyfluorescein (FAM)].

Name	Target microorganism	Target region in E.coli*	Length	Sequence	Labeling
Flavo285	*Flavobacterium* spp.	230	17	5′-GACCCCTACCCATCRTH-3′	CY3
FlavoP77	*F. psychrophilum*	138	22	5′-AGTGTGTTGATGCCAACTCACT-3′	FAM
FlavoP477	*F. psychrophilum*	532	19	5′-ACTTATCTGGCCGCCTACG-3′	FAM

*E.coli, GenBank sequence HM371196.

The probes were synthesized by Microsynth (Balgach, Switzerland).

FISH Conditions

Each putative *Flavobacterium* colony was resuspended in 200 µl of sterile water. Ten microliters were added in a well of ten-well immunofluorescence microscopy slides (bioMérieux, Geneva, Switzerland), air-dried and dehydrated sequentially in 50%, 70%, and 96% ethanol during 3 min for each condition. To determine stringent hybridization conditions, a formamide series was carried out with a pure culture of *F. psychrophilum*. The best results were obtained at a formamide concentration of 30%. 10 µl hybridization solution (0.9 M sodium chloride, 20 mM Tris/HCl pH 7, 30% formamide, water, 0.01% SDS) containing 50 ng of the oligonucleotide probe were added to each well and the sample incubated for 12 to 16 hours in an isotonically equilibrated, humid Falcon tube (Greiner bio-one, Verridia, Switzerland) at 46°C. After the incubation step, the slides were kept at 48°C for 20 min in 50 ml of washing solution (150 mM sodium chloride, 100 mM Tris/HCl pH 7, 5 mM EDTA pH 8, 0.01% SDS, water up to 50 ml), rinsed with distilled water, air-dried, and stained with an aqueous solution of 4′-6-diamidino-2-phenylindole (DAPI) (Fluka, Switzerland) for 7 min (10 µl ml^{-1}); after DAPI staining, slides were rinsed again with distilled water, air dried and mounted with Citifluor (Citifluor Ltd., London, UK). Slides were screened for fluorescence using an Axiolab microscope (ZEISS, Switzerland) equipped with filters for FITC (excitation 494 nm; emission 518 nm), Cy3 (excitation 562 nm; emission 576 nm) and DAPI (excitation 360 nm; emission 456 nm). *Flavobacterium psychrophilum* (DSM 3660), environmental samples of *Flavobacterium* spp. and *Chryseobacterium* spp. were used as controls.

Quantification of Bacteria

Optical Density (OD_{595}) of pure *F. psychrophilum* bacterial suspension (n = 10) was adjusted at 0.3 (±0.02) with a Perkin Elmer spectrophotometer (Perkin Elmer UV/VIS Spectrometer Lambda 2S, Waltham, MA). DNA was extracted from 1 ml of suspension with the QIAGEN tissue and blood kit (QIAGEN). The total amount of DNA was quantified with a Nanodrop spectrophotometer (ND1000, Witec, Switzerland) and divided by 3.137×10^{-6} ng [the weight of one *F. psychrophilum* genome (genome size 2′861′988 bp [20])]. This yields the number of bacteria present in one ml of the starting OD suspension. Thus, an OD of 0.3 corresponds to $3\times10^{9}\pm7\times10^{8}$ cells.ml^{-1} [21].

To determine the limit of detection and evaluate the goodness of FISH for diagnostic purposes, we plated out aliquots of the bacterial suspensions and assessed their growth, as cultures are currently used in veterinary laboratories to assess the presence of the pathogen in fish samples.

Limits of Detection for Suspensions from Pure Cultures

Pure cultures of *F. psychrophilum* grown on CYAM agar were adjusted in sterile water at 3×10^{9} cells.ml^{-1} (OD 0.3). Twenty serial two-fold dilutions were prepared and 100 µl of each were plated on CYAM; 10 µl of each dilution were put on a ten-well immunofluorescence microscopy slide. *F. psychrophilum* (DSM3660) and water were used as positive and negative controls, respectively.

Limits of Detection in Fish Tissues

Serial dilutions of a stock suspension of *F. psychrophilum* (9.4×10^{7} cells.ml^{-1}) were used for the experiment. For each serial dilution one spleen and one *F. psychrophilum* isolate were used. In a 200 µl Eppendorf, 10 µl of ground spleen were seeded with 10 µl of a bacterial suspension: 10 µl of the final suspension were placed on a ten-well immunofluorescence microscopy slide and 10 µl were plated on CYAM medium. A 1:32 dilution of the stock suspension was used as a positive control and a mix of 10 µl water and 10 µl of spleen was used as a negative control.

Diagnosis of Putative Infections by FISH

During 2011–2012, fish samples from Swiss fish farms were collected periodically to check for the systemic infection by *F. psychrophilum*. In addition, each potential infection reported by fish farmers was screened by FISH with the Pan-Flavo and *F. psychrophilum* probes to check for the presence of the pathogen on skin and spleen tissues.

The entire spleen and a sample of skin mucus were homogenized individually in 200 µl of sterilized water. 10 µl of each homogenate were added to a ten-well immunofluorescence microscopy slide and 100 µl were plated on CYAM agar medium for control.

Detection of *Flavobacterium* and *F. psychrophilum* in Paraffin Embedded Tissues

Serial sections of paraffin-embedded tissues from diseased fish were prepared and one section was stained with Giemsa. Pretreatment of FISH staining followed the protocol of Ridderstrale et al. [22]. Briefly, slides were heated at 65°C for 1 hour, immersed in 0.2 M HCl for 15 min, rinsed with water, incubated in 0.01 citrate buffer (pH 6.0) at 100°C for 90 s and at 90°C for 7 min. Slides were then immersed in 70% and 100% ethanol at 4°C for 3 min each, washed in standard saline citrate (SSC, 2X) and incubated in 0.5 mg/ml pepsin (Merck, Switzerland) in NaCl 0.9% (pH 2) for 20 min at 37°C. At the end, samples were dehydrated in 70% and 100% ethanol during 2 min each. FISH was carried out using the same method described for pure cultures.

Statistical Methods

Sensitivity (SE), specificity (SP), positive predictive values (PPV) and negative predictive values (NPV) for all probes were calculated using DAG-Stat.xls [23,24,25]. Alignments and phylogenetic tree construction were carried out using MEGA version 4 [26]. The limit of detection for pure culture and for spiked spleen was defined as the fifth percentile of all analyzed positive and negative samples. Receiver operating characteristic (ROC) analysis was done using SPSS version 17.0 (SPSS Inc., Chicago, IL, USA).

Results

Simultaneous Use of *F. psychrophilum* (FlavoP77 and FlavoP477) and Pan-Flavo (Flavo285) Probes

Two conserved regions within the *F. psychrophilum* 16S rDNA, with species-specific sequences, were chosen and tested individually. Results from a first experiment carried out on 10 *F. psychrophilum* strains in duplicates for each probe were not reproducible due to a too low fluorescence and an immediate loss of signal: therefore, in a second step a combination of both *F. psychrophilum* probes as well as a combination of the two *F. psychrophilum* probes with the Pan-Flavo probe were tested. Stable and accurate results were obtained using the two *F. psychrophilum* probes; the addition of the Pan-Flavo probe was, however, crucial to obtain the optimal fluorescence at which these bacteria can be easily seen through the microscope. To test whether or not the improved staining results, with the combination of the three probes, is due to a potentially cumulative fluorescence caused by the simultaneous presence of two types of fluorochromes (CY3 and FAM), we prepared a helper oligonucleotide probe with the same sequence as the Pan-Flavo probe but without a fluorescent label: this led to the same results as with the two fluorochromes. Further tests were carried out with the three probes available using them simultaneously, with essentially the same outcome.

Tests performed on 352 isolates (50 strains of *F. psychrophilum*, 226 *Flavobacterium* spp. and 76 other bacterial species) demonstrated that the Pan-Flavo and *F. psychrophilum* probes were highly sensitive and specific (98% and 100% for Pan-Flavo probe and 100% and 98% for *F. psychrophilum* probes) (Table 2). PPV and NPV values were 100% and 95%, respectively, for the Pan-Flavo probe and 91% and 100% for the specific *F. psychrophilum* probes. The probes showed no recognizable cross-reactions with other bacterial species (Figure 1).

Only 4 out of 276 *Flavobacterium* sp. strains did not react with the Pan-Flavo probe. Each strain was tested twice: the first essay was negative and the second could not unequivocally identify the strains as *Flavobacterium* sp. No mismatches were present in the alignment of the probe with the target sequence, therefore we have no clear explanation for this result: the error is approximately 1% (4 wrong identifications over 352 total strains) and, in our opinion, may be ascribed to natural variations among the samples studied. 5 out of the 352 strains tested were erroneously identified as *F. psychrophilum* but 16 s rRNA gene sequencing showed that they belonged to *Flavobacterium* sp. other than *F. psychrophilum.*

Limits of Detection for Pure Culture Suspensions

The LOD was established by investigating serial two-fold dilutions. In 95% of the tested cases (15 strains in duplicate), the LOD for the Pan-Flavo and *F. psychrophilum* probes was 7.3×10^5 cells.ml^{-1} by FISH; the LOD of the cultural method was only 3×10^9 cells*ml^{-1} (93% of the tested strains). ROC analysis (Figure 2A) showed a statistically significant higher sensitivity of the FISH method compared to culture (areas under the curve (AUC) for the Pan-Flavo probe: 0.89; for the *F. psychrophilum* probe: 0.88; culture method: 0.84).

Limit of Detection in Fish Tissues

The LOD for the Pan-Flavo and *F. psychrophilum* probes applied to fish tissue samples was 2.9×10^6 cells.ml^{-1}. An even lower LOD, 1.5×10^6 cells.ml^{-1} was reached with the PanFlavo probe in 80%, and with the *F. psychrophilum* probes in 70% of the cases. LOD was 3×10^9 cells.ml^{-1}, with only 40% of positive cultures. According to the ROC analysis, FISH appears to be more sensitive than culture, both for the Pan-Flavo and *F. psychrophilum* probes (Figure 2B). Likewise, the AUC values of both the Pan-Flavo and the *F. psychrophilum* probes were higher than those of the culture method (0.79 for FISH vs. 0.59 for culture).

Diagnosis of the Disease

FISH was very successful in detecting and identifying *F. psychrophilum* from fresh samples. Diagnosis by FISH was available within 24 hours as compared to 4 to 10 days with the culture method. The *F. psychrophilum* probes detected the pathogen in 13 cases of BCW and RTFS (Figure 3). In 9 cases, the FISH-based diagnosis was confirmed by culture, while in 1 case no growth in culture was seen. The remaining 3 cases were repeated samplings from the same fish farm: confirmation of the infection was possible with culture only after a fourth sampling more than one month after diagnosis by FISH.

Detection of *Flavobacterium* and *F. psychrophilum* from Paraffin Embedded Tissues

Five (4 positive and 1 negative) samples were fixed in paraffin, and 3–4 sections of each bloc were cut and mounted on a slide. Out of 11 samples, 7 were correctly detected as positive, 2 were correctly detected as negative and 2 samples were false negatives.

Discussion

The probes designed in this study, specifically targeting the genus *Flavobacterium* and the pathogenic species *F. psychrophilum*, are highly sensitive and specific (98% and 100% for Pan-Flavo, 100% and 98% for the *F. psychrophilum* probes) and allow correct identification of *Flavobacterium* spp. and *F. psychrophilum* in culture. The same probes were also used successfully to screen fish tissues for the presence of *Flavobacterium* sp. or *F. psychrophilum.* Compared to currently used diagnostic methods, FISH was rapid, as the results were obtained within 24 hours, as compared with 5 to 10 days needed to culture the bacteria. Thus the application of FISH offers a valuable tool for the rapid detection of *Flavobacterium* spp. and in particular *F. psychrophilum* in fish tissue.

Combining the two *F. psychrophilum* probes with the Pan-Flavo probe was crucial for a reliable detection of *F. psychrophilum.* The need for multiple labeling to increase signal strength has already been described by other authors [27]. It is assumed that the second probe enhances the annealing of the diagnostic probe with its corresponding rRNA [28]. Generally, a "helper" probe targets the sequence of a region directly adjacent to the diagnostic probe site [29]: however, the Pan-Flavo probe which, in our case, acted as a helper, is almost equidistant to both *F. psychrophilum* specific probes. This was not expected and we hypothesize that the effect may be related to the tertiary structure of the target region.

No cross-reaction was observed between the Pan-Flavo probe or the *F. psychrophilum* probes with other taxonomically closely related species that might be present in environmental and clinical samples. In our evaluation of sensitivity and specificity, we deliberately chose species known to be part of the aquatic environmental microbiota and we did not test opportunistic human and animal pathogens closely related with *Flavobacterium* such as *Capnocytophaga,* found in the mammal oropharyngeal tract

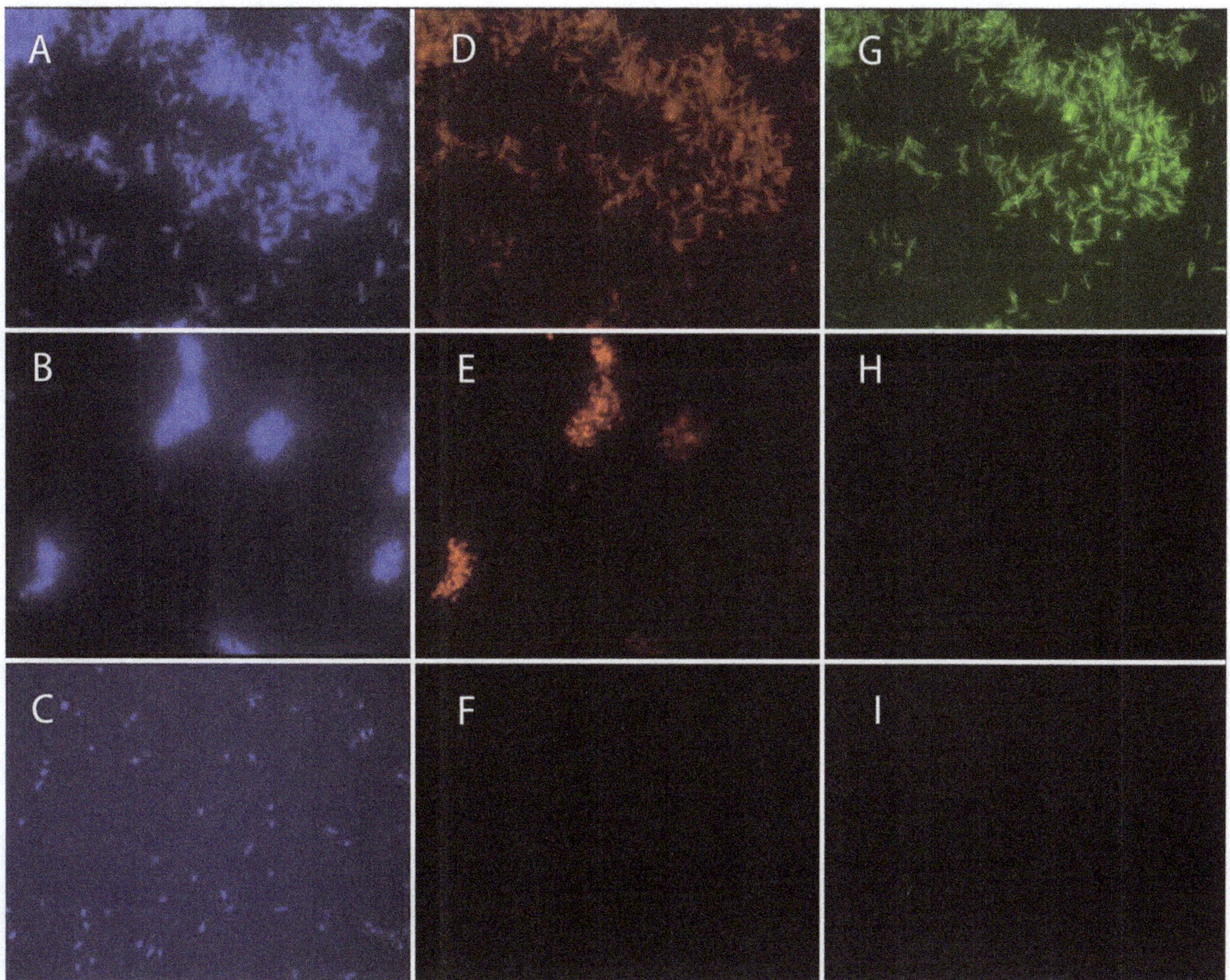

Figure 1. FISH assays of pure cultures. DAPI staining (A, B, C); Pan-Flavo probe (D, E, F); *F. psychrophilum* probes (G, H, I) (100x). *F. psychrophilum* (DSM3660) (A, D, G); *Flavobacterium* spp. (B, E, H); *Chryseobacterium* spp. (C, F, I).

Table 2. Agreement between FISH and 16S rDNA sequencing (SEQ, used as gold standard) in the experiments carried out with the Pan-Flavo (Flavo285) probe and the combination of two *F. psychrophilum* (FlavoP77, FlavoP477) probes.

Pan-Flavo	**FISH +**	**FISH −**	**Total**	
SEQ +	272	4	276	SE: 98%
SEQ −	0	76	76	SP: 100%
Total	272	80	352	
	PPV: 100%	NPV: 98%		
FlavoP77+ FlavoP477	FISH +	FISH −	Total	
SEQ +	50	0	50	SE: 100%
SEQ −	5	297	302	SP: 98%
Total	55	297	352	
	PPV: 91%	NPV: 100%		

SE: sensitivity; SP: specificity; PPV: positive predictive value, NPV: negative predictive value.

[30], or marine environment organisms such as *Tenacibaculum* [31]. Because of the particular ecological niche occupied by these species we do not expect them to be present in fish samples.

The minimal concentration of *F. psychrophilum* cells needed in a sample to yield a positive result by FISH is lower than for the culture method (7.3×10^5 cells ml^{-1} vs. 3×10^9 cells ml^{-1} for water, and 2.9×10^6 vs. 3×10^9 cells ml^{-1} for spleens). ROC analysis confirmed that FISH is more sensitive than culture (AUC for FISH 0.89 vs. 0.84 for culture with suspension of pure cultures and 0.79 for FISH vs. 0.59 for culture with spiked spleens). FISH also yielded reproducible results within and between isolates. This is not the case for the culture method, which showed variability even for one and the same isolate, with growth not always being reproducible.

In medical microbiology, FISH is frequently used as a cheap, easy and rapid method to identify pathogens directly in blood cultures; in these settings LODs are quite low, being approximately 1000 microorganism per ml ([32]). In our study, we detected the bacteria in spleen homogenates, a more difficult diagnostic matrix than blood. Indeed, for all three probes, the LOD for spiked spleens was higher than for pure culture suspensions. The LOD in spleens was $2.9*10^6$ cells/ml, mostly

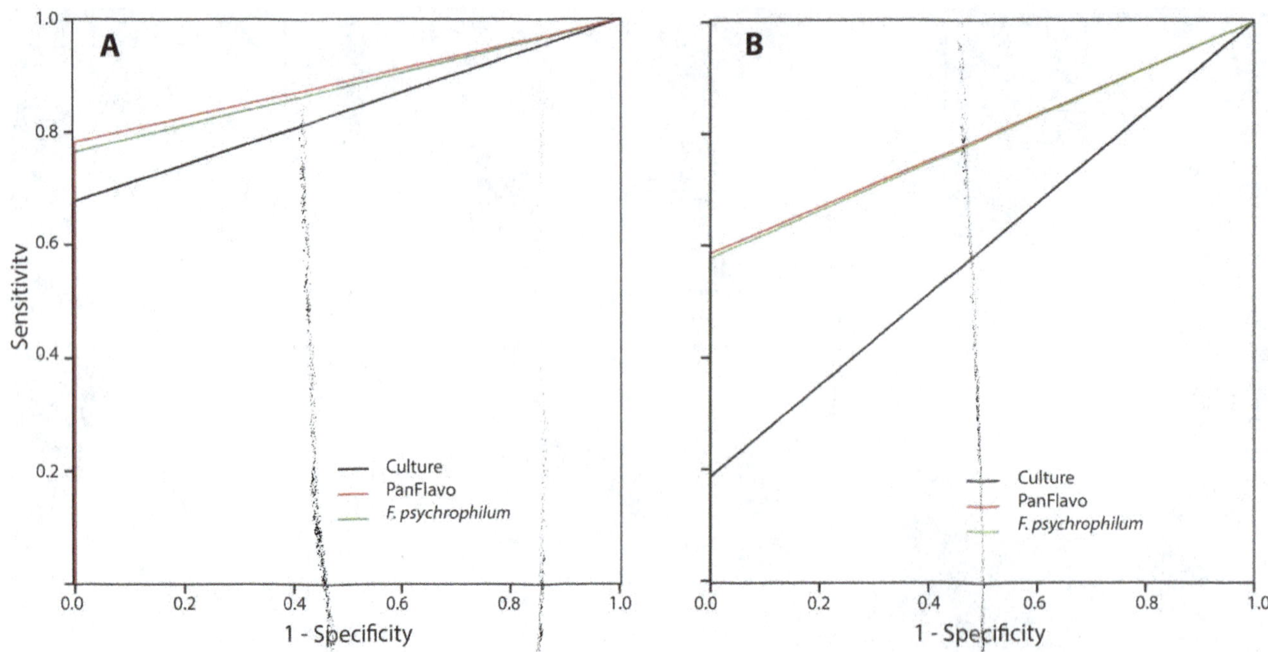

Figure 2. ROC curves for cell suspension of pure strains; area under the curve (AUC) for FISH: 0.89, for culture method: 0.79. (A). ROC curves for spiked spleens; AUC for FISH: 0.84, for culture method: 0.6 (B).

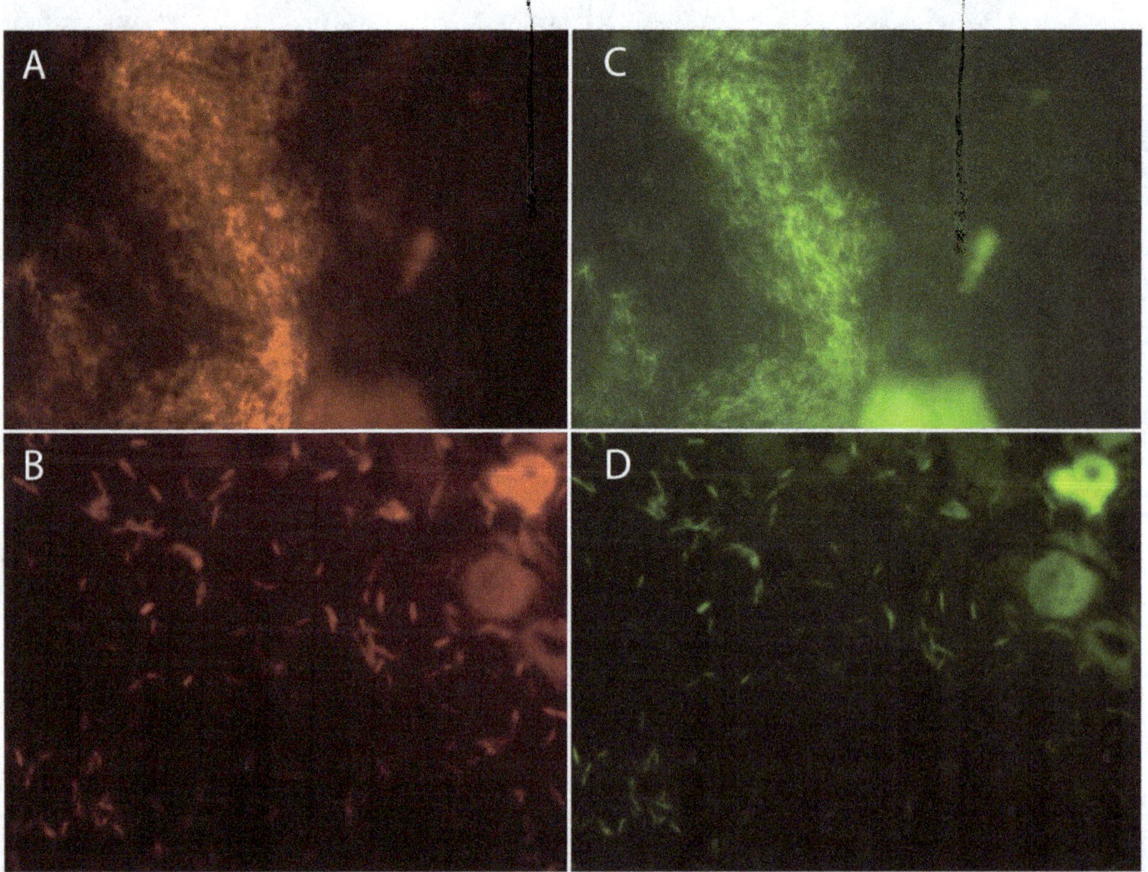

Figure 3. FISH assay on infected fish tissues. Pan-Flavo probe (A, B); *F. psychrophilum* probes (C, D). *F. psychrophilum* on skin (A, C) and *F. psychrophilum* in a spleen (B, D).

because of a rather high background fluorescence probably caused by the presence of muscular tissue and collagen. This is in agreement with a study by Marquardt and Wold [33] who used Raman spectroscopy to quantify collagen, fat and pigments such as carotenoids that were reportedly highly autofluorescent.

FISH detected *F. psychrophilum* within 24 hours in all infected samples: fresh samples of spleen and mucus are particularly well suited for analysis. The rapid diagnosis by FISH allows starting a timely and adequate treatment of the infection and could thus lead to better results in fish survival. FISH shows also a great potential for use on fixed tissues in retrospective studies of infections by *Flavobacterium*. Results, however, may be difficult to interpret due to the high background fluorescence of tissues. We had only four confirmed cases of *F. psychrophilum* infection available to test the method. In three of these cases the pathogen could be detected in spleen and liver tissues; while in one case the detection of *F. psychrophilum* was not possible. This could be explained by an inhomogeneous distribution of the infection in the tissues studied or by a bacterial count below LOD. On the other hand, the high background fluorescence could lead to false negative results when screening tissue sections.

FISH is an easy, fast and non-labour intensive technique. It does not require particular technical skills and is already used in many different fields such as clinical, veterinary, food and environmental microbiology [10,32,34,35,36]. Here we describe for the first time the successful use of FISH probes for the detection of *Flavobacterium* spp. and *F. psychrophilum* in environmental and tissue samples. The method described allows a fast and reliable qualitative detection of *Flavobacterium* spp. and *F. psychrophilum* in potentially infected tissues. While the method is particularly convenient in the diagnostic field, it does not replace culture, which is still needed for antibiotic sensitivity testing and other physiological studies.

Acknowledgments

We thank Joyce Rigozzi, AnnaPaola Caminada and Nadia Ruggeri-Bernardi for their excellent technical assistance. Dr. Jean-François Bernardet (Institut National de la Recherche Agronomique, Unité de Virologie et Immunologie Moléculaires, Jouy-en-Josas, France), Dr. Marc Lawrence (College of Veterinary Medicine Mississippi State, MS, USA) and Dr. Tom Wiklund (Laboratory of Aquatic Pathobiology, Environmental and Marine Biology, Department of Biosciences, Åbo Akademi University, Finland) provided us with well-characterized *Flavobacterium* isolates. The critical review of the manuscript by Dr. Cristina Fragoso and the proof-reading of the manuscript by Julie Guidotti are gratefully acknowledged.

Author Contributions

Conceived and designed the experiments: NS BP OP. Performed the experiments: NS. Analyzed the data: NS OP. Contributed reagents/materials/analysis tools: TW. Wrote the paper: NS TW HS OP.

References

1. Bernardet JF, Kerouault B (1989) Phenotypic and genomic studies of "Cytophaga psychrophila" isolated from diseased rainbow trout (Oncorhynchus mykiss) in France. Appl Environ Microbiol 55: 1796–1800.
2. Decostere A, Ducatelle R, Haesebrouck F (2002) Flavobacterium columnare (Flexibacter columnaris) associated with severe gill necrosis in koi carp (Cyprinus carpio L). Vet Rec 150: 694–695.
3. Kirby AL, Rosenkrantz WS, Ghubash RM, Neradilek B, Polissar NL (2010) Evaluation of otoscope cone disinfection techniques and contamination level in small animal private practice. Vet Dermatol 21: 175–183.
4. Laguna-del Estal P, Garcia-Montero P, Agud-Fernandez M, Lopez-Cano Gomez M, Castaneda-Pastor A, et al. (2010) [Bacterial meningitis due to gram-negative bacilli in adults]. Rev Neurol 50: 458–462.
5. Cipriano RC, Ford LA, Teska JD (1995) Association of Cytophaga psychrophila with mortality among eyed eggs of Atlantic salmon (Salmo salar). J Wildl Dis 31: 166–171.
6. Kondo M, Kawai K, Okabe M, Nakano N, Oshima S (2003) Efficacy of oral vaccine against bacterial coldwater disease in ayu Plecoglossus altivelis. Dis Aquat Organ 55: 261–264.
7. Nematollahi A, Decostere A, Pasmans F, Haesebrouck F (2003) Flavobacterium psychrophilum infections in salmonid fish. J Fish Dis 26: 563–574.
8. Wiklund T, Madsen L, Bruun MS, Dalsgaard I (2000) Detection of Flavobacterium psychrophilum from fish tissue and water samples by PCR amplification. J Appl Microbiol 88: 299–307.
9. Bottari B, Ercolini D, Gatti M, Neviani E (2006) Application of FISH technology for microbiological analysis: current state and prospects. Appl Microbiol Biotechnol 73: 485–494.
10. Werckenthin C, Gey A, Straubinger RK, Poppert S (2011) Rapid identification of the animal pathogens Streptococcus uberis and Arcanobacterium pyogenes by fluorescence in situ hybridization (FISH). Vet Microbiol 156: 330–335.
11. Manz W, Amann R, Ludwig W, Vancanneyt M, Schleifer KH (1996) Application of a suite of 16S rRNA-specific oligonucleotide probes designed to investigate bacteria of the phylum Cytophaga-Flavobacter-Bacteroides in the natural environment. Microbiology 142: 1097–1106.
12. Weller R, Glockner FO, Amann R (2000) 16S rRNA-targeted oligonucleotide probes for the in situ detection of members of the phylum Cytophaga-Flavobacterium-Bacteroides. Syst Appl Microbiol 23: 107–114.
13. Bernardet JF, Nakagawa Y, Holmes B (2002) Proposed minimal standards for describing new taxa of the family Flavobacteriaceae and emended description of the family. Int J Syst Evol Microbiol 52: 1049–1070.
14. Mellmann A, Cloud J, Maier T, Keckevoet U, Ramminger I, et al. (2008) Evaluation of matrix-assisted laser desorption ionization-time-of-flight mass spectrometry in comparison to 16S rRNA gene sequencing for species identification of nonfermenting bacteria. J Clin Microbiol 46: 1946–1954.
15. Benagli C, Rossi V, Dolina M, Tonolla M, Petrini O (2011) Matrix-assisted laser desorption ionization-time of flight mass spectrometry for the identification of clinically relevant bacteria. PLoS One 6: e16424.
16. Lane DJ (1991) 16S/23S rRNA sequencing: Chichester: Wiley & Sons. 115–175. p.
17. Fuchs BM, Wallner G, Beisker W, Schwippl I, Ludwig W, et al. (1998) Flow cytometric analysis of the in situ accessibility of Escherichia coli 16S rRNA for fluorescently labeled oligonucleotide probes. Appl Environ Microbiol 64: 4973–4982.
18. Cole JR, Chai B, Farris RJ, Wang Q, Kulam-Syed-Mohideen AS, et al. (2007) The ribosomal database project (RDP-II): introducing myRDP space and quality controlled public data. Nucleic Acids Res 35: D169–172.
19. Cole JR, Wang Q, Cardenas E, Fish J, Chai B, et al. (2009) The Ribosomal Database Project: improved alignments and new tools for rRNA analysis. Nucleic Acids Res 37: D141–145.
20. Duchaud E, Boussaha M, Loux V, Bernardet JF, Michel C, et al. (2007) Complete genome sequence of the fish pathogen Flavobacterium psychrophilum. Nat Biotechnol 25: 763–769.
21. Yun JJ, Heisler LE, Hwang, II, Wilkins O, Lau SK, et al. (2006) Genomic DNA functions as a universal external standard in quantitative real-time PCR. Nucleic Acids Res 34: e85.
22. Ridderstrale KK, Grushko TA, Kim HJ, Olopade OI (2005) Single-day FISH procedure for paraffin-embedded tissue sections using a microwave oven. Biotechniques 39: 316, 318, 320.
23. Kraemer HC (1992) Evaluating Medical Tests. Newbury Park, CA.
24. Mackinnon A (2000) A spreadsheet for the calculation of comprehensive statistics for the assessment of diagnostic tests and inter-rater agreement. Computers in Biology and Medicine 30: 127–134.
25. McKenzie D, Vida S, Mackinnon AJ, Onghena P, Clarke D (1997) Accurate confidence intervals for measures of test performance. Psychiatry Research 69: 207–209.
26. Tamura K, Dudley J, Nei M, Kumar S (2007) MEGA4: Molecular Evolutionary Genetics Analysis (MEGA) software version 4.0. Mol Biol Evol 24: 1596–1599.
27. Amann RI, Ludwig W, Schleifer KH (1995) Phylogenetic identification and in situ detection of individual microbial cells without cultivation. Microbiol Rev 59: 143–169.
28. Fuchs BM, Zubkov MV, Sahm K, Burkill PH, Amann R (2000) Changes in community composition during dilution cultures of marine bacterioplankton as assessed by flow cytometric and molecular biological techniques. Environ Microbiol 2: 191–201.
29. Fuchs BM, Glockner FO, Wulf J, Amann R (2000) Unlabeled helper oligonucleotides increase the in situ accessibility to 16S rRNA of fluorescently labeled oligonucleotide probes. Appl Environ Microbiol 66: 3603–3607.

30. Ciantar M, Newman HN, Wilson M, Spratt DA (2005) Molecular identification of Capnocytophaga spp. via 16S rRNA PCR-restriction fragment length polymorphism analysis. J Clin Microbiol 43: 1894–1901.
31. Bowman JP (2006) The Marine Clade of the Family Flavobacteriaceae: The Genera Aequorivita, Arenibacter, Cellulophaga, Croceibacter, Formosa, Gelidibacter, Gillisia, Maribacter, Mesonia, Muricauda, Polaribacter, Psychroflexus, Psychroserpens, Robiginitalea, Salegentibacter, Tenacibaculum, Ulvibacter, Vitellibacter and Zobellia. Procaryotes 7: 677–694.
32. Kempf VA, Trebesius K, Autenrieth IB (2000) Fluorescent In situ hybridization allows rapid identification of microorganisms in blood cultures. J Clin Microbiol 38: 830–838.
33. Marquardt BJ, Wold JP (2004) Raman analysis of fish: a potential method for rapid quality screening. LWT - Food Science and Technology 37: 1–8.
34. Angelidis AS, Tirodimos I, Bobos M, Kalamaki MS, Papageorgiou DK, et al. Detection of Helicobacter pylori in raw bovine milk by fluorescence in situ hybridization (FISH). Int J Food Microbiol 151: 252–256.
35. Nilsson SK, Hulspas R, Weier HU, Quesenberry PJ (1996) In situ detection of individual transplanted bone marrow cells using FISH on sections of paraffin-embedded whole murine femurs. J Histochem Cytochem 44: 1069–1074.
36. Rigby S, Procop GW, Haase G, Wilson D, Hall G, et al. (2002) Fluorescence in situ hybridization with peptide nucleic acid probes for rapid identification of Candida albicans directly from blood culture bottles. J Clin Microbiol 40: 2182–2186.

18

Reconstruction of *Danio rerio* Metabolic Model Accounting for Subcellular Compartmentalisation

Michaël Bekaert*

Institute of Aquaculture, University of Stirling, Stirling, Scotland, United Kingdom

Abstract

Plant and microbial metabolic engineering is commonly used in the production of functional foods and quality trait improvement. Computational model-based approaches have been used in this important endeavour. However, to date, fish metabolic models have only been scarcely and partially developed, in marked contrast to their prominent success in metabolic engineering. In this study we present the reconstruction of fully compartmentalised models of the *Danio rerio* (zebrafish) on a global scale. This reconstruction involves extraction of known biochemical reactions in *D. rerio* for both primary and secondary metabolism and the implementation of methods for determining subcellular localisation and assignment of enzymes. The reconstructed model (ZebraGEM) is amenable for constraint-based modelling analysis, and accounts for 4,988 genes coding for 2,406 gene-associated reactions and only 418 non-gene-associated reactions. A set of computational validations (i.e., simulations of known metabolic functionalities and experimental data) strongly testifies to the predictive ability of the model. Overall, the reconstructed model is expected to lay down the foundations for computational-based rational design of fish metabolic engineering in aquaculture.

Editor: Ramani Ramchandran, Medical College of Wisconsin, United States of America

Funding: This work was supported by the Marine Alliance for Science and Technology for Scotland (MASTS). The funders had no role in study design, data collection and analysis, decision to publish, or preparation of the manuscript.

Competing Interests: The author has declared that no competing interests exist.

* E-mail: michael.bekaert@stir.ac.uk

Introduction

High blood levels of long-chain omega-3 fatty acids (n-3 FAs), eicosapentaenoic acid (EPA, 20:5) and docosahexaenoic acid (DHA, 22:6), decrease risk of human cardiovascular disease events [1]. Intake of fatty, cold-water fish such as salmon, mackerel, and sardines, or fish oil capsules are the most effective methods of increasing EPA and DHA levels in the blood [2]. Fish metabolic engineering predominantly involves fisheries, however the fish farming industry has been criticised for being a net consumer of marine resources, in the form of fishmeals and fish oils used in feeds, despite the efforts made to replace them with alternatives, such as vegetable proteins and oils [3].

Current challenges in using fish as factories for public health nutrition and nutraceuticals require predesigned and efficient strategies for metabolic engineering [4]. Currently, fish metabolic engineering mostly involves trial-and-error approaches, without the utilisation of computational modelling procedures to rationally design genetic modifications. The marginal contribution played by metabolic modelling in fish until now stands in marked contrast to its prominent success in plant and microbial metabolic engineering [4–6]. The reconstruction process is well established for metabolic networks [7]. Once assembled, the reconstruction can be readily converted into a mathematical format by adding balances (*e.g.*, mass-balance constraints), steady-state assumptions [8]. The resulting model is condition-specific and can be used for phenotype simulations using various constraint-based reconstruction and analysis methods [8,9]. This approach has proven successful for numerous microorganisms [10] and eukaryotes for addressing various biological and biotechnological questions, such as the analysis of knowledge gaps [11], simulation of phenotype traits [12], analysis of evolution of metabolic networks [13,14] and metabolic engineering applications [15]. These analyses solely rely on simple physical-chemical constraints, thus overcoming the problem of missing enzyme kinetic data [5]. Numerous applications for large-scale plant and microbial networks have proven to be highly successful in predicting metabolic phenotypes in metabolic engineering and many other applications [5,16]. The reconstruction of metabolic network models for multicellular eukaryotes is significantly more challenging than that of bacteria, because of the larger size of the networks, the subcellular compartmentalisation of metabolic processes, and the considerable variation in tissue-specific metabolic activity [17].

A significant step forward in fish metabolic modelling was made with the publication of large-scale MetaFishNet model analysing high throughput expression data, as well as a framework for future systems studies [18]. This genome-wide model offered a significant expansion of the Kyoto Encyclopedia of Genes and Genomes (KEGG) zebrafish model [19]. In this work we present a computational reconstruction of genome-scale, subcellular compartmentalised metabolic network model for *D. rerio* (ZebraGEM), which relied on genomic and biochemical data extracted from various databases and literature. The validity of the model reconstruction steps is demonstrated via computational tests (cross-validation tests and simulations of known metabolic functionalities).

In this study, we present the reconstruction of the global *D. rerio* metabolic map. ZebraGEM is a comprehensive literature-based genome-scale metabolic reconstruction using manual curation, including extensive gap-assessment and filling. ZebraGEM has 9 compartments (cytoplasm, endoplasmic reticulum, extracellular space, Golgi apparatus, lysosome, mitochondria, mitochondrial intermembrane space, nucleus and peroxisome), 18 interfaces (surface orientations) and accounts for the functions of 4,988 proteins, 1,553 metabolites, and 2,824 metabolic and transport reactions. This network reconstruction was transformed into an *in silico* model and validated through the simulation of known metabolic functions found in a several tissue types.

Results and Discussion

Global Model Reconstruction

We adopted a reaction-centered view of *D. rerio* metabolism, where each metabolic gene is assigned to one or more reactions. We used Enzyme Commission numbers [20] to identify an initial set of metabolic genes from the KEGG database [19], National Center for Biotechnology Information's Gene [21], and Reactome Database [22]. These genes were mapped to a rudimentary network of 1,374 metabolic enzymes and 1,550 reactions (see Methods). In addition we collected zerbafish-specific reactions from MetaFishNET [18]. The final *D. reria* metabolic model (ZebraGEM) was then reconstructed based on biochemical and physiological knowledge by surveying literature and mining databases. We evaluated more than 15 years of biological evidence from more than 1,800 primary literature articles, reviews, and biochemical textbooks. The reconstruction was almost entirely constructed from zebrafish-specific data and includes many reactions directly extracted from the literature that are not described in any chart or database. Furthermore, it represents carefully formulated metabolites and reactions, which account for known reaction stoichiometry, substrate/cofactor specificity, and directionality, as well as overall conservation of mass and metabolite ionisation states at pH 7.2. Unambiguous metabolite names were verified using Chemical Identifier Resolver [23].

Compartmentalised Model Reconstruction

To extend the genome-scale *D. rerio* model reconstructed in the above step and to account for subcellular compartmentalisation of metabolic processes, we considered the known sub-localisation of the reactions taken from literature articles and reviews. Manual literature-based reconstruction ensured that the network components and their interactions were based on direct biochemical and subcellular evidence and reflected the current knowledge of *D. rerio* metabolism (see Methods).

The analysis resulted in a compartmentalised model Figure 1), with each enzyme localised between eight intracellular spaces (cytoplasm, endoplasmic reticulum, Golgi apparatus, lysosome, mitochondria, mitochondrial intermembrane space, nucleus and peroxisome) and the extracellular space Figure 1A. Furthermore, for each enzyme, the precise sub-cellular location and surface orientation on the membrane site or interface (when a reaction other than a transport is associated with the membrane; *e.g.*, endoplasmic reticulum membrane, cytosolic side) was identified.

Gap Filling

First, we decided to test whether the model was able to produce biomass by optimising the corresponding reaction accounting for all known biomass precursors required for cell replication. The non-compartmentalised draft model was able to produce biomass without any modifications. In contrast, the compartmentalised draft model was unable to produce biomass. Therefore, we sought to gap fill the draft compartmentalised to identify transporters needed to be added to the model for production of all biomass components. We searched for the minimal number of transporters necessary to add to the model in order to enable the production of biomass. The results of the algorithm were checked manually for two criteria: (i) the result did not suggest a reversible reaction for a known irreversible reaction and (ii) the added reaction transports a metabolite from a compartment where it was produced to a compartment where it was used as reactant.

Compartmentalised Model

A detailed list of the transporters is provided in Table S1. A small but significant percentage of the reactions are occurring in multiple compartments, in accordance with the experimental data used for the model reconstruction 20%, 3%, and 2% of the reactions were localised to two, three, or a higher number of compartments, respectively; This described 2,286 biochemical reactions, 2,824 compartment-specific reaction (with associated

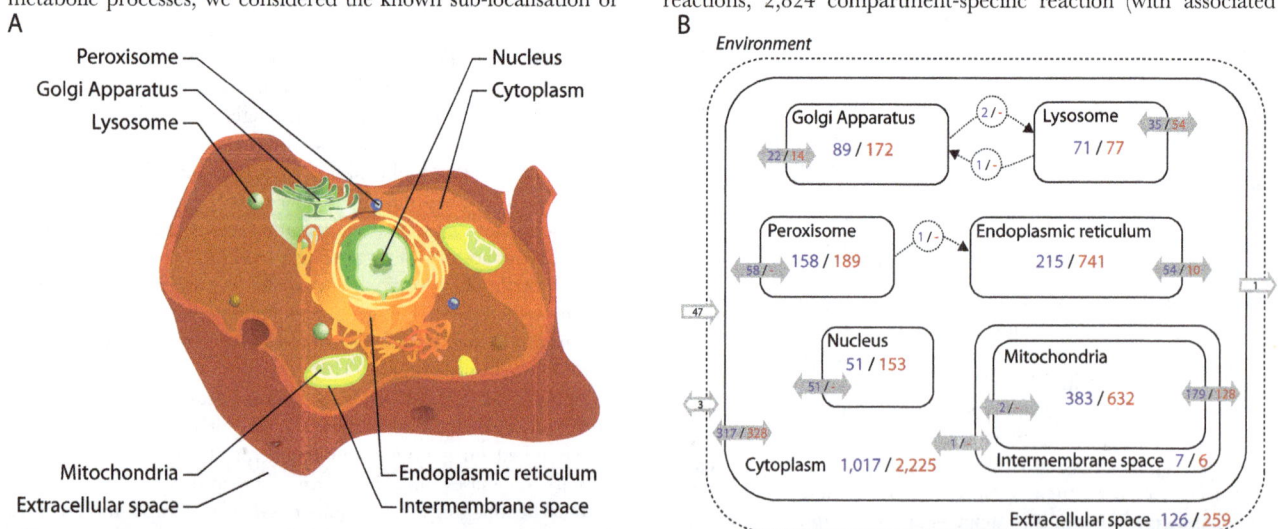

A

Peroxisome
Golgi Apparatus
Lysosome
Nucleus
Cytoplasm
Mitochondria
Extracellular space
Endoplasmic reticulum
Intermembrane space

B

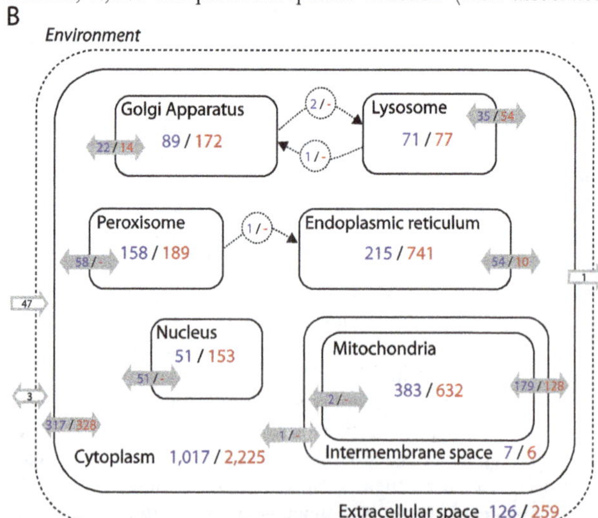

Figure 1. Overview of the metabolic network of *Danio rerio*, ZebraGEM. (A) Schematic of the 9 different compartments considered and their interfaces. **(B)** Break down of the reactions distribution (in blue) and associated genes (in red) for each compartment. Environmental exchanges are shown in grey-border arrows, biomass export as output, uptakes as input and oxygen, water and carbon dioxide freely exchange as doubled arrows. Solid-grey arrows denote reactions across membranes, with number of reactions and genes respectively indicated. Circles details vesicular transports.

enzymes) and 2,910 compartment-specific, surface-specific reactions (*e.g.*, one cytosolic reaction can occur both on the cytosolic side of the Golgi apparatus and the endoplasmic reticulum membranes). However, even if some reactions occur in more than one compartment, enzymes are reaction- and compartment-specific; No enzyme was described in more than one compartment. Table S2 describes the main pathways, their compartment, and the reaction count.

Figure 1B outlines the model by summarising the reaction and gene content for each subcellular compartmentalisation and metabolite exchanges between compartmentalisation and with the environment. ZebraGEM contains 9 different compartments Figure 2A. 18 interfaces were used for transport or exchange with the environment, and orientate membrane-associated reaction sides Figure 2B. There are 47 nutrient uptakes, while oxygen, water and carbon dioxide can freely exchange with the environment. The synthesis of 42 biomass components is summarised in the biomass reaction (Table S3). 1,553 different metabolites have associations with 2,824 reactions, 4,988 genes/enzymes. Only 418 reactions remain without any known enzyme. About 25% of all reactions involve transport between compartments. The resulting network has a density of 0.024, a diameter of 17 and average path length of 3.49. The entire content of *D. rerio* ZebraGEM is freely available Systems Biology Markup Language format [24] as EBI BioModel: MODEL1204120000. To date, it represents the most comprehencise metabolic networks. Compared to ZebrGEM, KEGG database presents only 1,550 individual reactions without compartment data and MetaFishNET 2,523 unique reactions without enzyme association and only a partial compartmentalisation.

Figures 2A & 3A outlines ZebraGEM model as it reports all reactions distributed in the 9 different compartments and their links (shared metabolites). Figures 2B & 3B details the reactions distributed in the 18 different interfaces (lumen and membrane organelles). We subsequently used the metabolic network model to examine classes of metabolites that are uniquely assigned to a particular compartment, either in their lumen or the membranes Figures 3. The extensive transport of metabolites between compartments makes such analysis at the network-scale of particular interest.

Model Validation

Following previous model reconstructions [25–27], we further validated the global model reconstruction using a set of simulations of known metabolic functions, such as amino acids, as well as secondary metabolites biosynthesis and degradation (Table S4). All simulations were successful, demonstrating that this model is indeed functional (see Methods).

Experimental Support

Examination of experimentally characterised pathway activity was used to validate the global metabolism model. For example, the *de novo* biosynthesis of taurine, our model suggests that the only path able to synthesis the taurine is controled by cysteine sulfinic acid decarboxylase. This finding is in accordance with the recently reported critical role of cysteine sulfinic acid decarboxylase in *D. rerio* taurine homeostasis and cardiac development [28]. Likewise, a large portion of the unique metabolites detected in the endoplasmic reticulum were fatty acid-related, including sterols, sphingolipids, and CoAs [29]. In the peroxisome, CoA metabolites were prominent [30–33].

Conclusion

System biology approaches to evolution are attractive as they bring the promise of quantitative analyses of trait evolution and behaviour, integrating phenomena from the molecular to the ecosystem level. As a result, they allow us to describe precise hypotheses as to how traits can be successively used for metabolic engineering.

Several limitations should be noted: (i) As the approach hinges upon different molecular data, its accuracy depends on the quality of the latter. (ii) Several assumptions are used in the reconstruction process, such as the assumption of minimal addition of reactions and directionality relaxation to resolve network gaps, and the

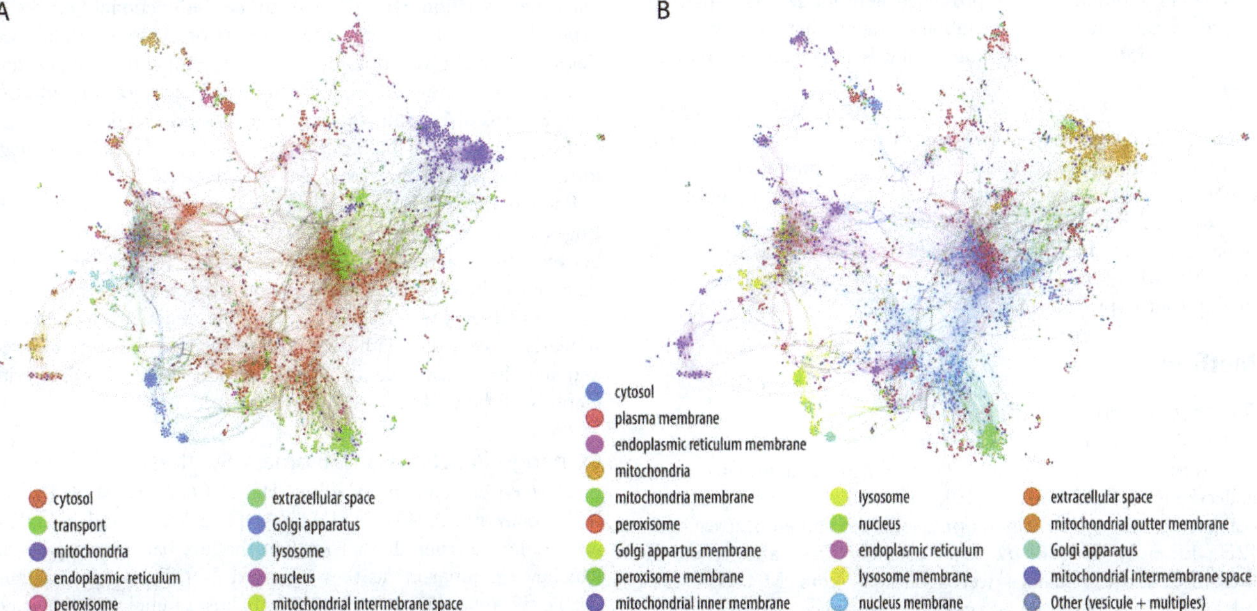

Figure 2. ZebraGEM model (EBI BioModel: MODEL1204120000). (**A**) Compartmentalised network. Each compartment is colour-coded. Nodes represent metabolic reactions, while edges links shared metabolites accounting for the reaction directionality, from product to reactant. (**B**) Surface orientation and interface details. The location of each reaction (excluding transporters) within the compartments and their membranes is detailed.

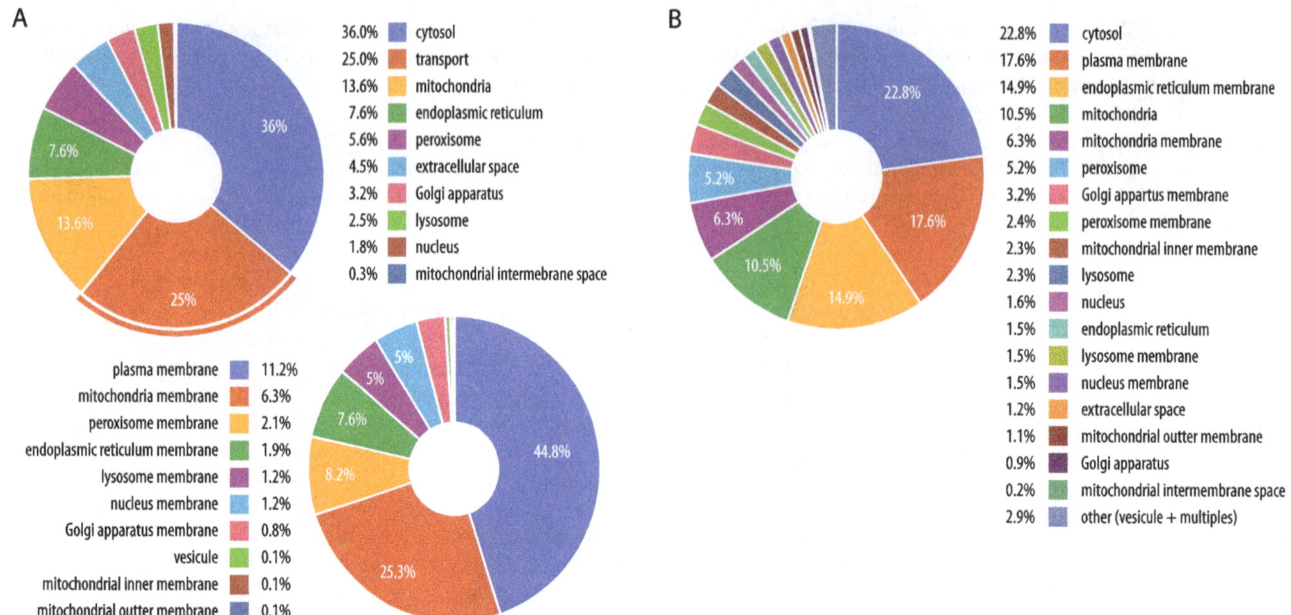

Figure 3. Subcellular localisation of metabolic reactions. (**A**) Distribution of the reaction per compartments or transporters. Transport compartment are detailed by membrane location. (**B**) Details of the reaction distribution (transporters and none transporters) per interfaces (lumen and membrane organelles).

assumption of minimal number of metabolite transmembrane transport used in the localisation assignment. These assumptions, although commonly used for similar purposes [34,35], may in some cases be oversimplified. (iii) The resulting models do not explicitly account for transcriptional and metabolic regulation; data that is still unavailable for most *D. rerio* metabolic functions, and thus cannot accurately predict some of the organism's functions. It should be noted, however, that further experimental validation should be performed to reaffirm the predictions.

A number of future refinements and extensions will improve the quality of the resulting network models. Such reconstructed models, in addition to those presented here for *D. rerio*, open up future opportunities for metabolic engineering of increased production rates of various nutraceuticals important for human and fish health.

In this work we present the reconstruction of a fully compartmentalised model of *D. reria*, a model fish and present a new feature in metabolic modelling: surface orientation. Despite its evolutionary divergence to farm fish, the genetic make-up of the *D. reria* model fish is the first step toward fish metabolic engineering, making its study highly useful to other fish, as well as to the fish industry, where metabolic engineering is commonly used for quality-trait improvement.

Methods

Reconstruction Procedure

An initial component list was assembled as described in the text (see Results and Discussion). Putative gene assignments were collected for KEGG build 39 [19], and verified based on evidence collected from genome annotation databases, namely EntrezGene [21], Ensembl 64 [36] *D. rerio* assembly Zv9, and scientific literature. Extra reactions were collected from MetaFishNET release 1.9 [18]. Substrate and cofactor preferences were identified from the literature and NCI/CADD Chemical Identifier Resolver beta 4 [23] or ChemSpider [37]. Metabolite formulae and charges were calculated based on their ionisation state at pH 7.2, which was assumed to be constant across all compartments. Reaction directionality was determined from thermodynamic data or inferred from empiric data and textbooks. For clarity, we have used only the apparent primary specificity of each enzyme in our network. Compartmentalisation was determined from protein localisation data; sequence targeting signals, and indirect physiological evidence. Transport reactions were entirely reconstructed based on literature reports and biochemistry textbooks because the current annotation of transporters is not sufficiently specific with regard to substrates and mechanisms. Each reaction in our network was manually curated with regard to reaction stoichiometry, subcellular localisation and directionality. By default each particular reaction was defined to be bidirectional, except if explicitly stated otherwise by the literature or the source databases. Based on the controlled nomenclature, the function of a reaction was used to generate descriptive statistics. The entire content of *D. rerio* ZebraGEM, with the references used to describe each reaction, is freely available Systems Biology Markup Language format [24] as EBI BioModel: MODEL1204120000.

Each mapped reaction is defined as a node in our network. Edges between these nodes are defined by shared metabolites between the reactions. The network is directed: for irreversible reactions if the product of one reaction is a reactant in the second, we defined a directed edge. Reversible reactions are treated similarly, except that both directions of the reaction were allowed and handled independently. The networks were visualised with Gephi v0.8 beta [38].

Exchange Reactions and Biomass Synthesis

Thirteen currency metabolites (H^+, H_2O, ATP, ADP, P_i, PP_i, Na^+, coenzyme A, O_2, NAD^+, NADH, $NADP^+$, and NADPH) were removed from all analyses (but the flux balance analyses) in whichever compartments they occurred [39]. Exchange reactions (Table S2, type 'import') allow metabolites to enter and/or leave the metabolic network and were formulated considering the minimum nutritional physiology of the cell. According to published dietary requirement from farm fish [40], the network considers 10 essential amino acids, 2 lipids, 9 carbohydrates, 6 macrominerals, 6 microminerals, 4 fat-soluble vitamins and 10 water-soluble vitamins nutrient uptakes. Likewise, four lipids can be used in fish nutrition; linolenic and linoleic acid, mainly for freshwater fish and eicosapentaenoic and docosahexaenoic acid for seawater fish; since the *D. rerio* in a freshwater fish and the model allows the synthesis of eicosapentaenoic and docosahexaenoic acid, only linolenic and linoleic acid were included. Oxygen, water and carbon dioxide can freely exchange with the environment. Metabolic network biomass is composed of all amino acids (protein), nucleic acids (DNA, RNA) and lipids, including triacylglycerol and cholesterol (Table S2, type 'biomass').

Gap Filling

Gap filling was done by importing the required metabolites from one compartment to another in order to allow all reactions to occur, and subsequently to produce all biomass components. Every metabolite that could not be produced or consumed was manually examined to identify possible reactions describing its degradation, production, or transport. Each added transporter was manually checked for applicability (such as the feasibility of the suggested directionality) and for being reported in at least one other species (fish when available, mammalian otherwise) before been included. We searched for the minimal number of transporters necessary to add to the model in order to enable the production of biomass.

Flux Balance Analysis

We used the Systems Biology Research Tool v2.0.0 [41] to perform flux balance analysis on the networks [34]. Exchange reactions were added to enable uptake and secretion of extracellular metabolites for the purpose of simulations (Table S5). The optimum flux distribution is here defined as the flux distribution that minimises uptakes for a fixed rate of biomass synthesis. The biomass composition (Table S2) was taken from the published dietary requirement from farm fish [40]. Note that exchange fluxes are negative for influx and positive for efflux by definition.

Functional Validation

Functional validation was performed by using flux balance analysis. The test simulations validates that the targeted metabolite could be synthetised from a particular source metabolite without violating the other model constraints (*i.e.*, reaction directions, model inputs).

Supporting Information

Table S1 Detailed list of the transporters.

Table S2 Detailed list of the pathways by compartment and the number of reaction involved.

Table S3 Detailed list the uptakes and biomass reactions.

Table S4 Simulations of known metabolic functions. Complete list of the 160 simulations used to validate ZebraGEM.

Table S5 Exchange reactions added to enable uptake and secretion of extracellular metabolites for the purpose of simulations.

Acknowledgments

The author thanks Margaret O' Toole for her help with manuscript preparation.

Author Contributions

Conceived and designed the experiments: MB. Performed the experiments: MB. Analyzed the data: MB. Contributed reagents/materials/analysis tools: MB. Wrote the paper: MB.

References

1. Harris WS, Miller M, Tighe AP, Davidson MH, Schaefer EJ (2008) Omega-3 fatty acids and coronary heart disease risk: clinical and mechanistic perspectives. Atherosclerosis 197: 12–24.
2. Harris K, Fleming J, Kris-Etherton P (2011) Challenges in estimating omega-3 fatty acid content of seafood from US nutrient databases: A salmon case study. Journal of Food Composition and Analysis 24: 1168–1173.
3. Bendiksen EÅ, Johnsen CA, Olsen HJ, Jobling M (2011) Sustainable aquafeeds: Progress towards reduced reliance upon marine ingredients in diets for farmed Atlantic salmon (*Salmo salar* L.). Aquaculture 314: 132–139.
4. Aharoni A, Jongsma MA, Bouwmeester HJ (2005) Volatile science? Metabolic engineering of terpenoids in plants. Trends in plant science 10: 594–602.
5. Feist AM, Palsson BØ (2008) The growing scope of applications of genome-scale metabolic reconstructions using *Escherichia coli*. Nature biotechnology 26: 659–667.
6. Burgard AP, Pharkya P, Maranas CD (2003) Optknock: a bilevel programming framework for identifying gene knockout strategies for microbial strain optimization. Biotechnology and bioengineering 84: 647–657.
7. Thiele I, Palsson BØ (2010) A protocol for generating a high-quality genome-scale metabolic reconstruction. Nature protocols 5: 93–121.
8. Price ND, Reed JL, Palsson BØ (2004) Genome-scale models of microbial cells: evaluating the consequences of constraints. Nature reviews Microbiology 2: 886–897.
9. Becker SA, Feist AM, Mo ML, Hannum G, Palsson BO, et al. (2007) Quantitative prediction of cellular metabolism with constraint-based models: the COBRA Toolbox. Nature protocols 2: 727–738.
10. Henry CS, DeJongh M, Best AA, Frybarger PM, Linsay B, et al. (2010) High-throughput generation, optimization and analysis of genome-scale metabolic models. Nature biotechnology 28: 977–982.
11. Reed JL, Patel TR, Chen KH, Joyce AR, Applebee MK, et al. (2006) Systems approach to refining genome annotation. Proceedings of the National Academy of Sciences of the United States of America 103: 17480–17484.
12. Nogales J, Palsson BO, Thiele I (2008) A genome-scale metabolic reconstruction of *Pseudomonas putida* KT2440: iJN746 as a cell factory. BMC systems biology 2: 79.
13. Pal C, Papp B, Lercher MJ, Csermely P, Oliver SG, et al. (2006) Chance and necessity in the evolution of minimal metabolic networks. Nature 440: 667–670.
14. Zhang Y, Thiele I, Weekes D, Li Z, Jaroszewski L, et al. (2009) Three-dimensional structural view of the central metabolic network of *Thermotoga maritima*. Science 325: 1544–1549.
15. Park JH, Lee KH, Kim TY, Lee SY (2007) Metabolic engineering of *Escherichia coli* for the production of L-valine based on transcriptome analysis and *in silico* gene knockout simulation. Proceedings of the National Academy of Sciences of the United States of America 104: 7797–7802.
16. Oberhardt MA, Palsson BØ, Papin JA (2009) Applications of genome-scale metabolic reconstructions. Molecular systems biology 5: 320.
17. Morais S, Pratoomyot J, Taggart JB, Bron JE, Guy DR, et al. (2011) Genotype-specific responses in Atlantic salmon (*Salmo salar*) subject to dietary fish oil replacement by vegetable oil: a liver transcriptomic analysis. BMC genomics 12: 255.
18. Li S, Pozhitkov A, Ryan RA, Manning CS, Brown-Peterson N, et al. (2010) Constructing a fish metabolic network model. Genome biology 11: R115.
19. Kanehisa M, Goto S, Sato Y, Furumichi M, Tanabe M (2012) KEGG for integration and interpretation of large-scale molecular data sets. Nucleic acids research 40: D109–114.
20. Webb EC (1992) Enzyme nomenclature 1992: recommendations of the Nomenclature Committee of the International Union of Biochemistry and Molecular Biology on the nomenclature and classification of enzymes: Published for the International Union of Biochemistry and Molecular Biology by Academic Press.
21. Maglott D, Ostell J, Pruitt KD, Tatusova T (2011) Entrez Gene: gene-centered information at NCBI. Nucleic acids research 39: D52–57.

22. Haw RA, Croft D, Yung CK, Ndegwa N, D'Eustachio P, et al. (2011) The Reactome BioMart. Database : the journal of biological databases and curation 2011: bar031.
23. Sitzmann M, Ihlenfeldt W-D, Nicklaus MC. NCI/CADD Chemical Identifier Resolver: Indexind and Analysis of Available Chemistry Space; 2011 June 5–9 2011; June 5–9 2011, Noordwijkerhout, The Netherlands.
24. Hucka M, Finney A, Sauro HM, Bolouri H, Doyle JC, et al. (2003) The systems biology markup language (SBML): a medium for representation and exchange of biochemical network models. Bioinformatics 19: 524–531.
25. Duarte NC, Becker SA, Jamshidi N, Thiele I, Mo ML, et al. (2007) Global reconstruction of the human metabolic network based on genomic and bibliomic data. Proceedings of the National Academy of Sciences of the United States of America 104: 1777–1782.
26. Jerby L, Shlomi T, Ruppin E (2010) Computational reconstruction of tissue-specific metabolic models: application to human liver metabolism. Molecular systems biology 6: 401.
27. Mintz-Oron S, Meir S, Malitsky S, Ruppin E, Aharoni A, et al. (2012) Reconstruction of Arabidopsis metabolic network models accounting for subcellular compartmentalization and tissue-specificity. Proceedings of the National Academy of Sciences of the United States of America 109: 339–344.
28. Chang YC, Ding ST, Lee YH, Wang YC, Huang MF, et al. (2012) Taurine homeostasis requires de novo synthesis via cysteine sulfinic acid decarboxylase during zebrafish early embryogenesis. Amino acids.
29. Melville DB, Knapik EW (2011) Traffic jams in fish bones: ER-to-Golgi protein transport during zebrafish development. Cell adhesion & migration 5: 114–118.
30. Tseng YC, Chen RD, Lucassen M, Schmidt MM, Dringen R, et al. (2011) Exploring uncoupling proteins and antioxidant mechanisms under acute cold exposure in brains of fish. PloS One 6: e18180.
31. Monroig O, Rotllant J, Cerda-Reverter JM, Dick JR, Figueras A, et al. (2010) Expression and role of Elovl4 elongases in biosynthesis of very long-chain fatty acids during zebrafish *Danio rerio* early embryonic development. Biochimica et biophysica acta 1801: 1145–1154.
32. Krysko O, Stevens M, Langenberg T, Fransen M, Espeel M, et al. (2010) Peroxisomes in zebrafish: distribution pattern and knockdown studies. Histochemistry and cell biology 134: 39–51.
33. Song Y, Selak MA, Watson CT, Coutts C, Scherer PC, et al. (2009) Mechanisms underlying metabolic and neural defects in zebrafish and human multiple acyl-CoA dehydrogenase deficiency (MADD). PloS One 4: e8329.
34. Orth JD, Palsson BØ (2010) Systematizing the generation of missing metabolic knowledge. Biotechnology and bioengineering 107: 403–412.
35. Mintz-Oron S, Aharoni A, Ruppin E, Shlomi T (2009) Network-based prediction of metabolic enzymes' subcellular localization. Bioinformatics 25: i247–252.
36. Flicek P, Amode MR, Barrell D, Beal K, Brent S, et al. (2011) Ensembl 2011. Nucleic acids research 39: D800–806.
37. Williams AJ, Tkachenko V, Golotvin S, Kidd R, McCann G (2010) ChemSpider - building a foundation for the semantic web by hosting a crowd sourced databasing platform for chemistry. Journal of Cheminformatics 2: O16-O16.
38. Bastian M, Heymann S, Jacomy M (2009) Gephi: An Open Source Software for Exploring and Manipulating Networks. Third International AAAI Conference on Weblogs and Social Media. Sn Jose, CA: AAAI Publications. pp. 361–362.
39. Huss M, Holme P (2007) Currency and commodity metabolites: their identification and relation to the modularity of metabolic networks. IET systems biology 1: 280–285.
40. Subcommittee on Fish Nutrition, National Research Council (1993) Nutrient Requirements of Fish. Washington, D.C.: The National Academies Press.
41. Wright J, Wagner A (2008) The Systems Biology Research Tool: evolvable open-source software. BMC systems biology 2: 55.

19

Space-Time Modelling of the Spread of Salmon Lice between and within Norwegian Marine Salmon Farms

Magne Aldrin[1,2]*, Bård Storvik[1], Anja Bråthen Kristoffersen[3,4], Peder Andreas Jansen[3]

1 Norwegian Computing Center, Oslo, Norway, 2 Department of Mathematics, University of Oslo, Oslo, Norway, 3 Norwegian Veterinary Institute, Oslo, Norway, 4 Department of Informatics, University of Oslo, Oslo, Norway

Abstract

Parasitic salmon lice are potentially harmful to salmonid hosts and farm produced lice pose a threat to wild salmonids. To control salmon lice infections in Norwegian salmonid farming, numbers of lice are regularly counted and lice abundance is reported from all salmonid farms every month. We have developed a stochastic space-time model where monthly lice abundance is modelled simultaneously for all farms. The set of farms is regarded as a network where the degree of contact between farms depends on their seaway distance. The expected lice abundance at each farm is modelled as a function of i) lice abundance in previous months at the same farm, ii) at neighbourhood farms, and iii) other, unspecified sources. In addition, the model includes explanatory variables such as seawater temperature and farm-numbers of fish. The model gives insight into factors that affect salmon lice abundance and contributing sources of infection. New findings in this study were that 66% of the expected salmon lice abundance was attributed to infection within farms, 28% was attributed to infection from neighbourhood farms and 6% to non-specified sources of infection. Furthermore, we present the relative risk of infection between neighbourhood farms as a function of seaway distance, which can be viewed as a between farm transmission kernel for salmon lice. The present modelling framework lays the foundation for development of future scenario simulation tools for examining the spread and abundance of salmon lice on farmed salmonids under different control regimes.

Editor: Martin Krkosek, University of Toronto, Canada

Funding: This work was funded by the Research Council of Norway project PREVENT, "Salmon lice-prevention and treatment," project number 199778/S40. The funders had no role in study design, data collection and analysis, decision to publish, or preparation of the manuscript.

Competing Interests: The authors have declared that no competing interests exist.

* E-mail: magne.aldrin@nr.no

Introduction

Infectious diseases constitute a constant problem in industrialised farming where there typically are both high densities of farms and high densities of animals within each farm. Disease outbreaks can have large economic consequences to farming industries, but can also have severe ecological effects if infections are spread to and impair the viability of wild animals. Several mathematical and statistical models within this field have been developed during the last decades and applied to as diverse diseases as foot-and-mouth disease, swine fever, bluetongue and infectious salmon anaemia [1–5]. In all these models, probabilities of infection relative to distance, often called transmission kernels [1,4], play an important role.

Salmon lice (*Lepeophtheirus salmonis*) are parasitic copepods that live on the skin surface of both wild and farmed salmonids. The parasite uses rasping mouthparts to feed on mucus, skin and underlying tissues of its host and thereby causes mechanical damage [6]. Possible effects of salmon farming on sea lice infections on wild stocks of salmonids, and hence the viability of such wild stocks, has evoked a large and contentious debate [7–9]. Nevertheless, the notion that salmon farming does affect local transmission of salmon lice to wild salmonids [9,10], as well as to farmed salmonids [11], seems well established. Recent studies also report that parasiticide-treatment of outwardly migrating salmon smolts significantly increases their marine survival compared to non-treated control smolts, suggesting that salmon lice induce mortality in wild salmonid hosts [12–14]. Due to the potential impact of salmon lice of farm origin on wild stocks of salmonids, salmon lice infections on farmed fish are strictly regulated in Norway [11]. To enforce these regulations, numbers of salmon lice are counted on samples of farmed salmonids at regular intervals from all actively producing marine fish farms each month. Salmon lice abundances from these counts, i.e. the average number of salmon lice per fish, are reported to a central data base [11]. Measures to control salmon lice infections, i.e. the application of medical treatments or the use of cleaner-fish to prey on lice, are also reported to this same data base.

The large spatio-temporal dataset covering salmon lice abundance on salmon farms, and efforts to control these infections, should give insight into factors that affect farm levels of salmon lice infections and the contributing sources for such infection. The aim of the present study was to develop a modelling framework that could: i) disentangle different contributing sources of salmon lice infections; ii) estimate functional relationships between expected salmon lice infections and contributing factors, e.g. the between farm transmission kernel as a function of seaway distance between farms [1,4]; and iii) lay the foundation for a scenario simulation tool to examine the potential spread of salmon lice within and between salmon farms to assess the impact of control measures. We developed a stochastic space-time model where the monthly salmon lice abundances at all Norwegian salmonid farms are modelled simultaneously. The model is similar in flavour to previous models developed by the same group of authors on infectious salmon anaemia and other salmonid diseases where farms appear as infected or non-infected [5,15,16]. Salmon lice, however, differ from these other diseases in that most farms are infected more or less all the time, but salmon lice abundances vary both over time and between farms.

The set of salmonid farms is regarded as a network where the degree of contact between each pair of farms depend on the seaway distance between them. The observed number of lice in a given sample of fish is assumed to be distributed according to a zero-inflated negative binomial distribution. The expected salmon lice abundance at each farm is modelled as a function of i) the observed lice abundances in previous months at the same farm, ii) the observed lice abundances in previous months at neighbouring farms, and iii) other, unspecified sources of infection. In addition, the model includes explanatory variables such as seawater temperature and numbers of fish at each farm. The model was fitted to data covering all marine salmonid farms in Norway from June 2003 until the end of 2011.

In the following sections, we first describe the data and then the model. Then, we present the fitted model and other results, and finally we draw some conclusions.

Materials and Methods

Salmonid Farming and the Problem of Salmon Lice

During a 30 year period from 1980 to 2010, annual production of farmed salmonids, i.e. Atlantic salmon (*Salmo salar*) and rainbow trout (*Onchorhyncus mykiss*), has grown from marginal to around one million tonnes in Norway [17]. For comparison, the total annual yield of wild Atlantic salmon and sea trout (*Salmo trutta*) in Norway was estimated at 850–1400 tonnes between 2000 and 2008 [18]. The production of salmonids consists of a freshwater juvenile phase, followed by a marine grow out phase, which is the focus of this study. Marine salmonid farming is regulated through a system of legal concessions authorised by the Norwegian Directorate for Fisheries (DFF; http://www.fiskeridir.no). All farms are registered with a geographic reference in the Aquaculture register, which is available at DFF's website. For marine farms that actively produce salmonids, it is mandatory to report key statistics on their fish stocks, fish health related statisics and seawater temperature at 3 m depth, on a monthly basis.

The production of a salmonid cohort on a farm typically initiates by stocking juvenile smolts to net-pens on the farm in spring or in the autumn. The net-pens allow free water exchange with the surroundings. After stocking, the fishes are on-grown on the farm for roughly 1.5 years and then slaughtered. The production time varies slightly according to the weight of the fish at slaughter, seawater temperature etcetera. On a given farm, only fish of a given year class are produced. After slaughtering, it is mandatory to fallow the farm location for a short period before stocking a new cohort of salmonids. Occasionally fish may be moved from one farm location to another, in which case a new farm will initially report fish weights larger than juveniles.

Sea lice are naturally occurring parasitic copepods that spread by planktonic larvae. Farmed salmonids are infected by water flow through the net-pens. The parasites live on the surface of the fish and feed on mucus, skin and underlying tissues of its host [6]. Sea lice infections on farmed salmonids are dominated by the salmonid specific salmon louse, but may also be by the generalist parasite *Caligus elongatus*. However, regulations on reporting requirements regarding sea lice infections on farmed salmonids in Norway (see below) are only directed at the salmon louse. Hence, we term the parasitic infections for salmon lice in this paper. The salmon louse has a life cycle consisting of three planktonic larval stages, of which the third copepodite-stage may attach to the surface skin of a salmonid host by a frontal filament. Upon infection, the parasite develops through four consecutive stages of attached larvae, followed by two pre-adult stages of lice that may move about on the surface of the fish (mobile stages), and finally adult males and females that also are mobile. Sea lice infections on farmed salmonids are regulated through a system of maximum thresholds of abundance of mobile stages. To enforce these regulations, farmers are instructed to count salmon lice on farmed salmonids at regular intervals. Counts are reported once a month to a database as the mean number of lice per fish within two stage categories: i) adult females, and ii) all other mobile stages (pre-adults and adult males). The farmers are instructed to report the highest mean count of adult females during a given month. In the present model, we sum the two categories of salmon lice and term this salmon lice abundance. The mean count of salmon lice from a sample of 20 fish from one net pen (before August 2009) or the mean of samples of multiples of 10 fish from half of the net-pens (from August 2009) was reported. Details of the sampling, counting and reporting procedures for salmon lice infections on farmed fish are given in Jansen et al. ([11]; and the Electronic supplementary material therein).

Data

The present data include all 1401 Norwegian marine fish farms with standing stocks of either Atlantic salmon or rainbow trout (salmonids) in any month from February 2003 to December 2011. The data is an updated version of the data presented in Jansen et al. [11],who analysed data up to 2010 in a study based on a different modelling approach than the present.

The data from the first four months were only used to construct lagged explanatory variables, whereas the model was fitted to data from June 2003 and onwards. In the following description, we therefore only summarise data from the latter period. The number of active farms was then reduced to 1361. Each salmonid farm normally had several consecutive periods of production of fish populations, interrupted by periods of fallowing (no fish on the farm). The fish population within a production period is termed a cohort and the present data consists of 4 255 cohorts with a total of 58 623 farm-months of cohort production. Farms produced between one and nine consecutive cohorts. For each salmonid farm, the salmon lice abundance on a sample of fish is obligatory to report to the responsible Norwegian Food Safety Authority. We assumed that the number of sampled fish always was 20. The reported salmon lice abundances were therefore multiplied by 20 and implemented as the response variable in the present model. Hence, we ignored that the number of fish sampled from August 2009 could be both less and more than 20 (see previous section), but note that this did not affect the mean number of lice per fish. However, lice abundance data were missing for about 3% of the total number of months with farmed fish, giving 56 836 farm-months with reported salmon lice abundance. Salmon lice abundances amount to between 0 and 7 in 95% of the farm-months, with a mean of 1.1 lice per fish. The distribution of salmon lice abundance is profoundly skewed, being exactly 0 in 32% of the farm-months, whereas the highest reported abundance is 163 lice per fish. The black curves in Figure 1 show how the observed lice abundance vary over the data period averaged over all farms and for three selected farms. The lice counts have a clear seasonal pattern driven by the seawater temperature. The farm-wise panels also show how the fish cohorts can be distributed over

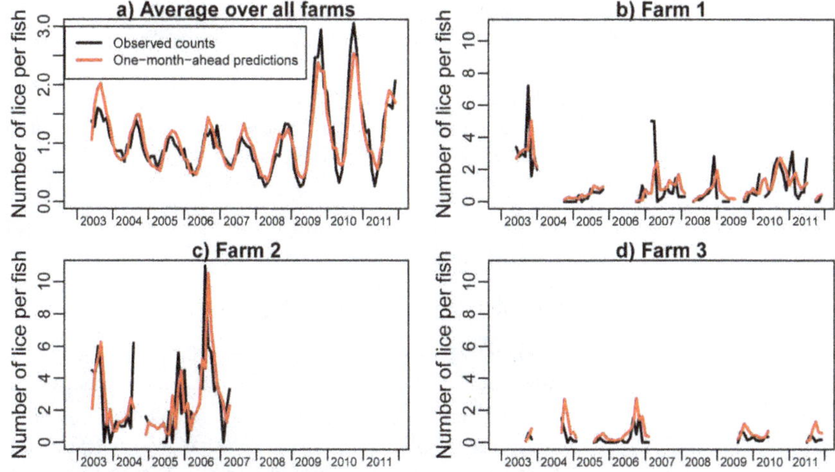

Figure 1. Time plots of observed (black) and one-month-ahead predictions (red) of salmon lice abundance averaged over all farms (panel a), and for three selected farms (panels b, c and d).

the data period. Farms 1 and 2 in the figure have periods where the farms are active, but the salmon lice counts are missing. This is shown by predictions (red line) but missing observations (no black line) in the time plots.

The geographic location of each farm and pairwise seaway distances between all farms were compiled from Jansen et al. [11]. The seaway distance between two farms is defined as the shortest distance through water. The various salmonid farms had between zero and 31 other salmonid farms within a seaway distance of 10 km (median five). Figure 2 shows the locations of all Norwegian salmonid farms that were actively producing salmonids at some month during the study period, with a closer look at the 37 farms that were active in the Sognefjorden area. Of the latter, 19 farms were actively producing in October 2011.

Monthly figures of several other variables were also given separately for each farm. Each farm could produce either Atlantic salmon or rainbow trout, or both species at the same time. We calculated the monthly proportion of Atlantic salmon biomass for each farm, which was 91% on average. The monthly average weight of fish was between 100 g and 5.8 kg for 95% of the farm-months, with a mean of 2.0 kg. The mean number of fish per farm was 530 000 (95% between 200 000 and 1.6 million). The mean seawater temperature at farms was 9.1°C (95% of all temperatures was between 3.4°C and 16.3°C). The seawater temperature was missing for 2.3% of the farm-months. Each missing temperature was imputed by a weighted mean of all observed temperatures the same month, with weights proportional to the inverse of the seaway distance to the current farm with the missing temperature. For about 4% of the farm-months, there were no fish at the same farm in the previous month in combination with the mean fish weight being less than 250 g. This indicates that the fish cohort was stocked for the first time on the given farm. Furthermore, for about 2% of the farm-months, there were no fish at the same farm in the previous month, but the mean fish weight was equal to or larger than 250 g. This indicates that the fish cohort had been relocated, i.e. moved from another marine farm. Medical salmon lice treatments were applied in 15% of the farm-months. Cleaner fish of the family *Labridae* were also applied at several farms [11], but we did not have sufficiently reliable data on the use of these, and we therefore ignored the use of cleaner fish in our modelling. Nor did we have sufficient data on salinity, which is known to affect the infection process of salmon lice [19].

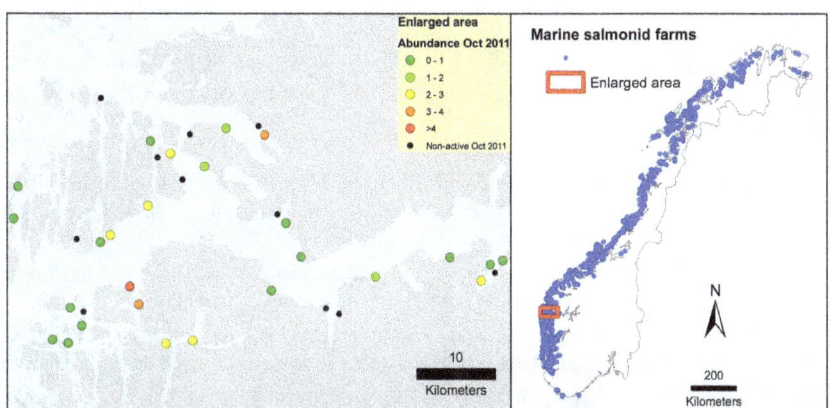

Figure 2. Location of salmonid farms. The left panel shows marine farms in the enlarged Sognefjorden area that were actively producing salmonids between June 2003 and December 2011, with observed salmon lice abundance indicated for those active in October 2011. The right panel shows marine farms that were active in the period June 2003 - December 2011 in the whole of Norway.

Overview of the Model

Here we first describe the main features of the present salmon lice model, and then go into more detail. The model includes all actively producing Norwegian marine salmonid farms simultaneously, but the counted number of salmon lice at a given farm in a given month is modelled conditionally upon the situation at the current farm of interest and all other farms in previous months.

Now, consider farm i at time or month t, and let μ_{it} denote the expected abundance of salmon lice. This expectation is modelled as a function of i) observed lice abundances in previous months at the current farm, ii) observed lice abundances in previous months at neighbouring farms, and iii) other factors such as seaway distances to neighbouring farms, seawater temperatures etcetera.

The response variable is constructed from the reported salmon lice abundances by multiplying it by the number of fish sampled, which is always assumed to be $n=20$, see previous section and Jansen et al. [11]. The response variable is denoted by y_{it}, and modelled as a zero-inflated, negative binomially distributed variable [20] with expectation $n\cdot\mu_{it}$. This means that the values of y_{it} tend to be zero more often than can be modelled by a negative binomial distribution alone. Let p^z_{it} denote the probability of excess zero observations in this compound distribution. These excess zeroes come in addition to those expected from the negative binomial part of the distribution. Thus, y_{it} comes from a negative binomial distribution with probability $1-p^z_{it}$, and has an excess zero value with probability p^z_{it}. Let further μ^{NB}_{it} denote the expectation in the negative binomial part of the distribution. Then, the expectation of y_{it} can be expressed as

$$\mathrm{E}(y_{it})=n\cdot\mu_{it}=(1-p^z_{it})\cdot\mu^{NB}_{it}. \tag{1}$$

Our main focus is on modelling the expected salmon lice abundance μ_{it}, and we give a detailed description of the model for μ_{it} in the next subsection. Furthermore, the probability of excess zero observations is modelled as a function of μ_{it} and a few other factors. Then, according to Eq. (1), the expectation in the negative binomial part of the compound distribution is given as $\mu^{NB}_{it}=n\cdot\mu_{it}/(1-p^z_{it})$. The negative binomial distribution has one parameter in addition to the expectation, here called R_{it}, and this is also modelled a function of μ_{it}. We present the sub-models for p^z_{it} and R_{it} in Appendix S1, since these only affect the shape of the distribution of y_{it}, which is not our focus here. The parameter R has often been denoted k in parasitological literature [21,22].

The motivation for using a negative binomial distribution begins with assuming that the number of lice counted on one fish, conditioned on the expected lice abundance, is Poisson distributed. However, we expect large variability from fish to fish, where typically some fish carry few salmon lice and some fish carry many. The negative binomial distribution is well suited to model such over-dispersed Poisson counts. Finally, if we sum the individual counts over $n=20$ fish, the total number of counted lice is also negative binomially distributed. However, when analysing the data, we found that there was an excess frequency of zeroes, which lead us to the zero-inflated, negative binomial distribution.

The zero-inflated, negative binomial distribution was also used by Jansen et al. [11], who modelled the same data as the present, although only updated up to December 2010. However, several other aspects of their approach differ from ours; i) they did not model all farms simultaneously in a network, ii) they ignored the lice counts at neighbouring farms, and iii) they had one model for the excess zeroes and another for the expectation in the negative binomial distribution, yielding a complex formula for the expected lice abundance, whereas we model the latter explicitly.

The Expected Lice Abundance

The model for the expected lice abundance at farm i in month t has the following additive-multiplicative form:

$$\mu_{it}=S_{it}\cdot\kappa^{susc}_{it}\cdot(\lambda^w_{it}+\lambda^d_{it}+\lambda^o_{it}). \tag{2}$$

The two multiplicative terms in Eq. (2) are:

- S_{it} is an "at-risk" indicator that is 1 when farm i is active (has a positive number of farmed fish) at month t and 0 otherwise.
- κ^{susc}_{it} is a factor proportional to the susceptibility of farm i. It depends on explanatory variables that characterise the conditions for the fish at farm i at month i. Some explanatory variables are common for all farms (e.g. season), whereas others are farm-specific (e.g. seawater temperature). The farm specific term has the form

$$\kappa^{susc}_{it}=\exp\Big(\sum_k\beta^{susc}_k x^{susc}_{ikt}\Big), \tag{3}$$

where x^{susc}_{ikt} denotes explanatory variables for farm i at month t and β^{susc}_k denotes corresponding regression coefficients. See Table 1 for a list of variables included in the final model.

The three additive terms represent three possible sources of lice infection:

1. λ^w_{it} represents infection within the current farm of interest.
2. λ^d_{it} represents infection from neighbouring farms, depending on among others the seaway distances to these farms and on their lice abundances.
3. λ^o_{it} represents infection from other, non-specified sources, for instance a reservoir of infection on free roaming salmonids.

We assume that the infection pressure from each source can be added. Details on how each of these terms are modelled are given in the following.

Infection within a farm is modelled as

$$\lambda^w_{it}=\left(y_{i(t-1)}/n+\sum_{l=2}^{l=L}\rho_l y'_{i(t-l)}/n\right)^{\alpha}\cdot S_{i(t-1)}, \tag{4}$$

where

- $y'_{i(t-l)}$ is equal to $y_{i(t-l)}$ if farm i has been active in all the months from $t-1$ to $t-l$, but is zero if the farm has been inactive in any of these months, and
- α is a positive parameter that allows for a non-linear dependency of the previous months' lice counts.
- $\rho_l, l=2,\ldots,L$ are parameters that account for the effect of previous lice abundances in sequential time steps. The term $(y_{i(t-1)}/n+\sum_{l=2}^{l=L}\rho_l y'_{i(t-l)}/n)$ divided by $1+\sum_{l=2}^{l=L}\rho_l$ is a weighted sum of the lagged observed lice abundances. Allowing for more than one lag may be useful, since the observed lice abundance is based on counts on small samples of fish, hence using several months may reduce the sampling uncertainty.

Table 1. Estimated parameters in the expected abundance μ_{it} with 95% confidence intervals for the selected model, with corresponding relative BIC values for selected parameters.

Parameter group	Variable name or parameter description	Farm specific variable	Parameter symbol	Est.	Lower	Upper	Relative BIC if deleted
misc.	Other sources		γ	0.076	0.068	0.084	
-"-	Lagged lice counts		ρ_2	0.076	0.062	0.090	
-"-	-"-		ρ_3	0.025	0.016	0.034	
-"-	-"-		ρ_4	0.022	0.013	0.031	39
-"-	Non-linear dependency		α	0.650	0.636	0.665	
-"-	Sea distance function		ϕ_0	−1.444	−1.605	−1.283	
-"-	-"-		ϕ_1	−0.351	−0.272	−0.430	
-"-	-"-		ϕ_2	0.568	0.478	0.658	
susc.	intercept	no	β_k^{susc}	−0.385	−0.439	−0.332	
-"-	$(t-103/2)$	-"-	-"-	$2.53\cdot10^{-3}$	$1.58\cdot10^{-3}$	$3.48\cdot10^{-3}$	
-"-	$(t-103/2)^2$	-"-	-"-	$5.44\cdot10^{-5}$	$4.04\cdot10^{-5}$	$6.84\cdot10^{-5}$	
-"-	$(t-103/2)^3$	-"-	-"-	$-1.50\cdot10^{-6}$	$-2.02\cdot10^{-6}$	$-0.99\cdot10^{-6}$	22
-"-	$(\text{temp}-9)$	yes	-"-	0.0979	0.0937	0.1020	
-"-	$(\text{temp}-9)^2$	-"-	-"-	−0.0047	−0.0056	−0.0038	91
-"-	(latitude-64)	-"-	-"-	0.0084	0.0042	0.0127	4
-"-	(temp-9)×(latitude-64)	-"-	-"-	0.0124	0.0112	0.0135	480
-"-	$\text{temp}_t-\text{temp}_{t-1}$	-"-	-"-	−0.0220	−0.0278	−0.0162	45
-"-	log(weight)	-"-	-"-	0.215	0.201	0.228	967
-"-	Stocked	-"-	-"-	−0.978	−1.115	−0.841	188
-"-	Relocated	-"-	-"-	0.218	0.101	0.336	4
-"-	Salmon proportion	-"-	-"-	0.130	0.083	0.176	20
inf.	log(number of fish)	yes	β_k^{inf}	0.258	0.191	0.326	57

misc.: Miscellaneous parameters.
susc.: Parameters related to the susceptible farm.
inf. : Parameters related to the infectious farm.
Est.: Estimate.
Lower: Lower bound of 95% confidence interval.
Lower: Upper bound of 95% confidence interval.

The term representing *infection from neighbouring farms* is summed over the separate contribution from all other farms and modelled as

$$\lambda_{it}^{d} = \sum_{j\neq i} \exp(\phi_0+\phi_1\cdot(d_{ij}^{\phi_2}-1)/\phi_2)\cdot\kappa_{j(t-1)}^{inf} \cdot\left(y_{j(t-1)}/n+\sum_{l=2}^{l=L}\rho_l y'_{j(t-l)}/n\right)^{\alpha}\cdot S_{j(t-1)}. \tag{5}$$

Here,

- ϕ_0 quantifies the importance of neighbouring infection compared to the other two sources of infection.
- d_{ij} is the seaway distance between farms i and j.
- ϕ_1 and ϕ_2 are parameters that reflect the effect of the seaway distances to the neighbouring farms. The transformation $(d_{ij}^{\phi_2}-1)/\phi_2$ is the well known Box-Cox transformation, which allows for many different shapes of the distance function (Figure 3). When ϕ_2 approaches zero, the Box-Cox transformation becomes $\log(d_{ij})$, where log here and elsewhere means the natural logarithm, and the distance function then becomes proportional to $d_{ij}^{\phi_1}$. When $\phi_2=1$, the distance function becomes proportional to the exponential function $\exp(\phi_1\cdot d_{ij})$.
- κ_{jt}^{inf} is a factor proportional to the infectiousness of farm j, depending on explanatory variables that characterise the neighbouring farm j, for instance the number of fish. The infectiousness term has the form

$$\kappa_{jt}^{inf} = \exp\left(\sum_k \beta_k^{inf} x_{jkt}^{inf}\right), \tag{6}$$

where, as before, x_{jkt}^{inf} denote explanatory variables and β_k^{inf} denote corresponding regression coefficients. See Table 1 for an overview of variables included in the final model and the

Results section for other variants of the model that were investigated.

Infection from other sources, λ^{o}_{it}, is currently modelled as a constant γ, and as such acts as an intercept term. This can, however, be modified to a more complex form, for instance by including functions of space and time.

Relative Contribution from Each Source

For each farm and each month, the expected lice abundance can be decomposed into the relative contribution from each of the three infection sources. Summing over all farms and months gives the mean relative contribution from each source. For instance,

$$r^{w} = (\sum_{t}\sum_{i} S_{it}\cdot\kappa^{susc}_{it}\cdot\lambda^{w}_{it})/(\sum_{t}\sum_{i}\mu_{it}) \qquad (7)$$

is the mean relative contribution within a farm. The mean relative contribution from neighbouring farms, r^{d}, and from other sources, r^{o}, are defined similarly.

Estimation

The unknown parameters consist of α, γ, ϕ_0, ϕ_1, ϕ_2, $\rho_2,\ldots,\rho_p$ and all β-s, joined into a parameter vector θ. The maximum likelihood estimates of θ was found by maximising the log likelihood of the data. An expression of the log likelihood is given in Appendix S1.

The log likelihood was maximised using the function *optim* in the statistical software *R*, using the method of Byrd et al. [23] for optimisation. Parameter uncertainties were based on the observed information matrix [24], and are reported as 95% Wald confidence intervals.

We investigated several variants of this model, with different explanatory variables included and with some parameters set to 0. For model selection, we used the Bayesian information criterion (BIC) [25,26]. This criterion balances the fit of the data to the number of parameters by minimising $-2ll(\theta)+\log(m)$, where $ll(\theta)$ is the log likelihood and m is the total number of observations and q the number of parameters. For some important explanatory variables, we considered second and third order terms as well as cross products between pairs of variables. We followed a strategy where main effects were included first, and second higher order effects or cross products were included if this improved the BIC value. Explanatory variables that were included in second or third order terms or cross products were centred to reduce correlations between the variables. We could alternatively have used Akaike's Information Criterion (AIC) [26,27] for model selection, which penalises the number of parameters less than BIC and therefore tends to give models with more parameters than when BIC is used. Choosing between BIC and AIC is primarily a matter of taste. We chose BIC because our dataset was large and hence we expected to end up with a rather complex optimal model also by using BIC.

Medical lice treatment was applied in 15% of the farm-months, which in principle should lead to reduced lice counts after treatments. It would indeed be useful if the model could be used to quantify the effect of such treatments. However, there is a dual relationship between treatment and lice counts; a high lice count will often induce a following treatment, whereas a treatment may result in lower lice counts. We don't know the actual dates for neither lice counts nor treatments, so it is not known for certain whether a treatment was applied before or after the corresponding lice count. Furthermore, we do not know what type of chemotherapeutic treatment that is used in each case. This makes

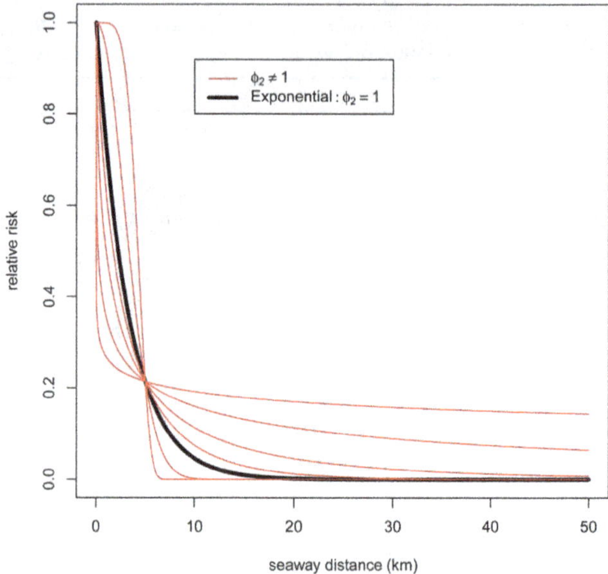

Figure 3. A selection of possible shapes for the relative effect of the seaway distance for various values of the ϕ_2 parameter.

it difficult to interpret the estimated effect of medical treatment. Therefore, we first consider a model without medical treatment as a factor, but then return to a model that includes medical treatment later on.

We estimated the model by fitting it to the observed lice counts from the 103 months between July 2003 and December 2011. We investigated models with up to $p=4$ lags of previous lice counts, so the four months of data from February to June 2003 were only used to construct these lagged lice counts. As mentioned in the Data Section, 3% of the lice counts were missing. It is reasonable to believe that these were missing at random in the context of Little and Rubin [28], and the corresponding farm-months were therefore ignored in the likelihood. However, for the lagged lice counts, each missing value was imputed by the last observed value within the same cohort, if any. If this was impossible, the missing value was imputed by the mean of all observed lice counts in the current month.

Results

We first consider a model without medical treatment. Then, we investigated several variants of the model and selected the model with the minimum BIC value. Figure 1 shows time plots of the estimated expected number of lice per fish and the corresponding observed lice abundance averaged over all farms and for three selected farms. In our model, the expected numbers of lice per fish are the same as one-month-ahead predictions, and there is therefore a tendency for the expected values being shifted to the right compared to the observations. The selected model is given in Table 1, with estimates and 95% confidence intervals for each coefficient that enters the model for the expected salmon lice abundance. Estimates for eight additional coefficients in the sub-models for the excess zero probability and R_{it} are given in Table A in Appendix S1. For some coefficients, the table also shows how much the BIC value would increase if this coefficient was excluded (set to 0) from the model, here called the relative (to the main model) BIC values. A high value of the relative BIC indicates that the corresponding variable improves the model fit significantly.

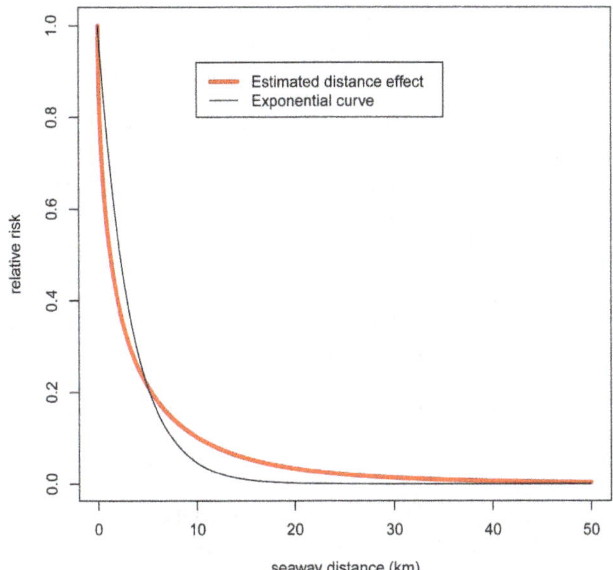

Figure 4. Estimated relative effect of the seaway distance.

Some effects that are handled by groups of explanatory variables, for instance the seawater temperature effect, and the relative BIC values for such effects are presented in the text below. The confidence intervals are rather tight, and all parameters are highly significantly different from natural reference values of 0 or 1 (the latter is relevant for the ϕ_2 parameter in the Box-Cox transformations, where a value of 1 implies linearity on the exponential scale, i.e. an exponential distance function). We have therefore not included any p values in the table.

In the following presentation, we go through the different parts of the model, starting at the top of Table 1. The parameter $\gamma = \lambda_{it}^{o}$, representing infection from other, non-specified sources (Eq. (2)), has to be interpreted relative to infection within and between farms. Infection from other, non-specified sources accounted for only 6% of the lice abundance, as calculated from Eq. (7). Infection from neighbourhood farms accounted for a further 28% of the lice abundance, whereas the remaining 66% was attributed to infection within farms. The confidence intervals for these proportions are all narrower than ± 1.5 %. Note that the latter includes both within farm infection of infectious copepodite-stage salmon lice and pre-adult and adult stages of lice that survive from previous counts.

All four lags of observed lice abundance (Eq. (4) and Eq. (5)) were selected. Deleting the last three lags would increase the BIC value by 688, so including these lags improves the fit considerably. However, the estimated values of ρ_2, ρ_3 and ρ_4 sum up to 0.12, so these lags only account for 11% (0.12/[1+0.12]) of the weighted observed lice abundance.

The estimated value of α ((Eq. (4) and Eq. (5)) was 0.65, which means that the expected lice abundance was not proportional to lagged observed lice abundance. Doubling the lagged observed lice abundance corresponds to a 57% increase in the expected lice abundance (since $2^{0.65} = 1.57$).

The infection pressure from neighbourhood farms decreased sharply by increasing seaway distance (Figure 4). A value equal to 1 for the parameter ϕ_2, which controls the Box-Cox transformation, would correspond to an exponential distance function, which we assumed in previous work on infectious diseases with less informative data [5,16]. However, ϕ_2 was estimated to be much lower than 1, which gives a much steeper curve than the exponential.

We continue by commenting on the effect of explanatory variables included in the factor κ_{it}^{susc} (Eq. (3)). All the variables included here were measured in the current month t, except the seawater temperature difference between month t and $t-1$. The κ_{it}^{susc} term includes a function of time common to all farms, represented by a third order polynomial. Deleting the time trend (represented by three terms) from the model would increase the BIC value with only 46, so the time trend is not among the most important factors for improving the model fit.

The factor κ_{it}^{susc} is also a function of several characteristics of the susceptible farm. The most important is the seawater temperature, which is represented by a second order polynomial of temperature, a cross product of temperature and latitude and a difference between the temperatures at month t and $t-1$. Deleting these temperature terms from the model would increase the BIC value with 2395, implying that water temperature is a major predictor of salmon lice abundance. Figure 5a) shows the estimated combined effects of the seawater temperature and the latitude for farms at 60° North (close to the city of Bergen) and 68° North (in the Lofoten area). The seawater temperature dependency was much stronger in northern Norway than in southern Norway. In addition, in southern Norway, the effect flattened out when the

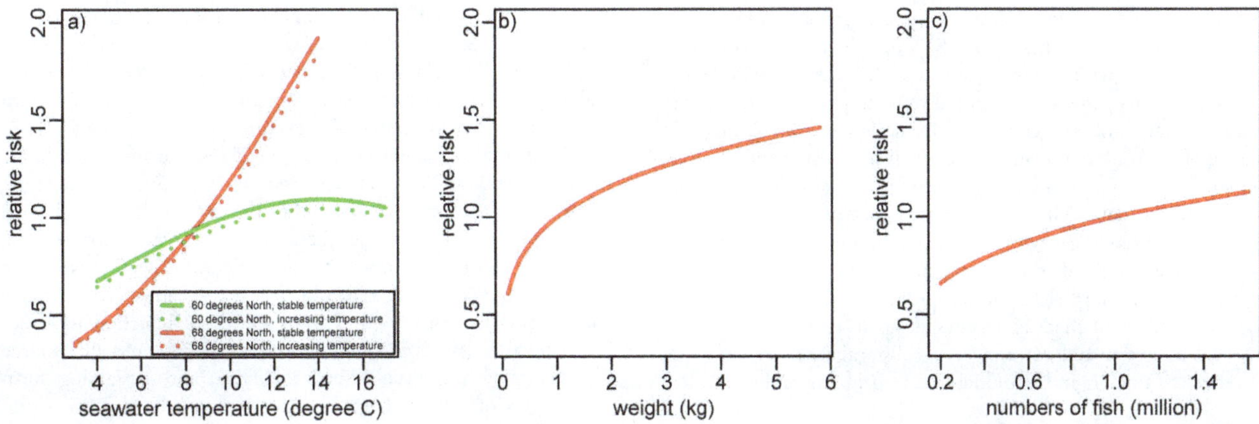

Figure 5. Estimated relative effects of the seawater temperature (a), of the mean fish weight (b) and of the number of fish at neighbouring farms (c).

seawater temperature reached about 14°C. There was also a smaller, but statistically significant, effect of the seawater temperature increase from the previous month, which reduced the expected lice abundance slightly. In practice, this means that the effect of the nominal seawater temperature in periods with increasing seawater temperatures (typically spring and summer) differs from that in periods with decreasing seawater temperatures (autumn and winter).

The mean fish weight at susceptible farms is a very important explanatory variable, with a relative BIC value of 967. Since it enters in Eq. (3) as the logarithm of the mean fish weight, it has the form $w_{it}^{\beta^{susc}}$ after taking the exponential, where w_{it} is the mean fish weight and β^{susc} the regression coefficient, which was estimated to be 0.215. Figure 5b) illustrates the effect of the mean fish weight.

When a farm was active in month t, but not in month $t-1$, the fish cohort at the farm was considered as stocked if the mean fish weight was less than 250 g and relocated otherwise. This was modelled by including corresponding indicator variables in Eq. (3). Remember that previous lice counts in this case is 0. If a fish cohort was stocked, its expected lice abundance was reduced by 62% (since $1-\exp(-0.978)=1-0.38=0.62$) compared to a farm that was active in the previous month, but with 0 observed lice abundance. On the other hand, if a fish cohort was relocated, the comparable expected lice abundance increased by 24% ($\exp(0.218)-1=1.24-1=0.24$). This is probably because relocated fish bring with them infection.

The expected lice abundance for a salmon cohort was 14% ($\exp(0.130)-1=1.14-1=0.14$) higher than for a cohort of rainbow trout, if they had the same lagged observed lice abundance and everything else being equal.

The number of fish was the only characteristic of the infectious farms that was included in the sub-model for infectiousness (Eq. (6)). The infection pressure from neighbourhood farms increased by increasing numbers of fish at such farms since $f_{jt}^{\beta^{inf}}=\exp(\beta^{inf}\cdot\log(f_{jt}))$, where f_{jt} is the number of fish at a neighbouring farm and β^{inf} the regression coefficient, which was estimated to be 0.258. Figure 5c) illustrates the effect of the number of fish at neighbouring farms.

We finally report the results for models including medical lice treatment. Extending Eq. (3) with an indicator variable for medical treatment in the current month improved the BIC value with 352. However, the estimated coefficient was 0.28 (95% confidence interval from 0.25 to 0.31), which is the opposite sign of the expected causal relationship. If we instead include an indicator variable for medical lice treatment in the previous month, the estimated coefficient was −0.34 (95% confidence interval from −.37 to −0.31), with an improvement in BIC value of 435. Hence medical lice treatment is potentially an important explanatory variable, but until we get more information on the timing and nature of medical treatments, this effect is not trustworthy. It does, however, make sense to include treatment in the previous month in the present data. The salmon farmers are instructed to report the highest abundance counted during a month and it is likely that this count is obtained before a potential medical treatment in a given month. Nevertheless, there is probably still some bias in the estimated effect of medical treatment. Inclusion or exclusion of medical treatment did not, however, substantially affect estimates of other parameters in the model (Table B in Appendix S1).

Discussion

The main structure of the present model is similar to models we have applied previously for viral diseases in salmonid farms [5,15,16] and is in addition inspired by models applied for other animal diseases [1–3], in that the distances to neighbouring farms as well as conditions at these farms are important factors for the infection pressure. However, the stochastic modelling differs, which is natural since the response data are different. The viral infection data are binary, i.e. a farm is considered to be infected or non-infected, whereas the salmon lice infection data consist of counts that may take positive values and that are available for every farm-month. This results in a comparatively informative dataset, which has given us the opportunity to account for many different factors and model some of them in detail. Our model also has similarities to models used for human infections. For instance, Held and Paul [29] modelled the monthly number of laboratory cases of influenza in 140 districts of Southern Germany by a spatio-temporal negative binomial regression model, taking into account infection within and between regions (termed epidemic components) and an additional external infection source (called endemic component).

The model can be seen as a generalisation of vector autoregressive models [30], which are widely used in areas such as econometrics: Let $\mathbf{y}_t$ denote the n^f-dimensional vector of salmon lice counts for all farms in month t, n^f being the number of farms. If the parameter α in Eq. (4) and Eq. (5) is exactly 1, the expected lice counts at month t is a weighted sum of the previous lice counts up to L lags back. This can be written as

$$\mathrm{E}(\mathbf{y}_t)=\gamma_{0t}+\sum_{l=1}^{l=L}\mathbf{\Phi}_{lt}\mathbf{y}_{t-l}, \tag{8}$$

where γ_{0t} is an n^f-dimensional vector of time-varying farm-specific intercepts representing infection from other, non-specified sources, whereas $\Phi_{lt}, l=1,\ldots,L$ are time-varying $(n^f\times n^f)$ matrices, where the diagonal represents within-farm infection and the off-diagonal elements represent between-farm infection.

The present modelling approach gives insight into the different contributing sources of salmon lice infection, as well as factors that affect lice abundance. Within farm infection was the dominating source of infection, accounting for 2/3 of the additive contribution to salmon lice abundance. The large within farm contribution to infection may partly be due to survival of pre-adult and adult lice over consecutive counts, since these stages are constituents of this internal source in addition to infectious copepodite-stage salmon lice produced within the farm stocks of salmonids. In contrast, the external sources of infection for a given farm constitute only infectious copepodite-stage salmon lice produced externally and transported passively with the water current. It is also important to note that the contribution from the different sources of infection will vary over time and space. As smolts are transferred from freshwater to marine waters, for example, they are free of salmon lice. Transmission of infectious copepodites will then only come from external sources. Hence, externally produced infection, which depends primarily on the density of salmon lice in neighbouring farms in the model, are important for seeding susceptible farm populations of salmonids with infection while the on-farm abundance of reproductive lice is low. The relative contribution from the within farm source of infection increases following the development of reproductive lice, and this source seems mainly responsible for amplifying and sustaining farm populations of salmon lice. The large contribution from within farm infection implies that the average farm primarily depends on itself in avoiding high levels of salmon louse infections. This, we argue, has important implications for the control and management of salmon lice. Due to the importance of self-inflicted infection, the farmer should be motivated to control infection levels of reproductive lice to minimise the production of infectious copepodites. From a management point of view, it is reasonable to hold the farms responsible for keeping levels of reproductive lice low, to minimise their contribution to local infection pressure to wild fish or neighbourhood salmon farms.

The external unknown source of infection accounted for only 6% of the additive contribution to salmon lice abundance. This source may represent infection originating from wild salmonids [9,31], or alternatively escaped farmed salmonids [32], but which we do not have information on in the present study. The external unknown source is modelled with fewer parameters than the other sources of infection, and it is possible that it would account for a larger proportion of the infection if a more flexible function was used. However, we did not find any logical reasons to do so. The low contribution from the unknown external source also agrees with earlier suggestions that farmed salmon must be the dominating host of salmon lice in Norwegian coastal waters due to the large population size compared to wild stocks [11,32].

Infection from external neighbourhood farms accounted for the final 28% of the additive contribution to salmon lice abundance. The contribution to a given farm by a neighbourhood farm in the model is shaped by the functional relationship between the relative risk and seaway distance between farms (Figure 4). Compared to the exponential function used in previous models [5,15,16], the present relative risk decreases steeply at low but increasing distance to neighbourhood farms, but levelled off at higher relative risks for intermediately distant neighbourhood farms. We interpret this to imply that i) very close farms on average interact intensively with respect to salmon lice infection, and ii) that the number of intermediately distant neighbourhood farms and their levels of infection are important for expected salmon lice abundance. The latter probably reflects the effect of being located in areas of intensive salmon farming [11]. The function describing the relationship between infection from neighbourhood farms and distance represents an expression of a transmission kernel for salmon lice transmission between neighbourhood farms that can be used to simulate the spread of infection between farms and assess the impact of different control measures. Such transmission kernels are key elements of models designed to examine scenarios for the spread and control of a range of different livestock diseases (e.g. [4]). To our knowledge, this is the first transmission kernel estimated for metazoan parasites from an extensive system of farm populations. It is worth noting, however, that the shape of the function describing the transmission kernel for salmon lice probably varies in space and time, for example influenced by the temperature dependent duration of the planktonic phase of the louse life cycle [33], and local hydrodynamics [34].

The large body of data used to parameterise the present model yield estimates with relatively tight confidence intervals. The reports of salmon lice abundance, however, are highly skewed and zero inflated, implying that modelled expected abundance of lice will have relatively large uncertainty while predicting individual lice reports. It is also worth noting that the farmers are instructed to report the highest count during a month such that the modelled expectations should tend to be overestimates of the true lice abundance at corresponding times. In addition, there is one general characteristic of the data that we suspect may be confounded by the regulation regime authorised by the legal authorities in Norway [11]. For the most influential factors in the model, effects seem relatively reduced for factor-values that imply expectations of high salmon lice abundance, i.e. effects are less than proportional to changes in the factors. This is apparent for example at especially high temperatures in the south, for large fish and for neighbourhood farms with large numbers of fish (Figure 5). Bearing in mind that the regulations on salmon lice infections focus on maximum legal thresholds of parasite abundances on farms, we anticipate that control efforts like the use of cleaner fish or medical treatments counter further increases in infection levels as lice abundances exceed legal thresholds. Since we cannot fully account for control efforts with the present resolution of data, this type of confounding related to legal thresholds of infection is to be expected. On top of this, we suspect that there may also be some underreporting of high salmon lice abundance. For example, the probability of finding a given salmon louse on a fish, i.e. the sensitivity of the counting procedure, probably decreases with increasing surface area to search on large fish. An implication of such potentially uncontrolled effects is that the model would be less predictive for high than for low lice abundance. Higher resolution of the data concerning timing and the nature of control measures, along with increased knowledge on the dynamic effects of different measures, are aspects that would improve the predictive capability of the present approach.

To conclude, we present a stochastic spatio-temporal model for salmon lice infection on farmed salmonids in Norway. We emphasise important insights that the model provides with respect to different sources of infection and how salmon lice abundance is affected by different environmental -, host related - and parasite related factors. In addition, we address control measures such as medical treatments and the use of cleaner fish, but point to shortcomings of the data to fully account for control measures at present. The present modelling framework lays the foundation for development of future scenario simulation tools for examining transmission and abundance of salmon lice on farmed salmonids under different control regimes. From 2012, the reporting procedures for salmon lice infections and medical treatments have changed from a monthly to a weekly frequency. With this increased time resolution we foresee that we can estimate more trustworthy effects of medical treatments and other control measures. This will benefit the prospects of applying the present model as a mathematical laboratory to investigate effects of complex and expensive actions before they are implemented in practice.

Acknowledgments

We thank Audun Stien for valuable comments to drafts of this article.

Author Contributions

Analyzed the data: MA BS. Contributed reagents/materials/analysis tools: ABK PAJ. Wrote the paper: MA BS ABK PAJ. Planned analysis methods and interpreted results: MA BS ABK PAJ.

References

1. Keeling M, Woolhouse M, Shaw D, Matthews L, Chase-Topping M, et al. (2001) Dynamics of the 2001 uk foot and mouth epidemic: stochastic dispersal in a heterogeneous landscape. Science 294: 813–817.
2. Diggle P (2006) Spatio-temporal point processes, partial likelihood, foot and mouth disease. Stat Methods Med Res 15: 325–336.

3. Höhle M (2009) Additive-multiplicative regression models for spatio-temporal epidemics. Biom J 51: 961–978.
4. Szmaragd C, Wilson A, Carpenter S, Wood J, Mellor P, et al. (2009) A modeling framework todescribe the transmission of bluetongue virus within and between farms in great britain. PLoS ONE 4: e7741.
5. Aldrin M, Lyngstad T, Kristoffersen A, Storvik B, Borgan Ø, et al. (2011) Modelling the spread of infectious salmon anaemia (isa) among salmon farms based on seaway distances between farmsand genetic relationships between isa virus isolates. J R Soc Interface 8: 1346–1356.
6. Costello M (2006) Ecology of sea lice parasitic on farmed and wild fish. Trends Parasitol Trends Parasitol 475–483.
7. Krkošek M, Ford J, Morton A, Lele S, Myers R, et al. (2007) Declining wild salmon populations in relation to parasites from farm salmon. Science 318: 1772–1775.
8. Krkošek M, Connors B, Morton A, Lewis M, Dill L, et al. (2011) Effects of parasites from salmon farms on productivity of wild salmon. Proc Natl Acad Sci U S A 108: 14700–14704.
9. Marty G, Saksida S, Quinn T (2010) Relationship of farm salmon, sea lice, and wild salmon populations. Proc Natl Acad Sci U S A 107: 22599–22604.
10. Krkošek M, Lewis M, Volpe J (2005) Transmission dynamics of parasitic sea lice from farm to wild salmon. Proc Biol Sci 272: 689–696.
11. Jansen P, Kristoffersen A, Viljugrein H, Jimenez D, Aldrin M, et al. (2012) Sea lice as a density dependent constraint to salmonid farming. Proc Biol Sci rspb20120084.
12. Krkošek M, Revie C, Gargan P, Skilbrei O, Finstad B, et al. (2012) Impact of parasites on salmon recruitment in the northeast atlantic ocean. Proc Biol Sci rspb20122359.
13. Skilbrei O, Finstad B, Urdal K, Bakke G, Kroglund F, et al. (2013) Impact of early salmon louse, *Lepeophtheirus salmon*, infestation and differences in survival and marine growth of sea-ranched atlantic salmon, Salmo salar l., smolts 1997–2009. J Fish Dis 36: 249–260.
14. Jackson D, Cotter D, Newell J, McEvoy S, O'Donohoe P, et al. (2013) Impact of *Lepeophtheirus salmonis* infestations on migrating atlantic salmon, *Salmo salar* l., smolts at eight locations in ireland with an analysis of lice-induced marine mortality. J Fish Dis 36: 273–281.
15. Scheel I, Aldrin M, Frigessi A, Jansen P (2007) A stochastic model for infectious salmon anemia (isa) in atlantic salmon farming. J R Soc Interface 4: 699–706.
16. Aldrin M, Storvik B, Frigessi A, Viljugrein H, Jansen P (2010) A stochastic model for the assessment of the transmission pathways of heart and skeleton muscle inammation, pancreas disease and infectious salmon anaemia in marine fish farms in norway. Prev Vet Med 93: 51–61.
17. Torrissen O, Olsen R, Toresen R, Hemre G, Tacon A, et al. (2011) Atlantic salmon (*Salmo salar* l.): the "super chicken" of the sea? Reviews in Fisheries Science 19: 257–278.
18. Anonymous (2009) Lakse- og sjøaurefiske 2008 (in Norwegian). Statistics Norway, Oslo. Noregs offsielle statistikk D 423. Available: http://www.ssb.no/1005.
19. Bricknell I, Dalesman S, O'Shea B, Pert C, Luntz A (2006) Effect of environmental salinity on sea lice *Lepeophtheirus salmonis* settlement success. Dis Aquat Organ 71: 201:2012.
20. Zuur A, Ieno E, Walker N, Saveliev A, Smith G (2009) Mixed Effects Models and Extensions in Ecology with R. Springer, 1st edition. 574 p. doi:10.1007/978-0-387-87458-6.
21. Grenfell B, Wilson K, Isham V, Boyd H, Dietz K (1995) Modelling patterns of parasite aggregation in natural populations: trichostrongylid namatode-ruminant interactions as a case study. Parasitology 111 (Suppl.): S135–S51.
22. Irvine R, Stien A, Halvorsen O, Langvatn R, Albon S (2000) Life-history strategies and population dynamics of abomasal nematodes in svalbard reindeer (Rangifer tarandus plathyrhynchus). Parasitology 120: 297–311.
23. Byrd R, Lu P, Nocedal J, Zhu C (1995) A limited memory algorithm for bound constrained optimization. SIAM Journal on Scientific Computing 16: 1190–1208.
24. Aalen O, Borgan Ø, Gjessing H (2008) Survival and event history analysis - A process point of view. Springer, New York.
25. Schwarz G (1978) Estimating the dimension of a model. The Annals of Statistics 6: 461–464.
26. Claeskens G, Hjort N (2008) Model selection and model averaging. Cambridge University Press, Cambridge.
27. Akaike H (1974) A new look at the statistical model identification. IEEE Transactions on Automatic Control 19: 716723.
28. Little R, Rubin D (1987) Statistical analysis with missing data. John Wiley & Sons, New York.
29. Held L, Paul M (2012) Modeling seasonality in space-time infectious disease surveillance data. Biom J 54: 824–843.
30. Lütkepohl H (2005) New introduction to multiple time series analysis. Springer, Berlin.
31. Bjørn P, Finstad B (2002) Salmon lice, *Lepeophtheirus salmonis* (krøyer), infestation in sympatric populations of arctic char, *Salvelinus alpinus* (l.), and sea trout, *Salmo trutta* (l.), in areas near and distant from salmon farms. ICES J Mar Sci 59: 131–139.
32. Heuch P, Mo T (2001) A model of salmon louse production in norway: effects of increasing salmon production and public management measures. Dis Aquat Organ 45: 145–152.
33. Stien A, Bjørn P, Heuch P, Elston D (2005) Population dynamics of salmon lice *Lepeophtheirus salmonis* on atlantic salmon and sea trout. Mar Ecol Prog Ser 290: 263–275.
34. Amundrud T, Murray A (2009) Modelling sea lice dispersion under varying environmental forcing in a scottish sea loch. J Fish Dis 32: 27–44.
35. Johnson N, Kotz S, Kemp A (1993) Univariate discrete distributions. John Wiley & Sons, Inc., New York.

Determining Vaccination Frequency in Farmed Rainbow Trout Using *Vibrio anguillarum* O1 Specific Serum Antibody Measurements

Lars Holten-Andersen[1,2]*, Inger Dalsgaard[2], Jørgen Nylén[3], Niels Lorenzen[4], Kurt Buchmann[1]

1 Department of Veterinary Disease Biology, Faculty of Health and Medical Sciences, University of Copenhagen, Frederiksberg C, Denmark, **2** Division of Veterinary Diagnostics and Research, National Veterinary Institute, Technical University of Denmark, Copenhagen, Denmark, **3** Aqua Denmark Unit, MSD Animal Health, Ballerup, Denmark, **4** Division of Poultry, Fish and Fur Animals, National Veterinary Institute, Technical University of Denmark, Aarhus N, Denmark

Abstract

Background: Despite vaccination with a commercial vaccine with a documented protective effect against *Vibrio anguillarum* O1 disease outbreaks caused by this bacterium have been registered among rainbow trout at Danish fish farms. The present study examined specific serum antibody levels as a valid marker for assessing vaccination status in a fish population. For this purpose a highly sensitive enzyme-linked immunosorbent assay (ELISA) was developed and used to evaluate sera from farmed rainbow trout vaccinated against *V. anguillarum* O1.

Study Design: Immune sera from rainbow trout immunised with an experimental vaccine based on inactivated *V. anguillarum* O1 bacterin in Freund's incomplete adjuvant were used for ELISA optimisation. Subsequently, sera from farmed rainbow trout vaccinated with a commercial vaccine against *V. anguillarum* were analysed with the ELISA. The measured serum antibody levels were compared with the vaccine status of the fish (vaccinated/unvaccinated) as evaluated through visual examination.

Results: Repeated immunisation with the experimental vaccine lead to increasing levels of specific serum antibodies in the vaccinated rainbow trout. The farmed rainbow trout responded with high antibody levels to a single injection with the commercial vaccine. However, the diversity in responses was more pronounced in the farmed fish. Primary visual examinations for vaccine status in rainbow trout from the commercial farm revealed a large pool of unvaccinated specimens (vaccination failure rate = 20%) among the otherwise vaccinated fish. Through serum analyses using the ELISA in a blinded set-up it was possible to separate samples collected from the farmed rainbow trout into vaccinated and unvaccinated fish.

Conclusions: Much attention has been devoted to development of new and more effective vaccines. Here we present a case from a Danish rainbow trout farm indicating that attention should also be directed to the vaccination procedure in order to secure high vaccination frequencies necessary for optimal protection with a reported effective vaccine.

Editor: Pierre Boudinot, INRA, France

Funding: This work was supported by a grant from the Danish Agency for Science, Technology and Innovation (MarinVac, DSF jnr 09-063102). The funders had no role in study design, data collection and analysis, decision to publish, or preparation of the manuscript.

Competing Interests: JN is an employee of MSD Animal Health. The commercial vaccine used in this project was provided by MSD Animal Health.

* E-mail: lhoa@life.ku.dk

Introduction

Salmonid aquaculture has increased three-fold since 1980 and aquaculture in general is by far the fastest growing sector of food animal production in the world. In Denmark, where the focus is on rainbow trout, the production is equally expected to grow over the next few years. However, it is also expected that the increase in production should not lead to a corresponding growth in consumption of antimicrobials. Hence, the use of commercial vaccines against bacterial infections has spread and vaccination is now taking place at all farms with marine net cages as part of the production line. The commercial vaccine used in Denmark for injection vaccination is an oil-adjuvanted construct containing whole-cell antigen preparations of *Aeromonas salmonicida* subsp. *salmonicida* and *Vibrio anguillarum* serovar O1 and O2. However, despite the use of such vaccines infections with *V. anguillarum* might occur at the farms. This marine bacterium is found regularly on marine rainbow trout farms in Denmark and can be a significant cause of mortality in the stocked fish [1,2]. While the level of protection from the vaccines may vary against *A. salmonicida* subsp. *salmonicida*, the vaccine efficacy against *V. anguillarum* is generally high [3]. Hence, we speculated that part of the explanation for the infections with *V. anguillarum* might be found in the vaccination procedure. To answer this question, we developed a highly sensitive and specific ELISA as a tool for monitoring vaccination frequencies in rainbow trout populations following large-scale vaccination procedures. With this assay we determined levels of anti-*V. anguillarum* antibodies in sera from farmed trout and compared these findings with visual examinations of the fish based on the Speilberg scale [4].

Results

Assay Characteristics

Coating antigen. Coating the microplates with sonicated bacteria compared with whole-cell antigen coating led to significantly different OD readings with serum (pooled, n = 20) from unvaccinated fish (control) diluted 1:1000 (P<0.0001), 1:5000 (P = 0.003) and 1:10000 (P = 0.009) in assay buffer. OD values for unvaccinated fish were lowest in microplates coated with sonicated material (Fig. 1). The two coatings also produced significantly different readouts with serum (pooled, n = 20) from three times immunised rainbow trout (formalin-killed *V. anguillarum* O1 in FIA) diluted 1:1000 (P = 0.012), 1:5000 (P = 0.0004) and 1:10000 (P = 0.0002). However, in the vaccinated fish the highest OD values were measured in microplates coated with sonicated bacteria. When antigen (*V. anguillarum* sonicate) was omitted from the coating buffer, reactions were not significantly different from blank wells (sample dilution buffer only; data not shown).

Dilution series and analytical signal. Representative titration curves based on pooled sera from unvaccinated rainbow trout and sera pooled from fish immunised one, two or three times are shown in Fig. 2. The plotted data produced sigmoid curves for the immunised fish, whereas serum from the naïve fish generated a linear horizontal curve showing a slight increase in background at the highest serum concentrations. The absorbance at 450 nm against the serum dilution factor exhibited a linear relationship in a narrow dilution range for all three immune sera with slopes increasing with the number of immunisations. The limit of detection for the *V. anguillarum* O1 assay was an OD of 0.066.

Intra- and interassay precision. The analysis of eight determinations of the same serum sample on one microplate gave a CV≤3.4% at 1:1000 and 1:5000 dilutions. The interassay CV was ≤6.3% based on duplicates of both dilutions of the control serum pool on individual plates run on 14 different days.

Stability The level of specific antibodies in the serum samples remained unchanged through ten cycles of freeze/thaw. In addition, keeping samples at room temperature for 24 h had no effect on antibody levels (data not shown).

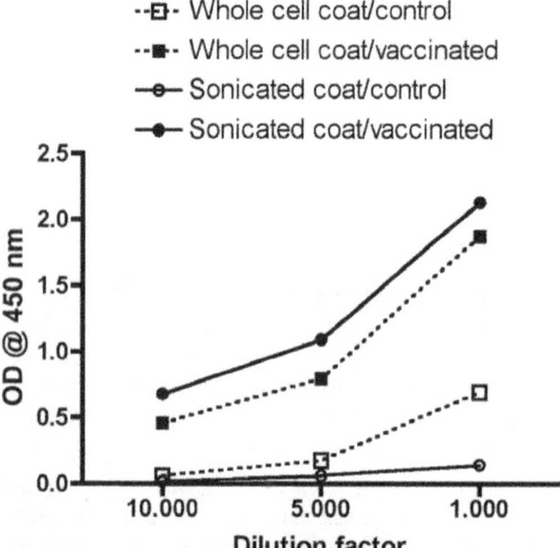

Figure 1. Performance of two different ELISAs coated with either whole cell or sonicated bacterial antigen. Serum pools from 20 unvaccinated (PBS injected, control) and 20 vaccinated (three immunisations) were analysed in 1:1000, 1:5000, and 1:10000 dilutions.

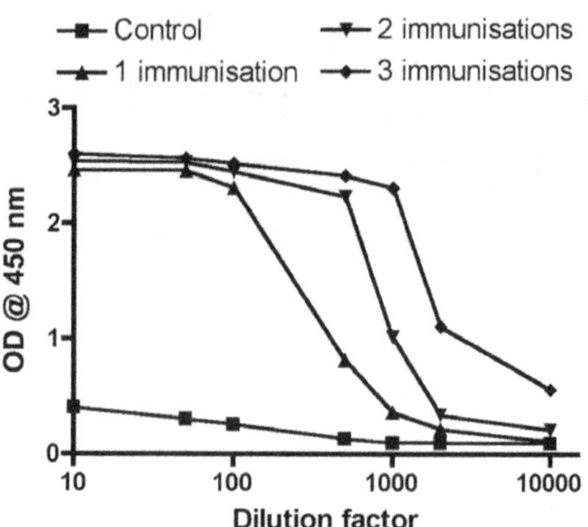

Figure 2. Titration curves based on pooled sera from 20 controls and 20 rainbow trout immunised i.p. one, two or three times with an experimental vaccine (formalin-killed *V. anguillarum* O1 in FIA).

Antibody Titre Increase for Each Immunisation

Sera from the fish that received three consecutive immunisations i.p. with a test vaccine were analysed individually for vaccine specific antibodies and Fig. 3 displays these data presented as OD values. As shown, there was no significant difference between control fish (0.10±0.0) and rainbow trout immunised once (0.24±0.14). However, following both the second (1.0±0.19) and third immunisation (2.64±0.23) a significant increase in *V. anguillarum* O1 specific antibodies was observed. Highest OD

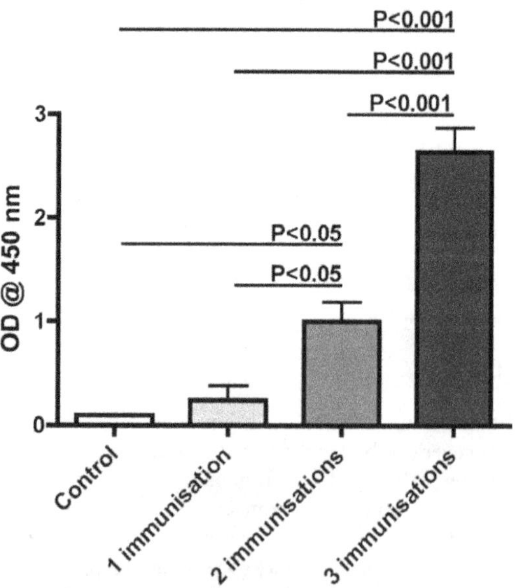

Figure 3. Antibody responses to repeated immunisations followed in 20 rainbow trout. Serum samples from 20 controls and 20 rainbow trout immunised i.p. one, two or three times with an experimental vaccine were analysed individually. All samples were diluted 1:1000.

values were found with serum from rainbow trout immunised three times.

Frequency of Unvaccinated Fish in a Population of Farmed Rainbow Trout

For a final evaluation of the *V. anguillarum* ELISA as a tool for determining vaccine status in farmed rainbow trout populations, blood samples were collected from 50 fish at a Danish sea farm. The fish were also examined for visual signs of previous vaccination. 15 months prior to this sampling 50 fish from the same batch of rainbow trout were randomly collected from the pond of supposedly vaccinated fish on the day of vaccination and visually examined for their vaccine status. The data from the visual examinations are presented in Table 1. There were found ten apparently unvaccinated fish (20%) out of the 50 examined specimens at both the fresh water and salt-water facility. This frequency of unvaccinated fish was confirmed with the ELISA (Fig. 4). OD values were significantly associated with the status as unvaccinated ($P<0.0001$) with an AUC of 1.00 (95% CI: 1.00–1.00). Moreover, the receiver operating characteristic (ROC) analysis showed 100% specificity and 100% sensitivity for the ELISA with an OD value cut-off set at 0.973. Based on these observations the 50 fish from the salt-water facility were divided into an unvaccinated and vaccinated group, and their data on size and condition are presented in Table 2. Although the unvaccinated fish were slightly longer compared to the vaccinated group ($P=0.041$), the weight ($P=0.12$) and average condition factor ($P=0.12$) of the two groups were not significantly different.

Discussion

Rainbow trout in culture suffer a variety of diseases, including bacterial, viral, parasitic and fungal infections. The majority of infections described are bacterial; including vibriosis caused by the extracellular bacterium *V. anguillarum*. Modern aquaculture takes prophylactic measures such as vaccination to combat serious infections, and with the continuous expansion of this industry there is an ongoing development of more effective ways to deliver vaccines to the fish [5]. In recent years refined machines for i.p. injections in fish have made automated vaccination a true alternative to manual vaccine delivery. However, the larger number of vaccinated fish per hour comes with an increased risk of vaccination failure if these machines are not handled optimally (personal observation). In Norway, being one of the largest producers of salmonids in aquaculture, the accepted vaccination failure rate at their fish farms is ≤2%. In Danish rainbow trout farming such a failure rate has yet to be agreed on.

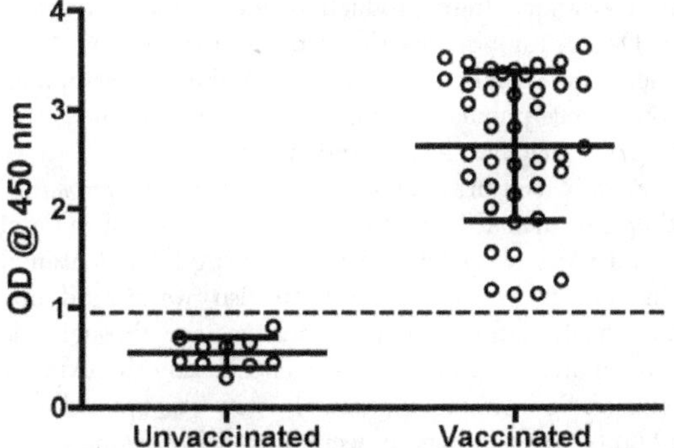

Figure 4. A scatter dot plot showing specific serum antibody responses from 50 farmed rainbow trout vaccinated once with a commercial vaccine against *V. anguillarum*. The dotted line represents the cut-off value for the ELISA at 0.97. All samples were diluted 1:1000.

The present study focused on the development and validation of an indirect ELISA for measuring anti-*V. anguillarum* O1 antibody levels as a marker for positive vaccination. The final assay was evaluated for its potential to determine the vaccination rate in a smaller population of rainbow trout subjected to automated vaccination against vibriosis at a Danish trout farm. Hence, the goal was to investigate and develop an assay useful for the farmers in determining the frequency of fish that accidentally have not received the vaccine after large-scale vaccination.

The development of a sensitive and specific ELISA requires special attention to a number of steps involved in the assay, here among the coating of the microplates. It has been described how optimising the coating process can lead to significantly better performance of the ELISA when the coating antigen is complex material like whole cell suspensions [6]. In our study, sonicated *V. anguillarum* O1 was found superior to whole cell *V. anguillarum* O1 as coating antigen in terms of significantly increased specific signal and lowered background. The reason for the better assay performance with sonicated coating material might simply be a larger number of exposed antigens on the bottom of the wells. Nonetheless, the result is a highly sensitive and specific ELISA, which is essential if the assay is to be used in screening for vaccination frequencies.

In the present study an experimental vaccine based on FIA and one inactivated bacterial isolate induced a weaker antibody response in rainbow trout compared to a commercial vaccine holding three bacterial isolates. However, the results were not directly comparable since fish stocks and rearing conditions differed. Moreover, there was a 15-month period between vaccination and serum sampling in case of the field-reared fish compared to 4½ months for the experimental fish. Nonetheless, the explanation for the observed difference could by a higher potency in multivalent vaccines [7]. Such an effect could possibly in part be attributed to a larger amount of bacterial immunomodulators in multivalent vaccines. Indeed, many of the most promising experimental adjuvant components are bacterially derived molecules that have been identified as highly conserved ligands for a range of immunologically important receptors on antigen presenting cells [8,9]. Finally, it may not be excluded that cross-reactive antibodies targeted at such conserved and thus shared epitopes between *V. anguillarum* and *A. salmonicida* have been induced by the *A. salmonicida* component in the commercial vaccine.

High levels of anti-*V. anguillarum* O1 antibodies were measured with this ELISA in sera from sea-farmed rainbow trout. In agreement with a previous report our analyses revealed a large variation in specific antibody levels among truly vaccinated fish [10]. Since it was outside the scope of this study to investigate protective qualities of the antibodies measured with the described ELISA it is unresolved whether these differences in titres reflect different levels of protection against vibriosis. However, such specific antibodies have been shown to confer protection in rainbow trout against vibriosis [7,11,12]. Moreover, it has been suggested that the protective immune response in rainbow trout against extracellular bacteria, like in higher vertebrates, is predominantly humoral [13]. Hence, we speculate that high numbers of unvaccinated fish with a very low antibody response against *V. anguillarum* O1 as reported here could be part of the explanation for the vibriosis outbreaks observed at Danish sea farms [2]. However, further studies including larger populations of

Table 1. Descriptive data for 100 commercially farmed rainbow trout.

	N fish	Length (cm)	Weight (g)	Condition factor	Speilberg score	% unvaccinated
Fresh water, February 2009	50	19.5±1.5	83.4±19.6	1.12±0.10	-	20
Salt water, May 2010	50	38.7±3.8	763.1±204.7	1.29±0.12	1.25±0.54	20

Condition factor: CF = $(W/L^3)\times100$; where W = weight (g) and L = total length (cm) [18].
Speilberg score is a measure of vaccine-induced side effects.

fish sampled at several farms are needed for conclusions on this theory.

There was not registered any outbreaks of vibriosis in the rainbow trout population during the study period. This might explain that the frequency of unvaccinated fish in the population had not decreased over time between the two sampling points. However, these outbreaks normally occur at the sea farms during summer when water temperatures are highest and hence would be expected to arise after the final sampling time point in the present study [1,2]. Thus, it remains to be investigated how the presence and frequency of unvaccinated fish affects the risk and seriousness of an outbreak of vibriosis among the farmed fish. Possibly, herd immunity could be provided to a certain level by the 80% vaccinated fish. In terms of analysis of the antibody response in sea reared rainbow trout it also should be determined how an infection with *V. anguillarum* affects the antibody profiles of the fish. However, this will not hamper the usability of the ELISA for screening purposes since the testing of vaccination status normally will be performed before the fish are transferred to environments, e.g. sea cages, potentially infected with *V. anguillarum* O1. Indeed, the ELISA revealed that apparently unvaccinated fish, although low, still had an antibody response against *V. anguillarum* O1, which might be the result of natural exposure to environmental *V. anguillarum* after transfer to the sea cages. Optimally, an unvaccinated control group should have been included in the field study for interpretation of potential exposure to environmental *V. anguillarum*. However, this was not feasible since all fish intended for sea cage rearing in Denmark should be vaccinated. A maximum water temperature of 15°C monitored at the sea farm during the test period would explain the lack of disease outbreak despite potential bacterial exposure. Distinguishing between non-specific reactivity and low level specific reactivity is a key problem in the development of ELISA [14,15]. To reduce the probability of false positive results (e.g. fish with detectable antibody levels induced by environmental bacterial exposure) it might be necessary, as in the present study, to set a relatively high cut-off value for the assay. However, this will also increase the probability of false negative results [10].

Table 2. Descriptive data for 50 commercially farmed rainbow trout divided into unvaccinated and vaccinated groups.

	N fish	Length (cm)	Weight (g)	Condition factor
Unvaccinated	10	40.9±3.2	852.1±173.9	1.24±0.10
Vaccinated	40	38.2±3.8	740.9±207.7	1.30±0.12

Length: $P = 0.041$, weight: $P = 0.12$, condition factor: $P = 0.12$.
Condition factor: CF = $(W/L^3)\times100$; where W = weight (g) and L = total length (cm) [18].

Apart from monitoring and improving efficacy of the vaccination procedure, knowledge on the immunological status of the actual fish population provided by the ELISA would help the farmer to handle vibriosis outbreaks in a more rational manner. As with other severe diseases vibriosis can lead to reduced net growth, increased need for treatment, and loss of fish dying from infection [16]. Thus, the ELISA presented here could be a useful tool for the fish farmer, and compared to visual examination it has the advantage that no fish needs to be killed in the process (sampling can be done on-site and the fish released).

Materials and Methods

Ethics Statement

The Committee for Animal Experimentation, Ministry of Justice, Copenhagen, Denmark, approved the study including the fish rearing and experimentation (license number 2006/561-1204), which was performed following the ethical guidelines listed in the license. All animal procedures were in agreement with the EU Directive 2010/63/EU for animal experiments?

Fish and Rearing Conditions

Rainbow trout (Skinderup strain, Jutland, Denmark) were hatched and reared under pathogen-free conditions (Danish Centre for Wild Salmon, Randers, Denmark), before they were brought to our experimental fish keeping facility. The pathogen-free status of the fish was confirmed in the laboratory. 40 rainbow trout (average weight 25 g/fish) were split into two groups of 20 fish each. The fish were kept in 200 L tanks (20 fish/tank) with bio-filters (Eheim, Germany) and maintained at a 12 h light and 12 h dark cycle in aerated (100% oxygen saturation) tap water at 12°C. One group was injected once intraperitoneally (i.p.) with PBS (control) while the other group received three consecutive immunisations i.p. with an experimental vaccine.

A commercial aquaculture production line considered representative for Danish rainbow trout farming comprising both fresh water and salt-water facilities was selected for the field sampling. The rainbow trout population for this production line was vaccinated i.p. by machine at the fresh water farm in February 2009. The vaccine comprised inactivated *Aeromonas salmonicida*, *Vibrio anguillarum* serovar O1 and O2 emulsified in a non-mineral oil adjuvant (FuruVac 5 Vibrio; Intervet Schering-Plough Animal Health). On the day of vaccination 50 fish were randomly collected for visual examination just after passage through the vaccination machine. First, the fish were checked on the exterior for needle entries. Second, the abdominal cavity was opened with a scissor and both cavity and organs were thoroughly examined for presence of injected vaccine. In case of no vaccine in the abdomen the fish was checked for incorrect vaccination in the dorsal part. A year later, the rainbow trout population was transferred to sea cages. On one day in May 2010 50 trout were randomly taken from the cages. The individual fish length and weight was registered and blood samples collected. Finally, the 50 rainbow trout were dissected and thoroughly examined for visible signs of previous vaccination (fibrinous adhesions, pigmentation/melanin deposits and/or vaccine residues).

Experimental Vaccine Preparation and Immunisation

48 h culture of the strain *V. anguillarum* O1 ATCC 43305 was inactivated with 0.9% formaldehyde for 2 h at room temperature, washed with PBS, adjusted to an optical density corresponding to 2×10^8 cells/mL PBS, and emulsified with an equal volume of Freund's incomplete adjuvant (FIA; Sigma, St. Louis, Mo.). Fish were immunised by i.p. injections with 0.1 mL of emulsified vaccine (1×10^7 cells/injection). Immunisations were given three times with 500 day degree intervals (six weeks).

Bacteria

The strain *Vibrio anguillarum* O1 6018/1 (ATCC 43305) was used for immunisations and coating of ELISA plates. Bacteria were grown with agitation for 48 h at 20°C in veal infusion broth with 1% NaCl and enumerated as colony forming units (CFU) on blood agar (blood agar base CM55, Oxoid, supplemented with 5% bovine blood). Stock cultures were maintained at −80°C in a broth culture supplemented with 15–20% glycerol.

Serum

Blood was collected as individual samples by caudal venipuncture and serum obtained by allowing the blood to clot overnight at 5°C followed by centrifugation (10.000 rpm for ten minutes without brakes). Serum samples were stored at −80°C in aliquots of 0.5 mL. Time points for blood collections from laboratory fish were one week prior to each immunisation and six weeks after third and final immunisation.

Field samples were collected likewise. Directly after sampling, the blood was put on ice and kept cool until arrival at the laboratory five hours later. At the laboratory blood samples were placed at 5°C. Serum was isolated the following morning and stored at −80°C until use.

ELISA for Determination of *V. anguillarum* O1 Specific Antibodies

For coating, each well of the microplates (flat-bottom 96-well plates, MaxiSorp™, Nunc) was filled with 100 µL coating buffer (C-3041, Sigma) containing 5 µg/mL of antigen (sonically disrupted *V. anguillarum* O1, Artek Sonic Dismembrator, Model 300, 60% for 3×5 min on ice; protein concentration determined with BCA Protein Assay, Thermo Scientific) and incubated overnight at 4°C. For comparison we also tested formalin inactivated whole-cell *V. anguillarum* O1 as coating antigen (5 µg/mL). After removal of coating solution, the wells were washed three times with 250 µL wash buffer (0.1% Tween 20 in PBS, pH 7.2) per well. For blocking residual protein binding capacity, PBS (200 µL) containing 10 g/L bovine serum albumin (BSA) was added to each well of the microplates, which were then incubated at room temperature for 1 h. After blocking, the microplates were washed three times, aspirated, sealed (microplate seal, Nunc) and stored at −20°C until use.

Serum samples were diluted in assay buffer (0.1% BSA and 0.1% Tween 20 in PBS, pH 7.2) and added to the microplates in duplicates with 50 µL per well. The microplates were sealed and placed at 4°C for overnight incubation. The next morning, the microplates were washed three times and 100 µL of a mouse anti-salmonid Ig antibody solution (MCA2182, AbD serotec, 1:400 dilution in assay buffer) was added. After 1 h incubation at room temperature, the microplates were washed three times, and 100 µl were added to each well of a rabbit F(ab')2 anti-mouse IgG:HRP solution (STAR13B, AbD serotech, 1:400 dilution in assay buffer). Optimal concentrations of the commercial antibodies were established through chessboard titration experiments (data not shown). The microplates were left for additional 1 h incubation at room temperature followed by three washes. Substrate solution was added (100 µl/well of TMB, Sigma). After 10 min of incubation, stop solution (100 µl 1N HCL) was added, and the absorbance at 450 nm was measured with a PowerWave 340 (BioTek).

The limit of detection for the assay was determined as the concentration corresponding to the signal three standard deviations above the mean for blank wells (sample dilution buffer only). The linearity of signal as a function of dilution was investigated in pooled serum samples (n = 20) diluted in sample dilution buffer ranging from 1:10–1:10000. Intra-assay coefficient of variation (CV) was determined for eight duplicates of 1:1000 and 1:5000 dilutions of a control serum pool on the same assay plate. The inter-assay CV was determined for duplicates of 1:1000 and 1:5000 dilutions of the control serum pool on individual plates run on 14 different days. The stability of the antibodies in the serum samples was tested when exposed to repeated steps of freeze and thaw (samples frozen and thawed in the range one to ten times) or keeping the samples at room temperature for 24 h.

Statistics

The Prism© software package (version 4.0 for Macintosh, GraphPad Software, Inc.) was used to manage data and for statistical analyses. Mean values and standard deviations were calculated and differences between means assessed by two-tailed unpaired t test. Sensitivity and specificity were summarised in a receiver operating characteristic (ROC) curve and area under the curve (AUC) given with 95% CI [17]. Moreover, the ROC analysis was used to identify an optimum cut-off value for the assay as the point giving the highest possible sensitivity (true-positive rate) in conjunction with the smallest false-positive fraction. The significance level was set at 0.05.

Acknowledgments

The authors would like to thank the Danish marine fish farmers for their cooperation. Also, we sincerely thank Kirsten Kaas and Lene Gertman, National Veterinary Institute, for skilful technical assistance.

Author Contributions

Conceived and designed the experiments: LHA NL KB. Performed the experiments: LHA JN ID. Analyzed the data: LHA. Contributed reagents/materials/analysis tools: LHA ID KB. Wrote the paper: LHA ID NL KB.

References

1. Larsen JL, Rasmussen HB, Dalsgaard I (1988) Study of Vibrio anguillarum strains from different sources with emphasis on ecological and pathobiological properties. Appl Environ Microbiol 54: 2264–2267.
2. Pedersen K, Skall HF, Lassen-Nielsen AM, Nielsen TF, Henriksen NH, et al. (2008) Surveillance of health status on eight marine rainbow trout, Oncorhynchus mykiss (Walbaum), farms in Denmark in 2006. J Fish Dis 31: 659–667.
3. Hastein T, Gudding R, Evensen O (2005) Bacterial vaccines for fish–an update of the current situation worldwide. Dev Biol (Basel) 121: 55–74.
4. Midtlyng PJ, Reitan LJ, Speilberg L (1996) Experimental studies on the efficacy and side-effects of intraperitoneal vaccination of Atlantic salmon (Salmo salar L) against furunculosis. Fish & Shellfish Immunology 6: 335–350.
5. Plant KP, Lapatra SE (2011) Advances in fish vaccine delivery. Dev Comp Immunol.

6. Douglas JT, Naka SO, Lee JW (1984) Development of an ELISA for detection of antibody in leprosy. Int J Lepr Other Mycobact Dis 52: 19–25.
7. Sun Y, Liu CS, Sun L (2011) A multivalent killed whole-cell vaccine induces effective protection against Edwardsiella tarda and Vibrio anguillarum. Fish Shellfish Immunol 31: 595–599.
8. Holten-Andersen L, Doherty TM, Korsholm KS, Andersen P (2004) Combination of the cationic surfactant dimethyl dioctadecyl ammonium bromide and synthetic mycobacterial cord factor as an efficient adjuvant for tuberculosis subunit vaccines. Infect Immun 72: 1608–1617.
9. Anderson DP (1997) Adjuvants and immunostimulants for enhancing vaccine potency in fish. Fish Vaccinology 90: 257–265.
10. Thorburn MA, Jansson EK (1988) Frequency-Distributions in Rainbow-Trout Populations of Absorbance Values from an Elisa for Vibrio-Anguillarum Antibodies. Diseases of Aquatic Organisms 5: 171–177.
11. Harrell LW, Etlinger HM, Hodgins HO (1975) Humoral factors important in resistance of salmonid fish to bacterial disease. I. Serum antibody protection of rainbow trout (*Salmo gairdneri*) against vibriosis. Aquaculture 6: 211–219.
12. Viele D, Kertsetter TH, Sullivan J (1980) Adoptive transfer of immunity against *Vibrio anguillarum* in rainbow trout, *Salmo gairdneri* Richardson, vaccinated by the immersion method. J Fish Biol 17: 379–386.
13. Palm RC Jr, Landolt ML, Busch RA (1998) Route of vaccine administration: effects on the specific humoral response in rainbow trout Oncorhynchus mykiss. Dis Aquat Organ 33: 157–166.
14. Korver K, Zeijlemaker WP, Schellekens PT, Vossen JM (1984) Measurement of primary in vivo IgM- and IgG-antibody response to KLH in humans: implications of pre-immune IgM binding in antigen-specific ELISA. J Immunol Methods 74: 241–251.
15. York JJ, Fahey KJ, Bagust TJ (1983) Development and evaluation of an ELISA for the detection of antibody to infectious laryngotracheitis virus in chickens. Avian Dis 27: 409–421.
16. Gudding R, Lillehaug A, Evensen O (1999) Recent developments in fish vaccinology. Vet Immunol Immunopathol 72: 203–212.
17. Hanley JA, McNeil BJ (1982) The meaning and use of the area under a receiver operating characteristic (ROC) curve. Radiology 143: 29–36.
18. Kheyrandish A, Abdoli A, Mostafavi H, Niksirat H, Naderi M, et al. (2010) Age and Growth of Brown Trout (*Salmo trutta*) in Six Rivers of the Southern Part of Caspian Basin. Am J Anim Vet Sci 5: 8–12.

Disease Resistance in Atlantic Salmon (*Salmo salar*): Coinfection of the Intracellular Bacterial Pathogen *Piscirickettsia salmonis* and the Sea Louse *Caligus rogercresseyi*

Jean Paul Lhorente[1], José A. Gallardo[2]*, Beatriz Villanueva[3], María J. Carabaño[3], Roberto Neira[1,4]

1 Aquainnovo S.A, Puerto Montt, Chile, **2** Pontificia Universidad Católica de Valparaíso, Valparaíso, Chile, **3** Departamento de Mejora Genética Animal, INIA, Madrid, Spain, **4** Departamento de Producción Animal, Facultad de Ciencias Agronómicas, Universidad de Chile, Santiago, Chile

Abstract

Background: Naturally occurring coinfections of pathogens have been reported in salmonids, but their consequences on disease resistance are unclear. We hypothesized that 1) coinfection of *Caligus rogercresseyi* reduces the resistance of Atlantic salmon to *Piscirickettsia salmonis*; and 2) coinfection resistance is a heritable trait that does not correlate with resistance to a single infection.

Methodology: In total, 1,634 pedigreed Atlantic salmon were exposed to a single infection (SI) of *P. salmonis* (primary pathogen) or coinfection with *C. rogercresseyi* (secondary pathogen). Low and high level of coinfection were evaluated (LC = 44 copepodites per fish; HC = 88 copepodites per fish). Survival and quantitative genetic analyses were performed to determine the resistance to the single infection and coinfections.

Main Findings: *C. rogercresseyi* significantly increased the mortality in fish infected with *P. salmonis* (SI mortality = 251/545; LC mortality = 544/544 and HC mortality = 545/545). Heritability estimates for resistance to *P. salmonis* were similar and of medium magnitude in all treatments ($h^2_{SI} = 0.23 \pm 0.07$; $h^2_{LC} = 0.17 \pm 0.08$; $h^2_{HC} = 0.24 \pm 0.07$). A large and significant genetic correlation with regard to resistance was observed between coinfection treatments (r_g LC-HC = 0.99 ± 0.01) but not between the single and coinfection treatments (r_g SI-LC = -0.14 ± 0.33; r_g SI-HC = 0.32 ± 0.34).

Conclusions/Significance: *C. rogercresseyi*, as a secondary pathogen, reduces the resistance of Atlantic salmon to the pathogen *P. salmonis*. Resistance to coinfection of *Piscirickettsia salmonis* and *Caligus rogercresseyi* in Atlantic salmon is a heritable trait. The absence of a genetic correlation between resistance to a single infection and resistance to coinfection indicates that different genes control these processes. Coinfection of different pathogens and resistance to coinfection needs to be considered in future research on salmon farming, selective breeding and conservation.

Editor: J. Stephen Dumler, The Johns Hopkins University School of Medicine, United States of America

Funding: This study was financed by INNOVA-CHILE of CORFO through the projects: 1) Consorcio empresarial de genética y desarrollo biotecnológico para la industria salmonera (N° 206-5047), and 2) Desarrollo de una nueva metodología para la identificación y selección de salmónidos genéticamente resistentes al ectoparásito Caligus rogercresseyi (07CN13PBT-61). This study is a collaborative work of different institutions belonging to "Red de genética e inmunología para el control de patógenos en Acuicultura" (RED CYTED N° 11RT0420). BV received funding from the Ministerio de Economía y Competitividad, Spain (project CGL2012-39861-C02-02). The funders had no role in study design, data collection and analysis, decision to publish, or preparation of the manuscript.

Competing Interests: Jean Paul Lhorente and Roberto Neira were employed in AQUAINNOVO when this research was performed.

* E-mail: jose.gallardo@ucv.cl

Introduction

Naturally occurring coinfections of pathogens have been reported in various salmonid species [1], [2]. However, the consequences for overall disease resistance remain unclear. Most studies have shown that presence of a primary pathogen reduces the resistance response to a secondary pathogen. For example, in rainbow trout, primary infestation of the parasite *Myxobolus cerebralis* suppresses the immune system, increasing mortality associated with a secondary infection by *Yersinia ruckeri* [3]. Similarly, Atlantic salmon infected with IPNV show significantly increased mortality as smolts when exposed to Furunculosis, Vibriosis and ISAv [4], [5]. Conversely, several studies have shown that coinfection does not necessarily increase mortality. For example, some studies on rainbow and brown trout showed that IPNV reduced the infection capacity of the hematopoietic necrosis virus (IHNV) and that the mortality of fish infected with both viruses was significantly lower than that observed in single challenges with each pathogen [6], [7], [8], [9]. Furthermore, acute coinfection of IPNv and ISAv in Atlantic salmon significantly reduced mortality compared with a single infection by ISAv [4].

Sea lice are among the most important sanitary problems in the global salmon aquaculture industry [10], further they have been linked to wild salmon and trout population declines around the world [11], [12], [13]. Salmonid infestation by sea lice may be associated with lethal or sub-lethal effects [14], [15], [16]. Sub-lethal effects may include stress [17], [18], loss of appetite, depression of the immune system and skin damage [19], [20], and therefore, it can contribute to increased susceptibility to other diseases [21], [22]. In agreement with this hypothesis, Mustafa et al. [22] reported that Atlantic salmon infected with the sea louse *Lepeophtheirus salmonis* showed increased susceptibility to a microsporidian parasite *(Loma salmonae)*. Recently, Nowak et al. [23], Bustos et al. [24] and Valdes-Donoso et al. [25] suggested, based on field studies on Atlantic salmon, that sea lice affect disease resistance to the amoeba *Neoparamoeba perurans* and to ISAv.

In Chile, caligidosis caused by *Caligus rogercresseyi* and piscirickettsiosis caused by *Piscirickettsia salmonis* have historically been the most important health problems in the salmon industry in the growth-out production phase [26], [27], [28], [10]. *C. rogercresseyi* is the only sea louse that affects the Chilean salmon industry [29], and annual losses attributed to this parasite are estimated at more than $178 million US [27], [10]. *P. salmonis*, an intracellular bacterium, was described in Chile at the end of the 1980s in the X region [30], [31], [32] and was initially found to be strongly associated with Atlantic salmon. It has now extended to other salmonid species farmed in Chile [33], producing mortality rates of up to 50% in fish in the sea growing stage, with monetary losses exceeding $100 million US per year [28]. Furthermore, *P. salmonis* has been reported in different countries and can infect rainbow trout (*Oncorhynchus mykiss*), cherry salmon (*Oncorhynchus masou*), chinook salmon (*Oncorhynchus tshawytscha*) and pink salmon (*Oncorhynchus gorbuscha*) [34].

Resistance of Atlantic salmon to *C. rogercresseyi* and *P. salmonis* has recently been studied in single-infection challenges in laboratory conditions by Yañez et al. [35] and Lhorente et al. [36]. Lhorente et al. [36] reported heritability of resistance of Atlantic salmon to the sea louse *C. rogercresseyi* of low (0.03–0.06) and medium (0.22–0.34) magnitudes for the mobile and sessile stages of the parasite, respectively. Other studies performed on different sea lice species confirmed that Atlantic salmon have a heritable defensive mechanism against *Lepeophtheirus salmonis* [37], [38], [39] and *Caligus elongatus* [40]. Additionally, it has been shown that both wild and farmed Atlantic salmon have a heritable defensive mechanism against various bacteria [41], [42], [43], [44], including *P. salmonis* [45]. For instance, Yañez et al. [35] reported heritability of resistance of Atlantic salmon to *P. salmonis* ranging from 0.11 to 0.41.

In this study, we hypothesized that coinfection with the sea louse *C. rogercresseyi* reduces the resistance of Atlantic salmon to *P. salmonis* because of the well-documented stress and depression of the immune system produced by sea lice infection in salmonids. Additionally, we hypothesized that coinfection resistance is a heritable trait that does not correlate with resistance to a single infection of *P. salmonis* because salmonid defense mechanisms against bacteria and parasites are substantially different [46]. We present experimental evidence supporting both hypotheses for the interaction between Atlantic salmon, *P. salmonis* and *C. rogercresseyi*.

Materials and Methods

Ethics Statement

This study was carried out in in accordance with the guide to the care and use of experimental animals of the Canadian Council on Animal Care. The protocol was approved by the Bioethical committee of the Pontificia Universidad Católica de Valparaíso (N° 10/2013). The animals were anaesthetized with benzocaine prior to the various handling processes and marking. Euthanasia was performed using an overdose of anesthesia. All efforts were made to provide the best growing conditions and to minimize suffering.

Fish

In total, 1,634 Atlantic salmon smolt fish of 103 g average body weight from 15 full-sib families (corresponding to seven paternal half-sib families) of the AquaChile Genetic Program were available for this study. The families originated from a nested mating design (15 females and seven males), in which one male fertilized the eggs of at least two females. Eggs from each family were produced during the spawning season of 2009. The fish were fed regularly with a commercial diet and individually tagged in April 2010 at an average weight of 5 g (SD = 8.0 g). Then, they were transferred as smolts in February 2011 to the Aquadvice S.A. experimental station located at the Quillaipe sector of Puerto Montt (Chile). A health check by PCR was performed prior to transfer to verify that the fish were free of viral (*IPNv* and *ISAv*) and bacterial pathogens (*P. salmonis*, *Renibacterium salmoninarum*, *Vibrio sp* and *Flavobacteria sp*). At the experimental station, the fish underwent a three-week acclimation period under seawater conditions (salinity of 33% and a temperature of 12°C).

Experimental Design

The fish were exposed to three different infection scenarios with three replicates (tanks) to simulate single infection by *P. salmonis* and coinfection by *P. salmonis* with two different levels of infection pressure of the parasite *C. rogercresseyi*. In all coinfection's experiments *P. salmonis* was used as primary pathogen and *C. rogercressyi* was used as a secondary pathogen. Both levels of infection pressure were established using information from previous experiments [47] and setup to experimental fish size to ensure successful but differential settlement between treatments and no mortality associated to single sea lice infection [47,48]. Sea lice used to produce copepodites were pathogen free (*P. salmonis* and ISAv). Fish from the 15 full-sib families were equally distributed in nine tanks of 0.72 m^3 such that the same number of fish per family was used in each of the following treatments:

a) Single infection (SI). The fish were infected with *P. salmonis* using a cohabitation challenge test in an environment free of *C. rogercresseyi.*

b) Low pressure of coinfection (LC). The fish were infected with *P. salmonis* using a cohabitation challenge test and then infested with *C. rogercresseyi* using a low infestation pressure of 44 copepodites per fish.

c) High pressure of coinfection (HC). The fish were infected with *P. salmonis* using a cohabitation challenge test and then infested with *C. rogercresseyi* using a high infestation pressure of 88 copepodites per fish.

The cohabitation method used for the primary infection with *P. salmonis* produces a natural infestation and reduces the manipulation of experimental fish in comparison with intra-peritoneal injection. At the beginning of the challenge with *P. salmonis* (day zero), 78 fish with unknown pedigree, referred to here as infective fish, were infected by intra-peritoneal injection. A volume of 0.2 ml/fish (at $10^{6,2}$ TCID/ml) of a virulent strain of *P. salmonis,* isolated from Atlantic salmon and commercially available from ADL-Diagnostic Ltda., was injected to each infective fish. The infective fish were then placed into cohabitation with tagged fish to reach an abundance of 260 fish app. per tank and a density of 42 kg/m^3. The mortality of the infective fish was 100% at 30 days post-infection, those fish were not included in our evaluation of resistance. Resistance to *P. salmonis* was measured as a mortality

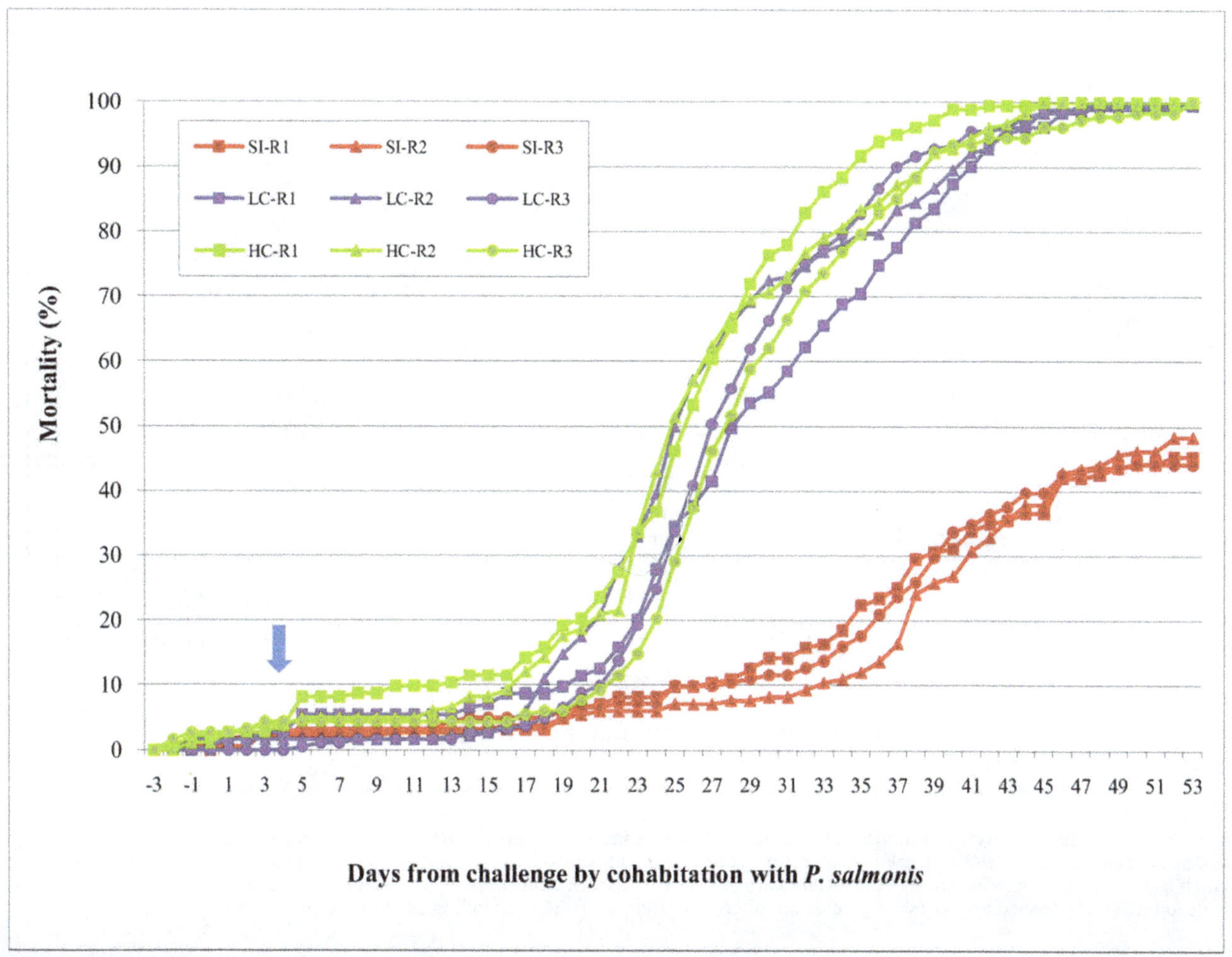

Figure 1. Mortality curves of Atlantic salmon coinfected with *P. salmonis* and *C. rogercresseyi*. The data show the cumulative mortality of three replicates (tanks) (R1–R3) for three different scenarios: single infection (SI) with *P. salmonis* and co-infection with two levels of infestation pressure of *C. rogercresseyi* (low pressure of coinfestation (LC) = 44 copepodites per fish; high pressure of coinfestation (HC) = 88 copepodites per fish). Each replicate had approximately 182 pedigreed fish that were free of disease. The arrow indicates the day of coinfection.

trait (dead/alive). The sexes, initial weights and final weights of all cohabitant tagged fish were recorded.

Four days post-infection, a secondary infection was conducted with *C. rogercresseyi* as described by Lhorente et al. [36]. Ten infective fish per tank were used to evaluate the effective burden of the parasite. We confirmed the lack of sea lice (i.e., chalimus I) in the SI treatment and an incremental response to settlement in both coinfection treatments (LC = 29±4.8 lice per fish; HC = 60±4.8 lice per fish).

Statistical Analysis

Kaplan-Meier survival curves were obtained using the software Survival kit v6 [49] to define the responses of Atlantic salmon to the different treatments. A chi-squared test was used to evaluate differences between treatments [49]. To estimate fixed effects and their interactions, an ANOVA analysis was conducted using a general lineal model (GLM) [50]. Mortality was defined at 30 days after infection with *P. salmonis* and treated as a normally distributed trait [43]. The GLM was:

$$\mathbf{y}_{ijklm} = \boldsymbol{\mu} + \mathbf{C}_i + \mathbf{S}_j + \mathbf{F}_k + \mathbf{T}(\mathbf{C})_{li} + \mathbf{CF}_{ik} + \mathbf{SF}_{jk} + \mathbf{CSF}_{ijk} + \mathbf{e}_{ijklm}$$

where $\boldsymbol{y_{ijklm}}$ is the fish mortality condition, $\boldsymbol{\mu}$ is the population mean, $\mathbf{C}_i$ is the treatment effect (SI, LC or HC), $\mathbf{S}_j$ is the sex (female or male), $\mathbf{F}_k$ is the full sib family effect (1,2,3,..., 15), $\mathbf{T(C)}_{li}$ is the lth tank within the ith treatment effect (C), $\mathbf{CF}_{ik}$ is the treatment by family interaction, $\mathbf{SF}_{jk}$ is the sex by family interaction, $\mathbf{CSF}_{ijk}$ is the treatment by sex by family interaction and $\mathbf{e}_{ijklm}$ is the random residual effect. The $\mathbf{CF}_{ik}$ interaction determines whether families resistant to SI are also resistant to LC or HC coinfection.

A quantitative genetic analysis of the resistance to *P. salmonis* was conducted for each infection treatment to compare estimates of the additive genetic variance for the mortality trait evaluated. A threshold model (TM) was assumed to estimate variance components [51]. This model assumes a normal underlying liability variable **l** determining the categorical outcomes of the test-period survival, such that $\mathbf{l}_{ijk} \leq 0$ corresponds to $\mathbf{Y}_{ijk} = 0$ and $\mathbf{l}_{ijk} > 0$ corresponds to $\mathbf{y}_{ijk} = 1$. The residual variance of **l** was assumed to be 1.

$$Pr(y_{ijk} = 1) = Pr(l_{ijk} > 0) = \Phi(\mathbf{w}'_i \theta)$$

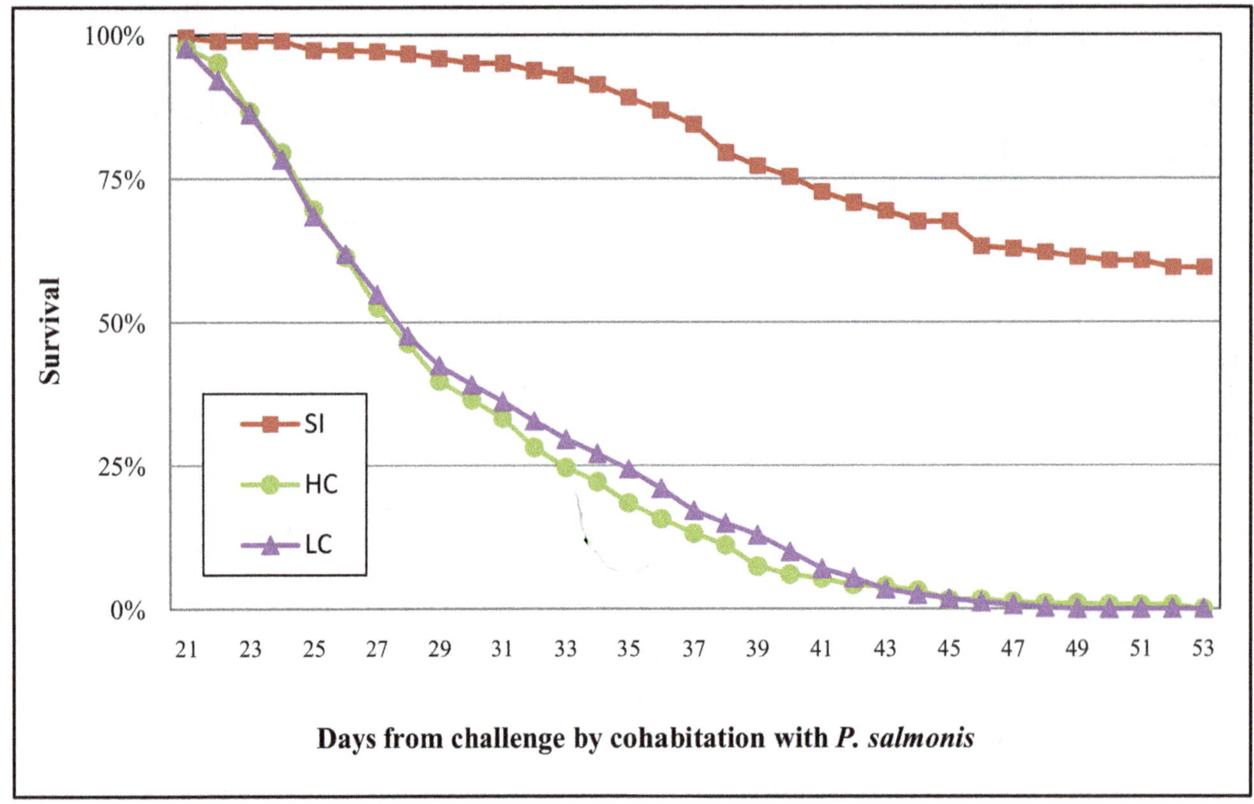

Figure 2. Kaplan-Meier survival function of Atlantic salmon coinfected with *P. salmonis* and *C. rogercresseyi*. The survival function represents the resistance of the Atlantic salmon (i.e., the proportion that had not died on each day following challenge) in three different treatments: single infection (SI) with *P. salmonis* and coinfection with two levels of infestation pressure of the sea louse *C. rogercresseyi* (low pressure of coinfection (LC) = 44 copepodites per fish; high pressure of co infestation (HC) = 88 copepodites per fish).

where Φ (*) corresponds to the standard normal distribution, w'_i is the incidence vector that links the data with the parameters that define the mean of the distribution of the liability indexed by the parameters in θ and θ contains the population mean, the additive genetic value and the significant (P<0.05) environmental effects (sex, challenge tank, and initial weight as covariate).

Because the same families were represented in the three infection treatments (SI, LC and HC), the genetic correlation of *P. salmonis* resistance in different infective scenarios measures the interaction that determines whether families resistant to SI are also resistant to LC or HC coinfection. A set of three bivariate linear model (LM) analyses was used to estimate the genetic correlation of resistance among the different treatments:

$$\mathbf{y} = \mathbf{Xb} + \mathbf{Za} + \mathbf{e}$$

where $\mathbf{y}$ is the observations vector for the proposed traits (SI, LC or HC), $\mathbf{X}$ is the design matrix, $\mathbf{b}$ is the vector of significant

Table 1. ANOVA results for mortality 30 days after *P. salmonis* infection.

Source of variation	fd	SS	SS%	MS	F	P-value
Infection treatment (I)	2	87.90	28.0%	43.940	310.29	0.0001
Sex (S)	1	1.75	0.6%	1.750	12.38	0.0004
Full-sib Family (F)	14	22.30	7.1%	1.590	11.25	0.0001
Tank within I (T)	6	3.72	1.2%	0.620	4.38	0.0002
I×F	28	12.94	4.1%	0.460	3.26	0.0001
S×F	2	0.53	0.2%	0.260	1.87	0.1547
S×I×F	42	6.37	2.0%	0.150	1.07	0.3517
Error	1267	179.46	57.0%			
Total	1362	314.98				

Table 2. Estimates of heritability (on diagonal), phenotypic (above diagonal) and genetic (below diagonal) correlations (± SE) in resistance of Atlantic salmon between single infection (SI) with *P. salmonis*, and co-infection with two incremental levels the of sea louse *C. rogercresseyi* (low pressure of coinfestation (LC) = 44 copepodites per fish; high pressure of coinfestation (HC) = 88 copepodites per fish).

	SI	LC	HC
SI	0.23±0.07*	−0.04±0.09^{ns}	0.06±0.07^{ns}
LC	−0.14±0.33^{ns}	0.17±0.08*	0.21±0.01*
HC	0.32±0.34^{ns}	0.99±0.01*	0.24±0.07*

ns: Not significantly different from zero, p>0.05; *; significantly different from zero, p<0.05.

(P<0.05) fixed effects within each test (sex, challenge tank and initial weight as a covariate), **Z** is the incidence matrix of the random effects, **a** is the breeding values vector and **e** is the residual error vector.

The covariance structure of random effects was:

$$Var = \begin{bmatrix} a_i \\ a_j \end{bmatrix} = \begin{bmatrix} A\sigma^2 a_{ii} & A\sigma^2 a_{ij} \\ A\sigma^2 a_{ij} & A\sigma^2 a_{jj} \end{bmatrix}; \; Var\begin{bmatrix} e_i \\ e_j \end{bmatrix} = \begin{bmatrix} I\sigma^2 e_{ii} & I\sigma^2 e_{ij} \\ I\sigma^2 e_{ij} & I\sigma^2 e_{jj} \end{bmatrix}$$

where $\boldsymbol{a_{i/j}}$ and $\boldsymbol{e_{i/j}}$ are the vectors of additive genetic and residual values for traits i/j, respectively; $\boldsymbol{A}$ is the additive genetic relationship matrix; $\boldsymbol{I}$ is the identify matrix; $\boldsymbol{\sigma^2 a_{ii/jj}}$ and $\boldsymbol{\sigma^2 e_{ii/jj}}$ are the variances of additive genetic and residual effects, respectively, for traits i/j and $\boldsymbol{\sigma^2 a_{ij}}$ and $\boldsymbol{\sigma^2 e_{ij}}$ are the covariances of additive genetic and residual effects, respectively, for the *ith* and *jth* traits.

Restricted Maximum Likelihood (REML) [52] and Asreml software [53] were used to solve the **TM** and **LM** models and obtain the genetic parameters.

Results

Development of Piscirickettsiosis with and without Coinfection with Sea Lice

The development of piscirickettsiosis was continuously recorded over 53 days until mortality reached 544/544 (100%) and 545/545 (100%) in both coinfection treatments (Figure 1). At that time, mortality in the group that received the single infection with *P. salmonis* only reached 251/545 (46%). At the beginning of the cohabitation challenge with *P. salmonis*, a small increase in mortality was observed (day 4). This increase was likely a consequence of the sea lice infection procedure and not a direct consequence of coinfection. At 14–16 days post-infection, mortality associated with piscirickettsiosis was observed in cohabitant fish, and this result was confirmed by PCR at the beginning of the outbreak in dead fish. The daily mortality rate was lower in single than in coinfection scenarios (Figure 1), but no differences were observed between the coinfection treatments.

Consistent with the mortality pattern previously described, Kaplan-Meier survival curves (Figure 2) confirmed that coinfection of *P. salmonis* and *C. rogercresseyi* significantly (p<0.05) reduced the survival of Atlantic salmon compared to single infection with *P. salmonis*. However, high parasite burden did not significantly reduce the survival of Atlantic salmon compared to low parasite burden (P>0.05).

Genetic Resistance of Atlantic Salmon to Single and Coinfection

As shown in Table 1, all main fixed effects significantly affected the mortality of Atlantic salmon challenged with *P. salmonis* and coinfected with *C. rogercresseyi*. The infection effect (I) was highly significant (P<0.001) and showed the highest relative value of the associated sum of squares over the total variability of mortality (28%). The effect of the full-sib family (F) and its interaction with the treatment effect (F×I) were also highly significant (P<0.001). These effects showed a relative influence of 7% and 4%, respectively, on the variability of mortality. Sex and tank within infection treatment effects also significantly affected mortality. However, their contributions to the overall variability of mortality were less than 2%.

Quantitative genetic parameters for the resistance of Atlantic salmon to *P. salmonis* for the three treatments of single and coinfection are shown in Table 2. The heritability of resistance was very similar between treatments and of medium magnitude (0.17–0.24). Resistance to the single infection with *P. salmonis* did not correlate phenotypically or genetically with resistance to *P. salmonis* upon coinfection with the sea louse *C. rogercresseyi* (Table 2). Conversely, a high and significant genetic correlation for resistance to *P. salmonis* was observed between the two coinfection treatments (r_g LC-HC = 0.99±0.01), providing solid evidence that these resistance values are measurements of the same trait.

Discussion

We have demonstrated for the first time that the sea louse *C. rogercresseyi*, as a secondary pathogen, significantly reduces the resistance of Atlantic salmon to the bacterium *P. salmonis*. The prevalence of *C. rogercresseyi* in Chilean salmon farms approaches 100% in some seasons (i.e., spring and summer) and geographic regions [27], [54]. Therefore, its effect on Atlantic salmon mortality may be higher than previously thought. In Atlantic salmon, coinfections of sea lice and other pathogens such as the amoeba *Neoparamoeba perurans* have also been reported in the USA [23] and in Chilean salmon farms [24]. Both of these studies suggested that sea lice may play an important role in the epidemiology of amoebic gill disease caused by *Neoparamoeba perurans* and/or in mortality of Atlantic salmon in sea farms. Similarly, Valdes-Donoso et al. [25] reported that most of the ISAV outbreaks between 2007 and 2009 in the Xth region of Chile were associated with high sea lice burdens. The reduced survival upon coinfection in Atlantic salmon might be explained by the direct skin damage caused by parasites that allows other pathogens to enter the fish [55]. Alternatively, it may result from the systemic effects of immunosuppression caused by sea lice [56].

Genetic variation in resistance to disease in salmonids has been reported for single infections of pathogens in Atlantic salmon [40], [57], [58], [36], [42], [35], rainbow trout [59], [60], [61], [62], [63], [64], Coho salmon [41] and brook charr [42]. However, genetic variation in the resistance to coinfection by two pathogens has not been previously estimated in salmonids. Using data from a farmed population of Atlantic salmon, we demonstrated genetic variation for resistance to *P. salmonis* upon coinfection with the sea louse *C. rogercresseyi*. Sea lice infections have been reported in salmon farms around the world [10], but coinfection with bacteria, viruses or parasites has been minimally investigated. Resistance to coinfection has two important implications for salmon breeding. First, if coinfection is common, selection for disease resistance to two or more pathogens, evaluated independently as proposed by Ødegård et al. [65], could be an inefficient method unless resistance to single and coinfection is positively correlated. Second, evaluation of resistance to two different pathogens could be performed in a simple assay, reducing costs associated with laboratory testing. Further studies are necessary to determine whether resistance to coinfection by *P. salmonis* and *C. rogercresseyi* or to coinfection by other pathogens relevant to salmon farming such as ISAV, *Aeromonas salmonicida* or *Neoparamoeba perurans* occurs in other populations of salmonids.

The genetic correlation for resistance among various Atlantic salmon pathogens has been described for some bacteria [66], [67], [68], [69], [70], [71] and viruses [69], [71], [72]. However, the genetic correlation for resistance between single and coinfection of two pathogens has not been previously estimated in salmonids. Our results strongly suggest that the resistance of Atlantic salmon to a single infection of *P. salmonis* and that to coinfection with the sea louse *C. rogercresseyi* are not genetically related. Therefore, we can infer that the best strategy for developing resistance to *P. salmonis* should consider coinfection with sea lice. However, further studies are necessary to establish whether the resistance to coinfection observed in experimental conditions correlates with higher survival rates in the field. A high genetic correlation for resistance between fresh and sea water has been described for other diseases such as furunculosis, sea lice and IPN [67], [38], [73].

Conclusion

Infection with the sea louse *C. rogercresseyi*, as a secondary pathogen, reduces the resistance of Atlantic salmon to the pathogen *P. salmonis*. Resistance to coinfection of *Piscirickettsia salmonis* and *Caligus rogercresseyi* in Atlantic salmon is a heritable trait. The absence of a genetic correlation between the resistance to single infection and that to coinfection indicates that different genes control these processes. Further studies are necessary to investigate the effects of coinfection when the sea louse is the primary pathogen. It is clear that coinfection of different pathogens and resistance to coinfection needs to be considered in future research on salmon farming, selective breeding and conservation.

Acknowledgments

The authors would like to thank Martín Hevia for his valuable contribution to the development of the experiments in Aquadvice S.A.

Author Contributions

Conceived and designed the experiments: JPL JAG RN. Performed the experiments: JPL JAG. Analyzed the data: JPL JAG. Contributed reagents/materials/analysis tools: BV MJC. Wrote the paper: JPL JAG. Contributed to designing the experiments and analyzing the data: BV MJC.

References

1. Mulcahy D, Fryer JL (1976). Double infection of rainbow trout fry with IHN and IPN viruses. Am Fish Soc/Fish Health Sect Newsl 5: 5–6.
2. Vilas MP, Rodríguez S, Perez S (1994) A case of coinfection of IPN and IHN virus in farmed rainbow trout in Spain. Bull Eur Assoc Fish Pathol 14 : 1–4.
3. Densmore CL, Ottinger CA, Blazer VS, Iwanowicz LR (2004) Immunomodulation and disease resistance in postyearling rainbow trout infected with *Myxobolus cerebralis*, the causative agent of whirling disease. J Aquat Anim Health 16: 73–82.
4. Johansen LH, Sommer AI (2001). Infectious pancreatic necrosis virus infection in Atlantic salmon *Salmo salar* post-smolt affects the outcome of secondary infections with infectious salmon anaemia virus or Vibrio salmonicida. Dis Aquat Organ 47: 109–117.
5. Johansen LH, Eggset G, Sommer AI (2009) Experimental IPN virus infection of Atlantic salmon parr; recurrence of IPN and effects on secondary bacterial infections in post-smolts. Aquaculture 290: 9–14.
6. Alonso M, Rodríguez Saint-Jean S, Pérez-Prieto SI (2003) Virulence of Infectious hematopoietic necrosis virus and Infectious pancreatic necrosis virus coinfection in rainbow trout (*Oncorhynchus mykiss*) and nucleotide sequence analysis of the IHNV glycoprotein gene. Arch Virol 148: 1507–1521.
7. Tafalla C, Rodríguez Saint-Jean S, Pérez-Prieto S (2006) Immunological consequences of the coinfection of brown trout (*Salmo trutta*) with infectious hematopoietic necrosis virus (IHNV) and infectious pancreatic necrosis virus (IPNV). Aquaculture 256: 15–22.
8. Byrne N, Castric J, Lamour F, Cabon J, Quentel C (2008) Study of the viral interference between infectious pancreatic necrosis virus (IPNV) and infectious haematopoietic necrosis virus (IHNV) in rainbow trout (*Oncorhynchus mykiss*). Fish Shellfish Immun, 24: 489–497.
9. Rodríguez Saint-Jean S, Pérez-Prieto S (2007) Effects of salmonid fish viruses on Mx gene expression and resistance to single or dual viral infections. Fish Shellfish Immun 23 (2): 390–400.
10. Costello MJ (2009) The global economic cost of sea lice to the salmonid farming industry. J Fish Dis 32: 115–118.
11. Costello MJ (2009) How sea lice from salmon farms may cause wild salmonid declines in Europe and North America and be a threat to fishes elsewhere. Proc Roy Soc B 276: 3385–3394.
12. Krkošek M, Ford JS, Morton A, Lele S, Myers RA, et al. (2007) Declining wild salmon populations in relation to parasites from farm salmon. Science 318: 1772–1775.
13. Skaala Ø, Kålås S, Borgstrøm R (2014) Evidence of salmon lice-induced mortality of anadromous brown trout (*Salmo trutta*) in the Hardangerfjord, Norway, Marine Biology Research, 10: 3, 279–288.
14. Costello MJ (2006) Ecology of sea lice parasitic on farmed and wild fish. Trends Parasitol 22: 475–483.
15. Connors BM, Hargreaves NB, Jones SRM, Dill LM (2010) Predation intensifies parasite exposure in a salmonid food chain. J Appl Ecol 47: 1365–1371.
16. Krkošek M, Connors B, Mages P, Peacock S, Ford H, et al. (2011) Fish farms, parasites, and predators: Implications for salmon population dynamics. Ecol Appl 21: 897–914.
17. Finstad B, Bjørn PA, Grimnes A, Hvidsten NA (2000) Laboratory and field investigations of salmon lice [*Lepeophtheirus salmonis* (Kroyer)] infectation on Atlantic salmon (*Salmo salar* L) post-smolts. Aquac Res 31: 795–803.
18. Fast MD, Muise DM, Easy RE, Ross NW, Johnson S (2006) The effects of *Lepeophtheirus salmonis* infections on the stress response and immunological status or Atlantic salmon (*Salmo salar*). Fish Shellfish Immun 21: 228–241.
19. MacKinnon BM (1998). Host factors important in sea lice infections. ICES J Mar Sci 55: 188–192.
20. Tully O, Nolan DT (2002) A review of the population biology and host-parasite interactions of the sea louse *Lepeophtheirus salmonis* (Copepoda: Caligidae). Parasitology 124: S165–S182.
21. Pike AW, Wadsworth SL (1999) Sea lice on salmonids: their biology and control. Adv Parasitol 44: 223–337.
22. Mustafa A, Speare DJ, Daley J, Conboy GA, Burka JF (2000) Enhanced susceptibility of seawater cultured rainbow trout, *Oncorhynchus mykiss* (Walbaum), to the microsporidian *Loma salmonae* during a primary infection with the sea louse, *Lepeophtheirus salmonis*. J Fish Dis 23: 337–341.
23. Nowak BF, Bryan JB, Jones S (2010) Do salmon lice, *Lepeophtheirus salmonis*, have a role in the epidemiology of amoebic gill disease caused by *Neoparamoeba perurans*?. J Fish Dis 33: 683–687.
24. Bustos PA, Young ND, Rozas MA, Bohle HM, Ildefonso RS, Morrison RN, Nowak BF (2011) Amoebic gill disease (AGD) in Atlantic salmon (*Salmo salar*) farmed in Chile. Aquaculture 310 (3–4): 281–288.
25. Valdes-Donoso P, Mardones FO, Jarpa M, Ulloa M, Carpenter TE, Perez AM (2013) Co-infection patterns of infectious salmon anaemia and sea lice in farmed Atlantic salmon, *Salmo salar* L., in southern Chile (2007–2009). J Fish Dis 36: 353–360.
26. Zagmutt-Vergara FJ, Carpenter TE, Harver TB, Hedrick RP (2005) Spatial and temporal variations in sea lice (Copepoda: Caligidae) infestations of three salmonid species farmed in net pens in Southern Chile. Dis Aquat Organ 64: 163–173.
27. Rozas M, Asencio G (2007) Evaluación de la Situacion Epidemiológica de la Caligiasis en Chile: Hacia una estrategia de control efectiva. *Salmociencia* 2 (1): 43–59.
28. Leal J, Woywood D (2007). Piscirickettsiosis en Chile: Avances y perspectivas para su control. Salmociencia 2: 34–42.
29. Boxshall G, Bravo S (2000) On the identity of common Caligus (Copepoda: Siphonostomatoida: Caligidae) from salmonid net pen systems in southern Chile. Contrib Zoo 69: 137–146.
30. Fryer JL, Lannan CN, Garcés LH, Larenas JJ, Smith PA (1990) Isolation of a rickettsiales-like organism from diseased Coho salmon in Chile. Fish Pathol 25: 107–114.

31. Cvitanich JD, Garate NO, Smith CE (1991) The isolation of a rickettsia-like organism causing disease and mortality in Chilean salmonids and its confirmation by Koch's postulate. J Fish Dis 14: 121–145.
32. Fryer JL, Lannan CN, Giovannoni SJ, Wood ND (1992) *Piscirickettsia salmonis* gen. nov., sp. nov., the causative agent of an epizootic disease in salmonid fishes. Int J Syst Bact 42: 120–126.
33. Gaggero A, Castro H, Sandino AM (1995) First isolation of *Piscirickettsia salmonis* from coho salmon, *Oncorhynchus kisutch* (Walbaum), and rainbow trout, *Oncorhynchus mykiss* (Walbaum), during the freshwater stage of their life cycle. J Fish Dis 18: 277–279.
34. Fryer JL, Hedrick RP (2003) *Piscirickettsia salmonis*: a Gram-negative intracellular bacterial pathogen of fish. J Fish Dis 26: 251–262.
35. Yáñez JM, Bangera R, Lhorente JP, Oyarzún M, Neira R (2013) Quantitative genetic variation of resistance against *Piscirickettsia salmonis* in Atlantic salmon (*Salmo salar*). Aquaculture 414: 155–159.
36. Lhorente JP, Gallardo JA, Villanueva B, Araya A, Torrealba DA, Toledo XE, Neira R (2012) Quantitative genetic basis for resistance to *Caligus rogercresseyi* sea lice in a breeding population of Atlantic salmon (*Salmo salar*). Aquaculture 324–325: 55–59.
37. Glover KA, Aasmundstad T, Nilsen F, Storset A, Skaala O (2005) Variation of Atlantic salmon families (*Salmo salar* L.) in susceptibility to the sea lice *Lepeophtheirus salmonis*and *Caligus elongatus*. Aquaculture 245: 19–30.
38. Kolstad K, Heuch PA, Gjerde B, Gjedrem T, Salte R (2005) Genetic variation in resistance of Atlantic salmon (*Salmo salar*) to the salmon louse *Lepeophtheirus salmonis*. Aquaculture 247: 145–151.
39. Gharbi K, Glover KA, Stone LC, MacDonald ES, Matthews L, Grimholt U, Stear MJ (2009) Genetic dissection of MHC-associated susceptibility to *Lepeophtheirus salmonis* in Atlantic salmon. BMC Genet 10: 20.
40. Mustafa A, MacKinnon BM (1999) Genetic variation in susceptibility of Atlantic salmon to the sea louse *Caligus elongatus* Nordmann, 1832. Can J Zool 77: 1332–1335.
41. Withler R, Evelyn T (1990) Genetic variation in resistance to bacterial kidney disease within and between two strains of coho salmon from British Columbia. Trans Am Fish Soc 119: 1003–1009.
42. Perry GML, Tarte P, Croisetiere S, Belhumeur P, Bernatchez L (2004) Genetic variance and covariance for 0+ brook charr (*Salvelinus fontinalis*) weight and survival time of furunculosis (*Aeromonas salmonicida*) exposure. Aquaculture 235: 263–271.
43. Ødegård J, Olesen I, Gjerde B, Klemetsdal G (2006) Evaluation of statistical models for genetic analysis of challenge test data on furunculosis resistance in Atlantic salmon (*Salmo salar*): prediction of field survival. Aquaculture 259: 116–123.
44. Holten-Andersen L, Dalsgaard I, Buchmann K (2012) Baltic salmon, *Salmo salar*, from Swedish river Lule Älv is more resistant to furunculosis compared to rainbow trout. Plos One: 7e29571.
45. Gómez D, Conejeros P, Consuegra S, Marshall SH (2011) MHC mediated resistance to *Piscirickettsia salmonis* in salmonids farmed in Chile. Aquaculture 318: 15–19.
46. Jones SRM (2011) Mechanisms of Resistance among Salmon to the Parasitic Copepod *Lepeophtheirus salmonis*. J Aquac Res Development S2: 003.
47. Araya A, Mancilla M, Lhorente JP, Neira R, Gallardo JA (2012) Experimental challenges of Atlantic salmon *Salmo salar* with incremental levels of copepodids of sea louse *Caligus rogercresseyi*: effects on infestation and early development. Aquac Res 43: 1904–1908.
48. González L, Carvajal J, George-Nascimento M (2000) Differential infectivity of Caligus flexispina (Copepoda:Caligidae) in three farmed salmonids in Chile. Aquaculture 183, 13–23.
49. Ducrocq V, Sölkner J, Mészáros G (2010) Survival Kit v6 - A software package for survival analysis. In: 9th World Congress on Genetics to Livestock Production, August 1–6, 2010, Leipzig, Germany.
50. SAS INSTITUTE INC, 1993. User's guide: Statistics, Versión 6.03. Edition. SAS Institute Inc., Cary, NC. 956 p.
51. Falconer DS, Mackay TFC. (1996) Introduction to quantitative genetics. London and New York, Longman Group Limited. Fourth Edition. 464 p.
52. Johnson DL, Thompson R (1995) Restricted maximum likelihood estimation of variance components for univariate animal models using sparse matrix techniques and a Quasi-Newton procedure. J Dairy Sci 78: 449–456.
53. Gilmour AR, Cullis BR, Welham SJ, Thompson R (1999). ASREML Reference Manual. 213 p.
54. Sernapesca (2012) Informe Sanitario Salmonicultura en Centros Marinos - Año 2012. Ministerio de Economía, Fomento y Reconstrucción, Valparaíso, Chile. Available at: http://www.sernapesca.cl (accessed November 15, 2013).
55. Smith PA, Pizarro P, Ojeda P, Contreras J, Oyanedel S, Larenas J (1999) Routes of entry of *Piscirickettsia salmonis* in rainbow trout *Oncorhynchus mykiss*. Dis Aquat Org 37: 165–172.
56. Tadiso TM, Krasnov A, Skugor S, Afanasyev S, Hordvik I, Nilsen F (2011) Gene expression analyses of immune responses in Atlantic salmon during early stages of infection by salmon louse (*Lepeophtheirus salmonis*) revealed bi-phasic responses coinciding with the copepod-chalimus transition. BMC Genomics 12: 141.
57. Taylor RS, Wynne JW, Kube PD, Elliott NG (2007) Genetic variation of resistance to amoebic gill disease in Atlantic salmon assessed in a challenge system. Aquaculture 272 S1: S94–S299.
58. Norris A, Foyle L, Ratcliff J (2008) Heritability of mortality in response to a natural páncreas disease (SPDV) challenge in Atlantic salmon, *Salmo salar* L., post-smolts on a West of Ireland sea site. J Fish Dis 31: 913–920.
59. Dorson M, Quillet E, Hollebecq MG, Torhy C, Chevassus B (1995) Selection of rainbow trout resistance to viral haemorrhagic septicaemia virus and transmission of resistance by gynogenesis. Vet Res 26: 361–368.
60. Henryon M, Jokumsen A, Berg P, Lund I, Pedersen PB, Olesen NJ, Slierendrecht WJ (2002) Genetic variation for growth rate, feed conversion efficiency, and disease resistance exists within a farmed population of rainbow trout. Aquaculture 209: 59–76. Erratum: Aquaculture 216: 387–388.
61. Henryon M, Peer Berg T, Olesen N, Kjaer T, Slierendrecht W, Jokumsen A, Lund I (2005) Selective breeding provides an approach to increase resistance of rainbow trout (*Onchorhynchus mykiss*) to the diseases, enteric redmouth disease, rainbow trout fry syndrome, and viral haemorrhagic septicaemia. Aquaculture 250: 621–636.
62. Weber GM, Vallejo RL, Lankford SE, Silverstein JT, Welch TJ (2008) Cortisol Response to a Crowding Stress: Heritability and Association with Disease Resistance to *Yersinia ruckeri* in Rainbow Trout. N Am J Aquacult 70: 425–433.
63. Silverstein JT, Vallejo RL, Palti Y, Leeds TD, Rexroad III CE, Welch TJ, Wiens GD, Ducrocq V (2009) Rainbow trout resistance to bacterial cold-water disease is moderately heritable and not adversely correlated with growth. J Anim Sci 87: 860–867.
64. Leeds TD, Silverstein JT, Weber GM, Vallejo RL, Palti Y, Rexroad III CE, Evenhuis J, Hadidi S, Welch TJ, Wiens GD (2010) Response to selection for bacterial cold water disease resistance in rainbow trout. J Anim Sci 88: 1936–1946.
65. Ødegård J, Baranski M, Gjerde B, Gjedrem T (2011) Methodology for genetic evaluation of disease resistance in aquaculture species: challenges and future prospects. Aquac Res 42: 103–114.
66. Gjedrem T, Gjoen HM (1995) Genetic variation in susceptibility of Atlantic salmon, *Salmo salar* L., to furunculosis, BKD and cold water vibriosis. Aquac Res 26: 129–134.
67. Gjøen HM, Refstie T, Ulla O, Gjerde B (1997). Genetic correlations between survival of Atlantic salmon in challenge and field tests. Aquaculture 158: 277–288.
68. Ødegård J, Olesen I, Gjerde B, Klemetsdal G (2007). Positive genetic correlation between resistance to bacterial (furunculosis) and viral (infectious salmon anemia) diseases in farmed Atlantic salmon (*Salmo salar*). Aquaculture 271: 173–177.
69. Kjoglum S, Henryon M, Assmundstad T, Korsgaard I (2008) Selective breeding can increase resistance of Atlantic salmon to furunculosis, infectious salmon anemia and infectious pancreatic necrosis. Aquac Res 39: 498–505.
70. Gjerde B, Evensen O, Bentsen HB, Storset A (2009) Genetic (co)variation of vaccine injuries and innate resistance to furunculosis (*Aeromona salmonicida*) and Infectious Salmon Anemia (ISA) in Atlantic salmon (*Salmo salar L*). Aquaculture 287: 52–58.
71. Drangsholt TMK, Gjerde B, Ødegård J, Finne-Fridell F, Evensen O, Bentsen HB (2012) Genetic correlations between disease resistance, vaccine-induced side effects and harvest body weight in Atlantic salmon (*Salmo salar*). Aquaculture 324–325: 55–59.
72. Guy DR, Bishop SC, Woolliams JA, Brotherstone S (2009) Genetic parameters for resistance to Infectious Pancreatic Necrosis in pedigreed Atlantic salmon (*Salmo salar*) post-smolts using a reduced animal model. Aquaculture 290 (3–4): 229–235.
73. Storset A, Strand C, Wetten M, Kjøglum S, Ramstad A (2007) Response to selection for resistance against infectious pancreatic necrosis in Atlantic salmon (*Salmo salar*, L.). Aquaculture 272 S1: S62–S68.

Permissions

All chapters in this book were first published in PLOS ONE, by The Public Library of Science; hereby published with permission under the Creative Commons Attribution License or equivalent. Every chapter published in this book has been scrutinized by our experts. Their significance has been extensively debated. The topics covered herein carry significant findings which will fuel the growth of the discipline. They may even be implemented as practical applications or may be referred to as a beginning point for another development.

The contributors of this book come from diverse backgrounds, making this book a truly international effort. This book will bring forth new frontiers with its revolutionizing research information and detailed analysis of the nascent developments around the world.

We would like to thank all the contributing authors for lending their expertise to make the book truly unique. They have played a crucial role in the development of this book. Without their invaluable contributions this book wouldn't have been possible. They have made vital efforts to compile up to date information on the varied aspects of this subject to make this book a valuable addition to the collection of many professionals and students.

This book was conceptualized with the vision of imparting up-to-date information and advanced data in this field. To ensure the same, a matchless editorial board was set up. Every individual on the board went through rigorous rounds of assessment to prove their worth. After which they invested a large part of their time researching and compiling the most relevant data for our readers.

The editorial board has been involved in producing this book since its inception. They have spent rigorous hours researching and exploring the diverse topics which have resulted in the successful publishing of this book. They have passed on their knowledge of decades through this book. To expedite this challenging task, the publisher supported the team at every step. A small team of assistant editors was also appointed to further simplify the editing procedure and attain best results for the readers.

Apart from the editorial board, the designing team has also invested a significant amount of their time in understanding the subject and creating the most relevant covers. They scrutinized every image to scout for the most suitable representation of the subject and create an appropriate cover for the book.

The publishing team has been an ardent support to the editorial, designing and production team. Their endless efforts to recruit the best for this project, has resulted in the accomplishment of this book. They are a veteran in the field of academics and their pool of knowledge is as vast as their experience in printing. Their expertise and guidance has proved useful at every step. Their uncompromising quality standards have made this book an exceptional effort. Their encouragement from time to time has been an inspiration for everyone.

The publisher and the editorial board hope that this book will prove to be a valuable piece of knowledge for researchers, students, practitioners and scholars across the globe.

List of Contributors

David Johansson, Jan Erik Fosseidengen, Lars Helge Stien, Tone Vågseth and Frode Oppedal
Institute of Marine Research, Matredal, Norway

Frida Laursen
Institute of Marine Research, Matredal, Norway
Department of Biology, University of Bergen, Bergen, Norway

Anders Fernö
Department of Biology, University of Bergen, Bergen, Norway

Pascal Klebert
Sintef Fisheries and Aquaculture, Trondheim, Norway

Francisco Ramírez and Manuela G. Forero
Estación Biológica de Doñana, Department de Biología de la Conservación, Sevilla, Spain

Joan Navarro
Institut de Ciéncies del Mar, Barcelona, Spain

Isabel Afán
Estacioń Biológica de Doñana, Laboratorio de SIG y Teledetección, Sevilla, Spain

Keith A. Hobson
Environment Canada, Saskatoon, Saskatchewan, Canada

Antonio Delgado
Instituto Andaluz de Ciencias de la Tierra, Granada, Spain

Elina Laanto and Jaana K. H. Bamford
Centre of Excellence in Biological Interactions, Universities of Jyväskylä and Helsinki, Finland
Department of Biological and Environmental Science and Nanoscience Center, University of Jyväskylä , Jyväskylä, Finland

Jouni Laakso
Centre of Excellence in Biological Interactions, Universities of Jyväskylä and Helsinki, Finland
Department of Biosciences, University of Helsinki, Helsinki, Finland

Lotta-Riina Sundberg
Centre of Excellence in Biological Interactions, Universities of Jyväskylä and Helsinki, Finland
Department of Biological and Environmental Science, University of Jyväskylä, Jyväskylä, Finland

Stefan W. Metz., Femke Feenstra., , Jan W. van Lent, Just M. Vlak and Gorben P. Pijlman
Laboratory of Virology, Wageningen University, Wageningen, The Netherlands

Stephane Villoing
Intervet Norbio, Bergen, Norway

Marielle C. van Hulten and Joseph Koumans
Intervet International BV, Boxmeer, The Netherlands

Åse Helen Garseth
Department of Health Surveillance, Norwegian Veterinary Institute, Trondheim, Norway
Department of Natural History, Norwegian University of Science and Technology University Museum, Trondheim, Norway

Torbjørn Ekrem
Department of Natural History, Norwegian University of Science and Technology University Museum, Trondheim, Norway

Eirik Biering
Department of Health Surveillance, Norwegian Veterinary Institute, Trondheim, Norway

Akiyuki Ozaki, Wataru Kai, Jun-ya Aoki, Yumi Kawabata and Kazuo Araki
National Research Institute of Aquaculture, Fisheries Research Agency, Nakatsuhamaura, Minamiise-cho, Watarai-gun, Mie, Japan

Kazunori Yoshida, Masahiro Nakagawa, Takurou Hotta and Tatsuo Tsuzaki
Seikai National Fisheries Research Institute, Fisheries Research Agency, Nunoura, Tamanoura-machi, Goto-shi, Nagasaki, Japan

Kanako Fuji, Satoshi Kubota, Junpei Suzuki, Kazuki Akita, Takashi Koyama, Nobuaki Okamoto and Takashi Sakamoto
Faculty of Marine Science, Tokyo University of Marine Science and Technology, Konan, Minato-ku, Tokyo, Japan

Maya Maria Mihályi Henriksen, Lone Madsen and Inger Dalsgaard
Technical University of Denmark, National Veterinary Institute, Bülowsvej 27, Frederiksberg C, Denmark

Weiling Zheng, Hongyan Xu, Siew Hong Lam and Zhiyuan Gong
Department of Biological Sciences, National University of Singapore, Singapore, Singapore

Huaien Luo and R. Krishna Murthy Karuturi
Computational and Systems Biology, Genome Institute of Singapore, Singapore, Singapore

Flavio F. Ribeiro and Jian G. Qin
School of Biological Sciences, Flinders University, Adelaide, South Australia, Australia

Mateus Maldonado Carriero, Márcia R. M. Silva and Antonio A. M. Maia
Departamento de Medicina Veterinária, Faculdade de Zootecnia e Engenharia de Alimentos, Universidade de São Paulo, Pirassununga, São Paulo, Brazil

Edson A. Adriano
Departamento de Ciências Biológicas, Universidade Federal de São Paulo, Diadema, São Paulo, Brazil
Departamento de Biologia Animal, Instituto de Biologia, Universidade Estadual de Campinas, Campinas, São Paulo, Brazil

Paulo S. Ceccarelli
Centro Nacional de Pesquisa e Conservação de Peixes Continentais, Instituto Chico Mendes de Conservação da Biodiversidade, Pirassununga, São Paulo, Brazil

Ilona Merikanto
Department of Biosciences, University of Helsinki, Helsinki, Finland
Department of Mental Health and Substance Abuse Services, National Institute for Health and Welfare, Helsinki, Finland

Jouni Laakso
Department of Biosciences, University of Helsinki, Helsinki, Finland
Centre of Excellence in Biological Interactions, Department of Biological and Environmental Science, University of Jyväskylä, Jyväskylä, Finland

Veijo Kaitala
Department of Biosciences, University of Helsinki, Helsinki, Finland

Windi Indra Muziasari, Katariina Pärnänen, Antti Karkman, Christina Lyra, Manu Tamminen and Marko Virta
Department of Food and Environmental Sciences, University of Helsinki, Helsinki, Finland

Satoru Suzuki
Centre for Marine Environmental Studies (CMES), Ehime University, Matsuyama, Ehime, Japan

Satoshi Managaki
Department of Environmental Sciences, Musashino University, Tokyo, Japan

Arijit Ganguly, Ranita Chakravorty, Dipak K. Mandal and Parimalendu Haldar
Department of Zoology, Visva-Bharati University, Santiniketan, West Bengal, India

Angshuman Sarkar
Department of Statistics, Visva-Bharati University, Santiniketan, West Bengal, India

Julieta Ramos-Elorduy and Jose Manuel Pino Moreno
Departamento de Zoología, Universidad Nacional Autonoma de Mexico, Mexico City, Distrito Federal, México

Bertrand H. Lemasson
Department of Biology and Ecology Center, Utah State University (USU), Logan, Utah, United States of America
Environmental Laboratory, United States Army Engineer Research and Development Center, Santa Barbara, California, United States of America

James W. Haefner
Department of Biology and Ecology Center, Utah State University (USU), Logan, Utah, United States of America

Mark D. Bowen
Fisheries and Wildlife Resources Group, United States Bureau of Reclamation, Denver, Colorado, United States of America
Turnpenny Horsfield Associates, Ashurst, Southampton, Hampshire, United Kingdom

Marta Librán-Pérez, Cristina Velasco, Marcos A. López-Patiño, Jesús M. Míguez, José L. Soengas
Laboratorio de Fisioloxía Animal, Departamento de Bioloxía Funcional e Ciencias da Saúde, Facultade de Bioloxía, Universidade de Vigo, Vigo, Spain

Tim Dempster
Department of Zoology, University of Melbourne, Parkville, Victoria, Australia
SINTEF Fisheries and Aquaculture, Trondheim, Norway

Pablo Sanchez-Jerez, Damian Fernandez-Jover and Just Bayle-Sempere
Department of Marine Sciences and Applied Biology, University of Alicante, Alicante, Spain

Rune Nilsen and Pal-Arne Bjørn
NOFIMA, Tromsø, Norway

Ingebrigt Uglem
Norwegian Institute for Nature Research, Trondheim, Norway

Nicole Strepparava and Orlando Petrini
Cantonal Institute of Microbiology, Bellinzona, Switzerland

Thomas Wahli and Helmut Segner
Centre for Fish and Wildlife Health, University of Bern, Bern, Switzerland

Bruno Polli
Cantonal Office of Hunting and Fisheries, Bellinzona, Switzerland

Michaël Bekaert
Institute of Aquaculture, University of Stirling, Stirling, Scotland, United Kingdom

Magne Aldrin
Norwegian Computing Center, Oslo, Norway
Department of Mathematics, University of Oslo, Oslo, Norway

Bård Storvik
Norwegian Computing Center, Oslo, Norway

Anja Bråthen Kristoffersen
Norwegian Veterinary Institute, Oslo, Norway
Department of Informatics, University of Oslo, Oslo, Norway

Peder Andreas Jansen
Norwegian Veterinary Institute, Oslo, Norway

Lars Holten-Andersen
Department of Veterinary Disease Biology, Faculty of Health and Medical Sciences, University of Copenhagen, Frederiksberg C, Denmark
Division of Veterinary Diagnostics and Research, National Veterinary Institute, Technical University of Denmark, Copenhagen, Denmark

Inger Dalsgaard
Division of Veterinary Diagnostics and Research, National Veterinary Institute, Technical University of Denmark, Copenhagen, Denmark

Jørgen Nylén
Aqua Denmark Unit, MSD Animal Health, Ballerup, Denmark

Niels Lorenzen
Division of Poultry, Fish and Fur Animals, National Veterinary Institute, Technical University of Denmark, Aarhus N, Denmark

Kurt Buchman
Department of Veterinary Disease Biology, Faculty of Health and Medical Sciences, University of Copenhagen, Frederiksberg C, Denmark

Jean Paul Lhorente
Aquainnovo S.A, Puerto Montt, Chile

José A. Gallardo
Pontificia Universidad Católica de Valparaíso, Valparaíso, Chile

Beatriz Villanueva and María J. Carabaño
Departamento de Mejora Gené tica Animal, INIA, Madrid, Spain

Roberto Neira
Aquainnovo S.A, Puerto Montt, Chile
Departamento de Producción Animal, Facultad de Ciencias Agronómicas, Universidad de Chile, Santiago, Chile

Index

www.ingramcontent.com/pod-product-compliance
Lightning Source LLC
Chambersburg PA
CBHW082012190326
41458CB00010B/3165

* 9 7 8 1 6 3 2 3 9 9 1 2 0 *